Applied Laser
Spectroscopy

W0245903

NATO ASI Series

Advanced Science Institutes Series

A series presenting the results of activities sponsored by the NATO Science Committee, which aims at the dissemination of advanced scientific and technological knowledge, with a view to strengthening links between scientific communities.

The series is published by an international board of publishers in conjunction with the NATO Scientific Affairs Division

A	**Life Sciences**	Plenum Publishing Corporation
B	**Physics**	New York and London
C	**Mathematical**	Kluwer Academic Publishers
	and Physical Sciences	Dordrecht, Boston, and London
D	**Behavioral and Social Sciences**	
E	**Applied Sciences**	
F	**Computer and Systems Sciences**	Springer-Verlag
G	**Ecological Sciences**	Berlin, Heidelberg, New York, London,
H	**Cell Biology**	Paris, and Tokyo

Recent Volumes in this Series

Volume 239—Kinetics of Ordering and Growth at Surfaces
edited by Max G. Lagally

Volume 240—Global Climate and Ecosystem Change
edited by Gordon J. MacDonald and Luigi Sertorio

Volume 241—Applied Laser Spectroscopy
edited by Wolfgang Demtröder and Massimo Inguscio

Volume 242—Light, Lasers, and Synchrotron Radiation: A Health Risk Assessment
edited by M. Grandolfo, A. Rindi, and D. H. Sliney

Volume 243—Davydov's Soliton Revisited: Self-Trapping of
Vibrational Energy in Protein
edited by Peter L. Christiansen and Alwyn C. Scott

Volume 244—Nonlinear Wave Processes in Excitable Media
edited by Arun V. Holden, Mario Markus, and Hans G. Othmer

Volume 245—Differential Geometric Methods in Theoretical Physics:
Physics and Geometry
edited by Ling-Lie Chau and Werner Nahm

Volume 246—Dynamics of Magnetic Fluctuations in
High-Temperature Superconductors
edited by George Reiter, Peter Horsch, and Gregory C. Psaltakis

Series B: Physics

Applied Laser Spectroscopy

Edited by
Wolfgang Demtröder
Kaiserslautern University
Kaiserslautern, Federal Republic of Germany

and
Massimo Inguscio
University of Florence and
European Laboratory for Nonlinear Spectroscopy (LENS)
Florence, Italy

Plenum Press
New York and London
Published in cooperation with NATO Scientific Affairs Division

Proceedings of a NATO Advanced Study Institute
on Applied Laser Spectroscopy,
held September 3-15, 1989,
in San Miniato (Pisa), Italy

Library of Congress Cataloging-in-Publication Data

NATO Advanced Study Institute on Applied Laser Spectroscopy (1989 :
  San Miniato, Italy)
    Applied laser spectroscopy
        p.   cm.
    ISBN-13: 978-1-4684-1344-1   e-ISBN-13: 978-1-4684-1342-7
    DOI: 10.1007/978-1-4684-1342-7

    1. Laser spectroscopy--Congresses. 2. Quantum electronics-
  -Congresses.
    QC454.L3N385   1989
    621.36'6--dc20   90-23596
                                                                 CIP

© 1990 Plenum Press, New York
Softcover reprint of the hardcover 1st edition 1990

A Division of Plenum Publishing Corporation
233 Spring Street, New York, N.Y. 10013

All rights reserved

No part of this book may be reproduced, stored in a retrieval system, or transmitted
in any form or by any means, electronic, mechanical, photocopying, microfilming,
recording, or otherwise, without written permission from the Publisher

PREFACE

This volume contains the lectures and seminars presented at the NATO Advanced Study Institute on "Applied Laser Spectroscopy" the fourteenth course of the Europhysics School of Quantum Electronics, held under the supervision of the Quantum Electronics Division of the European Physical Society. The Institute was held at Centro "I Cappuccini", San Miniato, Tuscany, Italy, September 3–15, 1989.

The Europhysics School of Quantum Electronics was started in 1970 with the aim of providing instruction for young researchers and advanced students already engaged in the area of quantum electronics or wishing to switch to this area from a different background. Presently the school is under the direction of Professors F.T. Arecchi and M Inguscio, University of Florence and Prof. H. Walther, University of Munich and has the headquarters at the National Institute of Optics (INO), Firenze, Italy. Each time the directors choose a subject of particular interest, alternating fundamental topics with technological ones, and ask colleagues specifically competent in a given area to take the scientific responsibility for that course.

The past courses were devoted to the following topics:

1. 1971: "Physical and Technical Measurements with Lasers"
2. 1972: "Nonlinear Optics and Short Pulses"
3. 1973: "Laser Frontiers: Short Wavelength and High Powers"
4. 1974: "Cooperative Phenomena in Multicomponent Systems
5. 1975: "Molecular Spectroscopy and Photochemistry with Lasers"
6. 1976: "Coherent Optical Engineering"
7. 1977: "Coherence in Spectroscopy and Modern Physics"
8. 1979: "Lasers in Biology and Medicine"
9. 1980: "Physical Processes in Laser Material Interactions"
10. 1981: "Advances in Laser Spectroscopy"
11. 1982: "Laser Applications to Chemistry"
12. 1984: "Physics of New Laser Sources"
13. 1987: "Instabilities and Chaos in Quantum Optics"

The objective of the ASI on "Applied Laser Spectroscopy" was to bring together researchers in fundamental and applied physics and chemistry in order to stimulate mutual interactions. A major aim was to illustrate how fascinating and useful interdisciplinary cooperation can be in a field where laser techniques are transferred from the highly specialized laboratories to a more general use.

The scientific organization of the course was taken care by F.T. Arecchi, University of Florence, S. Califano, University of Florence, W. Demtröder, University of Kaiserslautern, M. Inguscio, Director of the ASI, University of Naples now at Florence. G. Scoles, University of Princeton, J.P. Taran, ONERA, Chatillon, and H. Walther, University of Munich and MPI for Quantum Optics.

There were 20 invited lecturers, 15 seminar speakers and 85 other participants at the Institute. In addition, on Friday September 8th, we had a special one–day session including a round table on new coherent sources, a scientific visit to the European Laboratory for Nonlinear Spectroscopy (LENS) in Florence and a presentation of interdisciplinary laser spectroscopy projects related to this laboratory. On that occasion Prof. V. P. Chebotayev, Novosibirsk and Prof. V. S Letokhov, Moskow, guests of the University of Florence provided excellent lectures, which are reproduced as chapters in this volume.

The ASI was fuelled by the enthusiasm for the subject matter of all participants and was helped by the charming atmosphere of fine setting in a restored convent in the country side, very close to the small ancient city of San Miniato, in the very heart of Tuscany. An additional advantage of the location of the ASI was its nearness to the new European Laboratory for Nonlinear Spectroscopy (LENS) in Florence. We wish to express our appreciation to the NATO Scientific Affairs Division, whose financial support made the Institute possible. We also acknowledge the contributions of the following institutions:

    Consiglio Nazionale delle Ricerche (CNR)
    Ente Nazionale Energie Alternative (ENEA)
    US Department of the Air Force (London)
    Cassa di Risparmio di San Miniato
    Bruker Spectrospin Italia
    Contek S.R.L.
    Laser Optronics S.R.L.
    Laser Point S.R.L.
    Spectra Physics Italia

The National Science Foundation contributed with four travel grants for USA participants. Additional funding has been provided by NATO national offices for the participation of students from Turkey, Greece and Portugal.

We wish to thank Mrs. M. Petrone of INO, who, as a secretary of the Europhysics School of Quantum Electronics, significantly helped in the organization of this course. Together with the participants, we are grateful to Anna Chiara Arecchi for her competent and enthusiastic assistance during the course. We remember with great pleasure the "Vincenzo Galilei" Chorus of the Scuola Normale Superiore of Pisa, whose performance in the old S. Miniato Cathedral was particularly impressive.

W. Demtröder

M. Inguscio

CONTENTS

Introduction   xi

FUNDAMENTALS OF LASER SPECTROSCOPY AND QUANTUM OPTICS

Techniques of Laser Spectroscopy; their Advantages and Limitations   1
    W. Demtröder

Quantum Effects in Single–Atom and Single Photon Experiments   13
    G. Rempe, W. Schleich, M.O. Scully and H. Walther

Chaos and Complexity in Quantum Optics   31
    F.T. Arecchi

Synchronization of Atomic Quantum Transitions by Light Pulses   47
    V.P. Chebotayev and V.A. Ulybin

Amplification of Pulse Responses in Ensembles of Nonlinear   55
       Oscillators
    V.P. Chebotayev

SPECTROSCOPIC TECHNIQUES AND NEW COHERENT SOURCES

Narrowband Tunable VUV/XUV Radiation Generated   63
       by Third Order Frequency Mixing of Laser
       Radiation in Gases.
    A . Borsutzky, R. Brünger and R. Wallenstein

Laser and Synchrotron–Based Excitation Sources   69
       for Relaxation Studies
    S. Leach

Optical Parametric Oscillators of Bariumborate and   101
       Lithiumborate: New Sources for Powerful Tunable
       Laser Radiation in the Ultraviolet, Visible and
       Near Infrared
    A. Fix, T. Schröder, J. Nolting and R. Wallenstein

Two–Photon Transitions with Time Delayed Radiation Pulses   109
    A. Pasquarello and A. Quattropani

Diode Lasers and their Applications to Spectroscopy   117
    L. Hollberg

The CO—Overtone Laser: 127
    A Spectroscopic Source in a Most Interesting
    Wavelength Region
    W. Urban

Tunable Sideband—Laser Spectroscopy of Atoms and 137
    Molecules
    J. Legrand and P. Glorieux

High Resolution Far Infrared Spectroscopy 141
    L.R. Zink, M. Prevedelli, K.M. Evenson and M. Inguscio

LASER SPECTROSCOPY OF ATOMS AND MOLECULES

High-Resolution Laser Spectroscopy in the UV/VUV 149
    Spectral Region
    S. Svanberg

High Resolution Spectroscopy of Atomic Hydrogen: 167
    Measurement of the Rydberg Constant.
    L. Julien, F. Biraben, J.C. Garreau and M. Allegrini

Laser Spectroscopy of Atomic Discharges 173
    M. Inguscio

Infrared Laser Spectroscopy 189
    J.M. Brown

Opto—Thermal Spectroscopy 215
    D. Bassi, A. Boschetti and M. Scotoni

Multiphoton IR Spectroscopy 227
    J. Reuss and N. Dam

Light Induced Kinetic Effects in Alkali Vapors 241
    L. Moi

Anomalous Doppler Broadening in O_2—Noble Gases 249
    Radio—Frequency Discharges
    A. Sasso and G.M. Tino

The N_2^*—OCS System Microwave Spectroscopy Measurements 257
    of Vibrational Populations: The Perturbed
    Stationary State
    P.G. Favero, M.C. Righetti and L.B. Favero

INTERDISCIPLINARY APPLICATIONS OF LASER SPECTROSCOPY

A. Solid State and Cluster Physics

Phonon Lifetimes in Molecular Crystals 269
    S. Califano

Nonlinear Time Resolved Spectroscopy in Condensed Matter 293
    C. Flytzanis

Brioullin Gain Spectroscopy in Glasses and Crystals   307
G.W. Faris

Coherent Raman Spectroscopy: 313
  Techniques and Recent Applications
J.W. Nibler

B. Sensitive Detection of Isotopes, Combustion
Processes and Pollutants

Laser Detection of Very Rare Long–Lived Radioactive Isotopes   329
Yu. A. Kudryavtsev, V.S. Letokhov and V.V. Petrunin

Nuclear Ground–State Properties From Laser and   339
  Mass Spectroscopy
H.J. Kluge

Resonance Ionization Mass Spectroscopy for Trace Analysis   349
H.J. Kluge

CARS Spectroscopy and Applications   365
J.P. Taran

Laser Spectroscopy Applied to Combustion   393
A. D'Alessio and A. Cavaliere

Trace Gas Detection with Infrared Gas Lasers   403
J. Henningsen, A. Olafson and M. Hammerich

Environmental Monitoring Using Optical Techniques   417
S. Svanberg

Remote Detection of Atmospheric Pollutants Using   435
  Differential Absorption Lidar Techniques
J.P. Wolf, H.J. Kölsch, P. Riroux and L. Wöste

C. Application to Frequency Metrology and Stabilization

Laser Cooling of Atomic Beams and its Application   469
  to Frequency Standards
N. Beverini and F. Strumia

Possible Precision Far–Infrared Spectroscopy of   479
  Trapped Ions
G. Werth

Stable Ammonia Laser for Cooled Single Ion Frequency   487
  Standard
K.J. Siemsen, A.A. Madej and G. Magerl

Frequency Stabilization of Infrared Lasers: 491
  The Optogalvanic Effect in CO_2
L. Zink

Index   497

INTRODUCTION

Laser spectroscopy has spread into many areas of science and technology. Its applications are meanwhile so numerous, that a book with a limited size can only cover a selection of them. Many experts in different fields of laser spectroscopy have contributed to this volume. The various topics they selected from their current research work illustrates the close connection between basic research and possible applications.

Basic research in laser spectroscopy is not only necessary to solve fundamental problems related to the foundations of sciences but also to develop new coherent sources and novel spectroscopic techniques useful for applications. Of particular advantage is interdisciplinary research which helps to transfer knowledge from specialists in one field to researchers in other fields and which may induce new ideas by discussions and cooperation.This mutual interaction was one of the goals of the Nato Advanced Study Institute on "Applied Laser Spectroscopy" and the collection of the following contributions reflects clearly this intention.

The lectures in this volume are organized as follows:
The first five lectures cover basic aspects of different sensitive techniques in laser spectroscopy and their applications to fundamental problems in physics, such as quantum effects in single–atom and single–photon experiments or chaos and complexity in quantum optics. Lasers are good examples to study experimentally the transition from order to chaos as outlined in F.T. Arecchi's lecture. The invention of short laser pulses has not only brought great progress in time resolution but can be also utilized for high spectral resolution. This is shown in the lectures of Prof. Chebotayev.

The increasing importance of laser spectroscopy is due to the development of novel techniques and new coherent sources, a subject which is covered by the lectures comprised in the next section. The contributions range from new achievements in tunable VUV– and UV sources realized by frequency mixing techniques and free electron lasers to parametric oscillators covering the spectral region from the UV to the near infrared and are discussed in the lectures of R. Wallenstein and S. Leach. Of particular interest is the rapid progress of diode laser technology which allows the application of these cheap and handy sources to many problems in spectroscopy, as illustrated by the lecture of L. Hollberg.

In the infrared region the CO–laser with its numerous lines has proved to be a very useful tool for laser magnetic resonance spectroscopy as outlined in W. Urban's lecture. Another technique of infrared spectroscopy with fixed frequency lasers is the tunable sideband laser spectroscopy, covered by the contribution of Legrand and Glorieux.

The far infrared region, for a long time a frequency range above available microwave frequencies, but below existing widely tunable lasers, can now be covered by mixing techniques with fast MIM diodes. This interesting field is presented in the lecture by K. Evenson, a pioneer of this technique.

Detailed investigations of atoms and molecules with high resolution represent the main application domain of laser spectroscopy. An extensive section with 9 lectures is, therefore, devoted to this subject, ranging from high resolution spectroscopy of the hydrogen atom, light induced kinetic effects in alkali atoms to the study of more complex molecules.

The section starts with a review by S. Svanberg on atomic spectroscopy with lasers in the ultraviolet and VUV spectral range. The attainable accuracy is demonstrated by recent high resolution measurements of transitions in atomic hydrogen, reported by Julien et al. Many excited levels of larger atoms which are not populated at room temperatures are not yet known.

Novel spectroscopic technique, such as tunable diode opto—thermal and multiphoton IR—spectroscopy discussed in the articles by Brown, Bassi et. al. and Reuss and Dam have opened new possibilities for infrared molecular spectroscopy. This section closes with two special subjects on anomalous Doppler—broadening in discharges by Sasso and Tino and on perturbed vibrational populations in the N_2* OCS—system by Favero et al.

The largest number of lectures (15) comprised in Section V (16) are related to interdisciplinary applications of laser spectroscopy. They start with studies in solid state and cluster physics.

The interesting problem of phonon lifetimes in molecular crystals, addressed by S. Califano, the expert in this field and the nonlinear time resolved spectroscopy in condensed matter discussed by C. Flytzanis, anticipate sufficiently intense short laser pulses. Brioullin gain spectroscopy, presented by G.W. Faris and the different aspects of coherent Raman spectroscopy illustrate the usefulness of lasers in solid state physics and in the rapidly developing field of clusters, a transition regime between free molecules and the liquid or solid state.

The high sensitiviy of different techniques in laser spectroscopy has meanwhile been utilized in many different areas. This is illustrated by the contributions of subsection B. The lectures by V.S. Letokhov and H.J. Kluge on laser detection of very rare long—lived radioactive isotopes and on resonance ionization mass spectroscopy demonstrates the extreme sensitiviy, which reaches the "one atom—detection" limit, and which becomes extremely useful for trace analysis, radioactive dating techniques and fundamental research on short—lived unstable nuclei.

A field, where applications of laser spectroscopy gain increasing importance is the remote control of combustion processes, and environmental pollution. The lectures by J.P. Taran, A. D'Alessio and A. Cavaliere on applications of CARS, spontaneous Raman scattering and LIF to combustion illustrates the state of the art in this field. Environmental studies with infrared, visible and UV—spectroscopy are covered by the interesting lectures of J.Henningsen, S. Svanberg and L. Wöste.

The last subsection C deals with recent advances in high precision metrology and frequency stabilization. The possibility of laser cooling of atoms, discussed in the lecture of F. Strumia, increases the interaction time of neutral atoms with a light field, while trapping of ions in an ion trap allows to realize frequency standards with single ions. The lectures of G.Werth and K.J. Siemsen give examples of research in this area. For the establishment of a frequency chain from the cesium—standard to the visible range stable infrared lasers play a crucial role. The frequency stabilization of infrared lasers is discussed in the last lecture of this volume by L. Zink.

The variety of subjects in applied laser spectroscopy, covered by the following articles, may give an impression of the fascination and the importance of this fast growing field.

TECHNIQUES OF LASER SPECTROSCOPY:

THEIR ADVANTAGES AND LIMITATIONS

W. Demtröder

Fachbereich Physik, Universität Kaiserslautern
D–6750 Kaiserslautern, Germany

I Introduction

For all applications of spectroscopy to problems in science and technology intensities and wavelengths of absorbed or emitted radiation have to be measured. A major limitation in practical applications is imposed by the attainable detection sensitivity. The answer to the question:"How few atoms, molecules or photons can still be detected at a reasonable signal–to–noise ratio?" may be of crucial importance for the successful solution of a given problem.

For many applications furthermore a sufficiently high spectral and/or spatial resolution is required. These different demands of high sensitivity, high spectral and spatial resolution cannot always be fulfilled simultaneously and the spectroscopist has to look for an optimum compromise.

The development of various types of lasers or other coherent sources and their extensive use in spectroscopy has pushed both the attainable sensitivity and the possible resolution quite considerably to limits, which reach in favorable cases single atom detection [1, 2] and sub–Doppler spectral resolution.

This lecture gives a brief survey on some sensitive detection techniques in laser spectroscopy, their advantages compared with conventional techniques and their restrictions to limited spectral regions.

II Laser–Induced Fluorescence

When a light beam passes through an absorbing sample with pathlength L, the intensity I(L) of the transmitted light decreases to

$$I = I_0 \cdot e^{-\alpha L} \tag{1}$$

where the absorption coefficient $\alpha = N \cdot \sigma$ is the product of the density N of absorbing molecules and the absorption cross section σ of the absorbing transition. For sufficiently small absorptions ($\alpha \cdot L \ll 1$) the absorbed fraction ΔP of the incident power P_0 is

$$\frac{\Delta P}{P_0} = \alpha \cdot L.$$

Applied Laser Spectroscopy, Edited by W. Demtröder and
M. Inguscio, Plenum Press, New York, 1990

Because of unavoidable fluctuations of the incident power P_0 the sensitivity limit of measuring the difference ΔP between incident and transmitted power is for practical applications about $\Delta P / P_0 \geq 10^{-5}$, even if lock–in detection and reference cells are used. This limits the detectable density N_i of molecules absorbing on a transition $E_i \rightarrow E_k$ to

$$\alpha \cdot L = N_i \cdot \sigma_{ik} \cdot L \geq 10^{-5} \tag{3}$$

With typical absorption cross sections of $\sigma_{ik} = 10^{-16}$ cm^2 and an absorption pathlength of $L = 10$ cm we obtain $N_i \geq 10^{10}$ [cm^{-3}] which corresponds roughly to a partial pressure of 10^{-6} mb.

The absorption of a photon $h \cdot \nu$ results in the excitation of an atom or molecule from a thermally populated level $|E_i\rangle$ into an excited level $|E_k\rangle$. After a time τ_k, called its effective lifetime, the molecule returns to a lower level $|E_m\rangle$ by emitting a fluorescence photon or by suffering an inelastic (quenching) collision.

If the excited molecule does not suffer collisions during its lifetime τ the molecular quantum efficiency is $\eta_M = 1$, i.e. every absorbed laser photon produces a fluorescence photon which can be collected from the excitation volume by a collimating optics with a collection efficiency η_C and can be imaged onto a photon detector of quantum efficiency η_a. The rate n_a of absorbed laser photons then produces

$$n_{pe} = \eta_c \cdot \eta_a \cdot n_a \tag{4}$$

detected photoelectrons per second. From the measured rate n_{pe} we can deduce the density N_i of absorbing molecules, using (2) and (3) as

$$N_i = \alpha / \sigma_{ik} = \frac{n_{pe}}{n_L \cdot L \cdot \sigma_{ik} \cdot \eta_c \cdot \eta_a} \tag{5}$$

which shows that the incident flux n_L of laser photons and the efficiencies η_c, η_a should be large in order to reach a low limit for the detectable molecular density N_i. For a quantitative determination of N_i the absorption coefficient σ_{ik} has to be known.

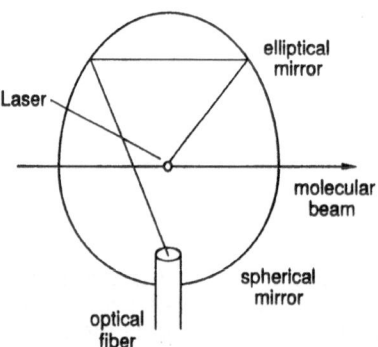

Fig. 1. High collection efficiency fluorescence optics which images the small excitation volume located in one focus of an ellipsoidal mirror into the second focus where an optical fiber bundle is placed.

The sensitivity of LIF detection can be best demonstrated for a small absorption pathlength L and a low density N_i, a situation, realized by laser spectroscopy in a collimated molecular beam. The small crossing volume of laser– and molecular beam can be effectively imaged onto the photon detector by the collection system shown in Fig. 1. The excitation volume is located in one focal point of an ellipsoidal mirror while the end of an optical fiber bundle is placed in the second focus. The fiber bundle images the light either directly to a photon detector or for spectrally resolved fluorescence measurements, onto the entrance slit of a monochromator. In the second case the fibers of the bundle are arranged in a rectangular cross section adapted to the size of the entrance slit.

The photo detector may be a photomultiplier or a recently developed silicon avalanche diode with a superior quantum efficiency $\eta_a \geq 0.6$ over the whole spectral range $0.4\ \mu m < \lambda < 0.95\ \mu m$. Because of its small sensitive area $A \leq 1\ mm^2$ the avalanche diode is in particular advantageous if only a small excitation volume has to be observed.

In case of atoms sometimes transitions $|i> \rightarrow |k>$ can be chosen on which the atom acts as a true two level system, which means that it returns after excitation only into the initial level $|i>$ and not into other levels. At sufficiently high laser intensities the atom with a spontaneous lifetime τ passing with a thermal velocity v through a laser beam of diameter d may then undergo $n = T/(2\tau)$ absorption–emission cycles during its transit time $T = d/v$.

This implies that it can emit $n_{Fl} \leq T/(2\tau)$ fluorescence photons (photon–burst). This, of course, considerably increases the detection sensitivity [3].

Example: $\tau = 10^{-8}s$, $d = 0.2$ cm, $v = 5 \cdot 10^4$ cm/s $\rightarrow T = 4 \cdot 10^{-6}s$
$N = T/2\tau = 200$.

Often the density of tracer molecules has to be determined in the presence of other gas constituents at high pressure. Examples are the measurements of small molecular concentrations in combustion processes.
The quantitative detection of these constituents with the LIF–method meets the following problem: The excited molecules are deactivated not only by emission of fluorescence but also by quenching collisions, and only the fraction

$$\eta_{Fl} = \frac{A_{ik}}{A_{ik} + \sigma_q \cdot N_B \cdot <v>} \tag{6}$$

of excited molecules emit fluorescence on a transition with a transition probability A_{ik} while all other excited molecules are quenched by collisions (quenching cross section σ_q) with other atoms or molecules at a density N_B and with a mean relative velocity $<v>$. The quenching cross section σ_q may differ by several orders of magnitude for different excited states and, therefore, the number of collected fluorescence photons is no longer a true measure of the density N_i in eq. (5).

A solution of this problem has been demonstrated by Andresen et al. [4]: Many excited molecular states have energy ranges where crossings with repulsive tentials occur which lead to predissociation. If the predissociation rate R_{PD} is much faster than the collisional quenching rate $\sigma_q N_B <v>$ then the fluorescence yield

$$\eta_{Fl} = \frac{A_{ik}}{A_{ik} + \sigma_a N_\beta \cdot v + R_{PD}} \approx \frac{A_{ik}}{R_{PD}} \tag{7}$$

becomes small ($R_{PD} >> A_{ik}$) but independent of quenching processes.

3

This technique of predissociation limited LIF has been successfully used to measure the concentration of OH, NO and CO—radicals within the combustion chamber of a car engine [5].

If the combustion chamber is illuminated with the enlarged beam of a tunable excimer laser (ArF, KrF and XeCl) the spectrally and spatially resolved detection of the LIF, which is possible by imaging the whole illuminated volume through interference filters onto a video camera with image intensifier, gives information about the spatial distribution of the different radicals in the combustion process at the time t where the laser is fired. Varying the time delay Δt between ignition of the combustion and laser illumination allows to observe the spatially resolved time evolution of the combustion process in detail [6].

III Resonant Multiphoton Ionisation

The most sensitive method to detect small concentrations of atoms or molecules is the resonant multiphoton ionization (REMPI) technique. In its simplest form only one tunable pulsed laser is required. When the laser is tuned to a resonance transition of the molecules to be detected, they are excited into an upper level $|k\rangle$. If the energy $E_K > IP/2$ is higher than one half of the ionization potential IP, a second photon of the same laser pulse can ionize the excited molecules (Fig. 2). Since the peak powers of pulsed dye lasers are sufficiently high to saturate a molecular transition, even without strong focussing, a volume of several cm^3 in a sample cell can be ionized with high efficiency [7].

More versatile is a two colour experiment where two different tunable lasers are used. The first laser is tuned to a selected transition $|i\rangle \rightarrow |k\rangle$ while the second laser can be tuned to a wavelength region where the ionization of the excited moledules is most efficient. If, for instance, the second laser excites molecular Rydberg states $|r\rangle$ above the ionization limit, these states can autoionize with nearly 100% ionization efficiency. The bound—bound transition $|k\rangle \rightarrow |r\rangle$ has a transition probability which might be 1—2 orders of magnitude larger than a bound—free transition [7].

If the ionization energy is larger than $h\nu_1 + h\nu_2$ two or more photons $n \cdot h\nu_2$ of the second laser are required for ionization. In such a case the two—colour

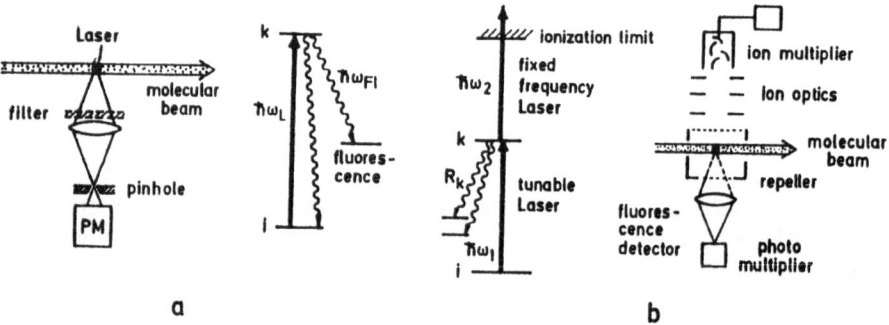

a   b

Fig. 2. Comparison of LIF detection and resonant two photon ionization of molecules in a molecular beam.

arrangement is clearly superior to the multiphoton ionization with a single laser, because both lasers can be tuned into resonance with the two transitions $|i> \rightarrow |k> \rightarrow |r>$ and a resonant two step excitation of a high lying molecular state $|r>$ is possible. This state is generally a Rydberg state which has a long spontaneous lifetime and a large dipole moment and can be, therefore, effectively ionized by a third photon $h\nu_1$ or $h\nu_2$. With a single laser the nonresonant two–photon excitation of $|r>$ has a much smaller cross section and furthermore the assignment of this state is more difficult than in case of two step excitation.

If a high spectral resolution is required, narrow band dye lasers can be used. However, for sub– Doppler resolution cw single mode dye lasers are superior. They have the additional advantage of continuous ionization while with pulsed lasers the molecules can be only detected during the laser pulse. Although the detection sensitivity can reach 100% during the pulse, the duty cycle of about 10^{-6} at a pulse repetition rate of 100Hz and a pulse duration of 10^{-8}s drops the time averaged detection sensitivity down to 10^{-6}.

Since cw lasers have much less power than pulsed lasers, the laser beams have to be tightly focussed in order to reach sufficient intensity for the second step $|k> \rightarrow IP$, while the first step $|i> \rightarrow |k>$ can be readily saturated even with moderate laser intensities. The molecules in the excited state $|k>$ with a typical lifetime of 10^{-8}s travel about 5 μm before they decay by fluorescence. They have to be ionized by the second laser before they radiate. It is, therefore, necessary to focus both laser beams down to diameters of about 5μm and to take care that the two beams overlap within this range.

A convenient experimental arrangement for resonant two–photon ionization in a molecular beam is shown in Fig. 3. The two laser beams are brought to the experiments in two optical fibers. The divergent light leaving the single mode fiber ends is collimated by a spherical lens and is then focussed into a focal line (5μm x 1 mm) by a cylindrical lens. Each molecule in the molecular beam has to pass through this two overlapping "sheets of light" and can be efficiently ionized.

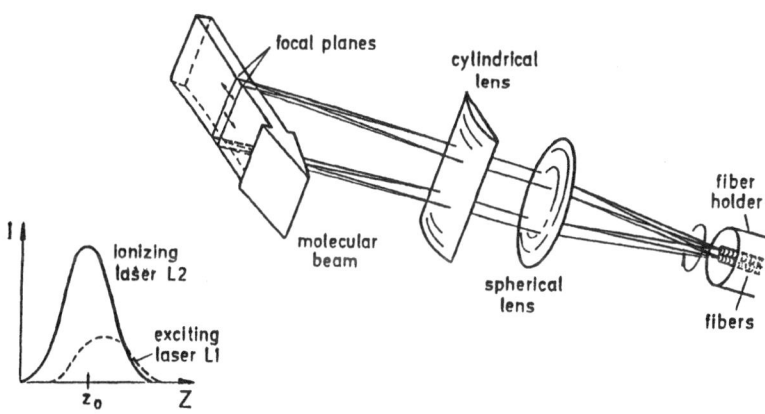

Fig. 3. Resonant two–photon ionization with two cw–lasers using two optical fibers and cylindrical focussing. The insert shows the optimum overlap of the two laser beam profiles.

IV Sensitive Detection Techniques in the Infrared

The two techniques LIF and REMPI are well suited in the visible and ultraviolet spectral range. In the infrared region of the vibrational and rotational transitions of molecules within their electronic ground state they loose sensitivity or selectivity, because of the following reasons:

1. The spontaneous lifetimes of vibrational levels in the electronic ground state become very long with typical values in the μs–ms–range. At low pressures the excited molecules may diffuse out of the observation volume before they emit a fluorescence photon. At higher pressures quenching collisions may deactivate the excited levels much faster than the fluorescence. In both cases the ratio of observed fluorescence photons to absorbed laser photons becomes small.

2. Infrared detectors have a sensitivity which is smaller by several orders of magnitude than that of photomultipliers.

3. The selectivity of the REMPI–method relies on the fact, that the excited state $|k>$ reached by absorption of the first photon is not thermally populated. This might not longer be true if $h \cdot \nu_1 \approx kT$.

Fortunately several other detection techniques have been developed which are particular adapted to the IR–region and which take advantage of the long spontaneous lifetime. We will briefly discuss two of them:

The first ist the opto–acoustic technique [8, 9], where the collisional energy transfer from the laser excited molecules to atoms or molecules of a background gas at higher pressure is utilized. The gas mixture is confined in a small cell which has a sensitive microphone placed on one of the inner walls (Fig. 4). The incident laser is tuned to an absorbing transition of the molecules under investigation and its intensity is periodically chopped at a frequency $f << 1/\tau_{eff}$, where the effective lifetime τ_{eff} of the excited molecules is mainly determined by collisional deactivation.

The periodical flux of absorption energy into the cell is converted by collisional energy transfer into a periodic modulation of translational energy, which results in acoustic waves in the cell. If the chopping frequency f is tuned to acoustic eigen resonances of the cell, standing acoustic waves are generated which are monitored by the microphone and transferred into an electric signal.

The method is not restricted to gases but can be also used for liquids and solids [10] and has found increasing applications. The sensitivity levels for absorbing gases range below the ppb–level, i.e. relative concentrations of 10^{-9} can still be detected.

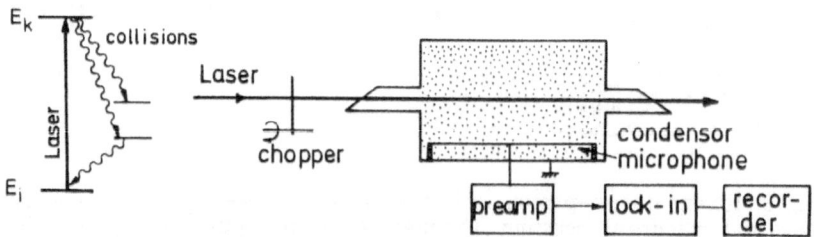

Fig. 4. Spectraphone for photoacoustic spectroscopy.

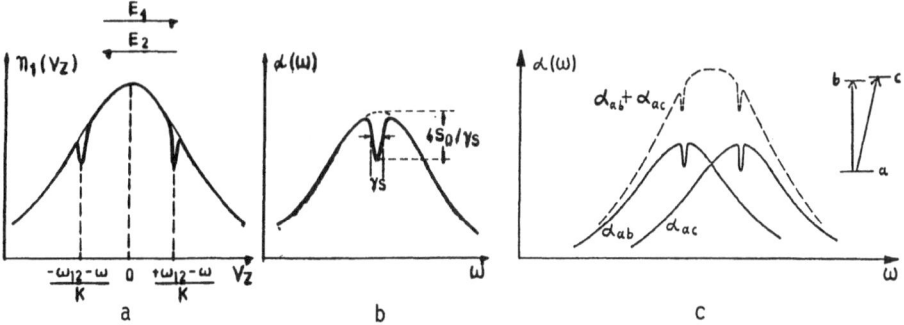

Fig. 6. Interaction of a monochromatic standing wave with a sample of molecules inside the laser resonator, a) hole burning in the population distribution $N(v_z)$ for $\omega_L \neq \omega_0$ b) absorption coefficient $\alpha(\omega_L)$ with the Lamb–dip for $\omega_L = \omega_0$
c) Resolution of the Lamb–dip of two closely lying transitions with overlapping Doppler–profiles.

The absorption coefficient $\alpha_{ik}(\omega) = \sigma_{ik}(\omega) \cdot N_i(v_z)$ decreases with increasing laser intensity because the lower level population N_i is depleted by absorption of laser photons (saturation) and becomes

$$N_i(v_z) = N_{i0}(1 - a \cdot I) \tag{9}$$

where the constant a depends on the transition probability of the pump transition.

Because for $\omega = \omega_0$ both waves interact with the same molecule, the intensity in (9) is twice as large as for $\omega \neq \omega_0$ and the absorption coefficient $\alpha_{ik}(\omega) = \sigma_{ik}(\omega) \cdot N_i(v_z)$ has a local minimum (Lamb–dip) at the center $\omega = \omega_0$ of the Doppler–broadened absorption profile (Fig. 6b).

This dip can be either monitored by observing the fluorescence $I_{Fl}(\omega)$ of the sample as a function of the laser frequency ω [14] or by measuring the laser output $I_L(\omega)$ which shows a local peak for $\omega = \omega_0$. The latter detection scheme is particularly sensitive if the laser is operated close above threshold since then even small changes of internal losses (due to absorption by the intracavity sample) cause already large intensity changes [15].

Saturation spectroscopy uses these narrow Lamb–dips which have a halfwidth γ determined by the homogeneous linewidth γ, for sub–Doppler high resolution spectroscopy. Even if the Doppler–broadened absorption line profiles overlap, their Lamb–dips may still be resolved (Fig. 6c).

In molecular beams saturation of molecular transitions is already achieved at moderate laser intensities because collisions which might refill the depleted level are absent. Since all molecules travel into the same direction the pump region can be spatially separated from the probe region. In Fig. 7 a typical experimental arangement is shown where the laser beam is split by the beam splitter BS into

The second technique is the opto–thermal spectroscopy which is mainly used for infrared laser spectroscopy in molecular beams. Here molecules, which are excited by an infrared laser impinge after a flight time T = D/v on a cooled bolometer at a distance D from the excitation region. If T is smaller than the lifetime τ of the excited molecules, their excitation energy is preserved and can be transferred to the bolometer when the molecules impinge on its surface.
This method is discussed in the lecture by Prof. Bassi in more detail.

V Absorption Techniques

In Section II we have seen that the sensitivity of conventional absorption spectroscopy is limited to values of $\alpha \cdot L > 10^{-5}$. There are several techniques to improve the sensitivity. One of them is the intracavity absorption where the absorbing probe is placed inside the laser resonator (Fig. 5).

When a multimode dye laser suffers at certain wavelengths λ_i extra losses, due to absorbing transitions of the sample inside the laser cavity, the modes with λ_i experience smaller net gain than the other modes which do not coincide with absorbing transitions. Due to gain competition between all modes within the homogeneous gain profile those modes with a higher net gain increase their intensity at the expense of the modes with lower net gain. This coupling effect enhances the effect of small absorption on the laser intensity $I(\lambda_i)$ at the absorbing wavelengths, which can even be completely extinguished at cw operation of a multimode cw laser. Also with pulsed dye lasers enhancement factors $M > 10^3$ can be obtained [11, 12].

With this method very weak absorptions on forbidden transitions and of samples at low pressures can still be measured [13]. The multimode laser output has to be spectrally dispersed in order to detect those wavelengths which are missing due to intracavity absorptions. The spectral resolution of this technique is, therefore, limited by that of the spectrometer.

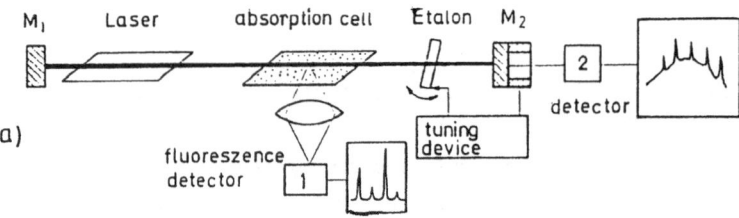

Fig. 5. Intra cavity absorption spectroscopy.

VI Sub–Doppler–Spectroscopy

For higher resolution single mode lasers can be used. The sample molecules inside the laser resonator are exposed to the monochromatic standing wave field of the laser, which can be composed of two counter propagating running waves $E = E_0 \cos (\omega t \pm kz)$. The absorption cross section $\sigma_{ik}(\omega)$ of a molecule with absorption frequency ω_0, which moves at a velocity $\vec{v}$ is given by

$$\sigma_{ik}(\omega) = \frac{(\gamma/2)^2}{(\omega_0 - \omega - kv_z)^2 + \gamma/2)^2} \tag{8}$$

where γ is the homogeneous linewidth of the molecular transition (natural linewidth + pressure broadening). The two counter propagating waves cause an opposite Doppler–shift $\pm kv_z$ and, therefore, they interact for $\omega \neq \omega_0$ with two different velocity classes $\pm (v_z \pm dv_z)$ of molecules (Fig. 7). Only for $\omega = \omega_0$ the Doppler–shift becomes zero and both waves interact simultaneously with the same two beams which cross the molecular beam perpendicularly. The laser–induced fluoresscence $I_{Fl}(\omega_L)$, collected from the second crossing point and recorded as a function of the laser frequency ω_L shows a Lamb–dip at the center of the absorption lines, which have already a reduced Doppler width due to the beam collimation (Fig. 8) The width of the Lamb–dip is determined by the homogeneous linewidth γ (natural linewidth + transit – time broadening + saturation broadening) and, in case of very narrow linewidths, also by frequency fluctuations of the laser.

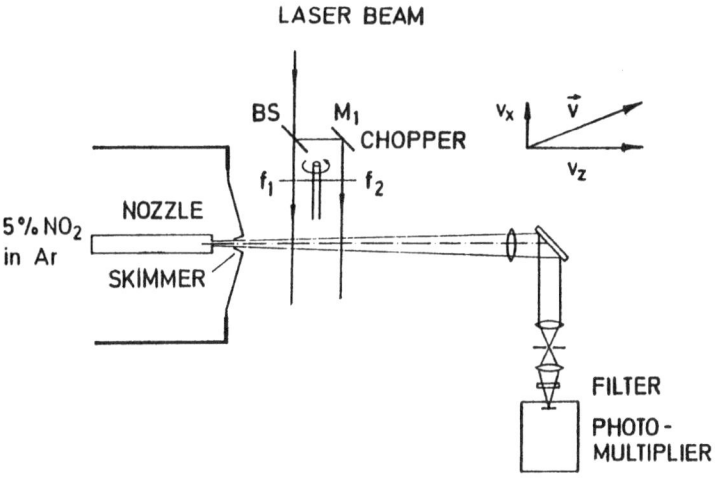

Fig. 7. Saturation Spectroscopy in a collimated molecular beam.

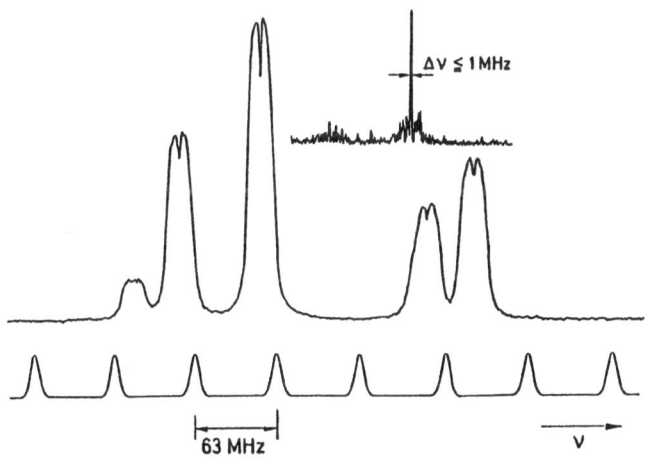

Fig. 8. Lamb–dips in the absorption lines of NO_2 measured by LIF with the arrangement of Fig. 7. The insert shows the elimination of the residual Doppler width by the technique of intermodulated fluorescence.

Another sensitive sub—Doppler technique is the polarization spectroscopy [16]. Here the sample is placed between two crossed polarizers. The laser output is split into a stronger linearly or circularly polarized pump beam and a weaker linearly polarized probe beam, which are send in opposite directions through the sample (Fig. 7). Without the pump beam the crossed analyzer blocks the probe beam and the detector receives only a weak background signal due to the imperfect extinction of the analyzer. If the laser wavelength is tuned to a molecular absorption line the molecules are partially orientated due to optical pumping by the polarized pump beam (i.e. the population of the degenerate m_y—levels is no longer equal). Such a sample of orientated molecules acts as a birefringent medium which tilts the plane of polarization of the probe laser beam when the probe leaser interducts with the optically pumped molecules. This happens only if the laser frequency a is tuned to the center ω_0 of the Doppler—broadened absorption line since for $\omega \neq \omega_0$ the two opposite laser beams interact with different velocity classes of molecules. Like saturation spectroscopy polarization spectroscopy represents a Doppler—free technique.

The high sensititivy of polarization spectroscopy is due to the following reasons:

1. Unlike saturation spectroscopy it is a "zero—background" method, because the transmitted signal is nearly zero with the pump beam off. High quality crossed polarizers have a residual transmission of less than 10^{-6}.

2. The system acts like an amplifier. A small tilt $\Theta/\!<\!<\!1$ of the polarization vector of the probe beam caused by the interaction with the oriented molecules generates a transmission signal of $\Theta \cdot I_{pr}$. Tilting angles of $\Theta = 10^{-7}$ can be still detected.

3. The pathlength L of the common interaction region of pump— and probe beam can be made much longer than, for instance, in LIF—experiments where L is limited to the region which still can be imaged onto the fluorescence detector.

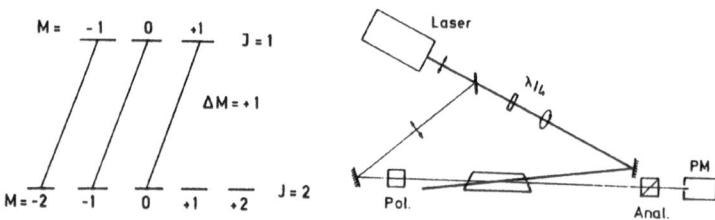

Fig. 9. Schematic arrangement for polarization spectroscopy.

VII Optical Double Resonance Techniques

The two—colour resonant two—photon ionization discussed in section III can be regarded as a special case of a more general optical double resonance method where one laser pumps a selected molecular transition and another radiation field which might be in the spectral range from radio—frequency waves to the ultraviolet region interacts simultaneously with the molecules. The possibilities for double–resonance experiments are numerous and all techniques discussed above can be combined with various double resonance methods.

A big advantage of double–resonance is due to the fact that single molecular levels can be selectively "labelled" even if many other neighbouring levels are thermally populated.

This labelling can be performed

   a) by a depletion or an increase of the level population by optical pumping.
   b) as an orientation or an alignment of a molecular level by pumping with polarized lasers. From the labelled level transitions can be induced to other levels by tunable radio–frequency, microwave, infrared, visible or UV–radiation.

These optical double resonance methods have several definite advantages:

   1. The assignment of molecular transitions is much easier than in other absorption spectra because all probe transitions start from a known level labelled by the pump laser. Due to selection rules the number of allowed probe transitions is restricted and therefore in most cases readily assigned.

   2. With single mode pump lasers only molecules of a definite velocity class $v_z \pm \Delta v_z$ are selected. Double resonance transitions between the depleted lower level and neighbouring level in the electronic ground state induced by rf or microwave transitions have a spectral resolution which may even surpass the natural line-width of the optical transitions. Therefore, hyperfine structure splittings in electronic ground states can be measured with extremely high resolution [17].

   3. By stepwise excitation high lying atomic and molecular Rydberg levels can be selectively excited. Double resonance experiments, where microwave radiation is used to induce transitions between molecular levels in Rydberg states has increased the accuracy considerably [18].

The combination of molecular beams with optical double–resonance techniques combines the advantages of adiabatic cooling in supersonic beams and collision–free conditions with the labelling of levels by a pump laser. This proves to be of particular importance for the assignment of perturbed molecular spectra.

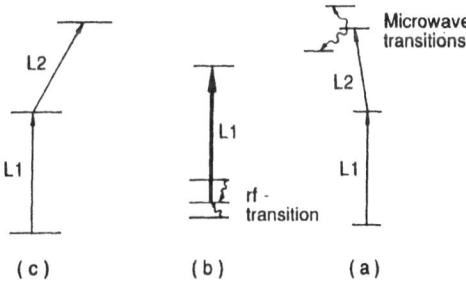

Fig. 10. Optical double resonances schemes. a) Stepwise excitation b) optical–rf double resonance spectroscopy of hfs–levels in the electronic groundstate c) optical–microwave double–resonance spectroscopy of Rydberg levels.

REFERENCES

1. G.S. Hurst, M.G. Payne: Principles and Applications of Resonance Ionization
   Spectroscopy
   (A. Hilger, Bristol 1988)
2. V.S. Letokhov: Laser Analytical Spectroscopy
   (A. Hilger Series in Optics, London 1986)
3. W. Demtröder: Laser Spectroscopy
   (Springer, Berlin, 1982)
4. P. Andresen, A. Bath, W. Gröger, H. W. Lülf:
   Appl. Opt. 27, 1 (1988)
5. W. Hentschel; VDI–Berichte 617, 347 (1986)
6. B. Raffel, J. Wolfrum; Z. Phys. Chemie, Neue Folge 161, 43 (1989)
7. M. Schwarz, R. Duchowicz, W. Demtröder, Ch. Jungen;
   J. Chem. Phys. 89, 5460 (1988)
8. Yoh–Han–Pao: Optoacoustic Spectroscopy and Detection
   (Academic Pres, New York 1977)
9. V.P. Zharov, V.S. Letokhov: Laser Optoacoustic Spectroscopy
   (Springer, Berlin 1986)
10. P. Hess, ed.: Photoacoustic, Photothermal and Photochemical Processes at
    Surfaces and in Thin Films (Topics in Current Physics, Vol 47,
    (Springer, Heidelberg 1987
11. H. Atmanspacher, H. Scheingraber, C.R. Vidal;
    Phys. Rev. A32, 254 (1985), Phys. Rev. 33, 1052 (1986)
12. T.D. Harris: Laser Intercavity Enhanced Spectroscopy
    in: D.S. Kliger, ed. Ultrasensitive Laser Spectroscopy
    (Academic Press, New York 1983)
13. V.M. Baev, A.Weiler, P.E. Toschek;
    J. de Physique Colloq. C7, Tome 48, 47–701 (1987)
14. M.S. Sorem, A.L. Schawlow;
    Opt. Commun. 5, 148 (1972)
15. T.W. Hänsch; Enrico Fermi School LXIV, p. 17 (1977)
    Ch. Borde: NASI on Advances in Laser Spectroscopy;
    T. Arrechi, F. Strumia and H. Walther, eds.
    (Plenum Press, London 1982)
16. M. Raab, G. Höning, W. Demtröder, C.R. Vidal;
    J. Chem. Phys. 76, 4370 (1984)
17. W.J. Childs, D.R. Cok, L.S. Goodman
    J. Chem. Phys. 76, 3993 (1982)
18. J. Boulmer, P. Camus, J.M. Gayne
    J. Phys. B, At. Mol. Phys. 20, L143 (1987)

QUANTUM EFFECTS IN SINGLE-ATOM AND SINGLE-PHOTON EXPERIMENTS

G. Rempe, W. Schleich, M.O. Scully[*] and H. Walther

Max-Planck-Institut für Quantenoptik and Sektion
Physik der Universität München, D-8046 Garching
bei München, FRG

INTRODUCTION

We review our recent work on the one-atom maser. We propose
and analyse an experiment based on this maser and designed to
probe the way in which the measurement process, that is, the
presence of a detector influences the investigated quantum
system. Phase transitions between chaotic and ordered
structures of ions stored in a Paul trap are analysed.

QUANTUM OPTICS IN THE REALM OF QUANTUM THEORY OF MEASUREMENT -- PROMOTER OF GEDANKEN EXPERIMENTS TO REAL WORLD EXPERIMENTS

The ingenious Einstein-Podolsky-Rosen Gedanken experiment[1]
aimed to illustrate in the most striking way the incomplete-
ness of quantum mechanics and Bohr's rebuttal[2], the concept
of elementary quantum phenomenon -- all just philosophical

---

[*]also at: Center for Advanced Studies, and Department of
Physics and Astronomy, University of New Mexico,
Albuquerque, N.M. 87131.

Applied Laser Spectroscopy, Edited by W. Demtröder and
M. Inguscio, Plenum Press, New York, 1990

talk which escapes experimental proof or disproof? No! Tools of laser spectroscopy, cascading atoms and fast switching Pockels cells bring to light the violation[3] of Bell's inequalities[4] in this situation and declare quantum mechanics the winner in this 50 year old saga[5].

"The past has no existence except as it is recorded in the present[6]". This intriguing statement summarizes in the most vivid way the alien features of quantum theory when applied to Young's double-slit experiment in a delayed choice mode[6,7]. In this domain it is fiber optics and quantum beat technology[8] which promote the delayed choice Gedanken experiment to real life[9-11] confirming that "no elementary quantum phenomenon is a phenomenon until it is a recorded phenomenon...[6]".

"Are there quantum jumps?[12]" How to design an experiment to answer this question put forward and discussed by Schrödinger for the first time during a stay in Copenhagen with Bohr in the late twenties and then more extensively in the fourties? How to see the discreteness of, that is, single photon emission events in a quasi-continuous beam of 10^8 photons? Couple the resonance transition of a single ion stored in a Paul-trap[13] to a longlived level driven by a second laser and interrupt so this photon stream whenever the electron is shelved in this state.[14] The outcome of recent trap experiments[14] along these lines provides the firm answer "yes" to Schrödinger's question.

Space permits us to choose only three out of a vast reservoir of quantum optics experiments which demonstrate in a striking way the enormous push experienced by the field of quantum theory of measurement from the techniques, approaches and tools of quantum optics. In the present article we continue this line of thought and introduce a new test of complementarity in quantum mechanics[15] using the most prominent representative of experimental quantum optics; the one-atom maser[16]. Moreover, we review our work[17,18] on ion traps aimed towards the creation of ion crystals whose crystal constant is either smaller than the wavelength of light or comparable to the de Broglie wavelength of the constituents of the crystal. In this regime new quantum effects are expected to make their appearance.

The article is organized as follows: In Sec. 2 we briefly summarize and highlight our work on the one-atom maser, and in particular, focus on the most recent improvements of this device. Section 3 is dedicated to the discussion of the non-classical radiation[19,20] generated by the one-atom maser[21,22]. We point out that this maser can create a state of well-defined photon number[23,24], a decisive ingredient of the new test of complementarity in quantum mechanics discussed in more detail in Sec. 4. In Sec. 5 we elaborate on the formation of regular structures of ions -- ion crystals -- stored in a Paul-trap. We conclude by summarizing our main results and by presenting an outlook on future experiments in Sec. 6.

A maser operation driven by a single atom in the cavity -just another example of a Gedanken experiment? Just another toy from the playground of a quantum optics theoretician? No, this device has been promoted from the wonderland of mathematics and theory to the real world of hardware and experiment. But what are the tools that make this amazing device work? This is the question addressed, discussed and elaborated on in this section.

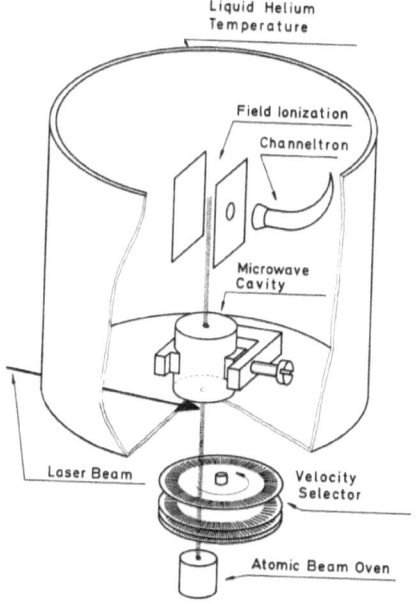

Fig. 1. Experimental setup of the one-atom maser

It was the enormous progress in constructing superconducting cavities together with the laser assisted preparation of highly excited atoms -- Rydberg atoms -- that has made the realization of such an one-atom maser possible. Rydberg atoms have quite remarkable properties[25] which make them ideal for such experiments: The probability of induced transitions between neighboring states of such a Rydberg atom scales as n^4, where n denotes the principle quantum number. Consequently a few photons are enough to saturate the transition between adjacent levels. Moreover, the spontaneous lifetime of a highly excited state is very large. We obtain a maser by injecting these Rydberg atoms into a superconducting cavity of a high quality factor. The injection rate is such that on the average there is less than one atom present inside the resonator at a time. A transition between two neighboring Rydberg levels is resonantly coupled to a mode of the cavity field. Due to the high quality factor of the cavity the radiation decay time is much larger than the characteristic time of the atom-field interaction, which is given by the inverse of the single photon Rabi frequency. Therefore it is possible to observe the dynamics[26] of the energy exchange between atom and field mode leading to collapse and revivals in the Rabi oscillations[27,28]. Moreover, a field is build up inside the cavity when the mean time between the atoms injected into the cavity is shorter than the cavity decay time.

The detailed experimental setup of the one-atom maser is shown in Fig. 1. A highly collimated beam of rubidium atoms passes through a Fizeau velocity selector. Before entering the superconducting cavity, the atoms are excited into the upper maser level $63p_{3/2}$ by the frequency doubled light of a cw ring dye laser. The laser frequency is stabilized onto the atomic transition $5s_{1/2} \rightarrow 63p_{3/2}$ which has a width determined by the laser linewidth and the transit time broadening corresponding to a total of a few MHz. In this way, it is possible to prepare a very stable beam of excited atoms. The ultraviolet light is linearly polarized parallel to the electric field of the cavity. Therefore only $\Delta m = 0$ transitions are excited by both the laser beam and the microwave field.

The superconducting niobium maser cavity is cooled down to a temperature of 0.5 K by means of a He3 cryostat. At such a low temperature the number of thermal photons is reduced to about 0.15 at a frequency of 21.5 GHz. The cryostat is carefully designed to prevent room temperature microwave photons from leaking into the cavity. This would considerably increase the temperature of the radiation field above the temperature of the cavity walls. The quality factor of the cavity is $3 \cdot 10^{10}$ corresponding to a photon storage time of about 0.2 s.

The cavity is carefully shielded against magnetic fields by several layers of cryoperm. In addition, three pairs of Helmholtz-coils are used to compensate the earth magnetic field to a value of several mG in a volume of $10 \times 4 \times 4$ cm^3. This is necessary in order to achieve the high quality factor and to prevent the different magnetic substates of the maser levels from mixing during the atom-field interaction time. Two maser transitions from the $63p_{3/2}$ level to the $61d_{3/2}$ and to the $61\ d_{5/2}$ level are studied.

To demonstrate maser operation, the cavity is tuned over the $63p_{3/2}-61d_{3/2}$ transition and the flux of atoms in the excited state is recorded simultaneously. Transitions from the initially prepared $63p_{3/2}$ state to the $61d_{3/2}$ level (21.50658 GHz) are detected by reduction of the electron count rate.

In the case of measurements at a cavity temperature of 0.5 K, shown in Fig. 2, a reduction of the $63p_{3/2}$ signal can be clearly seen for atomic fluxes as small as 1750 atoms/s. An increase in flux causes power broadening and a small shift. This shift is attributed to the ac Stark effect, caused pre-

dominantly by virtual transitions to neighboring Rydberg
levels. Over the range from 1750 to 28000 atoms/s the field
ionization signal at resonance is independent of the particle
flux which indicates that the transition is saturated. This,
and the observed power broadening show that there is a
multiple exchange of photons between Rydberg atoms and the
cavity field.

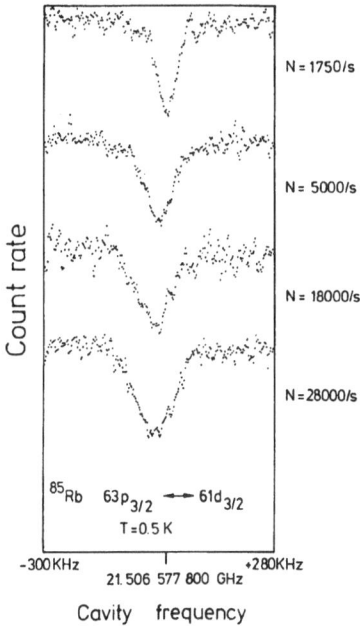

Fig. 2. Maser transition of the one-atom maser manifests
        itself in a decrease of atoms in excited state. Flux
        of excited atom, N, governs pump intensity. Power
        broadening of the resonance line indicates maser
        activity.

For an average transit time of the Rydberg atoms through the
cavity of 50 μs and a flux of 1750 atoms/s we obtain that
approximately 0.09 Rydberg atoms are in the cavity on the
average. According to Poisson statistics this implies that
more than 90 % of the events are due to single atoms. This
clearly demonstrates that single atoms are able to maintain a
continuous oscillation of the cavity with a mean number of
photons between unity and several hundreds.

THE ONE-ATOM-MASER -- A SOURCE OF NONCLASSICAL LIGHT

Send a stream of excited atoms through a superconducting
cavity with a single field mode coupled to a single transition
of the atom. An atom may deposit a photon in the cavity. The
next atom entering the cavity interacts with the field created
by all the earlier atoms and so on. This summarizes in brief
the underlying mechanism of the one-atom maser. But what are

the statistical properties of the electromagnetic field built up by this iterative process? This is the topic of the present section.

Electromagnetic radiation can show nonclassical properties[19,20], that is, properties that cannot be explained by classical probability theory. Loosely speaking we need to invoke "negative probabilities" to get deeper insight into these features. We know of essentially three phenomena which demonstrate the nonclassical character of light; photon antibunching[29], sub-Poissonian photon statistics[30] and squeezing[31]. Mostly methods of nonlinear optics are employed to generate nonclassical radiation. However, also the fluorescence light from a single atom caught in a trap exhibits nonclassical features[32,33].

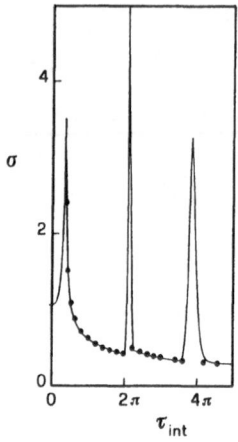

Fig. 3. Depending on the interaction time, τ_{int}, the normalize variance of the photon number distribution $\sigma = \Delta n/\langle n\rangle^{1/2}$, exhibits super-Poissonian and sub-Poissonian statistics, that is, $\sigma > 1$ and $\sigma < 1$ respectively. The solid line is taken from Ref. 21 whereas the dots are obtained by putting standard laser theory to use.

Another nonclassical light generator is the one-atom maser. We recall that the Fizeau velocity selector preselects the velocity of the atoms: Hence the interaction time is well-defined which leads to conditions usually not achievable in standard masers. Under appropriate conditions the number distribution of the photons in the cavity is sub-Poissonian[21,22] that is, narrower than a Poisson distribution as shown in Fig. 3. Even a number state that is, a state of well-defined photon number can be generated[23,24] using a cavity with a high enough quality factor and with no thermal photons in the cavity to begin with as depicted in Fig. 4. Both conditions can be fulfilled when the superconducting cavity is operated at very low temperatures, that is, at temperatures smaller than 0.5 K. In this case more interesting features such as trapping states of the cavity[34] make their appearance.

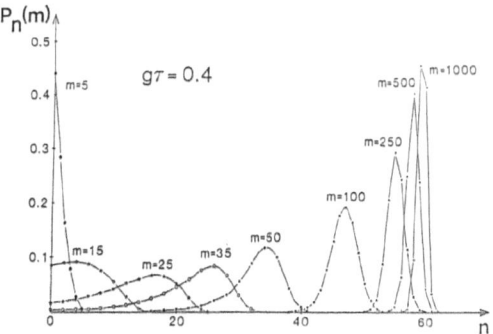

Fig. 4. Generation of a number state in a one-atom maser. Probability of obtaining n photons in the cavity after m atoms have passed. The curves are calculated for a fixed, normalized interaction time of 0.4.

Unfortunately, the measurement of the nonclassical photon statistics in the cavity is not that straightforward. The measurement process of the field invokes the coupling to a measuring device whereby losses lead inevitably to a destruction of the nonclassical properties. The ultimate technique to obtain information about the field employs the Rydberg atoms themselves: Measure the statistics via the dynamical behavior of the atoms in the radiation field, that is, via the collapse and the revivals of the Rabi oscillations, that is one possibility. However, a much more conclusive approach probes the population of the atoms in the upper and lower maser levels when they leave the cavity. In this case, the interaction time is kept constant. Moreover, this measurement is relatively easy since electric fields can be used to perform a selective ionization of the atoms. The detection sensitivity is sufficient so that the atomic statistics can be investigated. This technique maps the photon statistics of the field inside the cavity onto the atomic statistics. Experiments carried out along these lines have shown that the variance of the maser photon number distribution can show up to 70 % sub-Poissonian statistics[35].

We conclude this section by emphasizing that the one-atom maser is a unique device, and that for three reasons: (1) it is the first maser which sustains oscillations with less than one atom on average, (2) this setup allows to study in detail the conditions necessary to obtain nonclassical radiation especially sub-Poisson light and (3) it is possible to come to a complete physical understanding of the generation process of pure quantum light.

THE ONE-ATOM-MASER -- A NEW PROBE OF COMPLEMENTARITY IN QUANTUM MECHANICS

The electromagnetic field of a one-atom maser in a state of well-defined photon number, that is, in a number state? Yes! This state extremely fragile and impossible to couple outside of the cavity? Yes! But what is the use of such a nonclassical state? Test of quantum theory of measurement? Yes! But how? This is the crucial question discussed in the present section.

Complementarity[36], that is, the wave-particle duality of nature, lies at the heart of quantum mechanics: Matter sometimes displays wave-like properties manifesting themselves in interference phenomena, and at other times it displays particle-like behavior thus providing "which-path" information.

No other experiment illustrates this wave-particle duality in a more striking way than the classic Young's double-slit experiment[37]. Here we find it impossible to tell which slit light went through while observing an interference pattern. In other words, any attempt to gain which-path information disturbs the light so as to wash out the interference fringes. This point has been emphasized by N. Bohr in his rebuttal to Einstein's ingenious proposal of using recoiling slits[5], to obtain "which-path" information while still observing interference. The physical positions of the recoiling slits, Bohr argues, are only known to within the uncertainty principle. This error contributes a <u>random phase shift</u> to the light beams which destroys the interference pattern.

Such random-phase arguments, illustrating in a vivid way how the "which-path" information destroys the coherent-wave-like interference aspects of a given experimental setup, are appealing. Unfortunately, they are incomplete: In principle, and in practice, it is possible to design experiments which provide "which-path" information via detectors which <u>do not disturb</u> the system in any noticeable way. Such "Welcher Weg" - (German for "which-path") detectors have been recently considered within the context of studies involving spin coherence[38]. In the present section we describe a quantum optical experiment[15], which shows that the loss of coherence occasioned by "which-path" information, that is, by the presence of a "Welcher Weg"-detector, is due to the establishing of quantum correlations. It is in no way associated with large random-phase factors as in Einstein's recoiling slits. The two essential ingredients of this novel "Welcher Weg"-detector are the one-atom maser, discussed in the preceding sections, and the quantum beat concept[8].

In a quantum beat experiment[8] atoms are excited by a short laser pulse to a coherent superposition of the two excited states $|a\rangle$ and $|b\rangle$ separated by a frequency difference $\Delta\omega$. They are allowed to decay spontaneously to a lower level $|c\rangle$ as shown in Fig. 5. The spontaneously emitted radiation shows "beats", that is, temporal fringes. These beats are a result of the indistinguishability of the two "paths" of excitation $|c\rangle \rightarrow |a\rangle \rightarrow |c\rangle$ and $|c\rangle \rightarrow |b\rangle \rightarrow |c\rangle$ analogous to the interfering paths

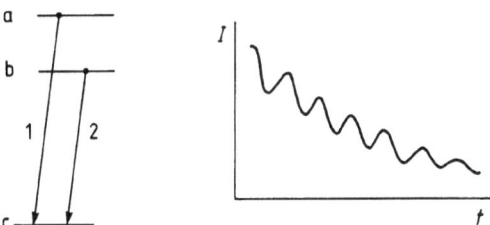

Fig. 5. Scheme for a quantum-beat experiment. Coherent superposition of upper levels a and b decay to ground state
c. The detector current shows a modulation in addition
to the usual exponential decay.

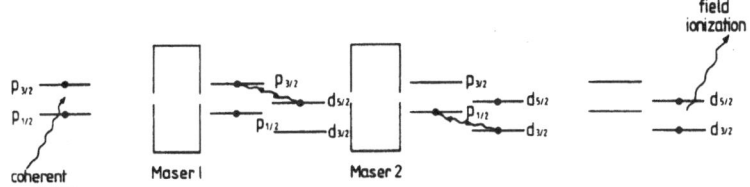

Fig. 6. A micromaser "Welcher Weg" (which-path) detector. In
passing through the first micromaser a Rb atom, prepared in a coherent superposition of $63p_{3/2}$ and $63p_{1/2}$
makes a transition from $63p_{3/2}$ to $61d_{5/2}$; and in the
second, from $63p_{1/2}$ to $61d_{3/2}$. Depending on the maser
state - coherent or number state - quantum beats
detected via field ionization do or do not occur.

of the light in the double-slit experiment. Hence the quantum
beat experiment represents only the interference side of the
coin of complementarity. To recognize its "Welcher Weg"-side
we consider the experimental arrangement of Fig. 6. Here we
depict the more complicated atomic configuration of a four-
level atom, and use two consecutive one-atom masers. We excite
a coherent superposition of the $63p_{3/2}$ and the $63p_{1/2}$ levels
of Rb^{85}. The velocity of the atoms is adjusted such that a
transition from $63p_{3/2}$ to $61d_{5/2}$ is guaranteed in the first
cavity; moreover, the second cavity induces the transition
from $63p_{1/2}$ to $61d_{3/2}$. The coherence transfered in this way
from the two initially excited states to the two d-states can
easily be detected by a field ionization quantum beat experiment[39]. But why does this experimental set-up provide
"Welcher Weg"-information? Where is the path-information?

The answers to these questions come to light in the ionization
current[16)]

$$I(t) = I_0 \, e^{-\gamma t} \left[1 + e^{i\Delta\omega t} \, \langle \Phi_1^{(f)}, \Phi_2^{(i)} | \Phi_1^{(i)}, \Phi_2^{(f)} \rangle + c.c. \right].$$ (1)

Here I_0 denotes the constant component of the current and
$\gamma_a = \gamma_b = \gamma$ are the natural linewidth of the two excited, $63p_{3/2}$
and $63p_{1/2}$ levels. The initial and final states of the
electromagnetic field in the j-th one-atom maser, are denoted
by $| \Phi_j^{(i)} \rangle$ and $| \Phi_j^{(f)} \rangle$, respectively. Equation (1) spot-
lights the crucial role of the field states in this play of
quantum beats "to be or not to be".

Let us prepare for example, both one-atom masers in coherent
states[40)] $| \Phi_j^{(i)} \rangle = | \alpha_j \rangle$ of large average photon number
$\langle m \rangle = | \alpha_j |^2 \gg 1$. The Poissonian photon distribution of such a
coherent state is very broad, $\Delta m \sim \alpha \gg 1$. Hence the two fields
are not changed much by the addition of a single photon
associated with the two corresponding transitions. We may
therefore write

$$| \Phi_j^{(f)} \rangle \cong | \alpha_j \rangle$$

which to a very good approximation yields

$$\langle \Phi_1^{(f)}, \Phi_2^{(i)} | \Phi_1^{(i)}, \Phi_2^{(f)} \rangle \cong \langle \alpha_1, \alpha_2 | \alpha_1, \alpha_2 \rangle = 1$$

Thus the interference cross term in Eq. (1) is present giving
so rize to quantum beats.

When we, however, prepare both maser fields in number
states[23;24)], $| n_j \rangle$, the situation is quite different. After
the transitions of the atom to the d-states, that is, after
emitting consecutively the two photons, the final states in
the two cavities read

$$| \Phi_j^{(f)} \rangle = | n_j + 1 \rangle$$

and hence

$$\langle \Phi_1^{(f)}, \Phi_2^{(i)} | \Phi_1^{(i)}, \Phi_2^{(f)} \rangle = \langle n_1 + 1, n_2 | n_1, n_2 + 1 \rangle = 0,$$

that is, the coherence cross term vanishes, and no quantum beats emerge.

On first sight this might seem a bit surprising when we recall that in the case of a coherent state the transitions did not destroy the coherent cross term, that is, it did not affect the temporal interference fringes. However, in the example of number states we can by simply "looking" at the one-atom maser state, tell which "path" the excitation took. When we find, for example, the first maser cavity in a state of n_1+1 photons we know that the internal atomic "path",

ground state $\rightarrow 63p_{3/2} \rightarrow 61d_{5/2} \rightarrow$ ionization

was followed. Likewise the knowledge of having n_2+1 photons in the second cavity allows us to deduce the path of excitation

ground state $\rightarrow 63p_{1/2} \rightarrow 61d_{3/2} \rightarrow$ ionization.

Hence the states of the electromagnetic field in the cavities play the role of the recoiling plate of the double-slit experiment in extracting which-path information. Moreover, it does not matter if we actually look at the states or not. Having the information stored there is enough to destroy the interference!

The use of Rydberg atoms in this experiment delivers us from the necessity of large photon numbers to produce the atomic transitions. In fact, the vacuum interaction, that is, spontaneous emission is sufficient to insure the transition. Thus we are relieved of preparing number states, $| n_j \rangle$, of large photon numbers and instead prepare our masers in the easiest number state of all, the vacuum.

One important condition, however, has to be fulfilled in this type of experiment. The beam density must be extremely low to guarantee that the cavities return to the vacuum state before the next atom enters. Otherwise a field different from the vacuum would build up in the cavities and the which-path information would be lost.

We conclude this section by emphasizing again that this new and potentially experimental example of wave-particle duality and observation in quantum mechanics displays a feature which makes it distinctly different from the Bohr-Einstein recoiling-slit experiment. In the latter the coherence, that is, the interference, is lost due to a phase disturbance of the light beams. In the present case, however, the loss of coherence is due to the correlation established between the system and the one-atom maser. Random-phase arguments never enter the discussion. We emphasize that the argument of the number state not having a well-defined phase is not relevant here; the important dynamics is due to the atomic transition. It is the fact that which-path information is made available which washes out the interference cross terms.

A single ion at rest, unperturbed by its environment and forced into this state for hours -- only a physicist's dream? No, it is reality - made possible by the combination of electromagnetic traps and laser technology. The Penning trap or the dynamical Paul-trap[41] developed in the thirties and the late fifties, respectively, give the experimenter a unique tool to isolate a single ion from its surroundings. Tunable lasers serve as a torch light and force the ion to fluoresce; simultaneously it is cooled[42] to Milli- or even Micro-Kelvin temperatures. An ion driven into saturation scatters roughly 10^8 photons per second leading to a high detection probability and at the same time to a reduction of the ion's kinetic energy via photon recoil.

We devote the present section to approach a fascinating aspect of trap physics: phase transitions of a few ions between an ordered, crystal-like state depicted in Fig. 7 and a chaotic cloud-like state. Here we concentrate on, review and highlight our own experimental efforts on this subject. For the exciting work performed at other laboratories around the world we refer to Ref. 18.

The first problem we face is to trap an ion. In contrast to neutral atoms, ions due to their charge can easily be influenced by electromagnetic fields. Apply a constant voltage, U_0, to a ring electrode with two end caps, such that the ring electrode is positively and the end-caps are negatively charged. Do we achieve a three-dimensional confinement? No! In such an electrostatic configuration a positively charged ion performs a stable motion in one direction and an unstable motion in the orthogonal direction. Hence the potential creating such a motion enjoys a saddle point at trap center. Would another, more sophisticated arrangement of electrodes achieve a three-dimensional confinement? No! The laws of electrostatics, more precisely, Poisson's equation prevent that.

Is there any way out of this one-way street? Yes! A trap solely based on electric fields -- however, now time dependent -- exists and is easily understandable from the saddle point potential. When the particle starts to roll down the unstable hillside, rotate the potential, so that the particle suddenly sees a rising rather than a decreasing potential and create so a dynamical binding. The rotation of the saddle point model translates itself into the electrical ring electrode - end-cap setup as an oscillatory exchange of the polarity of the electrodes, that is, into applying an alternating voltage, V_0. The equation of motion of an ion in such a situation is the Mathieu differential equation well-known in classical mechanics which depending on the two voltages U_0 and V_0 allows stable and unstable solutions. Under appropriate conditions we hence find dynamical, threedimensional confinement in such a Paul-trap.

So far we have learned how to keep an ion trapped. But how to get the ion into the trap? The most obvious solution: create it in there! Ionize for example a neutral atomic beam via electron collision at the trap center. Unfortunately the resulting trapped ion has a lot of kinetic energy rendering it useless for most of the applications such as spectroscopy. To extract kinetic energy from the ion, that is the immediate task -- radiation pressure[42] exerted by a laser is the tool to overcome this obstacle.

Let an atom or ion travel against an electromagnetic wave whose frequency is appropriately adjusted to be smaller than the atomic resonance transition frequency. Hence the energy of the photon is not sufficient to excite the atom. However, the ion can extract the missing energy from its motion thus reducing its kinetic energy. That is the most elementary explanation of laser cooling.

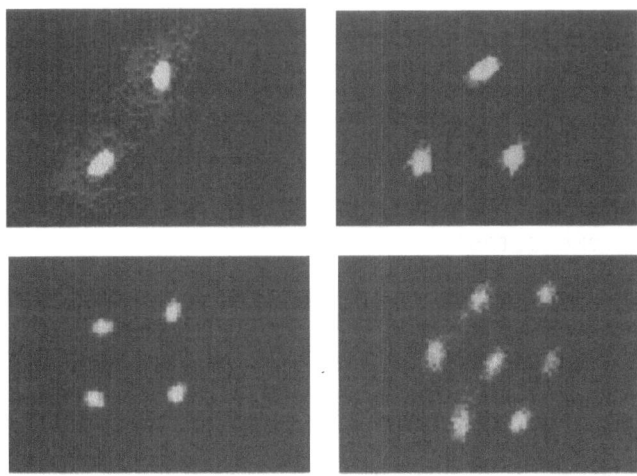

Fig. 7. Two, three, four and seven ions confined by the dyna-
        mical potential of a Paul-trap and crystallized to an
        ordered structure in a plane perpendicular to the
        symmetry axis of the trap. The average separation of
        the ions is 20 μm.

The heart of our experiment is a Paul-trap which is larger than most of the other traps used in laser experiments and which is described in great detail in Refs. 17 and 33. The fluorescence from the ions is observed through a hole in the upper end-cap. The large size of the trap affords a large solid angle for detecting the fluorescence radiation. This is done either by a photomultiplier, or by means of a photon-counting image system. For the observation of the ions, the cathode of the image system is placed in the image plane of a microscope objective attached to the trap. In this way the ion pictures of Figs. 7 have been obtained.

Detailed studies reveal that the behavior of the ions in the trap is governed by the trap-voltage V_0, the laser detuning and the laser power P. When we for example, increase the laser power while keeping Δ and V_0 constant the ions undergo a transition from a state of erratic motion to a situation in which the ions arrange themselves in regular structures similar to the electrons in Thomson's turn of the century raisin pudding-atom[43] -- or the Wigner-crystal[44]. In such a crystalline pattern the mutual ion-Coulomb repulsion is compensated by the external, dynamical trap potential. The two phases -- crystal-cloud -- can be verified experimentally by direct observation (see Fig. 7) with the help of the highly sensitive imaging system, theoretically by analyzing ion-trajectories of Monte Carlo simulations on a computer[17], and by discontinuous jumps in the excitation spectrum[17].

We conclude this section by noting that similar transitions to ordered structures can also be induced when we decrease the alternating voltage while keeping laser power and laser detuning constant, or when we alter the laser detuning. Moreover, there appears to be bistability and hysteresis associated with these phase transitions. Such a hysteresis behavior must be expected with laser-cooled ions because the cooling power of the laser is strongly dependent upon the details of the velocity distribution of the ions.

SUMMARY AND CONCLUSIONS

The laser -- a magic wand to visualize and manipulate a single ion caught in a miniature trap, a refrigerator to freeze an ion cloud into an ion crystal and an inducer of few body phase transitions between these two states -- that is in one pregnant sentence the summary of our studies on ions stored in a Paul-trap. But what is a possible further development? Investigation of collective effects? Yes! The distance between the constituents of the ion crystal depends on the trap para-meters: they can be changed as to force the ions closer together. Ion separations on the order of the wavelength of the resonance radiation seem to be achievable. Moreover, collective effects in the emission of the fluorescence photons from such a compact crystalline structure are predicted to make their appearance. New laser cooling techniques[45] allow to freeze the stored ions to temperatures in the range of Microkelvins. At those low temperatures the de Broglie wave-length of the ions is comparable to the distances in a crystal configuration- a novel structure with new interesting properties might arise.

Intriguing tests of the complementarity principle via the one-atom maser is another theme of the present paper. Send an atom prepared in a coherent superposition of two quantum levels through two consecutive one-atom masers. In this way, transfer the population to two neighboring levels. Is the initial coherence between the two atomic states as manifested in an oscillatory ionization signal -- temporal interference -preserved? To elucidate this question, imagine starting from coherent cavity fields with a large mean number of photons. In such a coherent state the fluctuations in photon number are large. Hence the two photons emitted in the transfer process, and caught in the cavities, do not change the field state significantly. One cannot deduce information concerning the transfer by studying the cavity fields. Hence the coherence is preserved giving rise to quantum beats. However, if the experiment is started from a state of well-defined photon number, it is possible to detect the increase, by one, in photon number induced by the population transfer. This destroys the coherence of the state, just as identifying which slit a photon passes through in the Young's double-slit experiment destroys the coherence and hence the interference fringes in that experiment.

References

1) A. Einstein, B. Podolsky and N. Rosen: Phys.Rev. 47:777 (1935).
2) N. Bohr, Nature 136:65 (1935); Phys.Rev. 48:696 (1935).
3) For a review on the pioneering experiments on the violation of Bell's inequalities see J.F. Clauser and A. Shimony, Rep. Prog. Phys. 41:1881 (1978), and in particular E.S. Fry and R.C. Thompson: Phys.Rev.Lett. 37:465 (1976); for the most recent experiments see A. Aspect, P. Grangier and G. Roger: Phys.Rev.Lett. 47:460 (1981); ibid 49:91 (1982); A. Aspect, J. Dalibard and G. Roger: Phys.Rev.Lett. 49:1804 (1982).
4) J.S. Bell, Physics 1:195 (1964).
5) For an excellent presentation of the Bohr-Einstein dialogue see chapter 1 in: J.A. Wheeler and W.H. Zurek: Quantum Theory and Measurement (Princeton University Press, Princeton, 1983) and in particular, the article by N. Bohr: Discussion with Einstein on Epistemological problems in Atomic Physics.
6) J.A. Wheeler: Mathematical Foundations of Quantum Theory (Academic Press, N.Y., 1978) p. 9; J.A. Wheeler: Some Strangeness in the Proportion (Addison-Wesley, Reading, 1980) p. 341; J.A. Wheeler: Problems in the Foundations of Physics, International School of Physics "Enrico Fermi", course LXXII (North-Holland, Amsterdam, 1979) p. 395; W.A. Miller and J.A. Wheeler: Proc. 2nd Int. Symp. Foundations of Quantum Mechanics, eds. S. Kamefuchi et al. (Phys. Soc. of Japan, Tokyo, 1984) p. 140.
7) The delayed-choice aspect has been first mentioned in connection with the Heisenberg microscope by C.F. von Weizsäcker, Zs. f. Physik 70:114 (1931); ibid 118:489 (1941).
8) S. Haroche in: High Resolution Laser Spectroscopy (Springer, New York, 1976) 253; see also W.W. Chow, M.O. Scully and J.O. Stoner: Phys.Rev. A11:1380 (1975).
9) T. Hellmuth, A.G. Zajonc and H. Walther: Proc. of Symposium on the Foundations of Modern Physics (World Scientific Publ. Corp., 1985); T. Hellmuth: Interferenzexperimente zum quantenmechanischen Meßprozeß: Eine Nachwahlversion des Youngschen Doppelspalt-Versuchs (Dissertation, Ludwig-Maximilians-Universität München, 1985); T. Hellmuth, H. Walther, A. Zajonc and W. Schleich: Phys. Rev. A35:2532 (1987); W. Schleich and H. Walther: Proc. 2nd Int. Symp. Foundations of Quantum Mechanics, eds. M. Namiki et al. (Phys. Soc. of Japan, Tokyo, 1986) p. 25.

10) W.C. Wickes, C.O. Alley and O. Jakubowicz, in ref. 5, p. 457; C.O. Alley, O. Jakubowicz, C.A. Steggerda and W.C. Wickes: Proc. 1st Int. Symp. Foundations of Quantum Mechanics, eds. S. Kamefuchi et al. (Phys. Soc. of Japan, Tokyo, 1984) p. 158; C.O. Alley, O.G. Jakubowicz and W.C. Wickes: Proc. 2nd Proc. Int. Symp. Foundations of Quantum Mechanics, eds. M. Namiki et al. (Phys. Soc. of Japan, Tokyo, 1986) p. 36.

11) For a recent delayed-choice experiment based on a single-photon state created by parametric fluorescence see J. Baldzuhn, E. Mohler and W. Martienssen, Z. Phys. B 77:347 (1989).

12) E. Schrödinger: The British Journal for the Philosophy of Science 3 (10) (1952).

13) W. Paul, O. Osberghaus and E. Fischer: Forschungs-berichte des Wirtschafts- und Verkehrsministeriums Nordrhein-Westfalen (1958) 415; E. Fischer: Z. Phys. 156:1 (1959); H.G. Dehmelt: Advances in Laser Spectroscopy, eds. F.T. Arecchi, F. Strumia and H. Walther (Plenum, New York, 1983) p. 153.

14) H.G. Dehmelt: Bull.Amer.Soc. 20:60 (1975); W. Nagourny, J. Sandberg and H.G. Dehmelt: Phys.Rev.Lett. 56:2797 (1986); Th. Sauter, W. Neuhauser, R. Blatt and P. Toschek, Phys.Rev.Lett. 57:1696 (1986); J.C. Bergquist, R.G. Hulet, W.M. Itano and D.J. Wineland, Phys.Rev.Lett. 57:1699 (1986); for a review see R. Brewer: Proc. 2nd Int. Symp. Foundations of Quantum Mechanics, eds. M. Namiki et al. (Phys. Soc. of Japan, Tokyo, 1987) p. 257.

15) M.O. Scully and H. Walther: Phys.Rev. A39:5229 (1989); see also W. Schleich and P.V.E. McClintock, Nature 339:257 (1989).

16) D. Meschede, H. Walther and G. Müller: Phys.Rev.Lett. 54:551 (1985); for a review see F. Diedrich, J. Krause, G. Rempe, M.O. Scully and H. Walther: IEEE J. Quantum Electron. 24:1314 (1988).

17) F. Diedrich, E. Peik, J.M. Chen, W. Quint and H. Walther, Phys.Rev.Lett. 59:2931 (1987); R. Blümel, J.M. Chen, E. Peik, W. Quint, W. Schleich, Y.R. Shen and H. Walther, Nature 334:309 (1988); W. Quint, W. Schleich and H. Walther, Physics World 2:30(8) (1989); La Recherche 20:1197 (1989); R. Blümel, C. Kappler, W. Quint and H. Walther; Phys.Rev. A40:808 (1989).

18) For related work on crystallization in ion traps, see D.J. Wineland, J.C. Bergquist, W.M. Itano, J.J. Bollinger and C.H. Manney, Phys.Rev.Lett. 59:2935 (1987); S.L. Gilbert, J.J. Bollinger and D.J. Wineland, Phys.Rev. Lett. 60 (1988):2022; J. Hoffnagle, R.G. DeVoe, L. Reyna and R.G. Brewer, Phys.Rev.Lett. 61:255 (1988); Th. Sauter, H. Gilhaus, I. Siemers, R. Blatt, W. Neuhauser and P.E. Toschek: Z. Phys. D 10:153 (1988); for a review see B.G. Levi, Physics Today 41:17 (9) (1988).

19) D.F. Walls: Nature 280:451 (1979); see also articles in: Photons and Quantum Fluctuations, edited by E.R. Pike and H. Walther (Hilger, Bristol, 1988).

20) D.F. Walls: Nature 306:141 (1983); ibid 324:210 (1986); see also the various articles in: Squeezed and Non-classical Light, edited by P. Tombesi and E.R. Pike (Plenum Press, New York, 1988).

21) P. Filipowicz, J. Javanainen and P. Meystre, Optics Commun. $\underline{58}$:327 (1986); Phys.Rev. $\underline{A34}$:3077 (1986).

22) L. Lugiato, M.O. Scully and H. Walther: Phys.Rev. $\underline{A36}$:740 (1987).

23) J. Krause, M.O. Scully and H. Walther, Phys.Rev. $\underline{A36}$:4547 (1987); J. Krause, M.O. Scully, T. Walther and H. Walther, Phys.Rev. $\underline{A39}$:1915 (1989).

24) P. Meystre: Optics Lett. $\underline{12}$ (1987) 669; P. Meystre: Squeezed and Nonclassical Light, edited by P. Tombesi and E.R. Pike (Plenum, New York, 1988) p. 115.

25) For a review see the following articles by S. Haroche and J.M. Raimond: Advances in Atomic and Molecular Physics, Vol. $\underline{20}$ (Academic Press, New York, 1985) p. 350; J.A. Gallas, G. Leuchs, H. Walther and H. Figger: Advances in Atomic and Molecular Physics, Vol. $\underline{20}$ (Academic Press, New York, 1985) p. 413.

26) E.T. Jaynes and F.W. Cummings, Proc. IEEE $\underline{51}$:89 (1963).

27) See for example: J.H. Eberly, N.B. Narozhny and J.J. Sanchez-Mondragon, Phys.Rev.Lett. $\underline{44}$:1323 (1980) and references therein.

28) G. Rempe, H. Walther and N. Klein: Phys.Rev.Lett. $\underline{58}$:353 (1987).

29) H.J. Kimble, M. Dagenais and L. Mandel, Phys.Rev.Lett. $\underline{39}$:691 (1977); Phys.Rev. $\underline{A18}$:201 (1978); J.D. Cresser, J. Häger, G. Leuchs, M. Rateike and H. Walther: Dissipative Systems in Quantum Optics (Springer, Berlin, 1982) p. 21.

30) M.C. Teich and B.E.A. Saleh: Progress in Optics XXVI, edited by E. Wolf (North Holland, Amsterdam, 1988) p. 1.

31) For a review on squeezed states of the radiation field see the two special issues of JOSA B $\underline{4}$ (10) (1987) and J. Mod. Optics $\underline{34}$ (6-7) (1987).

32) H.J. Carmichael and D.F. Walls, J. Phys. $\underline{9B}$:1199 L 43 (1976).

33) H. Walther and F. Diedrich, Phys.Rev.Lett. $\underline{58}$:203 (1987).

34) P. Meystre, G. Rempe and H. Walther, Optics Letters $\underline{13}$:1078 (1988).

35) G. Rempe, F. Schmidt-Kaler and H. Walther: to be published.

36) See for example, D. Bohm, Quantum Theory (Prentice Hall, Englewood Cliffs, 1951) or M. Jammer, The Philosophy of Quantum Mechanics (Wiley, New York, 1974).

37) A detailed analysis of Einstein's version of the double-slit experiment is given by W. Wootters and W. Zurek, Phys.Rev. D $\underline{19}$:473 (1979); see also Ref. 5.

38) B.-G. Englert, J. Schwinger and M.O. Scully: Found.Phys. $\underline{18}$:1045 (1988); J. Schwinger, M.O. Scully and B.-G. Englert, Z.Phys. D $\underline{10}$:135 (1988); M.O. Scully, B.-G. Englert and J. Schwinger, Phys.Rev. $\underline{A40}$:1775 (1989).

39) G. Leuchs, S. Smith and H. Walther: Laser Spectroscopy IV $\underline{21}$, edited by H. Walther and K.W. Rothe (Springer-Verlag, Berlin, 1979), p. 255; G. Leuchs and H. Walther: Z.Phys. A $\underline{293}$:93 (1979).

40) See for example M. Sargent, M.O. Scully and W. Lamb, Laser Physics (Addison-Wesley, Reading, MA, 1973).

41) For a review on the mechanism of Penning and Paul traps see for example H.G. Dehmelt: Advances in Atomic and Molecular Physics, Vol. $\underline{3}$ edited by D.R. Bates and I. Estermann (Academic, New York, 1967).

42) S. Stenholm, Rev. Mod. Phys. 58:699 (1986).
43) L. Föppl, J. reine angew. Math. 141:251 (1912).
44) E.P. Wigner, Trans. Faraday Soc. 34:678 (1938).
45) See for example A. Aspect, E. Arimondo, R. Kaiser,
     N. Vansteenkiste and C. Cohen-Tannoudji, Phys.Rev.Lett.
     61:826 (1988).

CHAOS AND COMPLEXITY IN QUANTUM OPTICS

F.T. Arecchi
Dept. of Physics University of Florence and
Istituto Nazionale di Ottica, Firenze

ABSTRACT

This is a review of how deterministic chaos enters quantum optics, introducing new features unexpected in the first twenty years of the laser era. After a general introduction to chaos in lasers, attention is focused to CO_2 lasers with feedback, which display the so called Shil'nikov chaos.

1. Introduction

Quantum optics from its beginning in 1960 with the first laser was considered as the physics of coherent and intrinsically stable radiation sources. Lamb's semiclassical theory[1] showed the role of the EM field in the cavity in ordering the phases of the induced atomic dipoles, thus giving rise to a macroscopic polarization and making possible a description in terms of very few collective variables. In the case of a single-mode laser and a homogeneous-gain line this meant just five coupled degrees of freedom, namely, a complex field amplitude E, a complex polarization P, and a population inversion ΔN. A corresponding quantum theory, even for the simplest model laser (the so called Dicke model, that is, a discrete collection of modes interacting with a finite number of two-level atoms) does not lead to a closed set of equations. However the interaction with other degrees of freedom acting as a thermal bath (atomic collisions, thermal radiation) provides truncation of high-order terms in the atom-field interation. The problem may be reduced to five coupled equations (the so-called Maxwell-Bloch equations) but now they are affected by noise sources to accound for the coupling with the thermal bath. Being stochastic, or Langevin equations, the corresponding solution in closed form refers to a suitable weight function or phase-space density. Anyway the average motion matches the semiclassical one, and fluctuations play a negligible role if one excludes the bifurcation points where there are changes of stability in the stationary branches. Leaving out the peculiar statistical phenomena which characterize the threshold points and which suggest a formal analogy with thermodynamic phase transitions,[2] the main point of interest is that a single-mode laser provides a highly stable or coherent radiation field. From the point of view of the associated information, the standard interferometric or spectroscopic measurements of classical optics, relying on average field values or on their first-order correlation functions, are insufficient. In order to characterize the statistical features of Quantum Optics it was necessary to make extensive use of photon statistics[3,4].

Coherence is equivalent to having a stable fixed point attractor and this does not depend on details of the nonlinear coupling, but on the number of relevant degrees of freedom. Since such a number depends on the time scales on which the output field is observed, coherence becomes a question of time scales. This is the reason why for some lasers coherence is a robust quality, persistent even in the presence of strong perturbations, whereas in other cases coherence is easily destroyed by the manipulations common in the laboratory use of lasers, such as modulation, feedback or injection from another laser.

Sect. 2 is a general presentation of low-dimensional chaos in lasers, including the description of the relevant measurements upon which any assessment on chaos has to rely. Sect. 3 describes features of a new type of chaos, the Shil'nikov chaos.

2. Deterministic chaos

Until recently the current point of view was that a few-body dynamics was fully predictable, and that only addition of noise sources, due to coupling with a thermal reservoir, could provide statistical fluctuations. Lack of long-time predictability, or turbulence, was considered as resulting from the interaction of a large number of degrees of freedom, as in a fluid above the critical Reynolds number (Landau-Hopf model of turbulence).

On the contrary, it is now known that even in systems with few degrees of freedom nonlinearities may give rise to expanding directions in phase space and this, together with the lack of precision in assigning initial conditions, is sufficient to induce a loss of predictability over long times.

This level of dynamical description was introduced by Poincaré, in dealing with the three-body problem in celestial mechanics. Already a three-body dynamic system is very different from the two-body problem since, in general, it may include an asymptotic instability. This means a divergence, exponential in time, of two phase space trajectories stemming from nearby initial points. The uniqueness theorem for solutions of differential systems always holds once one has assigned the coordinates of the initial point. However a fundamental difficulty arises. Only rational numbers can be assigned by a finite number of digits. A "precise" assignment of a real number requires an infinite acquisition time and an infinite memory capacity to store it, and neither of these two infinities is available to the physicist. Hence any initial condition implies a truncation. A whole range of initial conditions, even if small, is usually given and from within it trajectories may arise whose difference becomes sizeable after a given time, if there is an exponential divergence. As consequence predictions are in general limited in time and motions are complex, starting already from the three-body case. In fact, we know nowadays from very elementary topological considerations that a three-dimensional phase space corresponding to three coupled degrees of freedom is already sufficient to yield a positive Lyapunov exponent, and accordingly an expanding phase-space direction. This complexity is not due to coupling with a noise source as a thermal reservoir, but to sensitive dependence on initial conditions. It is called deterministic chaos.

2.1. Dynamical aspects of chaos

A dissipative system (i.e. with damping terms) does not conserve the phase-space volume. If we start with initial conditions confined in a hypersphere of radius ε , that is, with an initial phase volume

$$V_0 = \varepsilon^N$$

as time goes on, the sphere transforms into an ellipsoid with each axis modified by a time dependent factor. Its volume is

$$V_t = \varepsilon^N \exp\left(\Sigma_i \lambda_i t\right)$$

(λ_i: Lyapunov exponents). Since the volume has to contract, $V_t < V_0$, then

$$\Sigma_i \lambda_i < 0 .$$

We denote the sequence of λ exponents, starting from the smallest up to the highest as the Lyapunov spectrum. Let us consider for simplicity just the signs of nonzero λ_i, keeping the zero for $\lambda_i = 0$. We then describe a sequence of, negative, zero and positive λ_i as, e.g., $(- - 0 +)$. For N = 1, we have $(-)$ and a segment $V_0 = \varepsilon^1$ of initial conditions shrinks to a single point for $t \to \infty$, that is, the attractor is a fixed point. For N = 2 the system starts from $V_0 = \varepsilon^2$ and goes either to a fixed point $(- -)$, or to a limit cycle (-0). Chaotic motion $(-+)$ with $\lambda_+ < |\lambda_-|$ is forbidden in two dimensions by the Poincaré-Benedixon theorem. For N = 3, besides fixed point $(- - -)$, and limit cycle $(- - 0)$, we can have motion on a torus with two incommensurate frequencies $(- 0 0)$, but we can also have $(- 0 +)$, that is, a positive λ which gives an expanding direction along which we rapidly get uncertainty.
An example of chaotic motion is offered by the Lorenz model of hydrodynamic instabilities which corresponds to the following equations where the parameter values have been chosen in order to yield one positive Lyapunov exponent:

$$\dot{x} = -10x + 10y \ ,$$
$$\dot{y} = -y + 28x - xz \ ,$$
$$\dot{z} = -(8/3)z + xy \ . \tag{1}$$

The above considerations suggest the system will exhibit low-dimensional chaos, with the simplest phase-space topology allowing for the appearence of a positive Lyapunov exponent.

Focusing on these situations in quantum optics permits close comparison between experiments and theory. By purpose, I do not tackle the vast class of inhomogeneously broadened lasers, where it is extremely difficult to drive close correspondences between experiments and theory because of the large number of coupled degrees of freedom.

If we couple Maxwell equations with Schrödinger equations for N atoms confined in a cavity, and expand the field in cavity modes, keeping only the first mode E which goes unstable, this is coupled with the collective variables P and Δ describing the atomic polarization and population inversion as follows,

$$\dot{E} = -kE + gP \ ,$$
$$\dot{P} = -\gamma_\perp P + gE\Delta \ ,$$
$$\dot{\Delta} = -\gamma_\parallel(\Delta - \Delta_0) - 4g\,PE \ . \tag{2}$$

For simplicity, we consider the cavity frequency at resonance with the atomic resonance, so that we can take E and P as real variables and we have three coupled equations. Here, k, $\gamma_\perp$, $\gamma_\parallel$ are the loss rates for field, polarization and population, respectively, g is a coupling constant and Δ_0 is the population inversion which would be established by the pump mechanism in the atomic medium, in the absence of coupling. While the first equation comes from Maxwell equations, the two others imply the reduction of each atom to a two-level atom resonantly coupled with the field, that is, a description of each atom in a isospin space of spin 1/2. The last two equations are like Bloch equations which describe the spin procession in presence of a magnetic field.

The similarity of Maxwell-Bloch equations with Lorenz equations would suggest the easy appearence of chaotic instabilities in single-mode, homogeneous-line lasers. However, time-scale considerations rule out the full dynamics of (2) for most of the available lasers. Equations (1) have damping rates which lie within one order of magnitude of each other. On the contrary, in most lasers the three damping rates are wildly different from one another.

The following classification has been introduced

Class A (e.g., He-Ne, Ar, Kr, dye): $\gamma_\perp \simeq \gamma_\parallel \gg$ k. The last equations of (2) can be solved at equilibrium (adiabatic elimination procedure), thus the laser dynamics is described by one single nonlinear field equation. N = 1 means fixed point attractor, hence coherent emission.

34

Class B (e.g., ruby, Nd, CO): $\gamma_\perp \gg k \gtrsim \gamma_{\parallel}$. Only polarization is adiabatically eliminated (middle equation of (2)) and the dynamics is ruled by two rate equations for field and population. N = 2 allows also for periodic oscillations.

Class C (e.g., FIR lasers) $\gamma_{\parallel} \simeq \gamma_\perp \simeq k$. The complete set of (2) has to be used, hence Lorenz like chaos is feasible.

We have carried a series of experiments on the birth of deterministic chaos in CO_2 lasers (class B). In order to increase, by at least 1, the number of degrees of freedom, we have tested the following configurations:

i) Introduction of a time dependent parameter to make the system non autonomous. Precisely, an electro-optical modulator modulates the cavity losses at a frequency near the proper oscillation frequency Ω provided by a linear stability analysis, which for a CO_2 laser happens to lie in the 50-100 KHz range, providing easy and accurate sets of measurements.

ii) Injection of a signal from an external laser detuned with respect to main one, choosing the frequency difference near the above mentioned Ω. With respect to the external reference the laser field has two quadrature components which represent two dynamical variables. Hence we reach N = 3 and observe chaos.

iii) Use a bidirectional ring rather than a Fabry-Perot cavity. In the latter case the boundary conditions constrain the forward and backward waves, by phase relations on the mirror, to act as a single standing wave. In the former case forward and backward waves have just to fill the total ring length with an integer number of wavelengths but there are no mutual phase constrains, hence they act as two separate variables. Furthermore, when the field frequency is detuned with respect to the center of the gain line, a complex population grating arises from interference of the two counter-going waves, and as a result the dynamics becomes rather complex, requiring $N > 3$ dimensions.

iv) Add an overall feedback, besides that provided by the cavity mirrors, by modulating the losses with a signal provided by the output intensity. If the feedback has a time constant comparable with the population decay time, it provides a third equation sufficient ot yield chaos.

Notice that while methods (i), (ii) and (iv) require an external device, (iii) provides intrinsic chaos. In any case, since feedback, injection or modulation are currently used in laser applications, the evidence of chaotic regions puts a caution on the optimistic trust in the laser coherence.

2.2. Information aspects of chaos

Here we discuss what we measure to characterize chaos. We plot two of the three (or more) variables on a plane phase-space projection. This way, we build projections of phase space trajectories on an x-y oscilloscope. Simultaneously we can measure the power spectrum. In first experiment on a chaotic laser a sequence of subharmonic bifurcations was shown, which eventually leads to an intricated trajectory (strange attractor) and to a continuous power spectrum. But how can we discriminate between deterministic chaos and noise? After all, noise also would give a continuous spectrum, and the phase space point would fill ergodically part of the plane, thus densely covering a two-dimensional set.

In order to discriminate deterministic chaos from order as well as from random noise, we introduce two invariants of the motion, one static the other dynamic.

We partition the phase space into small boxes of linear size ε and give ith box a probability $p_i = M_i/M$ equal to fractional number of times it has been visited by the trajectory. This way, we build a Shannon information $I(\varepsilon)$, and with it an "information dimension" $D_1(\varepsilon)$ which is, in general, a fractional number, or a "fractal".

$$I(\varepsilon) = - \sum_i p_i \log p_i \; , \tag{3}$$

$$D_1(\varepsilon) = - \lim_{\varepsilon \to 0} \frac{I(\varepsilon)}{\log \varepsilon} \; . \tag{4}$$

To understand the meaning of a fractal, look up an operational definition of dimension. Let us compare three sets: (i) a segment of unit length; (ii) the Cantor set, built by taking out the middle one-third of the unit segment and repeating the operation on each fragment; (iii) the Koch curve, built by replacing the middle third with the other two sides of an equiteral triangle and repeating the operation "ad infinitum". At each stage of the partition, we cover each set with beads of suitable size not to lose in resolution (e.g., diameter 1/3 at the first partition) and count the number N for each set (at the first partition, we need 2 for the Cantor set, 3 for the segment, 4 for the Koch curve).

We define the fractal dimension as the ratio

$$D_0(\varepsilon) = \frac{\log N(\varepsilon)}{\log 1/\varepsilon} \; . \tag{5}$$

This definition is independent of the partition. Indeed, for the Cantor set and the Koch curve we have N = 2, ε = 1/3 and N = 4, ε = 1/3, respectively, at the first partition, yielding

$$D_0(\text{Cantor}) = \frac{\log 2}{\log 3} \simeq 0.63\dots, \quad \text{and}$$

$$D_0(\text{Koch}) = \frac{\log 4}{\log 3} \simeq 1.2618\dots.$$

At the second partition the number of necessary beads goes as N^2 and the diameter of each as ε^2, hence D_0 remains invariant.

Going back to the information dimension $D_1(\varepsilon)$ we see that we have replaced $\log N$ with $I(\varepsilon)$ which is an average (for p_i all equal, we recover $I(\varepsilon) = \log N$). Hence D_1 generalizes D_0 whenever the density of points is not uniform along the trajectory.

As D_0 was independent of the partition stage, similarly D_1 is an invariant, but static (time does not enter). It can be shown that $D_0 \ggeq D_1$, however for non pathological sets the difference is irrelevant. Let us refer for simplicity to an N = 3-dimensional phase space. If D = 0 (fixed point) or 1 (limit cycle) or 2 (torus) we have an ordered, or coherent, motion. In the order limit of random noise, fluctuations fill ergodically an N dimensional region of the space, hence D = 3. Deterministic chaos has to be in between, that is

$$2 < D < 3.$$

Hence, a fractal dimension is an indicator of chaos.

These features related to the topology of the attractor have a temporal counterpart in another invariant, which measures how information is dissipated in a motion to maintain knowledge of the system. To build this dynamic invariant, we partition both space and time in boxes of sizes ε and τ that we name $i_1, i_2, \dots i_d$ at each of the discrete times $\tau, 2\tau, \dots$ $d\tau$, and introduce the joint probability over the d time intervals,

$$p_{i_1 i_2 \dots i_d} \equiv \{x(t = \tau) \subset i_1; \dots; x(t = d\tau) \subset i_d\}$$

Correspondingly, we define a joint information

$$I_d(\varepsilon) = - \sum_{\{i_1 \dots i_d\}} p_{i_1 \dots i_d} \log p_{i_1 \dots i_d} \tag{6}$$

Then, by a limit operation, define the Kolmogorov entropy as the rate of information loss per unit time

$$K \equiv \lim_{\tau \to 0} \lim_{\varepsilon \to 0} \lim_{d \to \infty} \frac{1}{d\tau} \sum_{n=1}^{d} (I_{n+1} - I_n) = \lim \frac{1}{d\tau} I_d. \tag{7}$$

Now we have two indicators to gauge the difference among order, random noise (Brownian motion) and deterministic chaos. Referring to K, it is easily seen that

$$K = 0 \quad \text{for order (no information loss)};$$

$$K = \infty \quad \text{for random noise (total information loss)};$$

$$0 < K < \infty \quad \text{for deterministic chaos.}$$

The box counting method described above is impractical. It may require 10^6 points for a convergent numerical result. On the contrary, the following method introduced by Grassberger and Procaccia is applicable to only 10^3-10^4 independent data points. We generalize Shannon information defining the order-f information as

$$I_f(\varepsilon) = \frac{1}{1-f} \ln \sum_i p_i^f . \qquad (8)$$

For $f \to 1$ we recover the usual definition. Associated with I_f, there is an order-f dimension of the attractor

$$D_f = \lim_{\varepsilon \to 0} \frac{I_f(\varepsilon)}{\ln 1/\varepsilon} . \qquad (9)$$

For $f = 0$ and 1 we recover D_0 and D_1. Consider $f = 2$. The sum $\sum_i p_i^2$ is just the probability that a pair of random points on the attractor fall into the same box, that is, that two arbitrary points will have a distance less than ε. Calling this probability $C(\varepsilon)$, we expect thus

$$C(\varepsilon) = \lim_{N \to \infty} \frac{1}{N^2} \sum_{ij} \theta(\varepsilon - |x_i - x_j|) \simeq \varepsilon^{D_2} \qquad (10)$$

where $C(\varepsilon)$ is measured as the number of pairs (i,j) with a distance $|x_1 - x_j| < \varepsilon$. In (10), θ is the Heaviside step function.

Experimentally, we do not measure at each time the vector x(t) of phase space, but just one component $x_i(t)$ (for instance, just the light out of a laser).

However, in a nonlinear system, any component $x_k(t)$ will influence x_i at a later time (no normal mode transformation!). Hence, we can build an m-dimensional phase space $\xi(t)$ by just measuring one single component x_i at successive times and considering the m-fold as a single point in m space

$$\xi(t) \equiv \left[x_i(t), x_i(t + \tau), \ldots x_i(t + (m - 1)\tau) \right] .$$

(11)

As we evaluate the slope log C vs log $\mathcal{E}$ from our data, we can stop from increasing m when the slope shows saturation. The saturated slope is D_2.

3. Experimental characterization of Shil'nikov chaos by statistics of return times

The dynamic behavior of a single-mode CO_2 laser with feedback is characterized by global features in the phase space, related to the presence of three coexisting unstable fixed points. As a control parameter is monotonically increased, one can observe transitions from a Hopf bifurcation to a local chaos and eventually to regular spiking and Shil'nikov chaos.

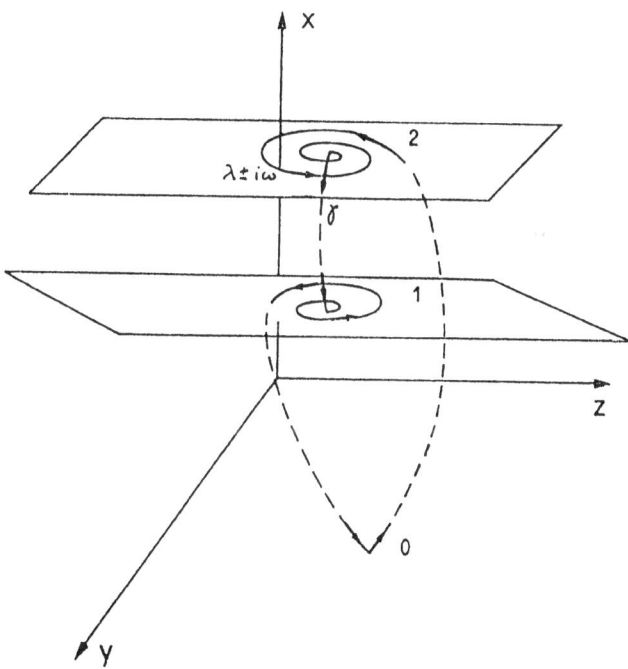

Fig. 1. Schematic view of a trajectory in the phase space when the dynamics are affected by all three unstable stationary points.

Furthermore, one can find evidence of competition among these different kinds of instability.[5] The phase-space trajectories are affected differently by each of the three unstable points, and by adjustment of the control parameters they can be characterized by the dominant role of only one, or a pair of them.

A linear stability analysis shows the local features at each fixed point. Precisely, point 0 (at zero intensity) is a saddle node with two stable directions and one unstable; point 1 has a plane unstable manifold with a focus and a stable third direction; point 2 has a stable manifold with a focus and an unstable third direction. Shil'nikov chaos is related to the saddle focus character of point 2. Around a saddle focus the motion consists of a contracting spiral $\exp(-\lambda t)\cos(\omega t)$ on the stable manifold and of an exponential expansion $\exp(\gamma t)$ along the unstable manifold. The presence of the other two unstable points ensures that the diverging flow is reinjected into the neighborhood of the saddle focus. Shil'nikov showed that for $|\lambda| < \gamma$ there exists a countable set of unstable trajectories close to the homoclinic one.[6] This structure of the flow is one of the simplest capable of generating chaotic behavior in many autonomous systems, such as the Lorenz equations[7] and the Belousov-Zhabotinski reaction.[8]

The temporal behavior of laser output intensity in this regime is characterized by pulses almost equal in shape but with chaotic recurrence times.[9] The regularity in shape means that the points at any Poincaré section are so closely packed that impossibly precise measurements of their position would be required if the relevant features of the motion were to be found. Instead, there is a large spread in the return times to a Poincaré section close to the unstable point. For this reason, the statistics of the return times appears to be the most appropriate characterization of Shil'nikov chaos.

Our experimental setup consists of a single mode CO_2 laser with an intracavity electro-optic modulator. A signal proportional to the laser output intensity is sent back to the electro-optic modulator[10]. Single mode CO_2 lasers have a dynamic behavior described by two coupled differential equations, one for the field amplitude and the other for the population inversion, the fast polarization being adiabatically eliminated from the complete set of Maxwell-Bloch equations. Thus, the presence of feedback introduces a third degree of freedom. When the feedback loop is so fast that it provides a practically instantly adapted loss coefficient, it does not modify the phase-space topology. On other hand, if the time scale of the feedback loop is of the same order as of the other two relevant variables, the system becomes three dimensional. With suitable normalizations such a system is described by three first-order differential equations for the laser intensity $x(t)$, the population inversion $y(t)$, and the modulation voltage $z(t)$ as follows:

$$x = -K_0 x \left[1 + \alpha \sin^2(z) - y \right],$$

$$y = -\gamma_{\parallel}(y + xy - A),$$
$$z = -\beta (z - B + \zeta x),$$

(12)

For a fixed pump A (fixed discharge current in the laser tube), our system has two control parameters: the bias voltage B applied to the electro-optic modulator and the gain r in the feedback loop.

From an experimental point of view we are able to visualize (x - z) phase-space projections, obtained by feeding onto a scope the photodetector signal proportional to the laser output intensity $x(t)$ and the feedback voltage $z(t)$. These phase-space projections consist of closed orbits visiting successively the neighborhoods of the three unstable stationary points 0,1, and 2.

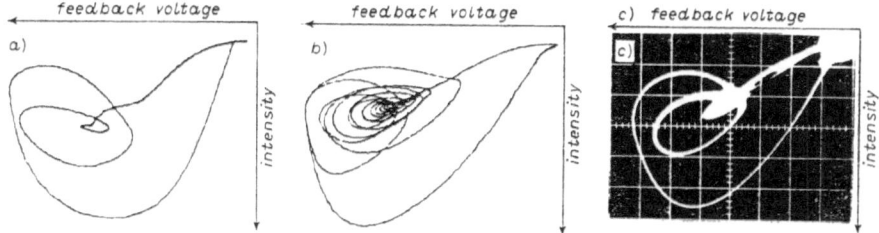

Fig. 2. Phase space projections x-z (laser intensity-feedback voltage). a) and b) are single orbits obtained by a digitizer, referring to the same parameters of Fig. 3a) and b), respectively. c) is the superposition of 30000 orbits of type a)

The local chaos around point 1, established at the end of a subharmonic sequence, has been characterized by standard methods as power spectra and correlation dimension measurements.[10]

However, the existence of a global behavior characterized by pulses with regular shapes but chaotic in their time of occurrence makes it significant to study the dynamics through measurements of return times to a Poincaré section. The measurements have been done by using a threshold circuit. An appropriate Poincaré section x = constant can be selected by adjusting the threshold level. This method permits us to distinguish among the different dynamical regimes. Adjusting the control parameters in order to have a dominance of the saddle focus 2, we obtain a motion consisting of a quasi-homoclinic orbit asymptotic to it (Fig. 3). In this regime, the laser output is characterized by pulses with regular shapes but chaotic in their recurrence. Based on such a consideration the iteration map of return times (τ_{i+1} versus τ_i) displays an extremely regular structure that we show below to be in close agreement with that arising from Shil'nikov theory of homoclinic chaos.

From a theoretical point of view, a homoclinic orbit asymptotic to a saddle focus can be modeled in terms of the following one-dimensional iteration map[11]:

$$\zeta_{n+1} = \zeta_n^{\lambda/\gamma} \cos[\omega/\gamma \ln(\zeta_n)] + \epsilon, \qquad (13)$$

where γ and $-\lambda \pm i\omega$ are the eigenvalues of the linearized flow at the saddle focus, ζ is the coordinate along the unstable manifold, and ϵ is the deviation along ζ from the homoclinic orbit at the Poincaré section in the neighborhood of the saddle point ($\epsilon = 0$ corresponds to the homoclinic condition).

If we build a small cubic box of unit side centered at the saddle focus and oriented along the eigenvectors ξ, η, and ζ, any tiny difference in the entrance coordinate along the expanding axis ζ will strongly influence the residence time inside the box and hence the spacing from the next reinjection.

Observing that most of the time is spent in the box around the saddle point, we relate the return time to the coordinate ζ of the unstable manifold by $\zeta = \zeta_0 \exp(\gamma \tau)$, thus obtaining an iteration map for the return times

$$\tau_{n+1} = -\ln[\exp(-\lambda/\gamma \tau_n)\cos(\omega/\gamma \tau_n) + \epsilon]$$
$$= -\ln[\varphi(\tau) + \epsilon], \qquad (14)$$

Comparison of Eqs.(13) and (14) shows the enhanced sensivity to fluctuations of the map with respect to the map.

Indeed, suppose that the offset ζ from homoclinicity is affected by a small amount of noise. The sensitivities of the two maps to such a noise are given, respectively, by $\partial\zeta/\partial\epsilon = 1$ and

$$\partial\tau/\partial\epsilon = [\varphi(\tau) + \epsilon]^{-1}. \qquad (15)$$

This sensitivity factor acts as a lever arm whenever $\varphi(\tau) + \epsilon$ becomes very small. Note the following: (1) This is not deterministic chaos; in fact, large fluctuations can be expected even for a regular dynamics, implying a fixed point τ^* . (2) It is not associated with the homoclinicity condition $\epsilon = 0$; in fact, for finite ϵ there may be a τ^* such that $\varphi(\tau^*) + \epsilon = 0$.

In figures 4 and 5 we show numerical and experimental iteration maps respectively. Fig. 5 also shows the spread related to the enhanced sensitivity to fluctuations.

To summarize, a CO_2 laser with feedback shows different dynamic regimes depending on the dominant role of one or two of three coexisting unstable stationary points. In particular, in the regime of Shil'nikov chaos the iteration maps of return times display a statistical spread owing to a transient fluctuation enhancement phenomenon peculiar to macroscopic systems, which is absent in low-dimensional chaotic dynamics.

In fact, the model description $\dot{x} = F(x)$ of a large system in terms of a low-dimensional dynamic variable x is just an ensemble-averaged description, and residual fluctuations on position x must be considered at some initial time, even though the successive evolution is accounted for by a deterministic law. In our case such a fluctuation is a stochastic spread $\delta\xi$ on the offset ξ of the position $\mathfrak{Z}$.

The same amount of $\delta\xi$ in Eqs.(13) and (14) leaves the $\mathfrak{Z}$ maps unaltered, while it strongly affects the τ maps, making them appear like the experimental data.

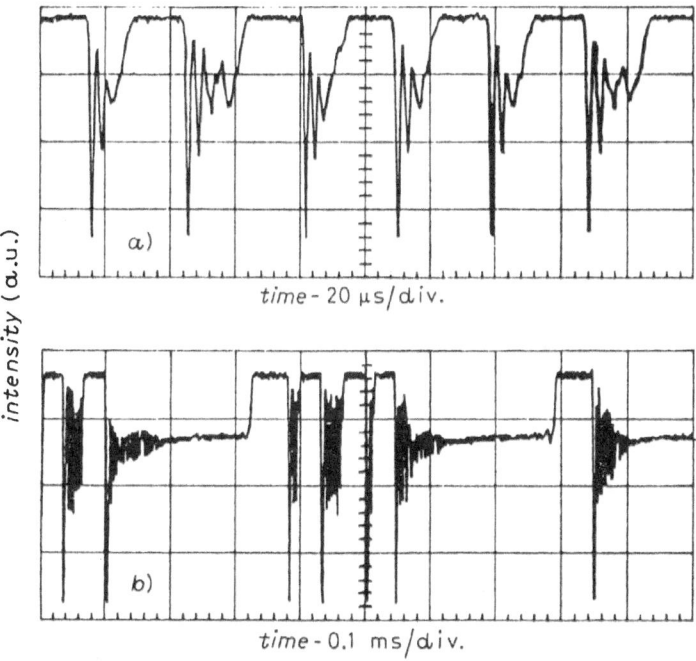

Fig. 3. Time plots of the intensity in the regime of Shil'nikov chaos. (a), (b) Refer to the same B value (B = 0.427) but two different gains of the feedback loop. (b) Shows long transients corresponding to a large number of small spirals around the saddle focus.

We have thus shown a fundamental difference between a small system, ruled by a few equations, and a large system, in which the corresponding low-dimensional dynamics is a contracted description in terms of macroscopic variables, which are ensemble averages over some initial spread.

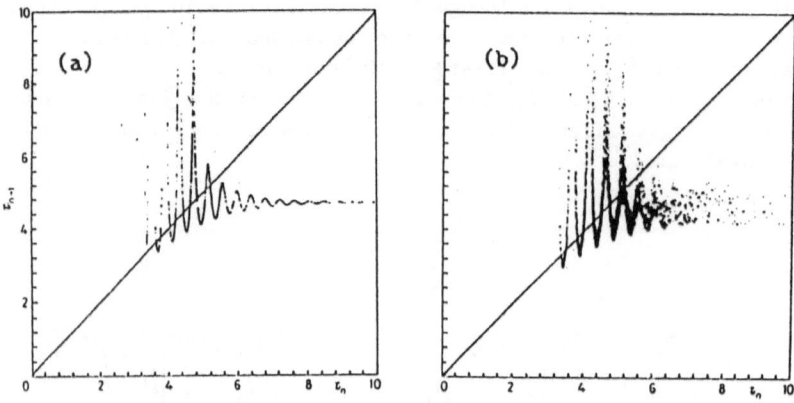

Fig. 4. Numerical iteration maps for Shil'nikov chaos. Paramater values: ω/γ = 13.0, λ/γ = 0.986, ε = 0.01. (a) and (b), τ maps without and with noise $\delta\varepsilon$ = 10^{-2}, respectively.

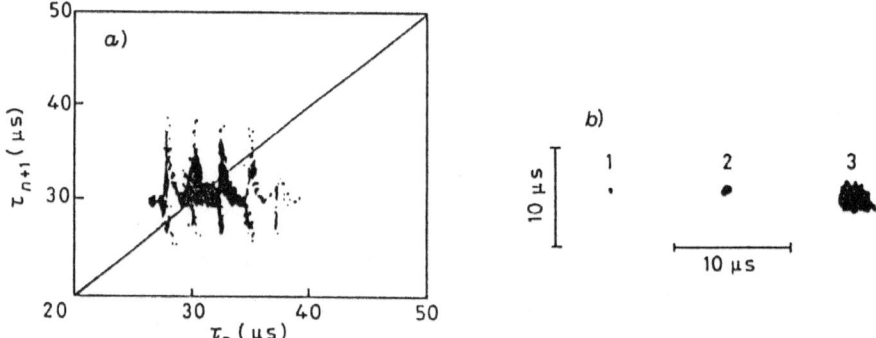

Fig. 5. Experimental iteration maps of the return times. (a) τ = 0.487 and B = 0.350. (b) Maps corresponding to regular periodic situations, namely, 1, an electronic oscillator; 2, the laser in a regular periodic regime; 3, the laser just at the onset of the instability but still with a regular period.

REFERENCES

1. Rather than giving specific references to well known fundamental papers on laser physics, I refer to a recent general review of mine plus References listed therein.

   F.T. Arecchi, in "Instabilities and Chaos in Quantum Optics", F.T. Arecchi and R.G. Harrison, eds., Vol. 34 of Springer Series in Synergetics (Springer-Verlag, Berlin, 1987), p. 9.

2. F.T. Arecchi in "Order and Fluctuations in Equilibrium and Nonequilibrium Statistical Mechanics" (Proc. XVII Solvay Conf. on Physics) ed. by G. Nicolis et al. (Wiley, New York 1981), p. 107.

3. R. J. Glauber in "Quantum Optics and Electronics", ed. by D. De Witt et al. (Gordon and Breach, New York 1965).

4. F.T. Arecchi in "Quantum Optics", ed. by R.J. Glauber (Academic, New York 1969).

5. F.T. Arecchi, R. Meucci, and W. Gadomski, Phys Rev. Lett. **58**, 2205 (1987).

6. L.P. Shil'nikov, Dokl. Akad, Nauk SSSR **160**, 558 (1965); L.P. Shil'nikov, Mat. Sb. **77**, 119, 461 (1968); **81**, 92, 1213 (1970).

7. P. Glendinning and C. Sparrow, J. Stat. Phys. **35**, 645 (1984); P. Gaspard, R. Kapral and G. Nicolis, J. Stat. Phys. **35**, 697 (1984).

8. F. Argoul, A. Arneodo, and P. Richetti, Phys. Lett. A **120**, 269 (1987).

9. F.T. Arecchi, A. Lapucci, R. Meucci, J.A. Roversi and P. Coullet, Europhys. Lett. **6**, 677 (1988).

10. F.T. Arecchi, W. Gadomski and R. Meucci, Phys. Rev. A **34**, 1617 (1986).

11. A. Arneodo, P.H. Coullet, E.A. Spiegel and C. Tresser, Physica **14D**, 327 (1985).

SYNCHRONIZATION OF ATOMIC QUANTUM TRANSITIONS BY LIGHT PULSES

V.P. Chebotayev and V.A. Ulybin

Institute of THERMOPHYSICS, Siberian Branch of the
USSR Academy of Sciences, Novosibirsk — 90, SU — 630090

ABSTRACT

A synchronization of atomic quantum transitions with natural Raman oscillations by ultrashort light pulses has been considered. This phenomenon may be observed if a perturbation pulse duration is smaller than the period of oscillations on the forbidden atomic transition. An accuracy of direct measurements of quantum transition times for trapped particles may be of the order of the ratio of two—photon transition frequency to the homogeneous width.

1. INTRODUCTION

Frequency measurements in atoms and molecules are currently made with an accuracy $10^{10} - 10^{14}$ by using methods of a super—high—resolution laser spectroscopy such as the saturated absorption, the Doppler—free two—photon absorption and other methods. The known methods are based on a resonant interaction between particles and an optical field, hence a high—monochromatic laser radiation is principally necessary in these methods. An observation of resonances is carried out for times which are much longer than oscillation periods of quantum transitions under investigation.

A new spectroscopy method based on using pairs of pulses of radiation with a linewidth which is broader than the measured quantum—transition frequency interval but with stable interpulse time has been considered in [1]. The duration of the interaction between a particle and the single pulse is shorter than the oscillation period of the quantum transition. Hence, the result of the atomic interaction with two short pulses is defined by the phase of free oscillations of a dipole moment at the time when the second pulse arrives. In other words, the particle makes the transition from one energy level into the other synchronouosly with the atomic oscillations. The synchronized quantum transitions are very accurately determined in time and may be applied to precise direct measurements of time and frequency.

Applied Laser Spectroscopy, Edited by W. Demtröder and
M. Inguscio, Plenum Press, New York, 1990

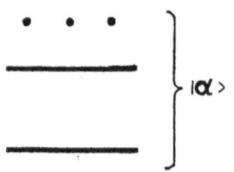

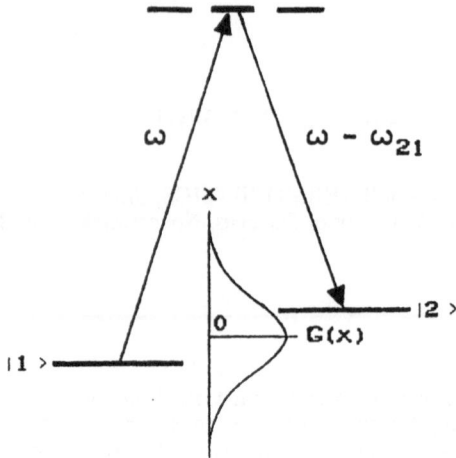

Fig. 1. Two photon Raman interaction between an atom and single ultrashort light pulse. $G(x)$ is the Fourier transformation of the pulse shape $|g(t)|^2$, x is a frequency.

2. QUALITATIVE ANALYSIS

The phenomenon of synchronization of quantum transitions is not so critical with respect to the nature of a pulse perturbation. Hence, we shall explain it by a simple example of an atomic interaction with two DC pulses of an electric field.

We write the last one in the form

$$E(t) = E\,g(t) + E'g(t-T)\,, \tag{1}$$

where $g(t)$ is the pulse shape, and T is the interpulse time. The field–induced dipole moment of a two–level atom may be written in the form

$$d(t) = a_{21}d_{21}^{*}\exp(-i\omega_{21}t) + \text{c.c.}\,, \tag{2}$$

where a_{21}, d_{21} and ω_{21} are the probability amplitude, the dipole moment and the frequency of the transition $|1> \longrightarrow |2>$, respectively (the atom is considered to be in the state $|1>$ when $t = -\infty$). Both $d(t)$ and d_{21} are assumed to be projections onto the field direction. According to [2] we have

$$a_{21} = (i/\hbar)d_{21}\int_{-\infty}^{t}dt'\,E(t')\exp(i\omega_{21}t')\,. \tag{3}$$

If the duration τ of the atomic interaction with a single pulse is much shorter than the period of atomic oscillations $2\pi/\omega_{21}$, then the function $g(t)$ in (1) and (3) may be replaced by $\tau\delta(t)$, where $\delta(t)$ is Dirac's delta–function. This means that we probe the atomic dipole moment at times $t = 0$ and $t = T$. For $t > T$ we find from (1) – (3) the dipole moment as the superposition of dipole moments excited at $t = 0$ and $t = T$

$$d(t) = id_{21}^* \tau\Omega_{21}\exp(-i\omega_{21}t)$$
$$x\, [1 + (E'/E)\exp(i\omega_{21}T)] + \text{c.c.} , \qquad (4)$$

where $\Omega_{21} = Ed_{21}/\hbar$, and the probability of the transition $|1> \longrightarrow |2>$

$$|a_{21}|^2 = (\tau \Omega_{21})^2 [1 + (E'/E)^2 + 2(E'/E) \cos\omega_{21}T] . \qquad (5)$$

As we see from (4) and (5), both $d(t)$ and $|a_{21}|^2$ have maxima at times T that are multiples of the period $2\pi/\omega_{21}$. Thus, when an atom interacts with two short pulses the quantum transition $|1> \longrightarrow |2>$ may be synchronized with its natural oscillations. The effect is also shown to be very distinctive with equal pulse amplitudes $(E' = E)$.

Instead of DC pulses of electric or magnetic fields, AC pulses may be used as well. The case of AC pulses is of interest for optical transitions. It is clear, that a carrier frequency of AC pulses may be of any value satisfying the inequality $\omega >> \tau^{-1}$, ω_{21}. There are only two pulse parameters of principal importance, i.e. a duration of a perturbation pulse should be shorter than an atomic–oscillation period, and the interpulse time should be stable and a multiple of the period.

The aim of this lecture is to consider the phenomenon of synchronization of quantum transitions by light pulses. At present, advanced methods of generating ultrashort laser pulses provide the possibility to obtain light pulses of $10 - 100$ fs duration [3]. This allows experiments for synchronization of IR and FIR quantum transitions and in particular the direct measurement of time with an accuracy of the order $10^{12} - 10^{13}$. As the light frequency $\omega >> \omega_{12}$, the excitation of the quantum transition is realized by a two–photon Raman process. We start the consideration with the analysis of the Raman interaction between an atom and a single light pulse of any intensity. This analysis is of general interest.

3.INTERACTION BETWEEN AN ATOM AND AN ULTRASHORT LIGHT

Let $|1>$ and $|2>$ be the ground and metastable atomic states to obey the selection rules for two–photon transition, see. Fig. 1. We write the light pulse in the form

$$E(t) = Eg(t) \exp(-i\omega t) + \text{c.c.} ,$$

where $2E$ is the amplitude of the electromagnetic wave pulse. The frequency ω is nonresonant to the intermediate transitions $|1> \longrightarrow |\alpha>$ and $|2> \longrightarrow |\alpha>$. The duration of the pulse is of the form

$$\tau = \int_{-\infty}^{\infty} dt\, g(t)^2$$

and obeys $\omega^{-1} << \tau \leq \omega_{21}^{-1}$. We assume that the light pulse is of symmetric shape $g(-t) = g(t)$.

The equations for the density matrix elements which describe the stimulated Raman scattering in the field $E(t)$ were reduced similarly [4] to the equations for a two–level atom in an effective nonoscillating field

$$\{d/dt + i\,[\omega_{21} - |g(t)|^2\Delta] + \Gamma\}\,\rho_{21} = (i/2)\,\Omega\,|g(t)|^2\,(\rho_{11} - \rho_{22})\,,$$

$$(6)$$

$$(d/dt + 2\,\Gamma)\,\rho_{22} = (-i/2)\,\Omega\,|g(t)|^2\,\rho_{21} + \text{c.c.}\,,$$

$\rho_{11} + \rho_{22} = 1$,

where $2\,\Gamma$ is the spontaneous decay rate of the upper state $|2\rangle$, $\Delta = E^2\,(D_{22} - D_{11})$ is the difference of the optical Stark shifts of the levels $|2\rangle$ and $|1\rangle$, $\Omega = 2E^2 D_{21}$ is the effective (two–photon) Rabi frequency, $D_{ik} = D_{ik}(\omega) + D_{ik}(-\omega)$, where $D_{ik}(\omega) = \Sigma_\alpha\, d_{i\alpha} d_{\alpha k}\hbar^{-2}\,(\omega_{\alpha 1} - \omega)^{-1}$ is a two–photon matrix element.

The following calculations will be carried out with $\Delta = 0$ in (6) because the Stark light shift Δ is small in comparison with the effective Rabi frequency Ω (for close atomic levels $|2\rangle$ and $|1\rangle$ $D_{22} - D_{11} << D_{21}$.

Before the interaction with the light pulse at time $t = 0$ the atom is in the ground state $|1\rangle$. Using the initial condition $\rho_{11}(-\infty) = 1$ we find the coherence and the upper–level population probability

$$\rho_{21}(t) = (i/2)\,\exp[-(\Gamma + i\omega_{21})t]\,\sin 2\Theta\,, \qquad (7)$$

$$\rho_{22}(t) = \exp(-2\,\Gamma t)\,\sin^2\Theta\,, \qquad (8)$$

where $t >> \tau$,

$$\Theta = (\Omega/2)\,G(\omega_{21})\,,$$

$$G(x) = \int_{-\infty}^{\infty} dt\,|g(t)|^2\,\exp(ixt)\,.$$

The parameter Θ determines the power of the pulse perturbation: the case of $\Theta << 1$ ($\Omega << \omega_{21} \lesssim \tau^{-1}$) corresponds to the weak perturbation and $\Theta \simeq 1$ ($\Omega \simeq \tau^{-1} \gtrsim \omega_{21}$) to the strong one. The function $G(\omega_{21})$ is the value of the Fourier transform of a light–pulse shape at frequency ω_{21}. It determines the atom–field interaction efficiency.

In the case of a Gaussian pulse $|g(t)|^2 = \pi^{-1/2}\exp(-t^2/\tau^2)$ we have $G(\omega_{21}) = \tau\exp[-(\omega_{21}\tau/2)^2]$. The atom–field interaction is shown to be efficient if $\tau \lesssim 2/\omega_{21}$. Increasing the pulse duration τ leads to a decrease of spectral components at $\omega \pm \omega_{21}$ in a light pulse, so the two–photon interaction becomes impossible.

For the rectangular light–pulse shape we have $G(\omega_{21}) = \sin(\omega_{21}\tau/2)\,/\,(\omega_{21}/2)$. The efficiency is high for $\tau = \pi(2n - 1)/\omega_{21}$, with n being integer here and elsewhere in the paper. The result of the atomic interaction with the light pulse of duration $\tau = \pi/\omega_{21}\,(n = 1)$ is the same as in the case of a long pulse ($n >> 1$). It means that only the sharp edges of the light pulse may give rise to the two–photon Raman process. The upper limit of τ is defined by the accuracy, required for the time localisation of the atomic transition $|1\rangle \longrightarrow |2\rangle$.

In this analysis we have neglected the one–photon excitation of the intermediate levels $|\alpha\rangle$, see Fig. 1. This process reduces the number of atoms involved in the two–photon process. For $\omega_{\alpha 1} - \omega >> \tau^{-1}$ the α–level population is of the form

$$\rho_{\alpha\alpha} = (\Omega_{\alpha 1}^2/2) \, \gamma_{\alpha 1} \, \tau (\omega_{\alpha 1} - \omega)^{-2}$$

where $\Omega_{\alpha i}$ and $\gamma_{\alpha i}$ $(i = 1, 2)$ are the one–photon Rabi frequency and the spontaneous decay rate of the transition $|\alpha\rangle \longrightarrow |i\rangle$, respectively. The population ρ_{22} induced by the two–photon process in a weak field is $\rho_{22} \approx \Theta^2$, (see (8)). So, the inequality $\rho_{22} \gg \Sigma_\alpha \rho_{\alpha\alpha}$ which corresponds to a suppression of the one–photon process, is reduced to

$$\Omega_{\alpha 2}^2 \gg \tau \, \gamma_{\alpha 1} \, G^2 \, (\omega_{21}) \sim \tau \, \gamma_{\alpha 1} \, \omega_{21}^2 \, ,$$

where α refers to the intermediate level which may be used to estimate the two–photon matrix element D_{21}.

To conclude this discussion we note the following: Firstly, the effect of the Stark light shift which we have neglected above may be shown to be distinct in the strong field alone and leads to the decrease in the dynamical variation of the coherence amplitude $|\rho_{21}|$ and the probability ρ_{22} in proportion to the ratios Ω/Ω_0 and $(\Omega/\Omega_0)^2$ respectively, with $\Omega_0 = (\Omega^2 + \Delta^2)^{1/2}$. Secondly, the formula (7) and (8) may be applied to the interaction between an atom and an ultrashort $(\tau \lesssim \omega_{21}^{-1})$ DC pulse of an electric field as well (for forbidden transitions). In this case, the frequency ω in the two–photon matrix elements $D_{ik}(\omega)$ is zero.

4. INTERACTION BETWEEN AN ATOM AND TWO LIGHT PULSES

The only physical parameter which describes the light field in equation (6) is the field power $\sim E^2 \, |g(t)|^2$. Therefore, the interaction between an atom and a pair of time–separated pulses will depend on the time delay T but not on the difference of their optical phases.

On the other hand, we have considered the excitation of an atomic natural oscillation with the frequency ω_{21} by one ultrashort light pulse, see (7). A second light pulse delayed by the time T (terminating pulse) will interact with the atom, when the phase of the oscillations will be equal within $\omega_{21}T$. It is the phase that determines the result of the atomic interaction with the pair of ultrashort pulses.

Using (7) and (8) at time $t = T$ as initial conditions for the interaction between the atom and the second light pulse we find from (6) the coherence and the upper–level population probability in the form

$$\rho_{21}(t) = (i/2) \sin 2\Theta \exp[-(\Gamma + i\omega_{21})(t - T)]$$

$$x \, \{1 + \exp[-(\Gamma + i\omega_{21})T] - 2 \sin^2\Theta \exp(-\Gamma T)$$

$$x \, [\exp(-\Gamma T) - \cos \omega_{21}T]\} \tag{9}$$

$$\rho_{22}(t) = \sin^2\Theta \exp[-2\Gamma(t - T)] \, [1 - \exp(-2\Gamma T) \cos 2\Theta$$

$$+ (1 - \cos 2\Theta) \exp(-\Gamma T) \cos\omega_{21}T] \, , \tag{10}$$

where $t - T \gg \tau$. We see that the density–matrix variation for the small time duration τ of the terminating–pulse is synchroneous with the atomic oscillations at the frequency ω_{21}. This is considered as the synchronized stimulated quantum transition. There is a peculiarity of the transitions induced by the weak and strong perturbations, hence we consider them separately.

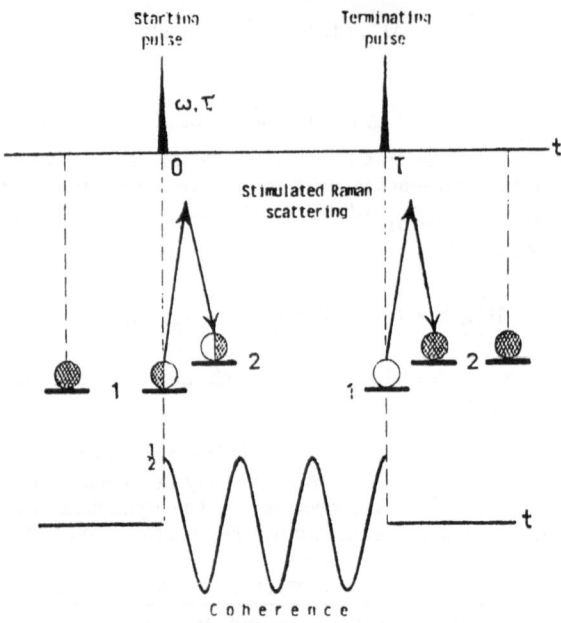

Fig. 2. Interaction between an atom and two
ultrashort light pulses ($\Theta = \pi/4$).

If the field is sufficiently weak to avoid saturation of the two–photon transition
$|1\rangle \longrightarrow |2\rangle$ ($\Theta << 1$), we see from (7) and (8) that the starting light pulse
induces the atomic coherence $\rho_{21} \sim E^2\tau$ and probability $\rho_{22} = |\rho_{21}|^2$. If the second
pulse delayed in time arrives synchroneously with the atomic oscillations
($T = 2\pi n/\omega_{21}$), then the amplitude of the oscillations is doubled (for $\Gamma T << 1$) and
the upper level population probabilities is quadrupled, see (9) and (10), due to the
interference of the probability amplitudes. If the delay time is equal to a half–integer of the atomic periods ($T = \pi(2n-1)/\omega_{21}$), the natural oscillations are annihilated and the atom returns into the ground state. The formulas (9), (10) for a
weak field are similar to (4), (5).

For the strong light field ($\Omega \simeq \tau^{-1} \gtrsim \omega_{21}$) we emphasize the case of a $\pi/4$–pulse
($\Theta = \pi/4$) which yields the maximum amplitude of the synchronized density
matrix variations. From (7) − (8) we see that the starting pulse induces the proba-
bility $\rho_{22} = 1/2$ and the atomic oscillations with the amplitude $|\rho_{21}| = 1/2$. The
terminating pulse arriving synchronously with the atomic oscillations
($T = 2\pi n/\omega_{21}$) doubles the upper–level population probability and returns the
atom into the ground state if $T = \pi(2n-1)/\omega_{21}$, with the natural oscillations
being annihilated in both cases, see Fig. 2.

5. ATOMIC ENSEMBLE, RAREFIED GAS

In order to find the upper–level population for an atomic ensemble we must
take into account the atomic motion and average the population probability $\langle\rho_{22}\rangle$
over coordinates and velocities. Atomic motion is accounted for by replacing the
value T   right   −   the   on   hand   side   of (10) by the interpulse time
$T_a = T + c^{-1}[z(T_a) - z_0]$ in the rest frame of the moving atom, where z_0 and
$z(T_a)$ being the atomic coordinates along the light wave propagation direction at
the times of the starting and terminating ultrashort pulses, respectively. The

ensemble–averaging of the upper–level population probability $<\rho_{22}>$ is reduced to averaging the oscillating factor $\cos\omega_{21}T_a$.

For a free atom $z(T_a) - z_0 \approx v_zT$, hence after the averaging over velocities v_z with the equilibrium distribution function we find

$$<\cos\omega_{21}T_a> = \cos\omega_{21}T \exp[-(T\delta/2)^2] , \qquad (11)$$

where $\delta = \omega_{21} v_0/c$ is the Doppler shift at the Raman frequency, with v_0 being the thermal velocity. We see that for rarefied gas the observation of the quantum transitions synchronized with the atomic oscillations at the frequency ω_{21} is limited by the time interval $T \lesssim \delta^{-1}$ which yields the accuracy of time measurements of order $(\tau\delta)^{-1} \simeq c/v_0$.

The influence of the Doppler effect may be decreased by elastic collisions in a sufficiently dense gas [5]. In this case the diffusion pathlength which a particle covers during the interpulse time should be of the order of the wave–length associated with the atomic transition $|1> \longrightarrow |2>$.

Note that in an atomic ensemble the spin–echo–like effect may be observed, i.e. the coherence $\rho_{21}(t)$ is seen to contain the term, which is in proportion to $\exp[-i\omega_{21}(t-2T)] \sin2\Theta \sin^2\Theta$, see (9).

6. ENSEMBLE OF TRAPPED ATOMS

During the interval between the starting and terminating pulses, an atom trapped in a finite volume may be displaced not more than the macroscopic oscillation amplitude v_{zmax}/ω_z, where ω_z is the frequency of atomic oscillations in a trap. Ensemble–averaging the factor $\cos\omega_{21}T_a$ we see that the parameter $\mu = \delta/\omega_2$ arises instead of $T\delta$ in (11). Furthermore, the Doppler effect does not limit the delay time T if the latter is a multiple of the macroscopic oscillation period. In this case the terminating pulse finds an atom at the same point as the starting pulse.

Let an atomic ensemble be trapped in a harmonic potential, with the principal axis of trap symmetry coinciding with the light direction z. The equation of motion of a single atom is of the form

$$z(t) = z_0 \cos\omega_zt + (v_{z0}/\omega_z) \sin\omega_zt .$$

Averaging over the initial coordinates z_0 and velocities v_{z0} we find

$$<\cos\omega_{21}T_a> = \cos\omega_{21}T\exp[-\mu^2 \sin^2(\omega_zT/2)] .$$

This confirms our above statements.

7. LASER SPECTROMETER SCHEME

In the method under consideration the light source has a very large linewidth $(\Delta\omega \simeq 1/\tau)$, and the value of its carrier frequency is unimportant. Only the stability of the pulse delay time is significant. There are at least two possibilites to realize the spectrometer. In the first case, the optical delay line may be used to form the terminating pulse. The delay time $T = L/c$ (where L is the delay line length) may be changed continuously with high accuracy. Unfortunately, the absolute accuracy of the delay time measurement is limited here by the length measurement accuracy. If the transition frequency is known, then according to the new definition of the meter, the delay time T and the length L may be directly measured. The second possibility is given by the laser spectrometer scheme shown in Fig. 3. This spectrometer allows one to measure the absolute values of both the delay time T and the transition frequency ω_{21}. The spectrometer is based on the use of ultrashort pulses generated by the forced—mode locked laser. The pulse repetition frequency is determined by the frequency ν of the RF generator, which controls the intracavity amplitude modulator. The optical pulse amplifier locked to the RF generator allows one to form the time—separated pulses. Their delay time will be a multiple of the laser interpulse time, i.e. $T = n\nu^{-1}$. Usually $\nu \sim 10^8$ Hz. Time tuning may be realized by tuning the frequency ν. Such a system may serve as a standard for time and frequency simultaneously. As the frequency ω_{21} is stable, it is possible to stabilize the delay time T and consequently the RF generator frequency ν.

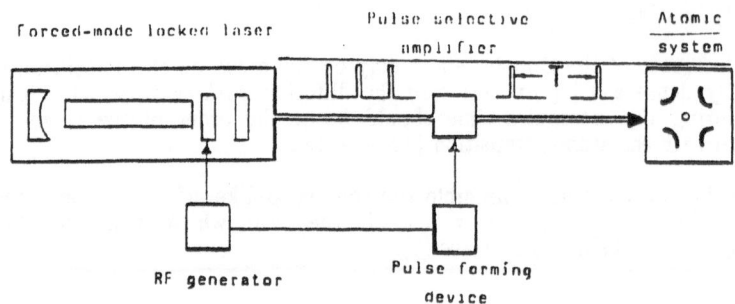

Conclusion

We have shown that for a single atom or an atomic ensemble the dynamics of the natural Raman oscillations may be investigated by the synchronization of atomic quantum transitions with these oscillations. This synchronization can be realized, for example, by time—separated ultrashort light pulses. The proposed effect may be used in super—high precision frequency measurements, developing new foundations for the time standards and magnetometers, the measurement of atomic spectroscopy constants, the selective excitation of atoms and molecules, etc. The possibility of using this phenomenon to develop the high—speed atomic memory systems is also of interest.

ACKNOWLEDGEMENTS

The authors are grateful to Prof. E.V. Baklanov for critical remarks, and Prof. L.S. Vasilenko and Dr. A.V. Shishayev for valuable discussions.

REFERENCES

1. V.P. Chebotayev: Pisma JETF, 49, 429 (1989)
2. L.D. Landau, E.M. Lifshitz: Quantum Mechanics. (Nauka, Moscow 1974), p. 177
3. S.A. Akhmanov, V.A. Vysloukh, A.S. Chirkin: Optics of Femtosecond Laser Pulses. (Nauka, Moscow 1988); W. Kaiser (ed.): Ultrashort Laser Pulses and Applications, Topics Appl. Phys. 60 (Springer, Berlin, Heidelberg 1988)
4. E.V. Baklanov, B.Ya. Dubetsky: Kvantovaya Elektronika, 5, 99 (1978)
5. R.H. Dicke: Phys. Rev., 89, 472 (1953)

AMPLIFICATION OF PULSE RESPONSES IN ENSEMBLES OF
NONLINEAR OSCILLATIONS

V.P. Chebotayev

Institute of THERMOPHYSICS, Siberian Branch of the
USSR Academy of Sciences, Novosibirsk — 90, SU — 630090

The response of an ensemble of oscillators to pulse excitation depends on the pulse parameters, the oscillation decay time, the degree of nonlinearity and so on. Due to frequency differences between the oscillators a dephasing of the oscillation takes place after the pulse. Therefore, the response decay time of the ensemble of oscillators is much shorter than the decay time of a single oscillator. In the case of nonlinear oscillation the dephasing may be compensated by the action of a second pulse with time delay T. Due to nonlinearity the phase locking of the oscillators occurs at the time t=2T. As a result a coherent nonlinear response as an "echo" [1] appears in the ensemble of oscillators.

1. Usually the response was considered in ensembles with a very large number of oscillators with a uniform frequency distribution. In that case the linear response is negligible. When the number of oscillators is small the linear response cannot be ignored [2]. In the particular case of an equidistant distribution of oscillator frequencies the linear response will be a pulse train at the constant interval $\tau = 2\pi/\Delta$, where Δ is the frequency difference between two adjacent oscillators. Both classical and quantum (Rydberg atom excited by a short light pulse) oscillator ensembles have been considered in [3]. The oscillator nonlinearity must be taken into account if the pulse amplitude increases. In this case responses were found at times $t = 2T + (n-1)\tau, t = n\tau - T$ $(n=1,2...)$. In the present paper the observation and the theoretical analysis of a new response is reported. A lot of attention is paid here to the amplification regime in which a weak initial response is increased after the action of an intensive second pulse. The pulse amplification may be considered as a result of collective nonlinear processes. The amplification depends on the observation time. When the decay rate λ is small, the pulse amplification may be large.

Another regime involving overlapping pulse responses is worth to be noted. When the pulses arising at times $t = (n-1)\tau + 2T$ and $t = n\tau - T$ are overlapped the whole picture of the responses is very sensitive to the delay time T. For example, the formation of the single pulse sequence with period τ was observed.

Applied Laser Spectroscopy, Edited by W. Demtröder and
M. Inguscio, Plenum Press, New York, 1990

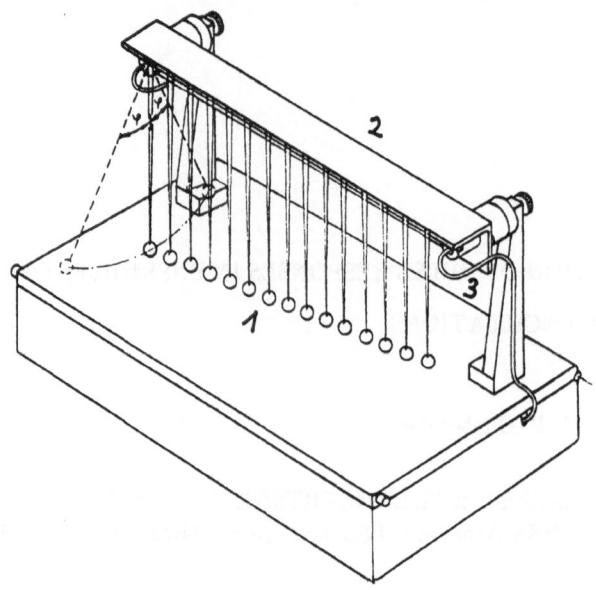

Fig. 1.   View of the experimental setup: 1—pendula,
2—pendula–carrying frame, 3—piezoceramic

2. Let us consider the response of an ensemble of nonlinear oscillators. To illustrate the conditions of the present experiment we shall consider an ensemble of pendula, with oscillation frequencies ω_i separated by the inverval Δ. As before [2, 3], we shall concentrate our attention on the value $\Psi(t)$ which corresponds to the sum of the oscillations of all pendula normalized to the number of oscillators N:

$$\Psi(t) = \frac{1}{N} \sum_{i=1}^{N} \psi_i(t) \qquad (1)$$

where $\Psi_i(t)$ describes the individual pendulum oscillations.
The equation describing a pendulum motion for a small amplitude has the form:

$$J_i(t) + 2\lambda \psi_i(t) + \omega_i^2 \psi_i(t) = \omega^2 \psi^3(t)/6 \qquad (2)$$

Assuming the initial conditions $\psi(0) = 0$ and $\psi(0) = A\omega$ at $t = 0$ the linear part of the solution for two excitation pulses at times $t = 0$ and $t = T$ has the form:

$$\psi_i(t) = Ae^{-\lambda t}\sin\omega_i t + Be^{-\lambda(t-T)}\sin\omega(t-T) \qquad (3)$$

where A and B are the pendulum oscillation amplitudes corresponding to the first and second excitation pulses respectively. In contrast to [2] the pulse duration is in our case much less than the oscillation period. The total respone $\Psi(t)$ is described by the expression:

$$\Psi(t) = F(t)Ae^{-\lambda t}\sin\omega t + F(t-T)Be^{-\lambda(t-T)}\sin\omega_i(t-T) \qquad (4)$$

where $\bar{\omega} = \omega_i + \Delta(N-1)/2$ $F(x) = \dfrac{1}{N} \dfrac{\sin(N\Delta x/2)}{\sin(\Delta x/2)}$

The function $F(x)$ describes the time behaviour of the response amplitude. According to (4) the ensemble response will represent two periodic pulse trains with a period $\tau = 2\pi/\Delta$. To account for the nonlinearity we substitute (3) into the right–hand side of Eq. (2). Then we obtain the classical equation with a decaying force (see Sec. 29 in [4]). The nonlinear part of the solution has the form (for $t > T$):

$$\phi_i^{nl} = \frac{1}{32\lambda}\, \omega_i \left[A^3\, (e^{-\lambda t} - e^{-3\lambda t}) \cos\omega_i t \right.$$

$$+ (e^{-\lambda(t-T)} - e^{-3\lambda(t-T)})\, [2AB^2 e^{-\lambda T} \cos\omega_i t$$

$$+ (B^3 + 2A^2 B e^{-2\lambda T}) \cos\omega_i(t-T) + AB^2 e^{-\lambda T} \cos\omega_i(t-2T)$$

$$\left. + A^2 B e^{-2\lambda T} \cos\omega_i(t+T)] \right] \tag{5}$$

Substituting (5) into (1) we obtain the pulse trains at times $t = (n-1)\tau + 2T$, $t = n\tau - T$, $t = n\tau$ and $t = (n-1)\tau - T$. The first two trains correspond to echo–type responses; the last two trains are to be considered as corrections to the linear responses.

The intensities for the first two cases are given by

$$\phi_{i,2T} = \omega_i\, \frac{AB^2}{32\lambda}\, e^{-\lambda T}\, (e^{-\lambda(t-T)} - e^{-3\lambda(t-T)})$$

and

$$\phi_{i,-T} = \frac{A^2 B}{32\lambda}\, e^{-2\lambda T}\, (e^{-\lambda(t-T)} - e^{-3\lambda(t-T)})$$

Both the analysis and the experiment showed that the regime of applicability of Eq. (5) is limited. Allowing for the change in oscillation frequency widens the applicability range of the solution. Using the amplitude dependence of the oscillation frequency [4] the solution (2) for the first harmonic of the oscillation is

$$\phi_{i,}(t) = Ae^{-\lambda t} \sin \int_0^t \omega_i(t)dt - Be^{-\lambda(t-T)} \sin \int_T^t \omega_i(t)dt \tag{6}$$

The pendulum oscillation frequency is given (see Sect. 11 in [4]) by

$$\omega_i(t) = \omega_i(1 - A^2 e^{-2\lambda t}/16) \qquad \text{(for } t < T\text{)}.$$
$$\omega_i(t) = \omega_i(1 - C^2 e^{-2\omega(t-T)}/16) \qquad \text{(for } t > T\text{)} \tag{7}$$

where

$$C = \left[A^2 e^{-2\lambda T} + B^2 + 2ABe^{-\lambda T} \cos\phi_i(T) \right]^{1/2}$$

is the pendulum oscillation amplitude after the second pulse.

$\phi_i(T) = \phi_i T - \frac{A^2}{32} (1 - e^{-\lambda T})$ is the phase of the pendulum oscillation at time T. Substituting (7) into (6) we obtain at $t > T$

$$\phi_i(t) = Ae^{-\omega t} \sin\left[\omega_i t - \omega_i \frac{A^2}{32\lambda}(1 - e^{-2\lambda T}) - \omega_i \frac{C^2}{32\lambda}(1 - e^{-2\lambda(t-T)})\right]$$
$$+ Be^{-\lambda(t-T)} \sin\left[\omega_i(t-T) - \omega_i \frac{C^2}{32\lambda}(1 - e^{-2\lambda(t-T)})\right] \qquad (8)$$

Nonlinear addition to the phase of pendulum oscillation depends on amplitude and oscillation phase at the moment of the second pulse arrival. Retaining only the terms depending on this phase and neglecting the phase change up to time T, we shall write eq. (8) in the form

$$\phi_i(t) = Ae^{-\lambda t} \sin(\omega_i t - \gamma\cos\omega_i T) + Be^{-\lambda(t-T)} \sin[\omega_i(t-T) - \gamma\cos\omega_i T] \qquad (9)$$

where $\gamma = \frac{AB}{16\lambda}\omega\, e^{-\lambda T}(1 - e^{-2\lambda(t-T)})$.

Using the well known relation

$$e^{i\gamma\cos x} = \Sigma\, i^k J_k(\gamma)e^{ikx}. \qquad (10)$$

where $J_k(x)$ is the Bessel function of the k–order, it is possible to obtain the expansion of Eq. (8) in a series of harmonic components $k\omega t$ (k = 0, ±1, ±2, ...). After summation over all the oscillators we obtain

$$\phi(t) = \Sigma\phi_k F(t-kT)\sin[\overline{\omega}(t-kT) + \varphi_k] \qquad (11)$$

where $\phi_k = \left[A^2 e^{-2\lambda t} J_k^2(\gamma) + B^2 e^{-2\lambda(t-T)} J_{k-1}^2(\gamma)\right]^{1/2}$

is the response amplitude and φ_k the phase of the k–th harmonic.

According to (11) a large number of response sequences appears. First of all we shall consider the behaviour of linear responses under the amplification conditions. The response amplitudes at times $t = n\tau$ and $t = n\tau + T$ are given by

$$\phi_0 = \left[A^2 e^{-2\lambda t} J_0^2(\gamma) + B^2 e^{-2\lambda(t-T)} J_1^2(\gamma)\right]^{1/2}$$
$$\phi_1 = \left[A^2 e^{-2\lambda t} J_1^2(\gamma) + B^2 e^{-2\lambda(t-T)} J_0^2(\gamma)\right]^{1/2} k \qquad (12)$$

For A $\ll$ B and $\gamma \ll 1$ we have

$$\phi_0 = \frac{AB^2}{32\lambda} e^{-\lambda T}\left[1 - e^{-2\lambda(t-T)}\right] \qquad (13)$$

The response amplification ϕ_0/A is determined by the amplitude B and the Q–factor of the oscillator (ba the ratio ω/λ). The response has a maximum at $\gamma = 2.5$. Its maximum value does not depend on the amplitude A. It is close to the value $\phi_0 = 5\, Be^{-\lambda(t-T)}$. From (11) one can see that the echo–type response is similar to that for linear response.

Eq. (11) is valid when the responses do not overlap. With a large number of oscillators the overlapping can be avoided by a proper selection of the dealy time T. For a small number of oscillators one can hardly avoid overlapping. In this case the response $\phi(t)$ may be determined by direct summation over the responses from each oscillator.

3. The experiments were carried out using the method describing in [3] The ensemble consisted of 15 pendula with successive frequencies differing by .016 Hz (see Fig. 1). The mean oscillation frequency and oscillation decay rate λ were equal to 1.2 Hz and 0.004 sec^{-1} respectively. These results which are in close agreement with the computer simulations are given in Fig. 2 and Fig. 3. The well–defined nonlinear amplificaiton was observed both in the experiment and in the computer simulation. The experimental value of the amplification in Fig. 3 is in agreement with computed values. The general behaviour was also confirmed. With the oscillation amplitude $A = 0.1$ rad a large number of overlapping responses became apparent. In this case exact summation was used and the results depend very critical on the delay time T. Fig. 4 shows the experimental data. The experimental picture is in close agreement with the numerical calculation for a delay time $T = 12.67$ sec. Small changes of the delay time T alter the picture of the responses dramatically.

The effects discussed above appear in an analog way for optical pulse amplification. The main object of the present work has to develop a model of a pulsed gravitational radiation detector provided by a set of resonant elements.

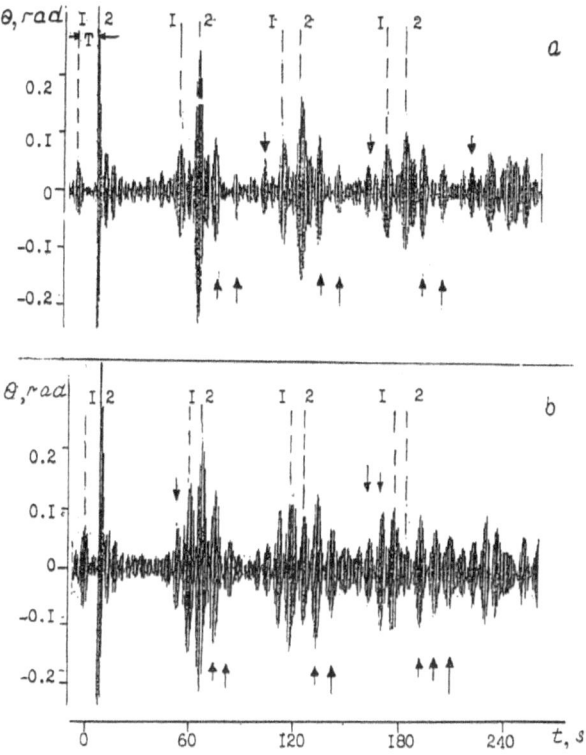

Fig. 2. Pendula ensemble responses for different amplitudes of the pulses. Numeral "1" and "2" correspond to linear response position at time $t = n\tau$ and $t = n\tau - T$. Arrows "↑" correspond to the pulse position at times $t = n\tau + 2T$, $t = n\tau + 3T$, $t = n\tau + 4T$; arrows "↓" – at times $t = (n+1)\tau - T$ and $t = (n+1)\tau - 2T$.

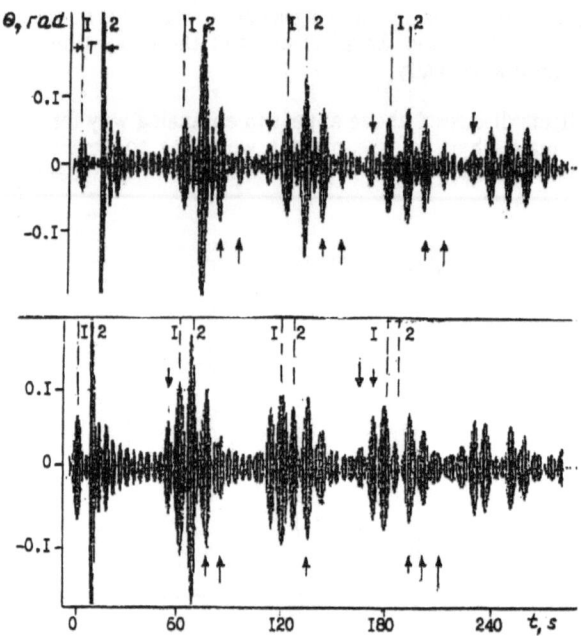

Fig. 3. Computer simulations of the responses for different
amplitudes of exciting pulses: a — A = 0.04 rad,
B = 0.3 rad, t = 11.3 sec, λ = 0.004 sec^{-1}; b — A = 0.07 rad,
B = 0.3 rad, T = 7.7 sec, λ = 0.004 sec^{-1}.

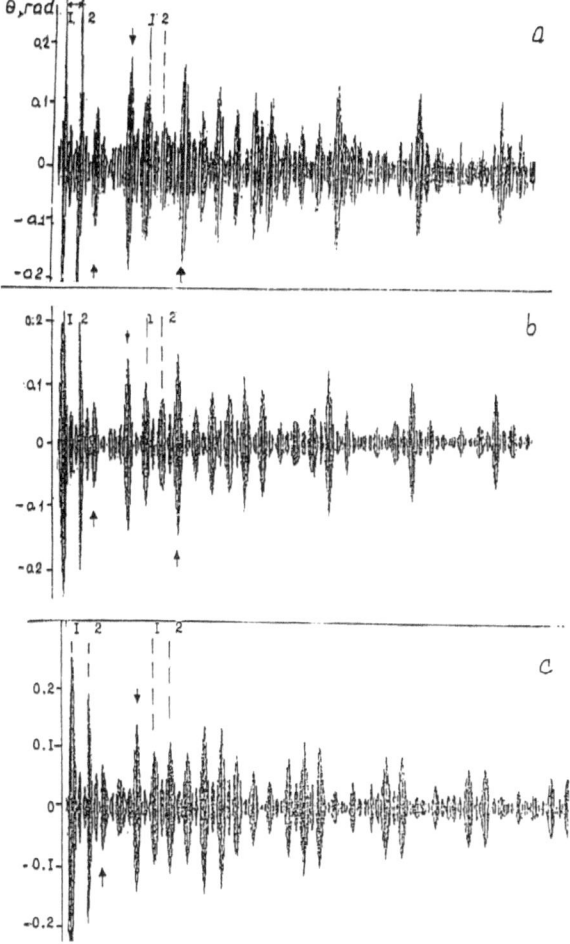

Fig. 4. Pendula ensemble responses: a – experiment; b, c – computer simulation for A = 0.26 rad, B = 0.21 rad, λ = 0.004 sec^{-1} and two values delay time T = 12.67 sec and t = 12.62 sec, respectively.

ACKNOWLEDGEMENT

The author would like to thank Prof. E.V.Baklanov, Dr. B.Ya. Dubetsky, Prof. W.R. Bennett, Jr., and Prof. R.K. Chang for their encouragement in carrying out this investigation.

REFERENCES

1. R.G. Brewer: in Conference on spectroscopy and Modern Physics, ed. by F.T. Arecchi, R. Bonifacio and M.O. Scully (Plenum Press, New York and London 1978) Serie B, Physics V. 37, p. 41.
2. V.P. Chebotayev, B.Ya. Dubetsky: Appl. Phys. B 31, 45, 1983.
3. B.Ya. Dubetsky, V.P. Chebotayev: Pizma v. ZETP, 41, 267, 1985.
4. L.D. Landau, E.M. Lifshits: Mechanics, M. Nauka, 1965.

NARROWBAND TUNABLE VUV/XUV RADIATION GENERATED BY THIRD-

ORDER FREQUENCY MIXING OF LASER RADIATION IN GASES

A. Borsutzky, R. Brünger and R. Wallenstein

Institut für Quantenoptik, Universität Hannover
3000 Hannover, FRG

Third-order frequency mixing in gases is a well established method for the generation of optical radiation in the spectral region of the vacuum ultraviolet (VUV) at wavelengths λ_{VUV}=100-200 nm and of the extreme ultraviolet (XUV) at λ_{XUV}=58-100 nm. The pioneering work in this field has been published more than ten years ago by Ward and New[1], Harris and Miles[2], Miles and Harris[3] and Kung et al[4]. During the past decade the results of a large number of theoretical and experimental investigations demonstrated that frequency mixing of powerful laser light generates intense VUV and XUV radiation with fixed or tunable frequency and high spectral brightness[5,6,7].

Nonresonant frequency tripling and sum- and difference frequency mixing ($\omega_{VUV}=2\omega_1\pm\omega_2$) generated in the rare gases Xe, Kr and Ar broadly tunable VUV radiation in the wavelength range λ_{VUV}=110-200 nm[8-15]. Third harmonic generation in Ar and Ne produced XUV light in spectral regions between 72 and 105 nm[16,17].

In these experiments laser pulse powers of 1 to 5 MW provided conversion efficiencies of 10^{-5} to 10^{-6}.

The efficiency could be increased by several orders of magnitude by using resonantly enhanced frequency conversions[18]. By tuning the laser frequency to a two-photon resonance, for example, the induced polarization is resonantly enhanced. The two-photon resonant conversion, which is usually of the type $\omega_{VUV}=2\omega_R+\omega_T$, where ω_R is tuned to a two-photon transition and ω_T is a variable frequency, provides conversion efficiencies of $\eta > 10^{-4}$ even at input powers of only a few kilowatts.

In the past the resonant frequency conversion has been investigated in metal vapors[19] (such as Sr, Mg, Cs, Ba, Hg and Zn) and in the rare gases Xe and Kr[20-22].

For experimental realization of the frequency mixing, rare gases are advantageous. Enclosed in a glas or metal cell (equipped with appropriate windows) these gases provide a nonlinear medium of homogenous, easily variable density. These gases are thus an appropriate medium for the

Applied Laser Spectroscopy, Edited by W. Demtröder and
M. Inguscio, Plenum Press, New York, 1990

construction of a reliable VUV light source useful for spectroscopy applications.

For spectroscopic applications parameters like the tuning range, the output power and the bandwidth are of special interest. In the following typical values of these parameters are summarized.

Tuning range

Phase-matching conditions between the generated VUV and the focused laser light restrict the tuning range of the sum frequency to spectral regions of negative mismatch ΔK, defined as the difference between the wavevectors of the generated radiation and the driving polarization[8,21].

The rare gases Ne, Ar, Kr and Xe provide the required negative dispersion at the high energy side of their resonance transitions in extended regions of the wavelength range λ_{VUV}=66-147 nm (see Fig.1). Above the transmission cut off of LiF (λ_{VUV}>105 nm) Kr and Xe are negative dispersive in the wavelength regions 110-116 nm (Kr), 113-117 nm(Xe), 117.2-119 nm(Xe), 120-135 nm(Kr), 126-129 nm(Xe) and 140-147 nm(Xe). At these wavelengths nonresonant third harmonic generation and sum frequency mixing has been investigated in detail[5,7]. Frequency tripling of the third harmonic of the Nd-YAG laser - that generates intense VUV at 118.3 nm - and the generation of radiation at $L\alpha$=121.6 nm are well-known examples.

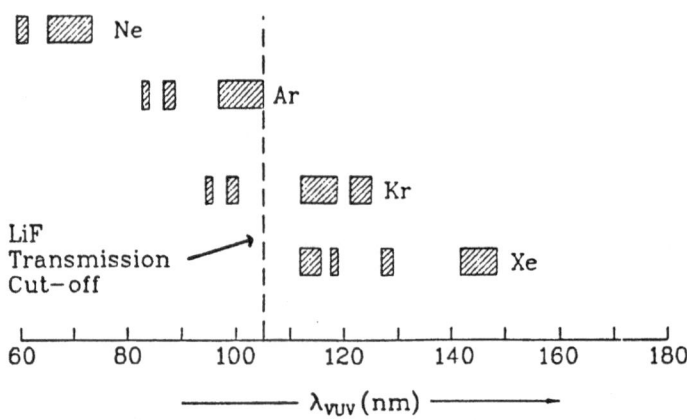

Fig. 1 Spectral regions with negative mismatch ΔK <0 for frequency tripling and sum-frequency mixing

In contrast to the sum, the difference frequency can be generated in a medium with positive or negative mismatch[8,21]. Since this conversion is not restricted by the dispersion of the medium it should be suited for the generation of VUV in the entire range between 115 and 200 nm.

This has been demonstrated by mixing the fundamental frequencies of the dye laser (ω_L) or of the Nd-YAG laser (ω_{IR}) with the UV radiation with the frequency ω_{UV}=2ω_L in the rare gases Xe, Kr and Ar[7,13,15]. Fig.2 displays the tuning ranges of the difference frequencies $\omega_{VUV}=2\omega_{UV}-\omega_{IR}$ (with $\omega_{UV}=2\omega_L$ and $\lambda_L=420-700$ nm) and $\omega_{VUV}=2\omega_{UV}-\omega_L$ (with $\omega_{UV}=2\omega_L$ and $\lambda_L=420-630$ nm).

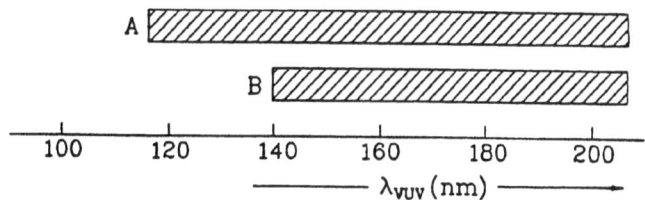

Fig. 2   Tuning range of the nonresonant difference frequency mixing in rare gases
A: $\omega_{VUV}=2\omega_{UV}-\omega_{IR}$ with $\omega_{UV}=2\omega_L$ and $\lambda_L=420-700$ nm
B: $\omega_{VUV}=2\omega_{UV}-\omega_L$ with $\omega_{UV}=2\omega_L$ and $\lambda_L=420-630$ nm
λ_L is the wavelength of the dye laser.

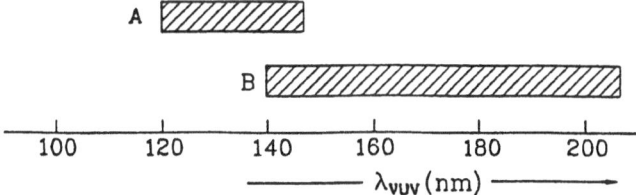

Fig. 3   Tuning range of the resonant frequency mixing in Kr:
$\omega_{VUV}=2\omega_R-\omega_T$ with $\lambda_R=212.5$ nm. A:$\omega_T=\omega_L$, B: $\omega_T=2\omega_L$
λ_L is the fundamental frequency of the dye laser
($\lambda_L=420-900$ nm).

It should be emphasized that this nonresonant frequency conversion requires only one dye laser pumped by the second or third harmonic of a Nd-YAG laser.

As the nonresonant difference frequency mixing the two-photon resonant difference frequency conversion produces widely tunable VUV. In Kr, for example, the difference frequency $\omega_{VUV}=2\omega_R-\omega_T$ of UV radiation at $\lambda_R=212.5$ nm (resonant with the 4p-5p [1/2,0] two-photon transition) and of tunable dye laser light at $\lambda_T=210-900$ nm generates radiation at $\lambda_{VUV}=120-215$ nm (see Fig.3).

It should be emphasized that the resonant sum-frequency $\omega_{VUV}=2\omega_R+\omega_T$ generates continuously tunable radiation at $\lambda_{XUV}=71.3-95.0$ nm[20]. In this case the frequency ω_T is tuned at $\lambda_T=217-900$ nm.

Output power

In nonresonant frequency conversion laser pulse powers of 1-5 MW provide conversion efficiencies of 10^{-5} to 10^{-6}. The power of the generated VUV light pulses is typically in the range of 1-20 W ($0.3-6 \times 10^{10}$ photons/pulse). The efficiency is limited by dielectric gas breakdown in the focus of the laser light and by nonlinear intensity-dependent changes of the refractive index[23].

The resonant frequency mixing usually provides conversion efficiencies of 10^{-2} to 10^{-4} depending on the resonant enhancement of the nonlinearity and on the laser power. With input pulse powers of the resonant UV radiation of only 100-200 KW the VUV output power generated in Kr is in the range of 0.2-2 KW ($6 \cdot 10^{11} - 6 \cdot 10^{12}$ photons/pulse)[20]. Higher input powers will further increase the VUV output by at least one order of magnitude.

Bandwidth

The spectral bandwidth of the VUV radiation is determined by the linewidth of the pulsed laser radiation.

The linewidth of commercial pulsed dye laser systems is typically 0.2-0.5 cm^{-1}. With additional line narrowing – provided by intracavity etalons – the spectral width is reduced to 0.02-0.1 cm^{-1}. For those conversions which use the radiation of the fundamental of the Nd-YAG laser injection-seeded systems may be used. The linewidth of these lasers is less than $3 \cdot 10^{-3}$ cm^{-1}. For narrowest bandwidths the radiation of cw dye lasers is amplified in pulsed dye laser amplifiers. The fourier-transform-limited linewidth of the amplified radiation is also less than $3 \cdot 10^{-3}$ cm^{-1}.

Depending on the used laser systems the bandwidth of the generated VUV radiation is on the order of 0.01-1.5 cm^{-1}.

Because of the small spectral width the spectral brightness of narrowband VUV generated by resonant frequency conversion, for example, is on the order of 10^{17} photons $sec^{-1} nm^{-1}$. This value surpasses the spectral brightness of synchrotron VUV light sources (which is typically 10^{14} photons $sec^{-1} nm^{-1}$) by orders of magnitude.

The discussed results demonstrate that nonlinear optical frequency conversion produce widely tunable VUV radiation. Because of the narrow spectral width and the high intensity the VUV light is a powerful tool for high resolution spectroscopy of atoms and molecules. This has been demonstrated, for example, by absorption spectroscopy, by excitation spectroscopy or by state selective resonant-excitation-ionization spectroscopy (see ref. 5-7 and references therein).

Today the number of spectroscopic applications of coherent laser-generated VUV light is still small. Because of the simple way of generation and the excellent spectral properties there is no doubt that in the future this radiation will be very useful for a large variety of spectroscopic applications.

References

1. G.H.C. New and J.F. Ward, Phys. Rev. Lett. 19, 556 (1967) J.F. Ward and G.H.C. New, Phys.Rev. 185, 57 (1969)

2. S.E. Harris and R.B. Miles, Appl. Phys. Lett. 19, 385 (1971)

3. R.B. Miles and S.E. Harris, IEEE J. Quantum Electron. QE-9, 470 (1973)

4. A.H. Kung, J.F. Young, G.C. Bjorklund and S.E. Harris, Phys. Rev. Lett. 29, 985 (1972)

5. W. Jamroz and B.P. Stoicheff, in 'Progress in Optics' (E. Wolf, ed., North Holland, Amsterdam 1983) vol 20, pp. 326-380

6. C.R. Vidal, in 'Tunable Lasers' (I.F. Mollenauer and J.C. White, eds. Springer Verlag, Heidelberg 1984)

7. R. Hilbig, G. Hilber, A. Lago, B. Wolff and R. Wallenstein, Comments At.Mol.Phys. 18, 157 (1986)

8. G.C. Bjorklund, IEEE J.Quantum Electron. QE-11, 287 (1975)

9. R. Mahon, T.J. McIlrath and D.W. Koopman, Appl. Phys. Lett. 33, 305 (1978)

10. D. Cotter, Optics Commun. 31, 397 (1979)

11. R. Wallenstein, Optics Commun. 33, 119 (1980)

12. R. Hilbig and R. Wallenstein, IEEE J. Quantum Electron. QE-17, 1566 (1981)

13. R. Hilbig, PhD Thesis, 1984 (to be published in Appl. Phys.)

14. R. Hilbig, G. Hilber, A. Timmermann and R. Wallenstein; 'Laser Techniques in the Extreme Ultraviolet', in Proc. AIP conf. vol. 119, 1 (1984)

15. R. Hilbig and R. Wallenstein, Appl. Optics 21, 913 (1982)

16. R. Hilbig and R. Wallenstein, Optics Commun. 44, 283 (1983)

17. R. Hilbig, A. Lago and R. Wallenstein, Optics Commun. 49, 297 (1984)

18. D.C. Hanna, M.A. Yuratich and D. Cotter 'Nonlinear Optics of Free Atoms and Molecules' (Berlin, Springer Verlag, 1979); M.A. Yuratich and D.C. Hanna, J. Phys. 89, 729 (1976)

19. see references no. 5,6 and 7 and references therein

20. G. Hilber, A. Lago and R. Wallenstein, JOSA B 4, 1753 (1987), and references therein.

21. G. Hilber, A. Lago and R. Wallenstein, Phys. Rev. A36, 3827 (1987)

22. G. Hilber, D.J. Brink, A. Lago and R. Wallenstein, Phys. Rev. A38, 6231 (1988)

23. H. Langer, H. Puell and H. Röhr, Opt. Commun. 34, 137 (1980)

LASER AND SYNCHROTRON-BASED EXCITATION SOURCES FOR RELAXATION STUDIES

Sydney Leach

Laboratoire de Photophysique Moléculaire du C.N.R.S.
Bâtiment 213, Université Paris-Sud
91405 - ORSAY, France
and
Département Atomes et Molécules en Astrophysique
Observatoire de Paris-Meudon
92190 - MEUDON, France

ABSTRACT

Source requirements for selective excitation and interrogation of physical and chemical processes are examined. Appropriate Laser and Synchrotron-based sources are characterised and described. The subjects discussed include UV and VUV lasers, synchrotron and undulator emission, harmonic generation, and free electron lasers of the Compton and Raman scattering types, as well as their associated equipment. Applications of free electron lasers are briefly considered.

I - INTRODUCTION

This review compares and contrasts laser and synchrotron sources for excitation and interrogation of physical and chemical processes. The main context is the use of such sources for intramolecular dynamics studies but includes other physical and chemical investigations. The discussion of synchrotron radiation leads to consider magnetic devices that provide ways of enhancing synchrotron radiation, including the generation of high energy harmonics of a fundamental frequency emitted by an internal or an external photon source. These magnetic devices also play a basic role in the free electron laser (FEL). The basic physics of free electron lasers are discussed and some results are presented. The status of electron accelerators and free electron laser devices operating or under construction in the world is considered and some comments made as to future expectations. Applications of FEL devices are briefly considered.

The general references [1-13] provide some introductory material on intramolecular dynamics [1,2] lasers [1-7], synchrotron radiation [1,8-10] and Free Electron Lasers [2,10-12]. A text book gives the situation of FEL devices in 1984 [11] with some emphasis on the so-called Raman-type FELs. A picture of world-wide FEL development and devices can be obtained from recent conference proceedings [13]. The present review is, in part, a shortened version of a more detailed presentation recently published [2]. Additional material brings the subject abreast of current activities.

A. Basic Photophysical Processes

Excited state intramolecular dynamics concerns the basic photophysical and chemical mechanisms of the acquisition and disposal of energy in
excited states of molecules, as viewed from the microscopic viewpoint [1].
Among the principal processes involving excited electronic states,
radiative decay is only one of several possible energy disposal mechanisms.
An outstanding goal of research is therefore the elucidation of a large
class of radiationless processes. These involve energy exchange between
and within electronic, vibrational and rotational degrees of freedom, e.g.
nonradiative coupling of electronic states and vibrational states, the
latter being often referred to as intramolecular vibrational redistribution (IVR) when it occurs within a particular electronic state. They also
include processes such as dissociation and ionization in their direct form
as well as the indirect processes of predissociation and autoionization.
The various intramolecular dynamics processes occur on characteristic
energy and time scales as discussed below.

B. Energy Scales

The energy regions for excited state photophysical processes in
molecules is indicated in figure 1, which also shows the present range of
synchrotron and laser sources. It is seen that the interesting energy
regions of particular processes are within the range 0.1 - 1000 eV.

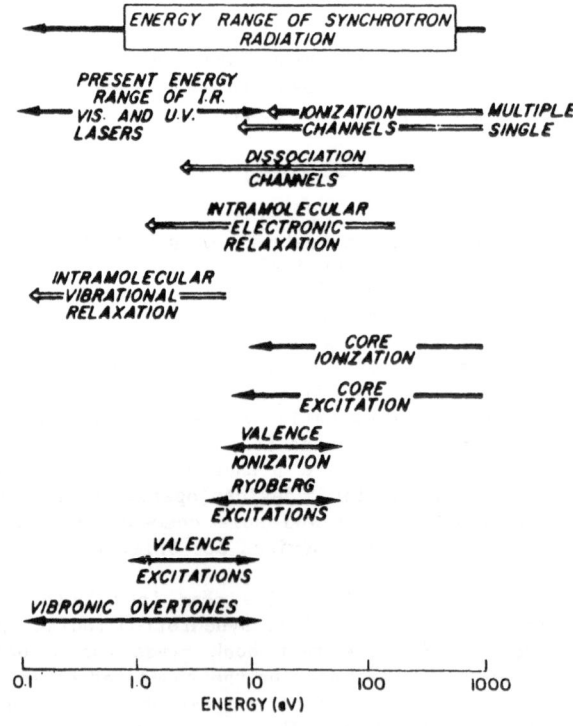

Fig. 1. Energy ranges of lasers, synchrotron radiation and of
intramolecular processes

As far as energy resolution is concerned, the characteristic values are determined by a number of parameters which include the following : Doppler broadening (10^{-3} - 10^{-1} cm^{-1}) ; Rotational spacing (10^{-2} - 10 cm^{-1}) ; Vibrational spacing (50 - 3000 cm^{-1}) ; Homogeneous decay width (e.g. for t = 10^{-14}, 10^{-12}, 10^{-8} s characteristic decay times, the homogeneous widths are 500, 5 and 0.0005 cm^{-1} respectively).

C. Time Scales

A non-exhaustive list of physico-chemical processes and their characteristic times is given Fig 2. The period of electronic motion provides the lower limit for the time scale of electronic autoionization, while the time scale for intramolecular nuclear motion determines the lower limit for a variety of processes which involve nuclear vibrations. The restrictions imposed by the Franck-Condon principle and its extensions for nonradiative intramolecular processes (which involve nonadiabatic coupling, bound state-continuum interactions and scattering effects) also influence the intramolecular relaxation rates.

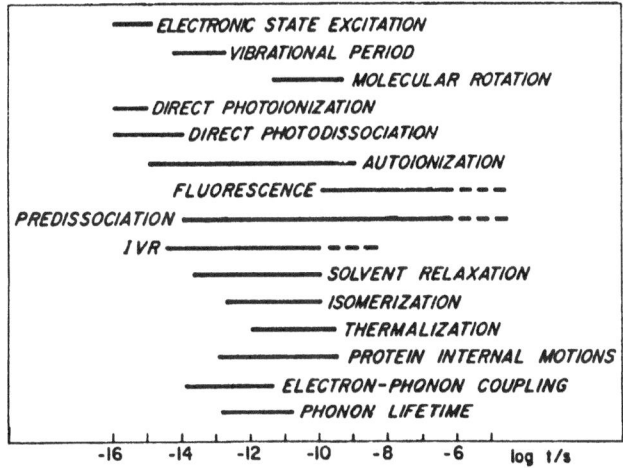

Fig. 2   Time scales for excitation and relaxation in physicochemistry processes.

D. Objectives and Current Situation

There are two major goals of studies in intramolecular (and intermolecular) dynamics. First, one has to identify and elucidate the various decay channels, the interactions between channels, and the sequence of photophysical (and photochemical) processes (non dissociative, dissociative, ionization) in excited states. Second, the phase relationships between excited states, as well as between the ground state and the excited doorway state, have to be explored in order to understand intramolecular interference effects, intramolecular dynamic processes, as well as medium perturbations of excited states.

The most extensive studies of intramolecular processes concern nondissociative electronic relaxation. This was already carried out with conventional light sources in the 1-5 eV energy range but the advent of laser sources and has led to much improved time-resolved experiments and studies of coherent optical effects. Relatively few direct studies have been conducted of electronic relaxation in higher (Rydberg) states. The study of intrastate vibrational energy redistribution (IVR), which can be carried out in the ground state in the energy range 0.1-2 eV, and above 1 eV for electronically excited states, has been vigorously pursued under the impact of recent theoretical models for understanding these phenomena. Direct detailed information on dissociation and ionization processes in polyatomic molecules is still rather fragmentary, since only few time-resolved observables and energy-resolved observables (except absorption and ionization cross sections) have been explored to date. This is, however, a rapidly developing field under the impetus of new excitation and interrogation techniques.

Current understanding of the various features and consequences of competitive and sequential processes in highly excited molecular states is still rather embryonic, especially for multiply charged molecular ion production and relaxation [14], [15]. Intramolecular dynamics is rather complex in highly excited molecular states. The interplay and interference between different excited channels govern much of molecular photochemistry and radiation chemistry. Many of these processes require optical excitation above ~ 5 eV.

III. IDEAL EXCITATION SOURCES FOR INTRAMOLECULAR DYNAMICS STUDIES

The study of excited state dynamics requires selective, well-defined excitation of the parent molecule. Thus, optical excitation is favoured over alternative techniques such as nonselective excitation by electron or ionic impact, fast neutral bombardment, or excitation by energetic (α, β, γ) particles. The basic characteristics of an ideal optical excitation source for studies of molecular dynamics should satisfy the various time and energy domains discussed above, and can be summarized as follows.

1) <u>Tunability</u> over a very broad 0.1 - 1000 eV energy range (cf fig. 1).

2) <u>Time resolution</u> : As seen from fig. 2 and the previous discussion, time scales for intramolecular dynamics vary from ~ 10^{-16} s to seconds.

3) <u>Energy resolution</u> : The spectral resolution for energy-resolved observables depends on the type of physical information required. Very high resolution ($< 10^{-2}$ cm^{-1}) is necessary for the study of the slow relaxation of a single rovibronic level. For many molecular relaxation processes, energy resolution in the 1 cm^{-1} range is usually sufficient, while for ultrafast processes 10 - 100 cm^{-1} is often adequate.

4) <u>Intensity requirements</u> : The intensity of the excitation source should be sufficient for time-resolved and/or energy resolved studies over the appropriate energy range. The precise requirements depend on many factors in the experimental set-up, in particular the sensitivity of detection methods and whether single or multiphoton processes are examined.

5) <u>Collimation</u> : A high degree of collimation is useful for maximizing the effective intensity (brightness) of excitation sources.

6) <u>Polarization</u> : Linear or circular polarization is very useful for studying angular distribution in fragmentation and ionization processes, as well as for energy transfer studies.

7) <u>Source stability</u> : High temporal, energetic and spatial stability is necessary for reliable high accuracy experiments.

8) <u>Temporal coherence</u> : This is important for interrogating coherent optical effects.
The sources that will be examined in this context are lasers, synchrotron radiation, and devices for enhancing synchrotron radiation including the free electron laser. The basic physics of synchrotron radiation will first be presented and a discussion given of the characteristics of present and projected synchrotron radiation facilities.

IV. SYNCHROTRON RADIATION SOURCES

A. Basic Physics of Synchrotron Radiation

An electron in an accelerating field will lose energy by radiation. This is the basis of synchrotron radiation and FEL devices, in which positive or negative electrons undergo radial acceleration, usually by magnetic constraint. In this case the radiation emitted can be considered as magnetobrehmsstrahlung. A low velocity electron radiates with a pattern close to that of a normal Hertzian electric dipole (Fig. 3a) but this is severely distorted at high electron velocity, due to relativistic effects. The zeros of the radiation pattern then occur at angles $\theta = (1-v^2/c^2)^{1/2}$ from the direction of motion. From the viewpoint of a stationary observer, the relativistic (Lorentz) transformation of the electron causes the power radiated to be projected into a very small foward cone of the order of a milliradian in angle, giving the radiation pattern indicated in fig. 3b.

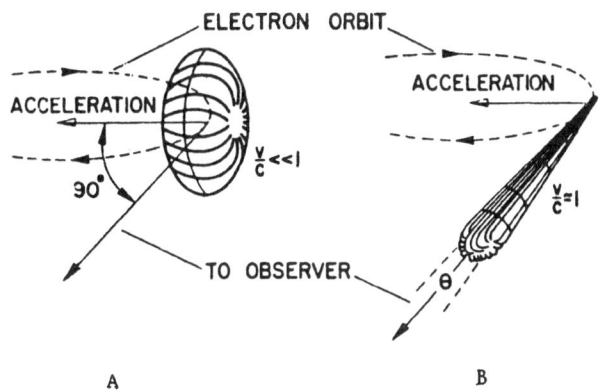

A   B

Fig. 3 Schematic radiation patterns for electrons in circular orbit (a) at low velocities ; (b) at relativistic velocities, $\theta = (1 - v^2/c^2)^{1/2}$.

The most useful synchrotron radiation device is based on the <u>electron storage ring</u>, (fig. 4). This contains a number of dipole "bending" magnets which force the electrons to undergo a circular trajectory where they radiate. Between the bending magnets are a number of straight sections where reside quadrupole magnets to focus the electron beam, and insertion devices such as undulators (see section VI). Electron beam correction may require additional, sextupole, magnet coils. Electrons, once injected into the magnet lattice, can have their energy modified or restored by a radiofrequency cavity.

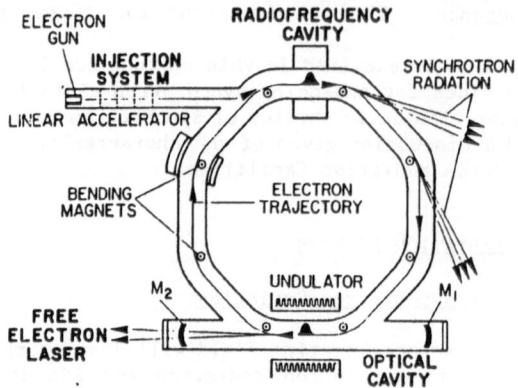

Fig. 4 Schematics of an electron storage ring and associated free electron laser. Two electron packets are represented. M1 and M2 are the optical cavity mirrors.

The basic relations between instrumental parameters and synchrotron radiation characteristics are as follows [16,17] (where the electron energy E is in GeV, the magnetic radius of curvature R is in meters, the electron current I in amps and the magnetic field B in Teslas) :

(i) total power radiated by a relativistic electron, for $(v/c) \approx 1$

$$P = \int I(\lambda,\Psi) \, d\lambda d\Psi = (2/3)(e^2c/R^2)(E/(m_0c^2))^4 \ldots\ldots\ldots\ldots\ldots(1)$$

(ii) energy loss per revolution, per electron

$$\delta E(keV) = 4\pi e^2 \gamma^4/3R = 88.47 \, E^4/R = 26.5 \, E^3 B \ldots\ldots\ldots\ldots\ldots(2)$$

where the relativistic factor $\gamma = E/m_0c^2 = 1957 \, E$

(iii) total power radiated
$$P_{tot}(kW) = 0.2654 \, E^3 BI \ldots\ldots\ldots\ldots\ldots\ldots\ldots\ldots\ldots\ldots\ldots(3)$$

(iv) critical wavelength
$$\lambda_c \, (\mathring{A}) = 4\pi R/3\gamma^3 = 5.59 \, R/E^3 = 18.64/BE^2 \ldots\ldots\ldots\ldots\ldots(4)$$

(v) characteristic (so-called critical) energy
$$\epsilon_c(eV) = 2218 \, E^3/R = 2.96 \times 10^{-7}\gamma^3/R$$
$$= 665.1 \, BE^2 \ldots\ldots\ldots\ldots\ldots\ldots\ldots\ldots\ldots\ldots\ldots(5)$$

(vi) mean emisssion angle $\theta \approx \gamma^{-1}$ $\ldots\ldots\ldots\ldots\ldots\ldots\ldots\ldots\ldots\ldots(6)$

Because of the Lorentz transformation, an observer at a particular angle in the orbit plane will detect synchrotron radiation emitted over an arc length R/γ in the bending magnet. The radiation pulse will have a duration $\Delta t \approx R/c\gamma^3$. The power spectrum comprises the Fourier components of the synchrotron radiation pulse and will contain harmonics of the orbit frequency ν up to $\omega = \nu\gamma^3$. The fundamental frequency is Doppler shifted from the MHz region into higher frequencies ; the harmonics blur into a continuum, so that the observed spectrum extends from the far infra-red to the ultraviolet and x-ray regions approximately up to the critical wavelength λ_c. The divergence of the resulting light, in the vertical plane, is comparable to that of a laser.

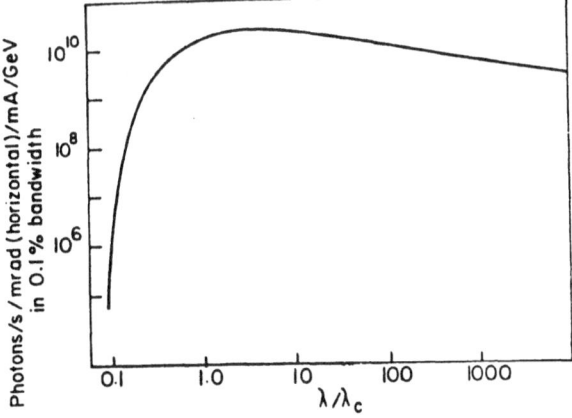

Fig. 5 Normalized radiation spectrum of an electron moving in a curved
trajectory, per GeV. λ_c - critical wavelength.

Fig. 5 gives the radiation spectrum of an electron moving in a
curved trajectory, per GeV. This is expressed in terms of the photon flux,
which is of practical interest, as a function of the radiation wavelength
divided by the critical wavelength. Its peak occurs at $\lambda/\lambda_c \approx 4$. Although
the flux falls off rapidly below 4 λ_c, it is still of useful intensity down
to $\approx 0.1 \lambda_c$.

The principal properties of synchrotron radiation of interest as a
spectroscopic source or excitation source in physicochemistry are as
follows [1] :

1) Tunability over a very broad energy range from below 0.1eV to above 1000
eV.

2) Pulsed time structure (ps-ns range) enabling time resolved experiments
    to be carried out.

3) High intensity (brightness).

4) High degree of spatial collimation (1 mrad).

5) Quasi-complete linear polarisation ; possibility of other polarisations.

6) Temporal and spatial stability.

Many of these properties carry over to, or are improved upon in,
insertion devices and FELs. It is thus seen that synchrotron radiation can
provide, in principle, most, but not all, of the characteristics of the
ideal source defined in section III.

B. Synchrotron Radiation Facilities

Characteristics of present-day and projected synchrotrons and
storage rings used or useful for synchrotron radiation are listed in Table
1, in order of increasing critical energy (decreasing critical wavelength)
[12]. A recent workshop on the construction and commissioning of dedicated
synchrotron radiation facilities [18] gives more detailed information on a
number of the facilities listed in table 1, as well as presentation of new
projects in several countries. Compact storage rings are also being

TABLE 1
Synchrotron Radiation Sources[a]

Machine Name	Location	E/(GeV)	I/mA	R/m	λ_c/Å	Remarks
N-100	Kharkov, USSR	0.10	50	0.5	2800	D
SURF II	Washigton, USA	0.25	25	0.84	300	D
TANTALUS I	Madison, USA	0.24	200	0.64	260	D
SOR RING	Tokyo, Japan	0.38	250	1.1	112	D
TERAS	Tsukuba, Japan	0.6	150		52	D
SIBERIA I	Moscow, USSR	0.45				UC ; D
MAX	Lund, Sweden	0.55	370	1.2	40.3	
ACO	Orsay, France	0.54	150	1.1	39	D
FIAN C-60	Moscow, USSR	0.67	10	1.6	30	S
UVSOR	Okazaki, Japan	0.75	500	2.2	29	D
NSLS I	Brookhaven, USA	0.75	500	1.9	25	D
HESYRL	Hefei, China	0.8	300	2.22	24	UC ; D
VEPP-2M	Novosibirsk, USSR	0.67	100	1.22	23	PD
BESSY	Berlin, FRG	0.8	500	1.83	20	D
SUPER ACO	Orsay, France	0.8	500	1.75	19.1	UC ; D
SRL	Stanford, USA	1.0	500	2.1	11.7	UC ; D
IN-ES	Tokyo, Japan	1.3	30	4.0	10.2	S
PAKHRA	Moscow, USSR	1.3	300	4.0	10.2	S
COSY	Berlin, FRG	0.63		0.44	9.8	UC ; D
SIRIUS	Tomsk, USSR	1.36	15	4.23	9.4	S
ADONE	Frascati, Italy	1.5	100	5.0	8.3	PD
BONN	Bonn, FRG	2.0	50	7.6	5.3	PD
ALADIN	Madison, USA	1.3	500	2.08	5.3	D
SRS	Daresbury, UK	2.0	370	5.56	3.9	D
DCI	Orsay, France	1.85	300	3.82	3.4	D
ELETTRA	Trieste, Italy	2.0	400		3.0	D
PHOTON FACTORY	Tsukuba, Japan	2.5	250	8.33	3.0	D
VEPP-3	Novosibirsk, USSR	2.25	100	6.15	3.0	D
NSL II	Brookhaven, USA	2.5	500	8.17	2.9	D
BEPC	Beijing, China	2.8	150	10.35	2.6	PD
SIBERIA II	Moscow, USSR	2.5	300	5.0	1.8	UC ; D
ARUS	Erevan, USSR	4.5	1.5	24.6	1.5	S
ELSA	Bonn, FRG	3.5	50	10.1	1.3	PD
SPEAR	Stanford, USA	4.0	100	12.7	1.1	PD
ESRF	Grenoble, France	5.0	200		0.9	UC ; D
DORIS II	Hamburg, FRG	5.0	50	12.1	0.54	PD
TRISTAN ACC.	Tsukuba, Japan	6-8				UC ; PAR
CESR	Ithaca, USA	8.0	100	32.5	0.35	PAR
VEPP-4	Novosibirsk, USSR	7.0	10	16.5	0.27	PAR
PETRA	Hamburg, FRG	18.0	18	192.0	0.18	POSS
PEP	Stanford, USA	18.0	10	165.5	0.16	PAR
TRISTAN	Tsukuba, Japan	30.0				UC ; PAR

[a]These are storage rings unless specified otherwise. D, dedicated ; PD, partially dedicated ; UC, under construction ; S, synchrotron ; PAR, parasitic mode ; and POSS, possible use for synchrotron SR.

developed, at Berlin (BESSY), the U.K. (Oxford Instruments) and Japan, mainly for industrial lithography.

V. LASER SOURCES

A. Introduction

There is a vast literature on lasers. Their development and characteristics can be advantageously followed in a number of Handbooks [3-5]. There are many workshops and conferences devoted to lasers, reported in the standard literature and in proceedings form. Two of the subjects of main concern here i.e. ultraviolet and VUV lasers [19,20], and short-pulsed lasers [6], have been the subject of recent reviews.

Laser generation of light involves the conversion of electrical, thermal, chemical, etc., energy into electromagnetic energy. Visible and U.V. lasers operate essentially as electron oscillators in which identical motions are induced in a large number of electrons. The initial energy state population is modified by pumping to higher levels by an external source. Laser action requires that sufficient population inversion occurs such that stimulated emission (i.e. $X^* + h\nu \rightarrow X + 2 h\nu$) dominates the spontaneous process $X^* \rightarrow X + h\nu$. Positive gain implies that the pump energy input exceeds the photon energy loss.

In a conventional laser the electrons are bound in atoms or molecules which themselves may be in the solid, liquid or gaseous state. The pump process creates a change of the state of motion of the bound electrons. The latter have, of course, discrete modes of motion, i.e. the energy levels are quantized. These constraints on electron motions lead to absorption or emission of photons over a restricted number of wavelengths. Thus, in conventional lasers, an initial limitation is the relatively small range of wavelengths of laser emission.

This section is devoted mainly to conventional lasers for the VUV, free electron lasers being discussed in section VII. However, it is worth mentioning at this point that since in free electron lasers the electrons are not bound, any state of electron motion or energy is theoretically possible in this case. Just as in the synchrotron radiation facilities, where free electrons are constrained magnetically to decelerate, leading to synchrotron radiation emission, the free electron laser involves acceleration and deceleration of unbound electrons as will be seen in detail later. This magnetobrehmsstrahlung corresponds to free-free transitions. Thus, in principle, a wide range of emission wavelengths is possible for a free electron laser.

B. Conventional Lasers for the VUV

The last few years have seen much progress in the development of coherent VUV radiation sources. Frequency doubling techniques using a visible laser and KDP (Potassium Dihydrogen Phosphate) and KPB (Potassium Pentaborate) anisotropic crystals enable one to continuously cover the region 210-330 nm. Recently β-Barium borate (β-Ba B_2 O_4) crystals have been developed in China as a superior nonlinear medium (second harmonic generation efficiency > 10 %) for harmonic generation. Excimer lasers are also useful in this and at shorter wavelength ranges. A number of frequency conversion techniques have been developed which extend the coherent emission range well into the VUV. These include the following :

1) Anti-Stokes Raman-shifting in H_2. Using a pump laser (frequency ω_p, e.g. an excimer laser or a dye laser), with H_2 as a Raman-active medium, one obtains stimulated emission at the first Stokes frequency $\omega_{-1} = \omega_p - 4155$ cm^{-1}. Four-wave mixing of ω_p and ω_{-1} generates blue-shifted radiation (First anti-Stokes) at $\omega_{+1} = \omega_p + 4155$ cm^{-1}. Higher order anti-Stokes radiation is generated at higher frequencies ω_n given by $\omega_n = (\omega_p + \omega_{n-1} - \omega_{-1}) = \omega_p + n \times 4155$ cm^{-1}, where n is an integer > 1. Pulse shortening occurs, giving pulses as short as ~ 3 ns. The lowest wavelength achieved is about 138 nm. The bandwidth is of the order of that of the pump laser.

2) Third-harmonic and fifth-harmonic Generation in rare gases and in metal vapours has been used to obtain VUV radiation. In these isotropic media the even order susceptibility tensors vanish but the odd order values are non-zero. The principal difficulties relate to phase-matching conditions which require negative dispersion of the medium. This limits the effective wavelength range of the radiation produced. Furthermore, the dispersion of the nonlinear medium regulates the spectral density of the VUV output since the latter depends on the spectral purity and the beam divergence of the laser source. These techniques provide VUV radiation over various limited wavelength intervals in the 60-120 nm range. Recent developments include the avoidance of window materials through the use of supersonic pulsed jets as the conversion medium. For example, 7th harmonic conversion of 248 nm radiation in a pulsed supersonic jet of He produced tunable coherent radiation at 35.5 nm [21]. Similarly, third harmonic generation in pulsed Ar or CO supersonic jets gave rise to useful radiation in the 80-130 nm region [22].

3) Four-wave Sum and Difference Frequency mixing techniques provide powerful methods to generate tunable VUV radiation [23]. Valid techniques include the sum and difference frequency mixing of pulsed dye laser radiation in rare gases (e.g. Kr, Xe) and metal vapours (Mg, Hg, Sr, Ba, Cs,...). The phase-matching condition is easily met in the difference frequency mixing technique. Widely tunable radiation in the 60-200 nm region can be obtained, with relatively narrow spectral widths (0.01-1 cm^{-1}) and high spectral intensity (10^9 - 5 x 10^{13} photons/pulse). The frequency conversion methods may be non-resonant or resonant. The pulse power is several orders of magnitude greater in the resonant conversion case, due to resonant enhancement of the induced polarization.

C. Comparison of Laser and Synchrotron Radiation Sources

It is convenient at this stage to compare and contrast conventional lasers and synchrotron radiation (SR) as sources for studies of intramolecular dynamics. This can be summarized as follows.

1) SR sources will undoubtedly be superior to lasers with respect to tunability over the broad energy range required for studies of molecular dynamics.

2) The time resolution of current SR sources is superior to that of current VUV lasers, but future VUV lasers will achieve subpicosecond time resolution, as has been obtained in the visible and near U.V.

3) The effective energy resolution of many lasers is superior to that obtained by monochromatization of SR sources.

4) The photon intensity of current visible, U.V. and VUV lasers is greater than current SR sources, making possible multiphoton excitation by lasers but not by SR.

5) The polarization characteristics of both SR and laser sources are excellent for molecular dynamics studies.

6) The temporal coherence of lasers permits interrogation of coherent optical effects which cannot be carried out using SR.

Restricting attention to the VUV region and extrapolating a little into the future, it is clear that VUV lasers will be extremely useful for some photoselective studies carried out over narrow spectral regions in the 5-13 eV energy range for one-photon excitation and 15-30 eV for multiphoton excitation. These studies include :
- energy-resolved experiments under ultra-high spectral resolution
- time-resolved experiments with subpicosecond time resolution using ultrashort pulses from future VUV lasers
- studies of coherent optical effects
- studies of multiphoton excitation in the 15-30 eV range.

Of these VUV studies, only the energy-resolved experiments have begun to be carried out in a widespread fashion. Relatively little has been attempted on the other themes.

There are, however, a number of inherent limitations of VUV lasers, as compared to SR sources. For example, the output of VUV lasers consists of a single spectral line or is limited to tuning over a narrow energy range. Furthermore, their available energy range is relatively small i.e. 5-13 eV for one photon and 10-30 eV for multiphoton excitation.

The use of VUV lasers in the area of molecular dynamics cannot be considered as an alternative to synchrotron radiation excitation. Rather, the utilization of VUV laser sources will provide useful information concerning one-photon and multiphoton excitation, ultrashort decay times and coherent effects, all interrogated over a narrow energy range, or at various selected energies. Thus synchrotron and VUV laser sources supplement and complement each other for molecular dynamics studies.

To end this section a list is given of the principal areas of study of molecular physics requiring VUV sources [24]. The present use of VUV lasers and synchrotron radiation for such studies are indicated (in brackets) by VUVL and SR respectively.

- <u>Structure and dynamics of molecular excited states above 5eV</u>
  . High resolution spectroscopy (VUVL, SR)
  . Predissociation and autoionization of rotationally resolved states : cross- sections, line profiles, lifetimes (VUVL, SR)
  . Collisional complex spectroscopy (VUVL, SR)

- <u>Photodissociation of small molecules</u>
  . Selective excitation : rotational resolution, polarization effects (VUVL, SR)
  . Photofragment fluorescence : spectral resolution, alignment, orientation (VUVL, SR)
  . Laser induced fluorescence (LIF) characterization of dissociation products : kinetic energy (Doppler effects), internal energy, alignment, orientation (VUVL).

- <u>Reactive collisions</u>
  . Preparation of reactants by photofragmentation : state selected molecules, ions, atoms (VUVL, SR)

. Product characterization by LIF : internal states of energy, Doppler
  profiles (VUVL)
. Transition state spectroscopy, (VUVL)

- Photoionization
  . Single photon ionization (VUVL, SR)
  . Multiphoton ionization (VUVL)
  . Product ion state analysis : photoelectron spectroscopy, (VUVL, SR),
    photoelectron-photoion coincidences (SR)
  . Single photon multiple ionization (SR)

- Molecule/Surface interaction
  . Initial state selection (VUVL,SR)
  . Product characterization (VUVL)

VI. UNDULATOR EMISSION

    In this section a return is made to the area of synchrotron
radiation. The discussion concerns magnetic devices inserted in a
relativistic electron beam trajectory with the object of modifying and
enhancing a synchrotron radiation source.

A. Insertion Devices

    Insertion devices have two important characteristics which enhance
the value of simple bending magnet synchrotron radiation sources i.e.
tunability and increased intensity. Many of these devices pertain to
storage rings where the insertion is in a field-free straight section of
the ring lattice. Ideally, it is possible to thus insert a magnetic field of
zero field integral, with respect to the orbit coordinate (for example by
an arrangement of fields of alternate polarity), without any consequent
alteration of the closed orbit of the ring electrons. The local magnetic
field simply produces a transverse acceleration of the electrons with no
overall deflection or displacement. Any value of the local magnetic field
can be used, and in principle, insertion devices can operate independently
of the rest of the storage ring without affecting electron beam properties
or other experiments.

    The principal insertion devices [12] are the three-pole wavelength
shifter, the wiggler, which is basically a sequence of several magnets with
the field alternating in polarity, and the undulator or multipole wiggler
having a smaller electron deflection angle α than a normal wiggler. These
are described in detail elsewhere [12]. Attention will be paid here mainly to
the undulator and its derivative the optical klystron. Fields in the
storage ring magnets are generally smaller than 1.2 Teslas but wiggler
fields of 2 Teslas or more are possible using iron-core magnets and 5 Teslas
or more with superconducting magnets [25]. Samarium-cobalt permanent
magnets, which have high remanent fields, are now considered to be among
the best materials for wigglers and undulators [26].

B. Undulators and Optical Klystrons

1. Undulator Structure and Parameters

    A transverse undulator is a multipole device constituted by a
sequence of several magnets with the field alternating in polarity and in a
direction perpendicular to the electron orbit (fig. 6). The spectral

characteristics of an undulator, as of all insertion devices, are largely determined by the dimensionless deflection parameter, or field index :

$$K = \alpha\gamma = eB\lambda_w/2\pi \ m_o c = 0.0934 \ B_o\lambda_w \ \dots\dots\dots\dots \ (7)$$

where B_o (Teslas) is the maximum magnetic field on the electron trajectory and λ_w (mm) is the undulator period. K is thus the ratio between the maximum angle of electron deflection and the radiation emission angle $\theta \approx \gamma^{-1}$. K has a value of the order of unity for most undulators. The photon emission is greatly enhanced in intensity with respect to a single bending magnet. Coherent radiation can be produced by interference effects in undulators having adapted characteristics, as described later.

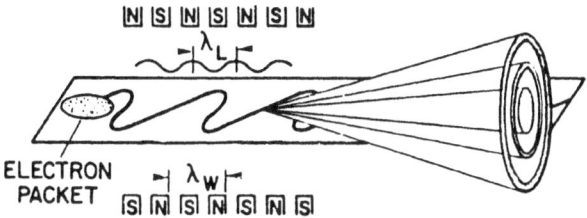

Fig. 6. Schematic representation of the passage of a relativistic electron packet in an undulator and radiation emitted at a particular point in the trajectory. λ_w = undulator period ; $\lambda_L = \lambda_w/ 2\gamma^2$ is the fundamental wavelength emitted (cf. Eq. (8)).

The synchrotron radiation emitted in the electron trajectory direction is linearly polarised, with the E vector in the direction of electron acceleration i.e. polarised in a plane perpendicular to the field. The small value of α means that radiation can be collected from practically the whole electron trajectory in the undulator.

Two undulator configurations are usually considered: (i) a transverse sinusoidal magnetic field with period λ_w and maximum field amplitude B_o, and (ii) a helical magnetic field in which the field amplitude remains constant but the direction of the field vector rotates about the axis of the undulator as a function of the distance along the axis.

2. Undulator Emission : The Resonance Condition

The wiggling electron in the undulator becomes equivalent to a dipole oscillator in the electron frame of reference. Since these are high velocity electrons, the electron sees the undulator relativistically contracted, so that it is forced to oscillate and emit dipole radiation of wavelength $\approx\lambda_w/\gamma$. The radiation emitted undergoes a Lorentz transformation in the laboratory frame, just as described in section V.A, the electromagnetic radiation being Doppler shifted to wavelength $\approx \lambda_w/2\gamma^2$ and projected forward in a small cone whose angular dimension is $\approx\gamma^{-1}$. The motion in the electron frame becomes more complex with increasing K, giving rise to the emission of harmonics. The fundamental (n = 1) or harmonic (n > 1) wavelength of this spontaneous emission is given by the important <u>resonance condition</u> [27-29] :

$$\lambda_n = (\lambda_w/2\gamma^2) \ (1 + cK^2 + \gamma^2\theta^2)/n \dotfill (8)$$

where θ = angle of observation with respect to mean electron trajectory. The constant c = 1/2 for a planar undulator, c = 1 for helical field undulators. The fraction of the power emitted in the fundamental is given by $(1 + K^2/2)^{-2}$, e.g. about 44 % for K = 1, but only about 10 % for K = 2.

Another perception of the properties of undulator radiation can be given in terms of interference between the synchrotron radiation emitted by the same electron at different points on its trajectory through the undulator. The resonance condition holds when the electron, which moves slower than the velocity of light, is overtaken by one period of the photon field as it passes through one period of the undulator. Destructive interference will develop after a number of periods but phase matching corresponding to the resonance condition will occur for some frequencies of the synchrotron radiation. There arises constructive interference contributions from different electrons that will be incoherent unless the electrons are structured in microbunches as will be discussed below. This process can be explained in terms of the ponderomotive force description of the interaction between the electromagnetic field and the charged particles [12,30].

Energy conservation implies that any energy lost or gained by the electron in the interaction is respectively given to or received from the radiation. Energy exchange will be zero at resonance, and close to zero elsewhere except when electron and light wave are slightly off resonance (cf.eqn.8). Thus, this process leads to absorption or to amplification of the electromagnetic wave. Since the electrons have varying velocities along their trajectory in the insertion device, those that are accelerated by the light wave interaction will catch up with those injected half a wavelength earlier that were decelerated. Therefore the electrons will have a tendency to group together in microbunches. As the electrons become grouped together at regular intervals within the electron packet, the emitted radiation acquires a <u>coherent</u> component whose relative importance will depend on the degree and the quality of the microbunches. As the coherent component increases, the tendency will be for the emitted amplitude to be proportional to N_e, i.e., for the intensity to vary as N_e^2.

The gain is relatively small but since the total gain is proportional to L^3 where L is the length of the undulator, the gain per unit length can be quite large at the further end of the undulator and so be useful for free electron laser devices.

Due to the relativistic Doppler effect the radiation is concentrated within horizontal and vertical half-angles $\theta_h \simeq (1+K^2/2)\gamma^{-1}$ and $\theta_v \simeq \gamma^{-1}$ respectively. The gain in power emitted in a solid angle $2\gamma^{-2}$ x2 γ^{-2}, as compared with normal bending magnets, is of the order of N, the number of undulations. An increase of the order of N^2 in spectral brilliance can be obtained for a well collimated beam. The spectral width W will depend on a number of factors : a) finite number of poles, since W is proportional to $(nN)^{-1}$; this can be considered as homogeneous broadening ; b) angular spread ; c) electron beam aperture ; d) acceptance angle θ, since W is a function of $\gamma^2\theta^2$. In practice, $\Delta\lambda/\lambda$ is of the order of 1 %, so that the use of undulator emission for selective excitation requires further monochromatization.

The fundamental wavelength can be changed either by modifying K, i.e. by varying the undulator gap, of by observing off axis, i.e. θ variation. The range of tunability can be calculated from eqn.8 by putting in suitable practical values of the parameters. The spectral effects of varying K and the angle θ are as follows. For a particular value of K, as θ increases the emitted wavelengths increase in value (blue to red) giving rise to a "rainbow" pattern. Successive patterns correspond to <u>different harmonics</u>. For a given value of θ the emitted frequency decreases with increase in K.

The harmonic content of undulator emission increases with magnetic field amplitude. The form of the <u>spectrum</u> depends on K, which is independent of the electron energy, and on the parameter Σ, which takes into account the angular spread of the electron (σ') and the acceptance angle of the detector (Ω) :

$$\Sigma = [(\gamma\sigma')^2 + \gamma^2 - \Omega/\pi]^{1/2} \quad \dots\dots\dots\dots\dots (9)$$

For K $\ll$ 1 only one spectral harmonic is emitted, at $\lambda = \lambda_w/2\gamma^2$; its wavelength is independent of B_0 and the power emitted is proportional to B_0^2 ; the wavelength spread is small if the radiation is observed through a pinhole on the axis of the undulator. The spectrum includes many harmonics when K>1 ; under certain conditions, their envelope ressembles a normal synchrotron radiation spectrum [12].

The phase relation between the electric field components perpendicular and parallel to the insertion device's magnetic field will determine the ellipticity of the <u>polarization</u> of the radiation. This phase relation depends on the particular configuration of the magnetic field. Most undulators have a vertical magnetic field so that the polarization of the radiation is then in the ring orbit plane. Circular polarization can be obtained by using crossed undulators, asymmetric wigglers, or helicoidal undulators [31].

The <u>time structure</u> of the emission of an electron in an insertion device also depends on the value of K. The time structure of an actual electron bunch is that corresponding to its length and its repetition rate. The spectral and time structure properties of insertion devices carry over to Free Electron Lasers in which they play a part.

3. Optical Klystrons

As discussed above, electron microbunching in undulator devices creates coherent spontaneous emission (and also leads to more efficient harmonic generation, see later). The optical klystron, proposed by Vinokurov and Skrinsky [32], enables electron bunching be further enhanced. It consists of two undulators, respectively the "modulator" and the "radiator", separated by a dispersive section (fig 7) in which the transit time of the electrons depends on their energy. Thus slower electrons will catch up with faster electrons in the dispersive section, increasing further the micro-bunching process for a given undulator length. The maximum increase of the gain occurs when the modulator and radiator undulators are identical. The optical klystron is a good device for optimizing the small signal gain. A recent theoretical description of the optical klystron has been given by Elleaume [33]. A number of optical klystrons have been built or are projected.

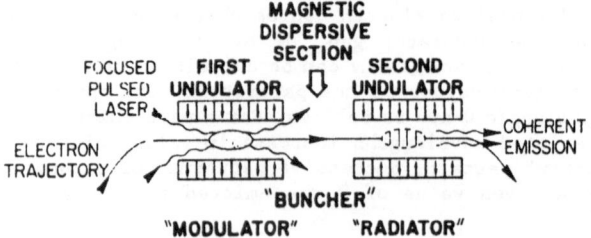

Fig. 7. Optical klystron. Schematics of undulator and dispersive sections. The focussed pulsed laser case is discussed in Sec. VI.C.

C. Harmonic Generation

Harmonic generation of radiation via insertion devices promises to have important applications in photophysics and photochemistry. Spatial bunching of electrons, as described in previous sections, serves not only to create coherent radiation but also to enhance the higher Fourier components in the emitted radiation. Harmonic generation may be "intrinsic" i.e. using spontaneous synchrotron radiation as a trigger, or "extrinsic" in which an external source (which may be a FEL) is used to enhance electron density modulation by focussing an external laser source into an undulator or an optical klystron [34]. In the latter case it is focussed into the modulator where it creates an energy modulation transformed into a spatial modulation of the electrons in the dispersive section (fig 7). The ensuing microbunches are separated by the external laser wavelength λ_ℓ. The Fourier transform in space of the electron density ρ (z,t) contains all harmonics of λ_ℓ, the intensity of each harmonic being dependent on the laser electric field. Light emitted by the electrons traversing the second undulator, or radiator, becomes coherent for these harmonics at wavelengths λ_ℓ/n. This technique can thus produce radiation in the vacuum ultraviolet region using an external laser source in the visible or near ultraviolet.

Up-conversion of this type is to be distinguished from laser harmonic generation obtained by non-linear optical devices. In the undulator case, the coherent output power is generated by loss of electron kinetic energy, i.e. an energy transfer process, rather than taken from the laser pump.

It has been estimated that this external laser technique should be able to create 10^{10}-10^{12} photons per pulse with modern storage rings operating at their nominal energy. From the Orsay experimental results [34] it is possible to foresee an extension of the spectral range to the VUV region. Indeed, recently [35] coherent radiation at λ_3 = 1773 A and λ_5 = 1064 A was observed in an experiment at Orsay in which radiation of λ_ℓ = 5320 A (0.4 J in 6ns, $\Delta\lambda$ = 0.35 A) from a frequency doubled Nd-YAG laser was focussed into the modulator of an optical klystron. These experiments require very precise superposition of external photon beam with the electron beam.

VII. FREE ELECTRON LASER

A. Introduction

As discussed in section V.A, free electron lasers (FEL) operate on free-free transitions i.e. by energy changes of <u>unbound</u> electrons, so that any state of electron motion or energy value may be involved. Photon absorption or emission is achieved by constraining the "free" electron to accelerate or decelerate its motion. The present theoretical situation with respect to free-free transitions of relativistic electrons has been summarized by Fedorov [36], by Marshall [11], and by Colson [37].

The basic elements of a FEL are a high velocity (relativistic) electron beam, a device enabling the electrons to undergo oscillatory motion so as to generate - and interact with - photons, and a resonant optical cavity, the latter having the same feedback function as in a conventional laser (fig. 4). The electron oscillations are usually produced by a suitably periodic magnetic field such as that existing in an undulator device. Some recent FELs have been conceived in which electron oscillation occurs by electrostatic means, or even by interaction with a high intensity light wave. The undulator field enables coupling to occur between the electrons traversing the undulator and the electromagnetic waves which they emit as a result of their oscillatory motion.

There are two main types of FEL : (i) the low electron density, high energy (E > 20 MeV) FEL in which electrons individually experience the undulator magnetic field and the photon fields, and (ii) the high electron density (usually k Amps current), low energy (E < 10 meV), FEL in which electrons interact collectively (plasma modes) with the undulator and photon fields. These types are distinguished by their electron scattering regimes, Compton for the low and Raman for the high electron density versions. The first is mainly useful from the XUV to the I.R. while the Raman FEL is of value for the cm to mm wavelength region.

B. Free Electron Laser Physics

1. Scattering Regimes

In the electron reference frame, the periodic undulator field appears as an electromagnetic field corresponding to virtual photons. It can thus amplify a real photon field travelling in the same direction as the relativistic electrons, the two fields being coupled by the electron charges. This is illustrated in fig 8 for the stimulated Compton scattering regime, which is a two-wave interaction process. An important parameter in the stimulated scattering of photons from high velocity electron beams is the ratio of the Debye length in the electron beam to the scattered wavelength. Stimulated Compton scattering occurs when this ratio is large. The resulting FEL is a low gain device. For small ratios, there is a modification in the nature of the scattering process, which is then best described as stimulated Raman scattering. In this case, along with the scattering interaction there develops a growing electrostatic oscillation at the electron beam plasma frequency ω_p. An important characteristic is that the growth rate of the scattering instability is much greater than in stimulated Compton scattering. In the pure Raman FEL case, and under appropriate conditions, e.g. cold beam, the gain can become exponential. In the low electron density Compton FEL the gain is linear in beam density N_e or in current I, but in the Raman case growth scales as $N_e^{1/4}$ or $I^{1/4}$. It is the high space charge of the electron beam which allows excitation of the beam space charge wave. Coupling between electromagnetic and electrostatic waves occurs via the undulator field.

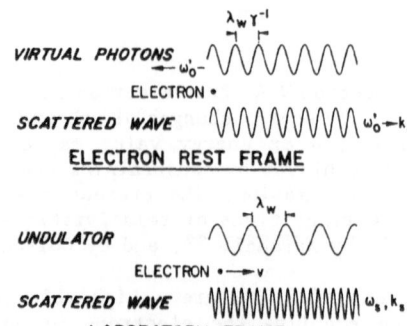

VIRTUAL PHOTONS ω_0^-

ELECTRON •

SCATTERED WAVE $\omega_0'-k_0'$

ELECTRON REST FRAME

UNDULATOR

ELECTRON •→ v

SCATTERED WAVE ω_s, k_s

LABORATORY FRAME

Fig. 8. Radiation emitted by an electron moving in an undulator. Stimulated Compton scattering regime illustrated for the laboratory frame and the electron rest frame.

The Raman-FEL gain process can be thought of as a convective three-wave parametric process in which the pump photon ω_0 decays into a plasmon ω_p and a scattered photon ω_s. In this parametric device the virtual photons corresponding to the undulator field act as the pump mode ω_0 and the space-charge wave ω_p plays the role of an idler wave mode that absorbs the energy remaining from inelastic Raman scattering. The scattered light is the signal mode. In this device the unstable slow space-charge wave, whose phase velocity is smaller than the electron velocity, picks up energy from the electron beam by an energetically favoured instability. The slow space-charge wave can then interact with electromagnetic radiation at a frequency defined by the FEL dispersion relation. This is discussed in detail by Marshall [11]. Since the Debye length is a function of $(T_e/N_e)^{1/2}$, where T_e is the electron temperature and N_e the electron density, a Raman FEL requires a cold dense electron beam. In general this leads to a lower wavelength limit of ~ 100 μm, so that Raman FELs are mainly microwave or mm/far infrared devices. If the beam is dense but not cold, collisionless Landau damping of the idler wave occurs, leading to a drop in gain. This particular situation is sometimes referred to as the warm-beam Compton regime.

Dense electron beams, for which resonance effects appear at plasma frequencies [38,39], cannot be achieved in storage rings. The present discussion will mainly consider the low density electron beam case.

2. Laser Characteristics

The three familar processes occurring in conventional lasers, i.e. spontaneous emission, stimulated emission, and amplification, take the following forms in FELs. The spontaneous emission wavelength is given by the resonance condition (eqn. 8). The "transition" is symbolised in fig. 9. Stimulated emission arises from the electron bunching process described earlier, where emission from a set of electrons with approximately the same phase gives rise to coherent addition of electromagnetic fields and

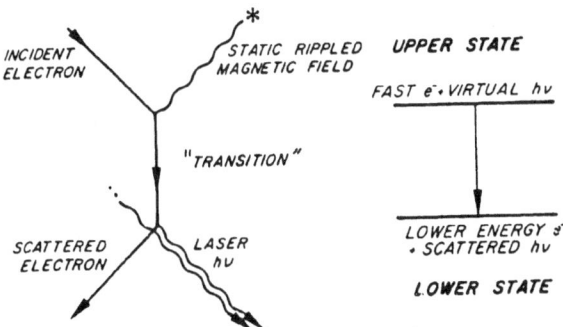

Fig. 9. Feynmann-type diagram of the stimulated Compton scattering process and the nature of upper and lower energy states in a FEL "transition".

constructive interferences. As mentioned earlier, <u>amplification</u> occurs via energy exchange between electrons and photons, the interaction between electrons and photons giving rise to acceleration or deceleration of the electrons according to the phase region in the undulator or optical klystron. Net energy transfer to photons leads to a concentration of electrons in the deceleration phases.

<u>Saturation</u> mechanisms limit the power output of undulators. The energy loss from an electron increases with increasing radiated power. At high power, the electron behaviour is perturbed, e.g. increase of angular dispersion, so that the resonance condition given by eqn. 8 is only valid in part of the undulator. The saturated power decreases with increasing N, however there is a trade-off between maximizing the gain and maximizing the output power since the maximum gain is proportional to N^3 (see later).

The FEL beam has spatial and time structures that reflect the electron beam structure. The radiation pulse length is proportional to the electron bunch length ℓ_e, and the radiation frequency width is given by $\delta \omega$ $= 2 \pi c / \ell_e$. Individual laser modes have linewidths which are approximately equal to the inverse of the correlation time. The practical limit to the correlation time, and therefore to the widths of individual lines in the spectrum, is set by mirror microphonics. The laser power $P_L = G_{max} \times P_{RF}$ where P_{RF} is the power given to the electron beam by the RF cavity.

Complete coherence of emission requires spatial (transverse) coherence, related to source divergence, and temporal (longitudinal) coherence, related to source monochromaticity. The undulator has a high spatial coherence, of the order of γ^{-1}, but the optical beam may have poor temporal coherence at the exit of the undulator. This can be improved by adding a monochromator.

The gain in a FEL is inversely proportional to the square of the linewidth of spontaneous radiation. The latter has two components (i) a homogeneous width due to the finite interaction time of electrons in the undulator, (ii) an inhomogeneous width due to the angular divergence and velocity spread of the electrons as well as to the variation with position of the strength of the periodic magnetic field. The maximum gain G_{max} is proportional to a number of factors :

$$G_{max} = k\rho_e B_o^2 \lambda_w^{3/2} \lambda^{3/2} N^2 L \ldots\ldots\ldots\ldots\ldots (10)$$

where k is a proportionality constant, $L = N\lambda_w$ is the undulator length and ρ_e is the electron density (assuming that the electron energy distribution is Gaussian). Non linear effects similar to self-focusing can occur in the case of a long insertion device and high gain. This can counteract the diffractive spreading of the light beam. Gain measurements on the Orsay FEL are described elsewhere [40],[41].

C. Some Free Electron Laser Experiments

1. Experiments

FEL experiments were carried out at Orsay with an optical klystron of 1.3 m length and 2 x 7 periods and an optical cavity 5.5 m long. Time structure and Q-switching studies were made [42],[43]. Orsay FEL performances in three sets of experiments [44] are listed in table 2.

TABLE 2

ORSAY Free Electron Laser Results

	Experiment 1 [45]	Experiment 2 [43]	Experiment 3 [44]
Wavelength tunability (Å)	6500 50 (100)	5750-6440(690)	4630-4860(230)
Smallest spectral bandwidth observed (Å)	3	1.7	2
Electron energy (MeV)	166-210	180-245	195-233
Laser lifetime after each injection of the ring (h)	1	3-4	2
Ring current (mA)	20-100	25-200	60-100
Max gain per pass for I 70 mA : (%)	0.18(200 MeV)	0.33(220 MeV)	0.35(220 MeV)
Mirrors	SiO_2/TiO_2 dielectric layers		
Round trip cavity losses (a) before irradiation (%)	0.02	0.04	0.13
(b) after the first irradiation of about 2 h(%)	0.07	0.13	0.20
Mirror transmission	3×10^{-5}	1×10^{-4}	8×10^{-5}
Average laser power intracavity (W)	13	20-30	15
exit power per mirror (mW)	0.4	2-3	1.2
total extracted power (mW)	20	60	60
Peak power (27 MHz pulsed laser) intracavity (MW)	0.05	0.5	0.3
exit power per mirror (W)	1.5	20-50	25

Laser operation necessitates excellent synchronism between the electron bunch revolution frequency in the storage ring and the light pulse round trip frequency in the optical resonator. It is better to fine tune by modification of the RF frequency rather than by mirror translation, so as to avoid backlash and misalignment of the cavity mirrors.

Gain measurements and laser operation over a wide spectral range (2400-6900 A) have very recently been made using an optical klystron installed on the VEPP-3 electron storage ring [46]. This is the first FEL operation in the ultraviolet region.

2. Temporal Structure of the Laser Emission

The FEL emits micropulses at the storage ring frequency. The width of the laser micropulse is certainly smaller than the electron bunch length but has not yet been measured. In the ACO case, where the electron pulses are 1ns long and the repetition period 37ns, the FEL emission macro-temporal structure has a pseudo-period of 10-20ms and a pulse duration of about 2-3ms. The laser time structure is thus similar to the relaxation oscillation of a conventional laser, but here the excursions of the electron beam energy spread about a mean value play a role equivalent to that of the excursions of an inverted population density.

It is difficult to achieve a perfect periodic signal since this requires perfect synchronisation, so that "natural" operation, less than perfect, is somewhat noisy. The noise can be reduced considerably by Q-switching, using either of two techniques peculiar to FELs. One is a "detuning" technique in which the optical gain is cancelled by varying slightly the RF frequency f by $(\Delta f/f)$ $\approx 10^{-6}$. The second method uses electric field modulation of the transverse position of the electron beam, causing modulation of the gain so that Q-switching occurs.

3. Laser Power

The FEL power output (table 2) is limited by the saturation mechanisms discussed in section VII.B.2. Saturation of the FEL gain occurs because the laser-induced electron energy spread creates inhomogeneous broadening effects and lengthening of the electron bunch. From measurements of bunch lengthening and energy spread [47], it was shown that for the Orsay FEL, bunch lengthening is relatively negligible as a saturation inducer since it decreases the gain by only 1 %, whereas inhomogeneous broadening is a much more important factor, being responsible for a 12 % variation of the gain.

The low power obtained with the FEL in the ACO storage ring is due to several limitations of this ring. It has a small diameter and a small space for insertion of an undulator or optical klystron (L = 1.3 m ; recall that the gain G is proportional to L^3). Furthermore, at the low beam energies used for FEL operations the synchrotron power emitted is relatively small and the electron density is low ($\rho_e \simeq 2 \times 10^{10}$ electrons/cm³). The beam quality could be improved by the use of positrons rather than electrons in the storage ring.

The intrinsic limitations of the ACO FEL have been mentioned earlier. Much improvement is expected with a FEL designed for operation with the Super-ACO storage ring constructed at Orsay. This is discussed in detail elsewhere [12]. With an optical klystron, and under the operating conditions of Super-ACO, the expected wavelength range for the FEL will be 1200 - 1700 A, with beam energies varying from 400 to 800 MeV. It will be possible to increase the photon energy by the generation of harmonics down to at least 300 A : pulses of the order of 10^8 photons have been estimated for $\lambda = 273$ A.

D. Electron Accelerators

The electrons entering the FEL should have optimum characteristics for maximizing the gain (cf eqn.10) and avoiding effects leading to line broadening. The desirable electron parameters can be summarized as follows : low emittance and energy spread, high peak current, proper bunch length in the micropulse, stability during the macropulse. Thus the feasibility of FELs is intimately linked to the quality of electron accelerator technology. It is only relatively recently that attention has been paid to the development of higher current electron beams, rather than the higher energy beams useful for collision physics. The most common accelerator devices, their typical beam energies and useful wavelength ranges of FELs using these as electron sources, are as follows :

1) Electron storage rings, $E > 200$ MeV, $\lambda = 500$ A-1 μm

2) Radiofrequency LINAC : $E > 20$ MeV, $\lambda = 1$-20 μm

3) Induction Linac : $E \approx 0.1$ - 50 MeV , $\lambda =$ mm-cm range

4) Van der Graaff electrostatic accelerator : $E < 10$ MeV, $\lambda = 1\mu$m-1mm

5) Classical Microtron : $E < 40$ MeV, $\lambda = 10$-500 μm

6) Racetrack Microtron : $E = 10$-500 MeV, $\lambda = 10$-500 μm

7) Pulse-Line : $E \approx 0.1$-5 MeV, $\lambda = 80$-3000 μm

E. Free Electron Laser Facilities

Table 3 lists Free Electron Laser Facilities and their typical parameter values. At the present time about a twenty of these FELs have operated successfully as lasers in the oscillator mode or as amplifiers of an external source. Others are in various stages of conception, design and achievement ; some of them may have gone into action before this text appears.

The FELs in Table 3 are grouped as a function of electron beam technology, which largely determines the spectral range for amplification and oscillation and whether they are repetitive or single pass devices. Storage rings inherently recirculate the electrons whereas the other beam sources operate in the pulsed mode, usually at a frequency of several Hz. Recirculation or beam recovery devices have been conceived and constructed for a number of the latter. Notes to table 3 give some additional information for particular FELs. Detailed accounts of each FEL can be obtained from the cited references. This compilation, although certainly incomplete, gives a picture of the recent FEL situation. Theoretical work on FELs, as well as other proposed FEL technologies or projects can be found in references [13a-13f].

The majority of the FEL devices in table 3 operate in the Compton regime. Those using the pulse-line electron beam technology correspond well to the high beam current, low energy limiting case of stimulated Raman scattering regime devices.

TABLE 3. Actual and Projected Free Electron lasers. Their Typical
Parameters and Beam Technologies (EBT).

Laboratory	Beam Energy (MeV)	Beam Current (amp)	Peak Power (oscillator mode)	Wavelength	Notes	Ref.
EBT : RF Linac						
Stanford Univ.	24	0.1		10.6 μm		[48]
Stanford Univ.	43	1.3	130 KW	3.3 μm	b	[48]
Los Alamos	20	10		10.6 μm		[49]
Los Alamos	20	30-60	10 MW	9-35 μm		[50,51]
MSNW/Boeing	20	5		10.6 μm		[52]
TRW	25	10		10.6 μm		[53]
TRW/Stanford	66	0.5-2.5	1.2 MW	1.6 μm		[54]
NRL	35	5	17 MW	16 μm		[13,55]
Boeing/Spectra T.	120	100	0.6 GW	0.5 μm	c	[56,57]
Glasgow	40-160	2.5		2-20 μm		[58,59]
Orsay	50	20	1 MW	1-15(50) μm		[60]
Stanford Univ.	43	240	5 MW	1.6-3.1 μm		[61,62]
EBT : Storage ring						
Orsay	240	2	13 W	0.5 μm	d	[63]
Orsay	240	2	50 W	0.5 μm		[43-45]
Frascati	600	10		0.5 μm		[55]
Brookhaven	500	108		0.35 μm	e	[55]
INP-Novosibirsk	350	6		0.24 μm	f	[46]
Stanford Univ.	1000	270	100 MW	< 1000 Å		[64]
Livermore	50			mm-infrared:visible		[13,55]
Berkeley	500-750		150 MW	400-1000 Å	g	[65]
EBT : Microtron						
Bell Labs	10-20	5	100 KW	100-400 μm		[66]
Frascati	20-35	0.6		10-35 μm		[67,68]
NBS	20-200	1.0	0.1-1 KW	100 μm-1500 Å	h	[69,70]
Dartmouth/Frascati	5	2		10-100 μm	i	[71]
			10-200 KW	375 μm-1mm	i	[72]
EBT : Induction Linac						
Livermore	3.6	850	1.9 GW	8.7 mm		[73,74]
NRL	0.7	200		8 mm		[13,55]
Livermore	5	400	80 MW	3-6 mm		[13,55]

(continued)

TABLE 3 (Continued)

Laboratory	Beam Energy (MeV)	Beam Current (amp)	Peak Power (oscillator mode)	Wavelength	Notes	Ref.
EBT : Pulse Line						
NRL	2	30,000		0.4 mm		[75]
NRL	1.35	1,500		4 mm		[75,76]
NRL/Columbia	1.2	25,000	75 MW	0.4 mm		[13,55]
Columbia Univ.	0.86	5,000	4 MW	1.7-3.5 mm		[77]
Columbia Univ.	0.9	10,000		0.6 mm		[13,55]
MIT	1	5,000		3 mm		[55]
Ecole Polytech. (Palaiseau)	1	2,000		2 mm		[55]
Osaka	0.6	40,000		4 mm		[78]
EBT : Electrostatic accelerator						
UCSB	3	2	8 KW	0.36 mm	j	[79,80]
UCSB	6	2		0.1-1 mm		[81]
UCSB	6	10,20	2 MW	703 ; 1.08 μm	k	[82]
UCSB	1			sub-mm	l	[83]
	25			infrared, visible	l	[83]
KMS Fusion	3				m	[84]
Rehovot	6	1		200 μm	n	[85]

[a] Peak current except for Pulse-Line accelerators

[b] First FEL operation (1977)

[c] Tapered wiggler, design goal 5% extraction of energy

[d] First FEL in visible (1983)

[e] Projected wavelength 3500 Å. Not achieved because of inadequate electron beam characteristics.

[f] Shortest wavelength operation 2400 Å

[g] By-pass FEL in storage ring

[i] Cerenkov laser

[j] pulses up to 50 μs duration

[k] 2-stage FEL

[l] Micro-undulator, gap $\cong$ 3mm

[m] Design for a 2-stage FEL with cylindrically summetric quasi-optical cavity for long (100-1000μm) and short (1-10 μm) wavelengths.

[n] Design to produce many harmonics of λ_i = 200 μm

F. Summary of FEL Properties and Applications

The properties of FELs which are considered to be possible with existing or foreseen improved technology can be summarized as follows. The operating wavelength is a continuous function of the electron energy and the magnetic field in the undulator ; the range is from the mm region to the vacuum ultraviolet, with $\lambda = 500$ A being a lower wavelength limit beyond which it would be difficult to achieve sufficient gain using an optical resonator. It should be possible to tune the emitted radiation over about one octave in the I.R., visible and U.V. regions. The best electron source varies with wavelength region : storage rings in the visible and U.V., microtrons in the I.R., and Van de Graaff machines in the mm region. Average powers greater than 1 KW have been obtained and much higher values, MW and even GW, are reached or contemplated with improved technology ; overall net energy conversion efficiencies of the order 1-30 % are achieved or expected. In the case of a FEL operated with an electron beam from an RF accelerator the optical cavity length must be set to a multiple of the electron bunch spacing ; the optical field in such a FEL is mode-locked by the periodic variation in electron current ; picosecond pulses can be obtained at high repetition rates (10 MHz - 1 GHz). All the usual laser techniques used to minimize linewidths can be applied to the FEL so that it should be possible to obtain individual lines with $\Delta\nu < 1$MHz. An important advantage of the FEL is the possibility for it to operate simultaneously at different frequencies, or with a set of independent undulators. Furthermore, a storage ring FEL also generates significant amounts of broadband incoherent synchrotron radiation in addition to the emitted (laser) coherent radiation ; this SR could be of interest for independent experiments or for use as a secondary source in conjunction with the laser.

Some possible uses for free electron lasers in spectroscopy, intramolecular dynamics and other studies and applications are summarized as follows.
1) _Far infrared_ : interesting wavelength range ; useful for saturation pumping, multiphonon processes, surface state spectroscopy, solid state spectroscopy and photophysics.
2) _U.V. - V.U.V._ : advantage can be taken of flexibility, power density, polarisation and wavelength range for studies on spectroscopy, multiphoton ionization, high energy-vibronic states, low density targets such a transients e.g. radicals, ions, muonium, positronium.
3) _Pump and probe experiments_ could be carried out by FEL + S.R., FEL + separate laser, or FEL operating in multifrequency mode.
4) _Study of laser assisted collisions and general_ photochemistry, in particular state and/or species selective photochemistry using the multifrequency resources. Selective purification ; laser initiated chain reactions ; laser isotope separation.
5) _Industrial applications_ of photophysical and photochemical processes, e.g. microfabrication of semiconductor circuits by lithography.
Some other scientific and technical applications are mentioned in ref. [13a].

Another suggested application of the FEL is as a driver for inertial confinement fusion [86]. There are also prospective medical applications of FELs. These reflect the desire to achieve specific effects on particular types of biological tissue without affecting neighbouring structures. The promise is in the areas of laser surgery and in photoradiation medicine. The FEL qualities of great interest to these ends are the possibilities of varying the power, wavelength, spot size and pulse width [87]. These are important qualities for situations where the skin or other tissue absorption depth is a critical factor.

Various actual or potential applications of FELs to condensed matter and biological physics have recently been described [88]. The direct excitation of far infrared vibrational modes in solids is a particularly promising application.

A sober, and sobering, assessment of the possibility of military application of FELs (and other lasers, as well as other possible directed energy weapons) has been given recently [89].

A note of caution is necessary in stressing (i) that the output of theoretical work on free-electron lasers still exceeds the experimental activity ; (ii) that much technological improvement is required before the FEL properties discussed in this and previous sections become completely effective. Feasibility studies are in progress in many FEL areas and the direction of the field and its real possibilities have become more clear over the past few years. Rapid progress is envisaged in FEL development under the impetus of scientific and publicity factors whose nature is evident from this summary section.

Finally, with respect to the initial context of this review, the comparison of sources for intramolecular dynamics and related studies, it can be concluded that free electron lasers must still undergo much development before becoming competitive with other laser sources but that they are already of some practical use in the infrared and mm wave regions. Possibly the immediately most promising area discussed in this review concerns the enhancement of synchrotron radiation by undulators and optical klystrons able to generate harmonics. This can provide a useful tunable source in the VUV region which is of interest for high energy optical excitation of molecules.

REFERENCES

1.   J. Jortner and S. Leach, J. Chim. Phys. Phys.-Chim. Biol. 77, 7 (1980)

2.   S. Leach, Vibrational Spectra and Structure 17B (H.D. Bist, J.R. Durig, J.F. Sullivan, eds.) Elsevier, Amsterdam, 1989, p. 187.

3.   Laser Handbook, (F.T. Arecchi and E.O. Schulz-Dubois, eds.) Vol. 1 and 2, North-Holland, Amsterdam, 1972.

4.   Laser Handbook Vol. 3 (M.L. Stitch ed.), North-Holland, Amsterdam, 1979.

5.   Laser Handbook, Vol. 4 (M.L. Stitch and M. Bass, eds.), North Holland, Amsterdam, 1985 ; Vol. 5, (M. Bass and M.L. Stitch, eds.) North Holland, Amsterdam, 1985.

6.   G.R. Fleming, Chemical Applications of Ultrafast Spectroscopy, Clarendon Press, Oxford, 1987.

7.   Frontiers of Laser Spectroscopy of Gases, (A.C.P. Alves, J.M. Brown and J.M. Hollas, eds.), NATO-ASI, Kluwer, Dordrecht, 1988.

8.   Synchrotron Radiation Research, (H. Winick and S. Doniach, eds.), Plenum, N.Y., 1980.

9. Perspectives of Synchrotron Radiation Applications to Molecular Dynamics and Photochemistry, (J. Jortner and S. Leach, eds.), J. Chim. Phys. Phys.- Chim. Biol. 77, 1-57 (1980)

10. G. Margaritondo, Introduction to Synchrotron Radiation, Oxford University Press, Oxford, 1988.

11. T.C. Marshall, Free Electron Lasers, MacMillan, N.Y. (1985)

12. S. Leach, in ref. 7, p.89

13. a) "Applications of Free Electron Lasers", Nucl. Inst. Meth. A 237, 371-433 (1985)
    b) I.E.E. J. Quant. Elec. QE-21, pp 804-1119 (1985)
    c) Int. Q. Electronics Conference (Abstracts) J. Opt. Soc. Am. 3, n° 8 (August 1986)
    d) Proc. 7th Int. Conf. Free Electron Lasers, Nucl. Inst. Meth.A250, 1-490 (1986)
    e) Proc. 8th Int. Conf. Free Electron Lasers, Nucl. Instr. Meth.A259, 1-316 (1987)
    f) Proc. 9th Int. Conf. Free Electron Lasers, Nucl. Instrum. Meth. A272, 1 (1988).

14. S. Leach, Acta Phys. Pol. A, 71, 671 (1987).

15. J.H.D. Eland, S.D. Price, J.C. Cheney, P. Lablanquie, I. Nenner and P.G. Fournier, Phil. Trans. R. Soc. London, A324, 247 (1988).

16. D. Ivanenko and I. Pomeranchuk, Phys. Rev. 65, 343 (1944)

17. J. Schwinger, Phys. Rev. 70, 798 (1946) ; 75, 1912 (1949) ; Proc. Natl. Acad. Sci. USA 40, 132 (1954)

18. "Construction and Commissioning of Dedicated Synchrotron Radiation Facilities", Proc. Workshop Brookhaven National Laboratory, Oct. 16-18, 1985 (ed. R.W. Klaffky), BNL 51959 (1986)

19. R. Wallenstein, in ref. 7, p. 53.

20. B.P. Stoicheff, in ref. 7, p. 63.

21. J. Bokor, P.H. Bucksbaum and R.R. Freeman, Opt. Letters 8, 217 (1983).

22. E.E. Marinero, C.T. Rettner, R.N. Zare and A. H. Kung, Chem. Phys. Lett. 95, 486 (1983).

23. R. Hilbig, G. Hilber, A. Lago, B. Wolff and R. Wallenstein, Comments At. Mol. Phys. 18, 157 (1986)

24. F. Rostas, XIII Summer School on Quantum Optics (eds J. Fiutak and J. Mizerski), World Scientific, Singapore (1986) p. 328.

25. J.E. Spencer and H. Winick, in Synchrotron Radiation Research, (ed. H. Winick and S. Doniach), Plenum, N.Y. (1980) p. 663

26. J.M. Ortega, C. Bazin, D.A.G. Deacon, C. Depautex and P. Elleaume, Nucl. Instr. Meth. A 206, 281 (1983)

27. H. Motz, *J. Appl. Phys.* 22, 527 (1951)

28. D.F. Alferov, Yu A. Bashmakov and E.G. Bessonov, *Sov. Phys.Tech. Phys.* 18, 1336 (1974)

29. A. Hofmann, *Nucl. Inst. Meth.* 152, 17 (1978)

30. L.D. Landau and E.M. Lifshitz, *Electrodynamics of Continuous Media*, Pergamon, Oxford (1966), p.242

31. *European Synchrotron Radiation Facility : Foundation Phase Report*, ESRF, Grenoble (1987)

32 - N.A. Vinokurov and A.N. Skrinsky, Preprint INP77-59, Novosibirsk (1977) ; N.A. Vinokurov, *Proc. 10th Int. Conf. High Energy Charged Particle Accelerators*, Serpukhov, vol.2, 454 (1977)

33. P. Elleaume, *Nucl. Instr. Meth.* A250, 220 (1986)

34. B. Girard, Y. Lapierre, J.M. Ortega, C. Bazin, M. Billardon, P. Elleaume, M. Bergher, M. Velghe and Y. Petroff, *Phys. Rev. Lett.*, 53, 2405 1984)

35. R. Prazeres, J.M. Ortega, C. Bazin, M. Bergher, M. Billardon, M. E. Couprie, H. Fang, M. Velghe and Y. Petroff, *Europhysics Letters* 4, 817 (1987)

36. M.V. Fedorov, *Prog. Quant. Electr.* 7, 73 (1981)

37. P. Elleaume, *Nucl. Instr. Meth.* A237, 28 (1985) ; W.B. Colson, *ibid* p.1

38. T. Kwan, J.M. Dawson and A.T. Lin, *Phys. Fluids* 20, 581 (1977)

39. D.B. Mc Dermott, T.C. Marshall, S.P. Schlesinger, R.K. Parker and V.L. Granatstein, *Phys. Rev. Lett.* 41, 1368 (1978)

40. D.A.G. Deacon, J.M.J. Madey, K.E. Robinson, C. Bazin, M. Billardon, P. Elleaume, Y. Farge, J.M. Ortega, Y. Petroff and M. Velghe, *I.E.E.E. Trans. Nucl. Sci.* NS-28-3142 (1981)

41. D.A.G. Deacon, K.E. Robinson, J.M.J. Madey, C. Bazin, M. Billardon, P. Elleaume, Y. Farge, J.M. Ortega, Y. Petroff and M. Velghe, *Opt. Commun.* 40, 373 (1982)

42. M. Billardon, P. Elleaume, J.M. Ortega, C. Bazin, M. Bergher, Y. Petroff and M. Velghe, *I.E.E.E. J. Quant. Elec.* QE-21, 805 (1985)

43. M. Billardon, P. Elleaume, J.M. Ortega, Y. Lapierre, Y. Petroff, M. Bergher, C. Bazin, and M. Marilleau, *Nucl. Instr. Meth.* A250, 26 (1986)

44. M. Billardon, P. Elleaume, J.M. Ortega, C. Bazin, M. Bergher, M.E. Couprie, Y.Lapierre, R. Prazeres, M. Velghe and Y. Petroff, *Europhysics Lett.* 3, 689 (1987)

45. M. Billardon, P. Elleaume, J.M. Ortega, C. Bazin, M. Bergher, M. Velghe, Y. Petroff, D.A.G. Deacon, K.E. Robinson and J.M.J. Madey, Phys. Rev. Lett. 51, 1652 (1983) ; P. Elleaume, J.M. Ortega, M. Billardon, C. Bazin, M. Bergher, M. Velghe, Y. Petroff, D. Deacon, K. Robinson and J. Madey, *J. Physique (Paris)* 45, 989 (1984)

46. I.B. Drobyazko, G.N. Kulipanov, V.N. Litvinenko, I.V. Pinayev, V.M. Popik, I.G. Silvestrov, A.N. Skrinsky, A.S. Sokolov and N.A. Vinokurov, Proc. Int. Congress Optical Sci. Eng., Paris, 1989, in press.

47. J.M. Ortega, P. Elleaume, M. Billardon, D.A.G. Deacon, B. Girard and Y. Lapierre, Nucl. Instr. Meth. A237, 254 (1985)

48. L.R. Elias, W.M. Fairbank, J.M.J. Madey, M.A. Schwettman and T.I Smith, Phys. Rev. Lett. 36, 717 (1976) ; D.A.G. Deacon, L.R. Elias, J.M.J. Madey, G.J. Ramian, H.A. Schwettman and T.I. Smith, Phys. Rev. Lett. 38, 892 (1977)

49. R.W. Warren, B.E. Newnam, L.M. Young, W.E. Stein, J.G. Winston and C.A. Brau, I.E.E.E. J. Quant. Elect. QE-19, 391 (1983)

50. B.E. Newnam, R.W. Warren, R.L. Sheffield, W.E. Stein, M.T. Lynch, J.S. Fraser, J.C. Goldstein, J.E. Sollid, T.A. Swann, J.M. Watson and C.A. Brau, I.E.E.E. J. Quant. Electr. QE-21, 867 (1985) ; J.M. Watson, Nucl. Instr. Meth. A250, 1 (1986)

51. J.C. Goldstein, B.E. Newnam, R.W. Warren and R.L. Sheffield, Nucl. Instr. Meth. A250, 4 (1986)

52. W.M. Grossman, J.M. Slater, D.C. Quimby, T.L. Churchill, J. Adamski, R.C. Kennedy and D.R. Shoffstall, Appl. Phys. Lett. 43, 745 (1983)

53. H. Boemher, M.Z. Caponi, J. Edighoffer, S. Fornaca, J. Much, G.R. Neil, B. Saur and C. Shih, Phys. Rev. Lett. 48, 141 (1982)

54. J.A. Edighoffer, G.R. Neil, C.E. Hess, T.I. Smith, S.W. Fornaca and H.A. Schwettman, Phys. Rev. Lett. 52, 344 (1984)

55. P. Sprangle and T. Coffey, Physics Today, (March 1984) p.44

56. J. Slater, T. Churchill, D. Quimby, K. Robinson, D. Shemwell, A. Valla, A.A. Vetter, J. Adamski, W. Gallagher, R. Kennedy, B. Robinson, D. Shoffstall, E. Tyson, A. Vetter and A. Yeremian, Nucl. Instr. Meth.A250, 228 (1986)

57. K.E. Robinson, T.L. Churchill, D.C. Quimby, D.M. Shemwell, J.M. Slater, A.S. Valla, A.A. Vetter, J. Adamski, T. Doering, W. Gallagher, R. Kennedy, B. Robinson, D. Shoffstall, E. Tyson, A. Vetter and A. Yeremian, Nucl. Instr. Meth. A259, 49 (1987)

58. W.A. Gillespie, P.F. Martin, M.W. Poole, G. Saxon, R.P. Walker, J.M. Reid, M.G. Kelliher, C.R. Pidgeon, S.D. Smith, W.J. Firth, D.A. Jaroszinski, D.M. Tratt, J.S. Mackay and M.F. Kimmitt, Nucl. Instr. Meth. A250, 233 (1986)

59. C.R. Pidgeon, D.A. Jaroszynski, D.M. Tratt, S.D. Smith, W.J. Firth, M.F. Kimmitt, C.W. Cheng, M.W. Poole, G. Saxon, R.P. Walker, J.S. Mackay, J.M. Reid, M.G. Kelliher, E.W. Laing, D.V. Land and W.A. Gillespie, Nuc. Instr. Meth. A259, 31 (1987)

60. J.M. Ortega, Y. Petroff, J.C. Bourdon, P. Brunet, J. Courau, J.L. Malglaive, P. Carlos and C. Hezard, LURE Report 1985-1987, Orsay, p. 479 (1987)

61. S.V. Benson, J.M.J. Madey, J. Schultz, M. Marc, W. Wadensweiler, G.A. Westenskow and M. Velghe, Nucl. Instr. Meth. A250, 39 (1986)

62. T.I. Smith, H.A. Schwettman, R. Rohatgi, Y. Lapierre and J. Edighoffer, Nucl. Instr. Meth. A259, 1 (1987)

63. M. Billardon, P. Elleaume, J.M. Ortega, C. Bazin, M. Bergher, M.E. Couprie, Y. Lapierre, Y. Petroff, R. Prazeres and M. Velghe, Nucl. Instr. Meth. A259, 72 (1987)

64. J. E. La Sala, D.A.G. Deacon and J.M.J. Madey, Nucl. Instrum. Meth. A, 250, 320 (1986).

65. M. Cornacchia, J. Bisognano, S. Chattopadhyay, A. Garren, K. Halbach, A. Jackson, K.J. Kim, H. Lancaster, J. Peterson, M.S. Zisman, C. Pellegrini and G. Vignola, Nucl. Instrum. Meth. A, 250, (1986).

66. E.D. Shaw, R.J. Chichester and S.C. Chen, Nucl. Instr. Meth. A250, 44 (1986)

67. U. Bizzarri, F. Ciocci, G. Dattoli, A. de Angelis, M. Ercolani, E. Fiorentino, G.P. Gallerano, T. Letardi, A. Marino, G. Messina, A. Renieri, E. Sabia and A. Vignati, J. Physique (Paris) 44, C1-313 (1983)

68. U. Bizzarri, F. Ciocci, G. Dattoli, A. de Angelis, G.P. Gallerano, I. Giabbai, G. Giordiano, T. Letardi, G. Messina, A. Mola, L. Piccardi, A. Renieri, E. Sabia, A. Vignati, E. Fiorentino and A. Marino, Nucl. Instr. Meth. A250, 254 (1986)

69. C.M. Tang, P. Sprangle, S. Penner, B.M. Kincaid and R.R. Freeman, Nucl. Instrum. Meth. A, 250, 278 (1986).

70. X.K. Mariyama, S. Penner, C. M. Tang and P. Sprangle, Nucl. Instrum. Meth. A, 259, 259 (1987).

71. J.E. Walsh, B. Johnson, C. Shaughnessy, F. Ciocci, G. Dattoli, A. de Angelis, A. Dipace, E. Fiorentino, G.P. Gallerano, T. Letardi, A. Renieri, E. Sabio, I. Giabbai and G. Giordiano, Nucl. Instr. Meth. A250, 308 (1986)

72. E.P. Garate, J. Walsh, C. Shaughnessy, B. Johnson and S. Moustaizis, Nucl. Instr. Meth. A259, 125 (1987)

73. T.J. Orzechowski, B.R. Anderson, J.C. Clark, W.M. Fawley, A.C. Paul, D. Prosnitz, E.T. Scharlemann, S.M. Yarema, D.B. Hopkins, A.M. Sessler and J.S. Wurtele, Phys. Rev. Lett., 57, 2172 (1986).

74. T.J. Orzechowski, B.R. Anderson, W.M. Fawley, D. Prosnitz, E.T. Scharlemann, S.M. Yarema, A.M. Sessler, D.B. Hopkins, A.C. Paul and J.S. Wurtele, Nucl. Instrum. Meth. A, 250, 144 (1986).

75. J.A. Pasour and S.H. Gold, I.E.E.E. J.Q. Electr. QE-21, 845 (1985)

76. J.A. Pasour, J. Mathew and C. Kapetanakos, Nucl. Instr. Meth. A259, 94 (1987)

77. J. Masud, Y.G. Yee, T.C. Marshall and S.P. Schlesinger, Nucl. Instr. Meth. A250, 342 (1986)

78. N. Ohigashi, K. Mima, S. Miyamoto, K. Imasaki, Y. Kitagawa, H. Fujita, S.I. Kuruma, S. Nakai and C. Yamanaka, Nucl. Instr. Meth. A259, 111 (1987)

79. A. Amir, L.R. Elias, D.J. Gregoire, J. Kotthaus, G.J. Ramian and A. Stern, Nucl. Instr. Meth. A250, 35 (1986)

80. L.R. Elias, R.J. Hu and G.J. Ramian, Nucl. Instr. Meth. A237, 203 (1985)

81. I. Kimel, L.R. Elias and G. Ramian, Nucl. Instrum. Meth. A, 250, 320 (1986).

82. J. Gallardo and L. Elias, Nucl. Instr. Meth. A250, 438 (1986)

83. G. Ramian, L. Elias and I. Kimel, Nucl. Instr. Meth. A250, 125 (1986)

84. S.B. Segull, M.S. Curtin and S.A. Von Laven, Nucl. Instr. Meth. A250, 316 (1986)

85. E. Jerby, A. Gover, S. Ruschin, H. Kleinman, I. Ben-Zvi, J.S. Sokolowski, S. Eckhouse, Y. Gorren and Y. Shiloh, Nucl. Instr. Meth. A259, 263 (1987)

86. R.A. Jong and E.T. Scharlemann, Nucl. Instrum. Meth. A, 259, 254 (1987).

88. Physics of Free-Electron Laser Applications, J. Opt. Soc. Am. B6, pp 972-1089 (1989).

89. N. Bloembergen, C.K.N. Patel, P. Azivonis, R.G. Clem, A. Hertzberg, T.H. Johnson, T. Marshall, R.B. Miller, W.E. Morrow, E.E. Salpeter, A.M. Sessler, J.D. Sullivan, J.C. Wyant, A. Yariv, R.N. Zare, A.J. Glass and L.C. Hebel, Rev. Mod. Phys., 59, n° 3, Part II, July 1987, pp. S1-S200.

OPTICAL PARAMETRIC OSCILLATORS OF BARIUMBORATE AND LITHIUMBORATE: NEW SOURCES FOR POWERFUL TUNABLE LASER RADIATION IN THE ULTRAVIOLET, VISIBLE AND NEAR INFRARED

A. Fix, T. Schröder, J. Nolting and R. Wallenstein

Institut für Quantenoptik
Universität Hannover, Hannover, FRG

Since the first demonstration of an optical parametric oscillator (OPO) by Giordmaine and Miller[1] in 1965 the OPO has been subject to detailed theoretical and experimental investigations[2,3]. The OPO is considered as a source of powerful, broadly tunable coherent radiation. The development and the scientific application has been hampered, however, by the scarcity of nonlinear optical materials with suitable optical and mechanical properties.

Recently new nonlinear materials - Bariumborate (BBO) and Lithiumborate (LBO) - became available. With their unique properties (i.e. high nonlinearity, wide transparency range and high damage threshold) BBO and LBO should be very useful materials for an OPO. These OPO's combine the advantages of an all solid-state tunable source with a wide tuning range in the ultraviolet, visible and near infrared, high peak and average power and high conversion efficiency.

Pumped by the harmonics of the Nd:YAG laser, the BBO-OPO as well as the LBO-OPO generate tunable laser radiation in the spectral range between 300 and 3000 nm. Fig. 1 shows the calculated tuning regions of the signal and idler waves as function of the phase-matching angle. The OPO's are pumped by 532 nm, 355 nm and 266 nm Nd:YAG laser radiation.

In a first experimental demonstration a BBO-OPO was pumped by the 532 nm second harmonic of a Nd:YAG laser[4]. The wavelength tuning (940-1220 nm) was limited by the used mirrors. The tuning range could be extended substantially into the UV and the near infrared (330 nm - 2550 nm) by pumping with the 355 nm third or the 266 nm fourth harmonic of a Nd:YAG laser or a 308 nm XeCl excimer laser. This was demonstrated in several experiments first reported in 1988 at the CLEO conference[5][6][7]. Details of these investigations have meanwhile been published[5][6][7].

In the experiments performed at the Stanford University[5] the BBO-OPO was pumped by single axial mode 355 nm third harmonic Nd:YAG laser radiation. The Nd:YAG pump laser with

Applied Laser Spectroscopy, Edited by W. Demtröder and
M. Inguscio, Plenum Press, New York, 1990

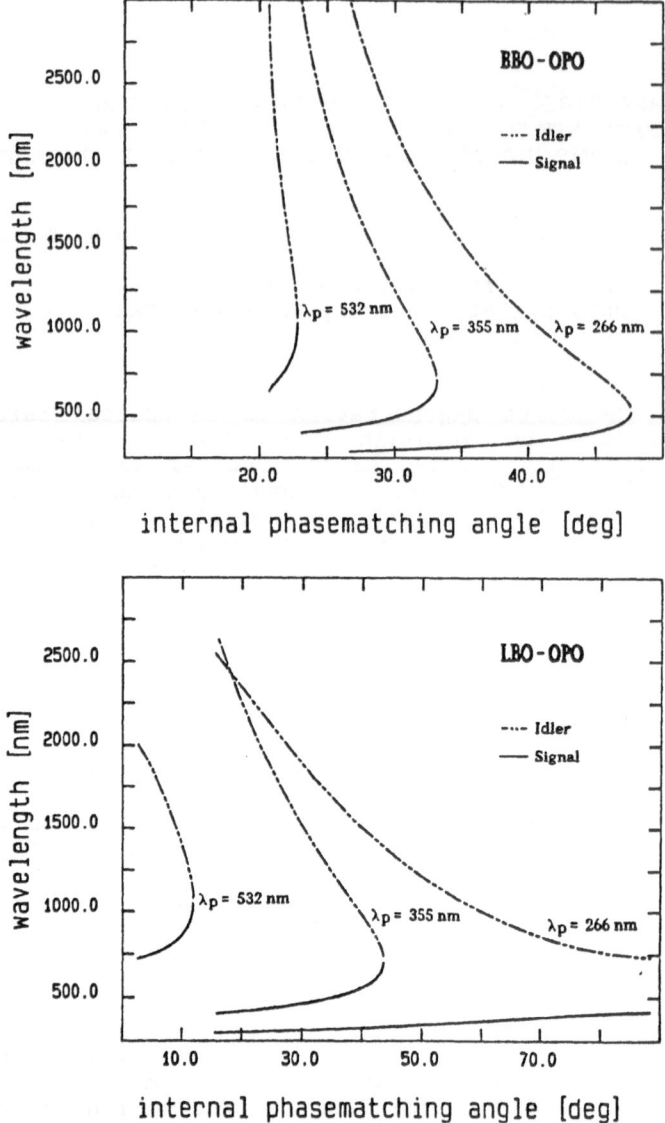

Fig.1: Wavelength of the signal and idler wave of the BBO-OPO and LBO-OPO as function of the phase-matching angle. The OPO's are pumped by the second, third or fourth harmonic of a Nd:YAG laser.

unstable resonator (Spectra Physics, Model DCR3) was injection seeded to obtain single-mode operation (Spectra Physics Model 6300 Injection Seeder). The doughnut shaped output beam was spatially filtered in the far field by a suitable pinhole and frequency tripled with KD*P. The 355 nm radiation with almost Gaussian intensity distribution (diameter d=2.8 nm) provided 6 ns long light pulses with an energy of 30 mJ/pulse and a repetition rate of 30 Hz.

The 3.2 cm long resonator consisted of two flat mirrors with a reflectivity of 98 percent (input mirror) and about 80 percent (output mirror). The transmission at 355 nm and at the idler wavelength exceeded 80 percent. The BBO crystals (size 6 x 6 x 12 mm^3) were cut at an angle of 25 and 35 degrees. Fig. 2 displays the measured wavelengths of the signal and idler radiation as function of the type I phasematching angle. The estimated experimental accuracy of each angle and wavelength measurement was about +1 degree and ±1 nm, respectively. Within these uncertainties the measured values were in good agreement with those calculated from the Sellmeier formula reported by Kato [8]. The observed tuning range extended from 412 nm to 710 nm (signal wave) and 710 nm to 2.6 μm (idler wave). This operating range corresponds almost to the maximum possible tuning range which is limited by the increasing absorption of BBO at wavelengths larger than 2.5 μm [9]. As shown in Fig. 2 the measured phasematching angle varies between 24.5 and 33.2 degrees.

Besides the wavelength tuning, parameters like threshold power, conversion efficiency and the spectral width of the generated light are of special importance.

Fig.3 displays the energy density at threshold (J_0) measured as function of wavelength with an experimental set up similar to the one described in ref. 5. The value of J_0 depends - as expected - on the mirror reflectivity . The minimum values of J_0 increase with decreasing wavelength from $J_0=0.12$ J/cm^2 to $J_0=0.19$ J/cm^2. The corresponding pulse energy is 5-7 mJ. The power density of 20 to 40 MW/cm^2 is well below the BBO damage threshold which is expected to be several GW/cm^2. Calculated values of J_0 - also shown in Fig.3 - are larger by a factor of about 2. This difference between the theoretical and experimental results might indicate that the previously measured nonlinear coefficent of BBO is low[10].

Fig.4 displays the measured energy conversion efficiencies as function of the ratio J_p/J_0, where J_p is the energy density of the pump radiation. With a 12 mm long crystal the conversion efficiency is close to 25 percent. For an 8 mm long crystal J_0 is larger by a factor of 2.2. At a ratio of $J_p/J_0=1.8$ - which is limited by the available energy $E_p=24mJ$ of the laser pulse - the conversion efficiency is less than 8 percent. The efficiency should increase with crystal length and larger values of J_p/J_0. However, walk-off between the pump and the generated radiation limits the useful BBO crystal length to about 25 mm.

For many applications narrowband operation is highly desirable. Pumping with a single-axial mode laser source[5] the output contained typically 6 axial modes of the 3.2 cm long OPO resonator indicating a linewidth of about 23 GHz.

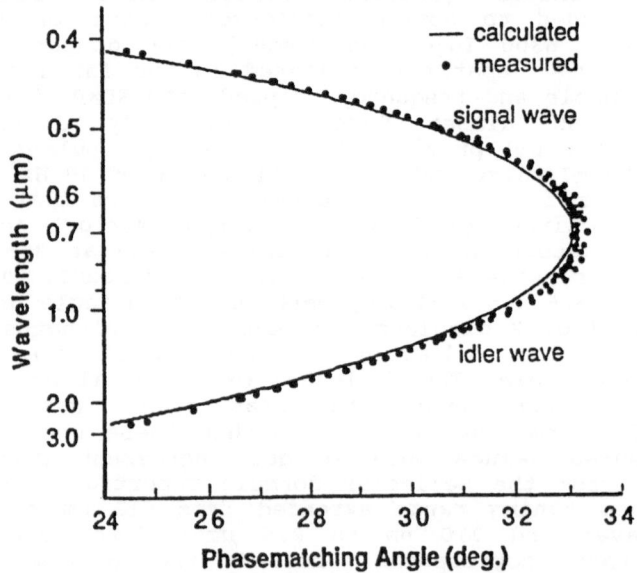

<u>Fig.2</u>. Measured and calculated wavelength of the signal and idler wave as function of phasematching angle of the BBO optical parametric oscillator (Ref.5).

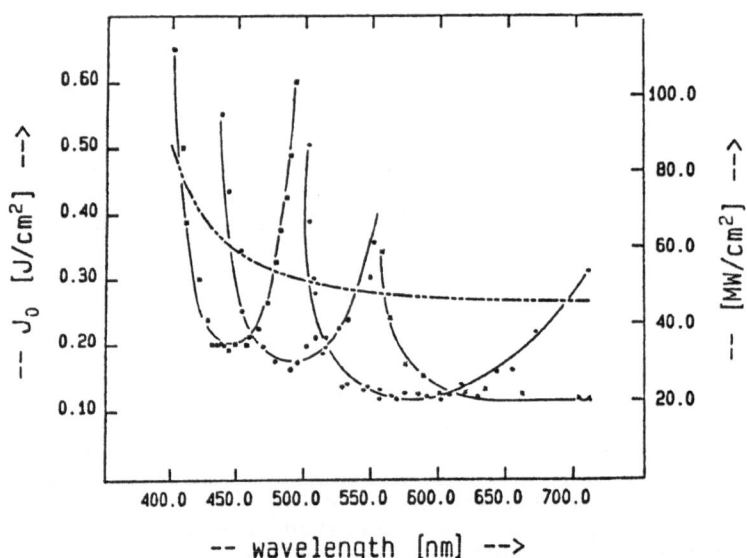

<u>Fig.3</u>. Energy density J_0 at threshold measured for four sets of mirrors. The beam diameter is 2.5 mm. The theoretical values of J_0 are calculated for a mirror reflectivity of 70 percent.

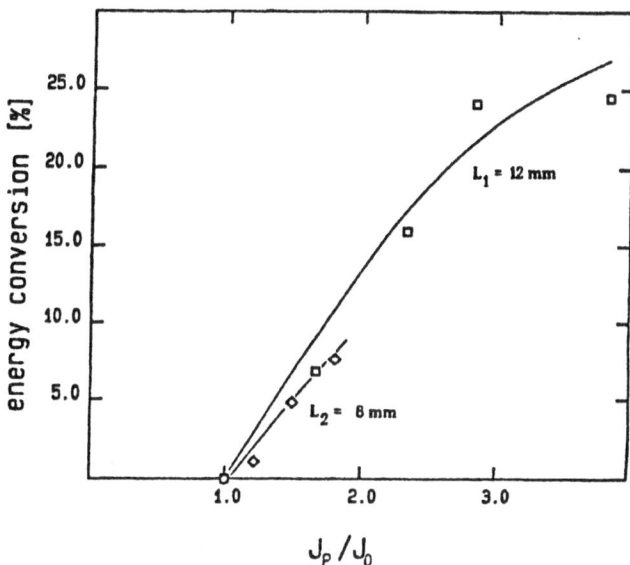

J_p / J_0

<u>Fig.4</u>. Total energy conversion measured as function of the ration J_p/J_0. J_0 is the energy density of the pump radiation. The length of the crystals are $L_1=$ 12 mm and $L_2=8$ mm. The corresponding values of J_0 are $J_{01}=0.12$ J/cm^2 and $J_{02} = 0.26$ J/cm^2. The wavelength of the signal wave is 620 nm.

Single mode operation was achieved at 532 nm or 1,06 μm (and at the corresponding idler and signal wavelength) by injection seeding with light of the second harmonic or the fundamental of the single-mode Nd:YAG pump laser. The injection seeding not only provided single-mode operation but also reduced the oscillator build-up time from 4 to 2.2 nsec. Because of the reduced build-up time the pulse length increased from 2.5 nsec to 4 nsec. Simultaneously the OPO threshold power decreased by a factor of 3.

In addition to the systems mentioned so far, synchronously pumped BBO-OPO devices have been investigated. With the second harmonic of a mode-locked pulsed Nd:YAG laser [11] or the third harmonic of a Nd:YAP system[12] the BBO-OPO produced 75 psec long light pulses tunable in the regions of 680 nm to 2.4 μm [11] or pulses of about 20 psec duration at 406 nm to 3.17 μm [12], respectively.

In addition to BBO LBO crystals are now available[13] in sufficiently large size and with high optical quality. As seen in Fig. 1 the LBO-OPO should be tunable in the same wavelength range as the BBO-OPO. In LBO the effective nonlinearity is smaller, however. At 710 nm, for example, d_{eff}(LBO)=0.39 d_{eff}(BBO). The values of J_0 of the LBO-OPO should thus be larger by a factor of 6.7. The values of the ratio of J_0(BBO)/J_0(LBO) obtained in first measurements with 12 mm long crystals at the wavelengths of 630 nm, 560 nm and 440 nm provide ratios of 4.7, 5.2, 3.2, respectively[14].

These measurements confirm the higher energy density at threshold of the LBO-OPO. The results are in agreement with the theoretical predictions.

In LBO the tuning rate is considerably smaller compared to BBO. Wavelength tuning in the range of 410-2600 nm requires, for example, a change of the type I phase-matching angle $\Delta\phi=28°$. In the UV tuning from 300 nm to 400 nm even requires a change by $\Delta\phi=55°$.

The possible advantages of the LBO-OPO have still to be investigated in particular with respect to narrow bandwidth operation and short pulse generation.

In summary these first experiments clearly demonstrate the advantages of the BBO and LBO-OPO's as tunable coherent light sources. The most attractive features are the high power capability, high conversion efficiency, and in particular the large tuning range. The results obtained so far are certainly very promising for the further development of these new tunable all solid-state sources, which might be in the near future the radiation source of choice for applications like pulsed laser spectroscopy.

References

(1) J.A. Giordmaine and R.C. Miller, Phys.Lett.14, 973 (1965)

(2) S.E. Harris, Proc. IEEE 57, 2096 (1969)

(3) R.L. Byer, in Treatise in Quantum Electronics, edited by H. Rabin and C.L. Tang (Academic, New York, 1973) pp.587-702 see also: Y.X. Fan and R.L. Byer in SPIE Proceedings Vol. 461, 27(1984)

(4) Y.X. Fan, R.C. Eckardt and R.L. Byer, Conference on laser and Electro-Optics (CLEO) 1986 (San Francisco), postdeadline paper ThT4

(5) Y.X. Fan, R.C. Eckardt, R.L. Byer, J. Nolting and R. Wallenstein, Conference on Lasers and Electro-Optics (CLEO) 1988 (Anaheim), postdeadline paper pd31; Appl.Phys.Lett.53, 2014 (1988)

(6) H. Komine, Conference on Lasers and Electro-Optics (CLEO) 1988 (Anaheim), postdeadline paper pd32; Opt.Lett.13, 643 (1988)

(7) L.K. Cheng, W.R. Bosenberg, D.C. Edelstein and C.L. Tang, Conference on lasers and Electro-Optics (CLEO) 1988 (Anaheim), postdeadline paper pd 33; Appl. Phys.Lett.54, 13 (1989)

(8) K. Kato, IEEE J.Quant.Electron., QE-22, 1013 (1986)

(9) D. Eimerl, L. Davis, S. Velsko, E.K. Graham and A. Zalkin, J.Appl. Phys. 62, 1968 (1987)

(10) C. Chen, B. Wu, A. Jiang, and C. You, Sci.Sin.Ser.B28, 235 (1985)

(11) L.J. Bromley, A. Guy and D.C. Hanna, Opt. Commun. 67,317 (1988)

(12) S. Burdulis, R. Grigonis, A. Piskarskas, G. Sinkevicius, V. Sirutkaitis, J. Nolting and R. Wallenstein (submitted for publication)

(13) Ch.Chen, Y. Wu, A. Jiang, B. Wu, G. You, R. Li and S. Liu JOSA B6, 616 (1989)

(14) A. Fix, T. Schröder, Ch. Huang and R. Wallenstein (to be published)

TWO-PHOTON TRANSITIONS WITH TIME-DELAYED RADIATION PULSES

Alfredo Pasquarello and Antonio Quattropani

Institut de Physique Théorique
Ecole Polytechnique Fédérale de Lausanne
CH-1015 Lausanne, Switzerland

INTRODUCTION

It is known that in order to have two-photon transitions in a system interacting with two monochromatic radiation beams the sum of the photon energies has to be equal to the energy difference between the initial and final level.[1] Although all states have to be taken into account as intermediate virtual states in the calculation of the transition rate, it is not necessary that the single photon energies be in resonance with an intermediate level in order to have a two-photon transition. In the case of near-resonance on an intermediate state the two-photon transition rate increases. This resonant enhancement has been observed by Bjorkholm and Liao.[2] In the case of resonance on an intermediate state another process consisting of two successive one-photon transitions (two-step transition) is expected to occur. Two-photon transitions are characterized by the fact that the two beams have to interact simultaneously (coherent absorption). On the other hand, two-step transitions are expected to occur also if the beams are separated in time. Moreover, in the latter case no enhancement in the transition rate is expected when the sum of the photon energies equals the energy difference between the initial and final level.

In this chapter, we study to what extent it is possible to separate the contributions to the transition rate due to these two processes. We calculate the transition rate in second-order perturbation theory. For simplicity, we consider a three-level system and suppose the lifetime of the levels to be long with respect to the length of the radiation pulses. For this system analytical results can be obtained which allow a better insight. We show that the result from perturbation theory contains both contributions. Besides the usual two-photon contribution,[1] we find additional contributions which are important when one of the two radiation beams is in resonance on the intermediate level. Assuming that two-step transitions be only dependent on the absolute values of the detunings, we are able to identify the enhancement due to the coherent process. By varying the time-delay between the two radiation pulses, we analyse how the transition occurs between a regime where coherent two-photon processes contribute to absorption (simultaneous pulses) to a regime where only two-step processes are present (time-delayed pulses).

FULL SOLUTION IN SECOND ORDER PERTURBATION THEORY

We consider a three-level system; the transitions occur from the ground-state ψ_1 to the final state ψ_2, ψ_3 being the intermediate level. We assume infinite lifetimes for these states.

Applied Laser Spectroscopy, Edited by W. Demtröder and
M. Inguscio, Plenum Press, New York, 1990

The system interacts with two laser fields,

$$\mathbf{E} = E_1 \cos(\Omega_1 t)\boldsymbol{\epsilon}_1 + E_2 \cos(\Omega_2 t)\boldsymbol{\epsilon}_2, \tag{1}$$

through the dipole interaction Hamiltonian

$$H_d = -\boldsymbol{\mu}\cdot\mathbf{E}. \tag{2}$$

The electric dipole matrix elements are given by

$$\begin{aligned}
\langle 3|\boldsymbol{\mu}\cdot\boldsymbol{\epsilon}_1|1\rangle &= \mu_{31}, \\
\langle 2|\boldsymbol{\mu}\cdot\boldsymbol{\epsilon}_2|3\rangle &= \mu_{23}, \\
\langle 1|\boldsymbol{\mu}|2\rangle &= 0.
\end{aligned} \tag{3}$$

Moreover, we assume that the frequency Ω_1 is in near-resonance with the 1-3 transition and Ω_2 in near-resonance with the 2-3 transition. The two frequencies are sufficiently different that each laser field induces only one transition rather than both. We adopt the definitions

$$\begin{aligned}
\omega_{ij} &= (E_i - E_j)/\hbar, \\
\alpha &= \mu_{13}E_1/2\hbar, \quad \beta = \mu_{23}E_2/2\hbar, \\
\Delta_1 &= \omega_{31} - \Omega_1, \quad \Delta_2 = \omega_{23} - \Omega_2.
\end{aligned} \tag{4}$$

Initially $(t = 0)$, the system is not excited and is found in level 1

$$\psi(t = 0) = \psi_1. \tag{5}$$

We now derive the usual two-photon formula looking for additional contributions in the case of resonance on the intermediate level. The population ρ_{22} of level 2 can be expressed in terms of the time dependent state $\psi(t)$

$$\rho_{22}(t) = |\langle \psi_2|\psi(t)\rangle|^2. \tag{6}$$

Studying the evolution in second-order time-dependent perturbation theory Eq. (6) becomes

$$\rho_{22}(t) = \left| \frac{1}{(i\hbar)^2} \int_0^t dt' \langle \psi_2|H_i|\psi_3\rangle \int_0^{t'} dt'' \langle \psi_3|H_i|\psi_1\rangle \right|^2, \tag{7}$$

where, in rotating wave approximation,

$$\begin{aligned}
\langle \psi_3|H_i|\psi_1\rangle &= \hbar\alpha e^{i\Delta_1 t}, \\
\langle \psi_2|H_i|\psi_3\rangle &= \hbar\beta e^{i\Delta_2 t}.
\end{aligned} \tag{8}$$

The final result is

$$\rho_{22}(t) = \alpha^2\beta^2 \left[-\frac{4\sin^2(\Delta_1 + \Delta_2)t/2}{(\Delta_1 + \Delta_2)^2\Delta_1\Delta_2} + \frac{4\sin^2\Delta_2 t/2}{(\Delta_1 + \Delta_2)\Delta_2^2\Delta_1} + \frac{4\sin^2\Delta_1 t/2}{(\Delta_1 + \Delta_2)\Delta_1^2\Delta_2} \right]. \tag{9}$$

The first term in Eq. (9) is the usual two-photon contribution. In addition we find two other terms, which are in resonance on the intermediate level, i.e. their contribution is important when both frequencies Ω_1 and Ω_2 are in resonance on the transitions 1-3 and 2-3, respectively.

We remark that if we define a transition rate by

$$W_{1\to 2} = \lim_{t\to\infty} \frac{\rho_{22}(t)}{t}, \tag{10}$$

we obtain

$$W_{1\to 2} = 2\pi\alpha^2\beta^2 \left[\frac{\delta(\Delta_1 + \Delta_2)}{\Delta_1^2} + \frac{\delta(\Delta_1)}{\Delta_2^2} + \frac{\delta(\Delta_2)}{\Delta_1^2} \right]. \tag{11}$$

110

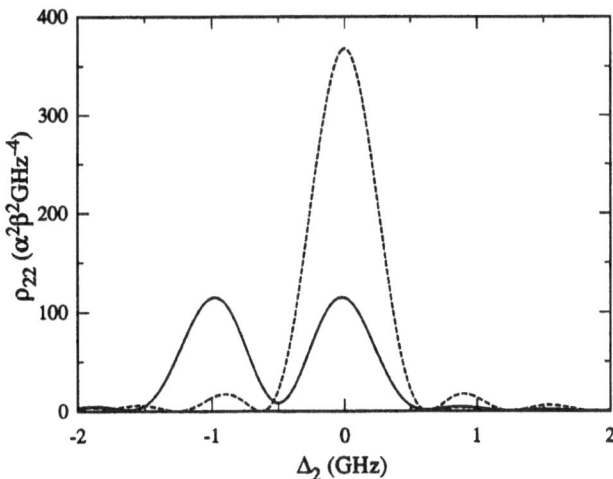

Figure 1. Population of level ψ_2 as a function of detuning Δ_2, $\Delta_1 = 1$ GHz being
fixed, for simultaneous pulses (solid line) and for pulses separated in
time (dashed line). The pulse length is 10 ns.

In the Δ_1-Δ_2 plane the two-photon resonance occurs on the line $\Delta_1 + \Delta_2 = 0$ and the
additional resonances on the lines $\Delta_1 = 0$ and $\Delta_2 = 0$. We note that in contrast with the
usual two-photon transition rate, represented by the first term in Eq. (11), the additional
terms in this equation give rise to contributions that are not energy conserving. It is evident
that in the limit of Eq. (10) the second-order of perturbation theory is insufficient to recover
the conservation of energy.

In Fig. 1, we show the population ρ_{22} of Eq. (9) as a function of detuning Δ_2, $\Delta_1 = 1$
GHz being fixed. These detunings correspond to a line in the Δ_1-Δ_2 plane which crosses the
two-photon resonance line at $\Delta_2 = -1$ GHz as well as one additional resonance line at $\Delta_2 = 0$
GHz. In correspondence of these resonances two peaks are found in Fig. 1. In this model the
peaks are found of equal height. However, if we consider effects such as finite lifetimes the
linewidth increases and turns out to be different for the two resonances. In fact the width
of the two-photon resonance will depend on broadening effects on the transition between the
initial and the final state whereas the additional resonance is broadened by effects that involve
the intermediate state ψ_3. In Fig. 1 the interaction time has been chosen to be of 10 ns. We
note that if shorter pulses are considered the two peaks merge into one broad peak.

We remark that qualitatively the same results are obtained, if T and Δ_1,Δ_2 are scaled
keeping the products $T\Delta_1$ and $T\Delta_2$ constant. However, the transition probability is proportional to the reciprocal of the fourth power of the detuning scaling parameter. So, if we
want to observe phenomena which occur at large values of $T\Delta$, it is preferable to have long
radiation pulses. Our model does not account for the finite lifetime of atomic levels. This
fact, in turn, limits the length T of the pulses.

TIME-DELAYED RADIATION PULSES

In this section, we consider the effect of a time-delay between the two laser pulses. We
consider, for simplicity, laser pulses of the same length T. We fix our time axis in such a way
that the first pulse, which induces the first transition 1-3, begins at $t = 0$ and the second at
$t = R$. If $R = 0$, the pulses are simultaneous and the results of the preceeding section are

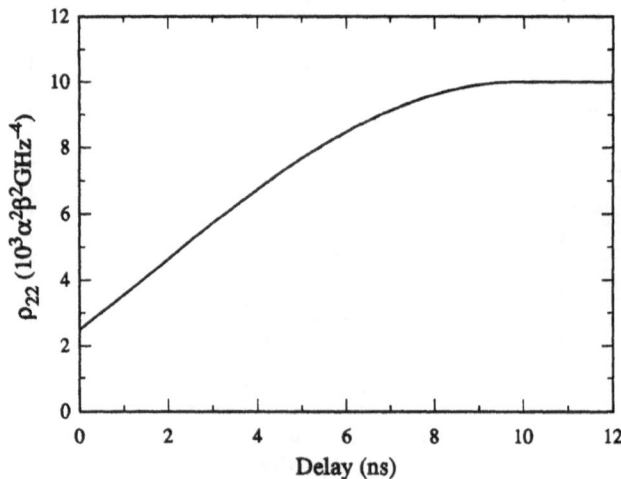

Figure 2. Population of level ψ_2 as a function of delay R for detunings $|\Delta_1| < 0.01$ GHz and $|\Delta_2| < 0.01$ GHz. The pulse length is 10 ns.

recovered. The population of the highest level at the end of the second pulse is now obtained by

$$\rho_{22}(R+T) = \left| \int_R^{R+T} dt' \, \beta e^{i\Delta_2 t'} \int_0^{t'} dt'' \, \theta(T-t'')\alpha e^{i\Delta_1 t''} \right|^2 \tag{12}$$

where θ is the unit step function. It can be shown that the result is symmetric for interchange of Δ_1 and Δ_2. For all $R > T$, the final population corresponds to succesive independent one-photon interactions of the system with the two pulses:

$$\rho_{22}(R+T) = \alpha^2\beta^2 \frac{4\sin^2(\Delta_1 T/2)}{\Delta_1^2} \frac{4\sin^2(\Delta_2 T/2)}{\Delta_2^2} \tag{13}$$

The result is independent on the time between the two pulses because the present model does not allow the excited states to decay. For $R > T$, the population depends only on the absolute values of the detunings.

In Fig. 1, the dashed curve is obtained for two pulses which are completely separated in time. We see that with respect to the result for simultaneous pulses the peak at $\Delta_2 = 0$ GHz has increased while the two-photon peak has disappeared. In fact, for $R > T$, coherent effects do not contribute to absorption. We note that the result for delayed pulses does not depend on the sign of Δ_2. In the case of simultaneous pulses a comparison of the population of the final level for $\Delta_2 = -1$ GHz with that for $\Delta_2 = 1$ GHz shows the enhancement due to the two-photon contribution. In fact in absence of coherent effects these populations are expected to be equal.

In Figs. 2, 3 and 4, we plot the population of the highest level at the end of the second pulse, $\rho_{22}(R+T)$, as a function of the delay R. We have only considered pulses with $T = 10$ ns, because of the described scaling properties. In Fig. 2, we have allowed Δ_1 and Δ_2 to vary between -0.01 GHz and 0.01 GHz. In the scale of the figure, the different curves cannot be distinguished. The population increases for increasing delays.

In Fig. 3, we take $\Delta_1 = 0.1$ GHz and three values for Δ_2: $\Delta_2 = -0.1$ GHz, which corresponds to the two-photon resonance, $\Delta_2 = 0$ on the additional resonance line and $\Delta_2 = 1$ GHz, which is not on a resonance line. The figure is qualitatively similar to the preceeding one. We notice that for small delays the curves which correspond to resonant situations are

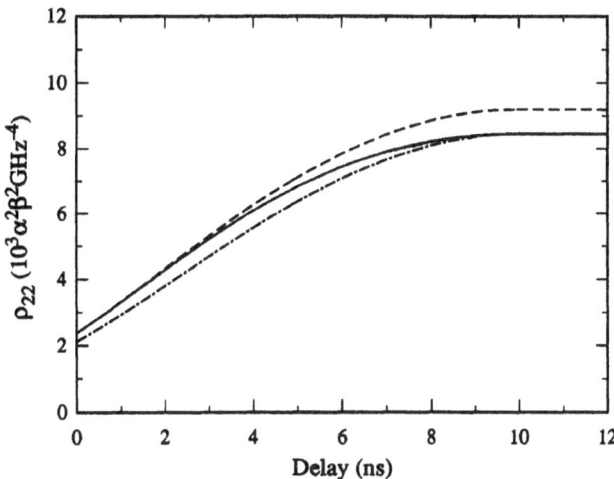

Figure 3. Population of level ψ_2 as a function of delay R for detunings $\Delta_1 = 0.1$ GHz, and $\Delta_2 = -0.1$ GHz (solid), $\Delta_2 = 0$ GHz (dashed) and $\Delta_2 = 0.1$ GHz (dotted-dashed). The pulse length is 10 ns.

higher. When the pulses are separated in time $(R > T)$, the two-photon curve and the off-resonance curve coincide as $\rho_{22}(R+T)$ of Eq. (13) only depends on the absolute value of the detunings which, in both cases, is 1 GHz. The curve, which corresponds to the additional resonance line, ends up higher because the detuning Δ_2 is smaller ($\Delta_2 = 0$).

In the three graphs of Fig. 4, we take $\Delta_1 = 1$ GHz and vary Δ_2 between -1 GHz and 1 GHz. In this case, great oscillations are observed in the population of the final level. The final population for pulses separated in time is largest on the additional resonance line, i.e. for $\Delta_2 = 0$. Because of the strong increase as a function of the delay, the oscillations are in this case less evident. In each graph we present two curves which correspond to opposite Δ_2. The result for $R > T$ is equal in the two cases as, in the case of independent one-photon transitions the final population does not depend on the sign of the detuning. For $R < T$, the result contains also the contribution of the coherent effects, which depend on the sign of Δ_2. As before, by comparing the two curves the effect due to coherent processes is evident. Strong coherent effects are observed when $\Delta_2 = -1$ GHz, which corresponds to the two-photon resonance. In this case, the population of the final level is large for simultaneous pulses and decreases about one order of magnitude when the pulses become separated in time.

CONCLUSIONS

We have analysed the contribution to absorption due to coherent two-photon transitions as compared with that due to two-step transitions in the case of near-resonance on an intermediate state. We have coupled a three-level system to the electromagnetic field in second-order perturbation theory. In the case of simultaneous radiation pulses we have found, besides the usual two-photon contribution, additional contributions which are important when one of the photon energies is resonant on the intermediate state. The enhancement due to coherent two-photon processes can be identified by comparing the transition rate in the case of opposite detunings to that in the case of equal detunings. In fact, two-photon transitions are resonant in the former but not in the latter case, whereas two-step transitions are expected to contribute in the same way in the two cases. By considering a delay between the radiation pulses

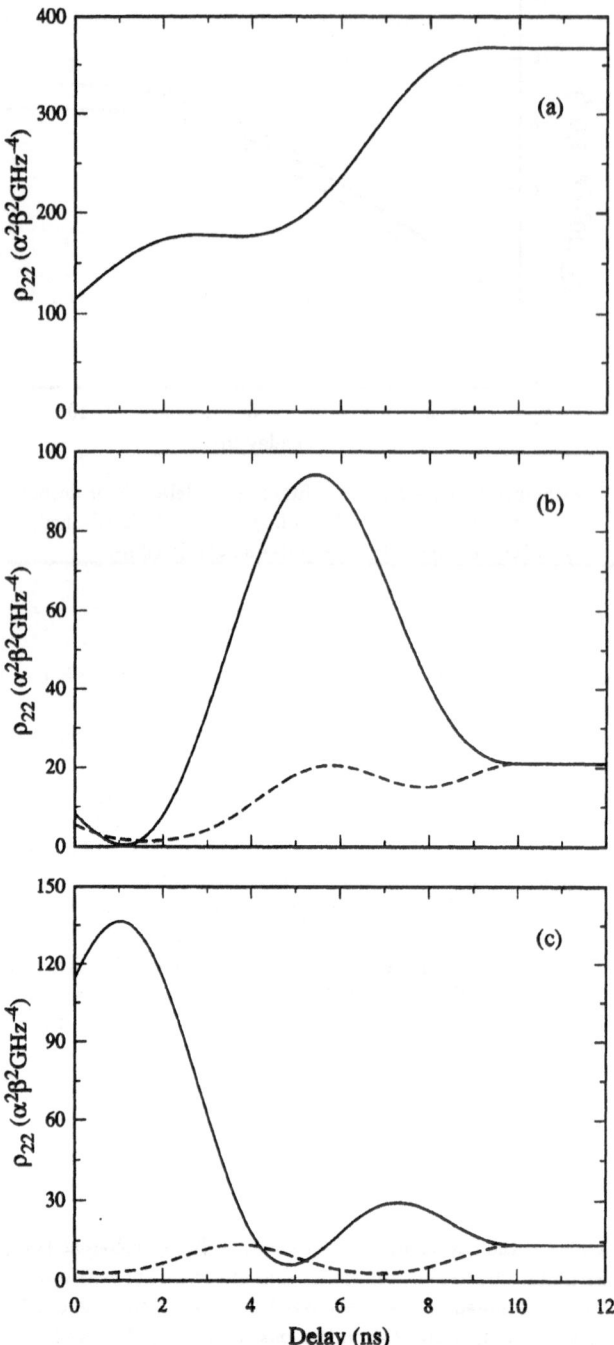

Figure 4. Population of level ψ_2 as a function of delay R for detunings $\Delta_1 = 1$ GHz and (a) $\Delta_2 = 0$ GHz, (b) $\Delta_2 = \pm 0.5$ GHz and (c) $\Delta_2 = \pm 1$ GHz. Solid lines and dashed lines represent positive and negative Δ_2 respectively. The pulse length is 10 ns.

it is possible to study how the coherent effects progessively vanish going from simultaneous pulses to pulses which are completely separated in time.

We have found that absorption as a function of delay is strongly dependent on the magnitude of the detunings. For small detunings we have shown that absorption increases for increasing delays. However, for large detunings, when the the product of T, the duration of the pulses, and the detunings is about 10, interesting oscillations are found. In order to verify these results experimentally, it is preferable to have an atomic system with states with long lifetime, as the results presented here have been obtained neglecting the loss of population due to spontaneous decay.

ACKNOWLEDGMENTS

We acknowledge interesting discussions with L. C. Andreani and S. Degironcoli.

REFERENCES

1. M. Göppert-Mayer, Ann. Phys. (Leipzig) **9**, 273 (1931).
2. J. E. Bjorkholm and P. F. Liao, Phys. Rev. Lett. **33**, 128 (1974).

DIODE LASERS AND THEIR APPLICATION TO SPECTROSCOPY

L. Hollberg

Time and Frequency Division
National Institute of Standards and Technology
Boulder, CO. 80303

Semiconductor diode lasers are emerging as important tools for the future of laser spectroscopy and precision optical measurements. The technology of diode lasers has advanced considerably in the past few years as has their availability. Because these lasers are practical, efficient and inexpensive they will open many research possibilities. Their low cost is particularly significant for laboratories with limited budgets. The impact of diode lasers on atomic and molecular spectroscopy will be profound, and is just beginning.

In this short summary of the characteristics of diode lasers and their application to spectroscopy we will focus attention on the room temperature semiconductor lasers operating in the near infrared and red regions of the spectrum. This means we will neglect completely the considerable amount of spectroscopic work that has been done with the cryogenically cooled lead-salt diode lasers that operate further in the infrared. The lasers we consider here are based on the mixed semiconductors of GaAs and InP and are produced primarily for commercial electronics applications including laser printers, compact disk players, and fiber-optics communications systems. Our goal is to take these commercial lasers and apply them to scientific and measurement applications. Some of the more general articles on the characteristics of diode lasers[1-4] and their application to spectroscopy[5-7] are included in the references.

Semiconductor diode lasers are very _small_, electrically efficient, tunable sources of laser radiation that can provide reasonable cw power levels. They also have the potential for very high resolution spectroscopy. The diode lasers are semiconductor devices with dimensions of about 125x300x250 μm with an active laser region of about .3x3x250 μm. The cleaved facets of the laser chip can serve as the mirrors for the laser's resonator because the index of refraction of the semiconductor is about 3.5 which gives a Fresnel reflectance of about 30%. Even with this small active region and mirror reflectance near 30% these lasers produce tens of milliwatts of cw optical power. The gain is obviously very high.

Contribution of NIST, not subject to copyright.

Applied Laser Spectroscopy, Edited by W. Demtröder and
M. Inguscio, Plenum Press, New York, 1990

The laser light is generated by the injection current which forces
carriers through the active region. The fluorescent output power of the
diodes increases gradually as the injection current is increased up to
the threshold current, at which point the diode begins to lase and output
power increases rapidly as a function of the injection current. A
typical laser might have a threshold current of 50 mA (with a diode
voltage drop of about 1.7 V) and a maximum output power per facet of 10
mW at 80 mA. Many of the new higher power lasers use high reflectance
coatings on the laser's back facet and reduced reflectance coatings on
the laser's output facet, so that all of the useful power comes out in
one direction.

The performance characteristics that one can obtain from commercial
diode lasers are diagrammed in fig. 1. Here we see the output powers
that are available as a function of wavelength, with lasers available in
four basic wavelength bands. The wavelengths are determined by the
bandgap of the material; thus the shortest wavelength lasers (670 nm) are
made from the semiconductor InGaAlP, those near 800 nm are made from
AlGaAs, and the longer wavelength lasers come from various compositions
of InGaAsP. The AlGaAs lasers generally have the best characteristics in
terms of power and spectral purity, and they have naturally seen more
applications in spectroscopy. We have used lasers from many regions of
this chart but that is not to imply that it is necessarily easy to obtain
lasers at any power and wavelength that one desires. Most of the lower
power lasers are readily available from distributors but wavelength
selection can be a problem. As a rule of thumb the price generally
increases with power and wavelength.

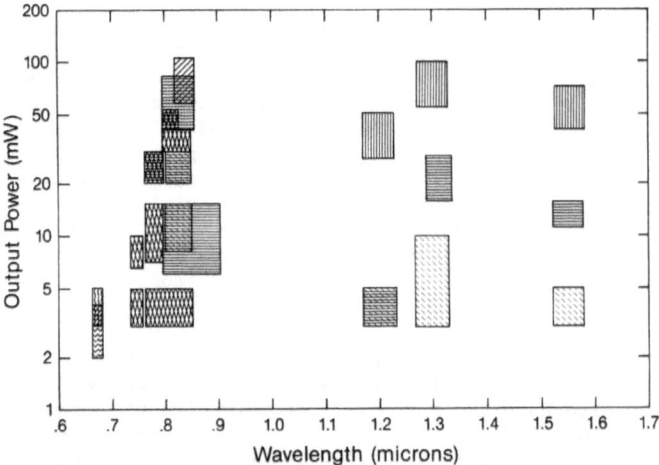

Fig. 1.   Output powers for commercial semiconductor diode lasers
           plotted as a function of wavelength. The hatched boxes
           represent the distribution of lasers as advertised by various
           manufacturers.

A given laser's wavelength can be tuned easily with temperature and injection current. For example the tuning coefficients for AlGaAs lasers (for small temperature changes) are approximately +0.07 nm/K (or -30 GHz/K) and +0.007 nm/mA (or -3 GHz/mA) respectively. The tuning with the injection current is mainly due to heating of the semiconductor junction although there is a contribution due to changes in the carrier density. For larger temperature changes the laser's frequency will jump progressively from one longitudinal mode to the next. When one averages over these mode jumps and the linear tuning regions near each mode the net tuning rate with temperature is about 0.1 nm/K. But because of the mode jumps the spectral coverage is not complete. The pattern of mode jumps is usually irregular and often shows hysteresis, so the probability that a given laser will reach a specific wavelength within its tuning range is only about 50%. With a Peltier element for temperature control, a typical laser will cover (incompletely) a wavelength range of about 8 nm. For spectroscopic applications it is obviously wise to have several diode lasers available. In practice the problem is not very serious because diode lasers are inexpensive (some as low as US $20) and in addition we can use optical feedback techniques to achieve almost complete spectral coverage within a wavelength range of about 20 nm.

Before applying diode lasers to spectroscopy we will find it useful to have some understanding of their noise characteristics. We need to consider both their amplitude and frequency noise. The amplitude noise of unmodified commercial diode lasers is broadband and extends from the audio frequency range to just above the laser's relaxation-oscillation frequency at a few gigahertz. The magnitude of the noise over this frequency range is typically 20 dB above the fundamental shot noise limit, and peaks to higher noise levels at low frequencies (below 100 kHz) and at the relaxation oscillation frequency (about 3 GHz). To scale these numbers in terms of fractional amplitude fluctuations we recall that the shot noise corresponds to the purely statistical fluctuations of randomly distributed photons, and thus varies as the square root of the number of photons. For example, the fractional power fluctuation due to shot noise for a 5 mW laser beam at 800 nm is about 10^{-8}. Thus we can expect a typical diode laser to have fractional power fluctuations of 10^{-6}, except at the lowest frequencies where the noise is higher. Generally the noise is that it decreases with an increase in injection current or a decrease in junction temperature. In fact, the amplitude noise on many diode lasers is low when compared to that of other spectroscopic laser sources such as dye lasers. This advantage of diode lasers allows high sensitivity absorption measurements to be made relatively easily.

The characteristics of the amplitude and frequency noise depend profoundly on any optical feedback that finds its way back into the laser. This is a disadvantage of diode lasers for some applications and care must be taken to avoid feedback. Feedback can be avoided by careful attention to optical layout, or when necessary by using some form of optical isolator (attenuator, quarter-wave plate and polarizer, or Faraday isolator).

The spectral properties of the frequency noise on diode lasers depends strongly on the type of laser. In fact, some commercial lasers do not even run on a single longitudinal mode, which makes them hard to use for spectroscopy. The longitudinal mode spacing for these very tiny lasers is about 0.3 nm ($\approx$140 GHz) and the bandwidth of the gain is about

40 nm. As the technology advances more and more lasers that operate on a single longitudinal mode are available. Although there are optical methods that can be used to force multimode lasers to operate on a single mode, we will concentrate on the single mode lasers. The spectral linewidth of most commercial diode lasers varies from about 20 to 300 MHz. These linewidths provide a resolution that is adequate for some spectroscopic applications but not for others. The spectral character of the frequency noise that generates these broad linewidths is similar (and related) to the amplitude noise. That is, the spectral density of frequency fluctuations is broadband and extends out to tens of megahertz before it drops off significantly. There is then a broad flat region of frequency noise out to higher frequencies and again a peak in the noise at the relaxation oscillation frequency (at a few gigahertz). The frequency noise below 20 MHz is the most significant because it contributes the most to the linewidth.

One of the early surprises that people found when trying to use diode lasers for atomic spectroscopy was that the signal-to-noise ratio in some cases was not nearly as good as expected.[8] It turns out that the frequency noise on diode lasers can be converted to amplitude noise by atomic resonances. This then degrades the signal-to-noise ratio in absorption and fluorescence measurements. In using diode lasers for optical pumping in cesium atomic clocks, people discovered that the noise on the strong cycling transition (F = 4 to 5) was about 100 times worse than expected. The physics of this FM-to-AM noise conversion by the atomic resonance has been explained, at least in part, by the work of Zoller and collaborators.[9] In fig. 2 we show some experimental manifestation of this noise in fluorescence and absorption measurements. In fig. 2a we see the output from a very simple experiment; the diode laser's output was sent through a cesium cell and then onto a fast photodetector. As the laser was scanned across the Doppler broadened transition, the high frequency noise at the photodetector increased dramatically. Figure 2b shows the Doppler free fluorescence spectrum of the cesium, F = 4 to 5, transition taken in an atomic beam. Here again we see a large increase in the noise from the atomic signal, whereas the Fabry-Perot transmission fringe that was monitored simultaneously shows mainly frequency modulation and not the excess amplitude noise. Fortunately, this type of noise is insignificant on many transitions and can be eliminated when it is significant by spectrally narrowing the lasers as described below.

Two main approaches have been taken to reduce the linewidth of semiconductor diode lasers: one is fast electronic feedback and the other is some form of controlled optical feedback. The electronic feedback approach uses fairly traditional laser frequency control to feed back to the laser's injection current in order to stabilize the lasers frequency to a Fabry-Perot resonance. With this method it is very easy to precisely control the laser's center frequency, but it is not easy to reduce the diode laser's linewidth. The trick here is that the frequency noise extends to high Fourier frequencies so that very fast electronics are required in the servo-loop. A few groups have had some success in narrowing diode laser's linewidths with fast electronics.[10-12]

The other main approach to diode laser frequency control is to use some form of controlled optical feedback in order to narrow the laser's linewidth. The linewidth of a diode laser can be reduced simply by reflecting some of the laser's output back to the laser. This effectively extends the laser's resonator and creates a coupled cavity system with the original diode laser cavity coupled to an external reflector. This extended cavity system has a higher Q than the original diode laser cavity and thus produces a narrower linewidth. It is often

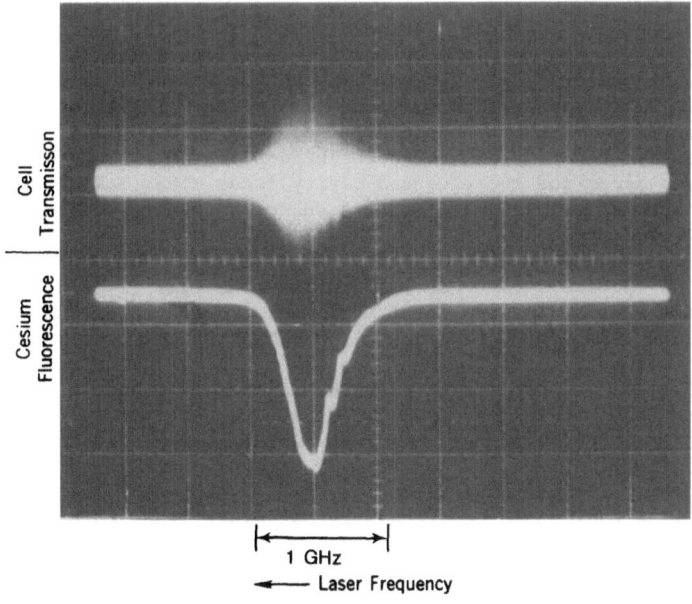

1 GHz

← Laser Frequency

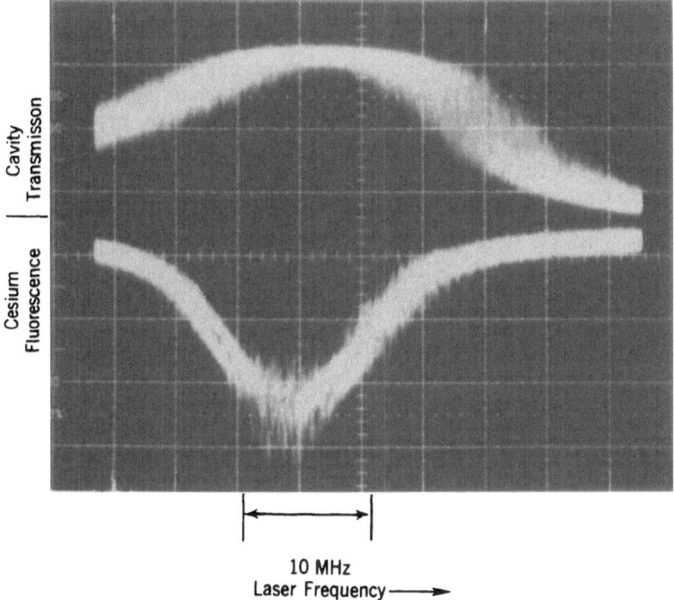

10 MHz
Laser Frequency →

Fig. 2.   Noise observed in cesium spectroscopy. In 2a the upper trace
shows excess high frequency noise on a diode laser beam that
has passed through an absorption cell. The lower trace in 2a
is the fluorescence signal detected simultaneously. In 2b
the lower trace shows excess noise in atomic beam
fluorescence from the F = 4 to 5 transition. Here the upper
trace is a Fabry-Perot transmission fringe that was taken
simultaneously.

advantageous to have some frequency selectivity in the optical feedback to force the laser to a specific single longitudinal mode. A wide variety of optical feedback systems has been developed to control the center frequency and narrow the linewidth of diode lasers. For example, the feedback can come from mirrors, fibers, gratings, etalons or combinations of these. With many of these systems it is necessary to have an antireflection (or reduced reflection) coating on the diode laser's output facet. This reduces the competition between the normal diode laser modes and the extended-cavity laser modes. We will look with a little more detail at two optical feedback systems that have been very useful for spectroscopy.

One of the ideas that has existed for some time is to use a grating to extend the diode laser's cavity and thus provide an extended cavity with frequency selective feedback.[13-19] This system works very well if the laser has some antireflection coating on its output facet. Such grating systems can be used to tune the laser's frequency in steps over a wavelength range of approximately 20 nm. Fortunately (and fairly recently) many of the commercial high-power lasers already have antireflection coatings and can be used directly with gratings without modification. The grating then allows one to dependably select a specific longitudinal mode of the original diode laser chip. Small changes in the diode's temperature can be used for gross tuning of these modes. With stable mechanical construction these grating extended cavities have linewidths of 0.1 to 1 MHz depending on the actual design. With the grating mounted on a PZT translator the laser's frequency can be scanned continuously over a range of about 1 GHz. If the diode laser's injection current is scanned synchronously with the PZT the continuous scanning range can be tens of GHz.

Another frequency stabilization method that we have used successfully is an optical-feedback lock[20-22] that is based on resonant optical feedback from a high-Q optical resonator. A typical system of this kind uses an unmodified commercial diode laser and weak optical feedback from a confocal Fabry-Perot cavity. In certain geometries the laser sees feedback from the Fabry-Perot only when the laser's frequency matches the resonance frequency of the Fabry-Perot. In this coupled cavity system the laser's oscillation frequency automatically locks to the cavity resonances and is thereby stabilized. These optical self-locking systems have achieved diode laser linewidths of a few kilohertz.

One of the very useful properties of diode lasers is that they can be modulated very efficiently and very rapidly via the laser's injection current. When the injection current is modulated, the laser's power and frequency are both modulated. In practice, for spectroscopic applications, the variation in power is small relative to the variation in frequency and can often be ignored. The modulation response of the diodes extends from DC to a few GHz and allows a number of applications that are much more difficult with other types of lasers. Some of these applications include rapid frequency scans (frequency chirps) which can be used for diode laser cooling of atoms,[23] rapid frequency jumps which can be useful for transient spectroscopy and optical pumping, and high frequency modulation which can be used for optical heterodyne spectroscopy. Unfortunately, the modulation capabilities are altered and usually degraded by the optical stabilization methods that are useful for narrowing the laser's linewidth. This is not surprising because in one case we are asking the laser's frequency to be very stable and in the other we are asking it to change easily and rapidly.

To be realistic about applying diode lasers to spectroscopy and precision measurements we must also recognize their limitations. One of these is that some of the special diode lasers we would like to use are very difficult to obtain. In addition, some of the lasers that we can get do not tune to the wavelength we need, and their tuning is irregular with many mode jumps. For high resolution spectroscopy the diode laser's broad linewidths (tens of megahertz) are a problem. The diode lasers are also particularly sensitive to optical feedback. For some applications it would be useful to have more power than is readily available from common diode lasers. This brief list of complaints contains most of the limitations that we face when trying to use diode lasers, but the problems are not very serious. Relatively simple techniques have been developed to deal with the tuning, linewidth, and optical feedback problems.

On the other hand, the advantages of semiconductor diode lasers far outweigh their limitations. For the most part the lasers are inexpensive, electrically efficient, readily available, and extremely easy to use. Typical powers of tens of milliwatts are more than enough for even nonlinear spectroscopy. The diode lasers are easily swept and modulated which allows unique applications. With a little bit of extra work the laser's frequency can be stabilized and one can achieve resolution capabilities of a few kilohertz. The available diode lasers are produced for commercial electronics, and we are just in the early phases of learning how to use them for scientific applications. The development and application of diode lasers for spectroscopy is somewhat reminiscent of the early days when the transistor began to take over some of the roles of the vacuum tube. The technology is changing rapidly and the future certainly promises higher powers, broader spectral coverage and better spectral purity.

REFERENCES

1. K. Petermann, Laser diode modulation and noise, Kluwer Academic Publishers, Dordrecht, (1988).

2. M. B. Panish, Heterostructure injection lasers, Proceedings of the IEEE 64, 1512 (1976).

3. G. H. B. Thompson, in Physics of semiconductor laser devices (John Wiley & Sons, 1980)

4. H. C. Casey, Jr. and M. B. Panish, Heterostructure lasers part A and B, New York, (1978).

5. J. C. Camparo, The diode laser in atomic physics, Contemp. Phys. 26, 443 (1985).

6. M. Ohtsu and T. Tako, Coherence in semiconductor lasers, Progress in optics XXV, E. Wolf, Ed., Elsevier Science Pub. B. V., (1988).

7. C. Wieman and L. Hollberg, Diode lasers and their application to spectroscopy, invited paper, submitted to Rev. Sci. Inst.

8. Avila G., De Clercq E., De Labachellerie M. and Cerez P., Microwave Ramsey Resonances from a Laser Diode Optically Pumped

Cesium Beam Resonator, IEEE Transactions on Instrumentation and Measurement, IM-34, 139, (1985).

9.   Th. Haslwanter, H. Tirsch, J. Cooper, and P. Zoller, Laser-noise-induced population fluctuations in two- and three-level systems, Phys. Rev. A 38, 5652 (1988).

10.   S. Saito, O. Nilsson, and Y. Yamamoto, Frequency modulation noise and linewidth reduction in a semiconductor laser by means of negative frequency feedback technique, Appl. Phys. Lett. 46, 3 (1985).

11.   M. Ohtsu and N. Tabuchi, Electrical feedback and its network analysis for linewidth reduction of a semiconductor laser, J. Lightwave Tech. 6, 357 (1988).

12.   H. R. Telle and B. Lipphardt, Efficient frequency noise reduction of GaAlAs laser diodes by negative electronic feedback, 436, in Frequency Standards and Metrology edited by A. DeMarchi (Springer-Verlag, Berlin, Heidelberg, (1989).

13.   R. P. Salathe, Diode lasers coupled to external resonators, Appl. Phys. 20, 1 (1979).

14.   M. Ito and T. Kimura, Oscillation properties of AlGaAs DH lasers with an external grating, IEEE J. Quantum Electron. QE-16, 69 (1980).

15.   M. W. Fleming and A. Mooradian, Spectral characteristics of external-cavity controlled semiconductor lasers, IEEE J. Quantum Electron. QE-17, 44 (1981)

16.   S. Saito, O. Nilsson, and Y. Yamamoto, Oscillation center frequency tuning, quantum FM noise, and direct frequency modulation characteristics in external grating loaded semiconductor lasers, IEEE J. Quantum Electron. QE-18, 961 (1982).

17.   E. M. Belenov, V. L. Velichanskii, A. S. Zibrov, V. V. Nikitin, V. A. Sautenkov, and A. V. Uskov, Methods for narrowing the emission line of an injection laser, Sov. J. Quantum Electron. 13, 792 (1983).

18.   De Labachelerie M. and Cerez P. An 850 nm semiconductor laser tunable over a 300 Å Range, Opt. Commun., 55, 174 (1985).

19.   Favre F., Le Guen D., Simon J.C., Landousies B., External-cavity semiconductor laser with 15 nm Continuous Tuning Range, IEEE Quant. Electron., QE-21, 1937, (1985).

20.   B. Dahmani, L. Hollberg, and R. Drullinger, Frequency stabilization of semiconductor lasers by resonant optical feedback, Opt. Lett. 12, 876 (1987).

21.   H. Li and H. R. Telle, Efficient frequency noise reduction of GaAlAs semiconductor lasers by optical feedback from an external high finesse resonator, IEEE J. Quantum Electron. 25, 257 (1989).

22.   Ph. Laurent, A. Clairon, and Ch. Breant, Frequency noise analysis
       of optically self-locked diode lasers, IEEE J. Quantum
       Electron. 25, 1131 (1989).

23.   R. N. Watts and C. E. Wieman, Manipulating atomic velocities using
       diode lasers, Opt. Lett. 11, 291 (1986).

THE CO-OVERTONE LASER A SPECTROSCOPIC SOURCE IN A MOST

INTERESTING WAVELENGTH REGION

Wolfgang Urban
Institut für Angewandte Physik
Universität Bonn, D 5300 Bonn 1

1. INTRODUCTION

For spectroscopic applications a continuously tunable laser source is certainly the most desirable solution; however, fixed frequency laser may be much better lasers, as far as intensity, spectral purity and reliability are concerned. If one can achieve tunability on the molecular side, for example by Zeeman, -Stark or Doppler-effect, a stepwise tunable laser, such as the a molecular gas laser with its rotational manyfold may be a very good compromise.

The most common spectroscopic method in this context is certainly the so called laser magnetic resonance (LMR) technique, that has first been developed by K.M. Evenson et al in the FIR [1] but has been readily applied in the medium infrared using CO_2-lasers and CO-lasers. [2] These lasers can be used to study transitions between 900 and 1100 cm^{-1} (CO_2) or 1200 and 2100 cm^{-1} (CO). Recently we were able the extend the range of the CO-laser considerably by getting a liquid nitrogen cooled plasma to lase on overtone transitions $\Delta v=2$ between 2500 and 3500 cm^{-1} [3]. For a part of this range which is particularly interesting for spectroscopic applications no other powerful laser has been available so far. The lasing regions of the various lasers are indicated in Fig. 1. After a brief review concerning the gain mechanism in the CO-laser plasma, some of the particular problems and possibilities of the CO-overtone laser will be discussed.

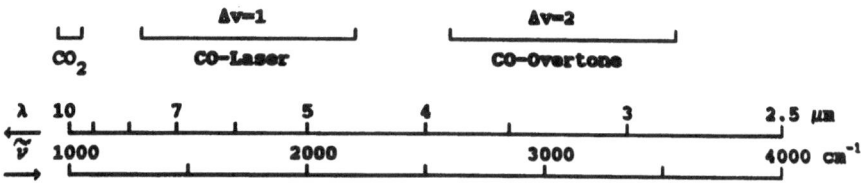

Fig. 1 Spectral regions for the various gas lasers
CO_2, CO, CO-Overtone, used in IR-Spectroscopy.

Applied Laser Spectroscopy, Edited by W. Demtröder and
M. Inguscio, Plenum Press, New York, 1990

2. THE PARTIAL INVERSION MECHANISM

There is one fundamental difference between the pumping mechanism of the CO_2-laser and the CO-laser. In the CO_2-laser, the upper lasing state is directly populated by resonant energy transfer from a different species. Corresponding to the He-Ne system, the CO_2-laser is a N_2-CO_2-laser. The main energy is available from vibrationally excited N_2-molecules and leads to the population of one particular CO_2-vibrational level from where lasing transitions occur into lower, unpopulated vibrational states. Thus (total) inversion is easily achieved. There are two types of vibrational transitions showing each rotational manyfolds with fully developed P- and R-branches.

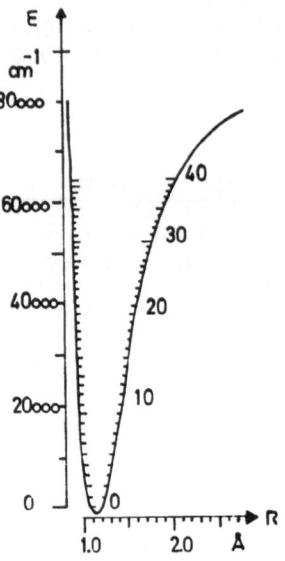

Fig. 2
The potential of the $X^1\Sigma$ state in carbon monoxide. The vibrational levels up to v=40 are indicated.

On the other hand, CO has only one type of vibrational transitions. In its $X^1\Sigma$ electronic ground state it has a very deep, slightly anharmonic potential (Fig.2). From the lower vibrational states populated in discharge producing the laser plasma, the energy is pumped by collisional transitions into the vibrational levels. Due to the particular situation for relaxation processes, the vibrational degree of freedom is heated up very fast, whereas the rotational degree is closely coupled to the translational temperature. The heating up of the vibration by so called VV-transfer is the more effective, the cooler the translational temperature in the discharge. This phenomenon has first been described theoretically by

Treanor, Rich and Rehm [4] and eversince is known as "Treanor pumping". Nevertheless, no "total inversion" is achieved in the CO-plasma, as has been experimentally verified [5]. Evaluation of the fluorescence data obtained from a liquid nitrogen cooled CO-laser plasma shows, that the higher vibrational state is always less populated than the lower adjacent state. However, there is a wide range of vibrational states, where the ratio of N_{v+1}/N_v is quite close to unity (0.95) and nearly constant. This is called the "plateau region" (Fig. 3).

The CO-laser exists, no doubt, but its spectrum consists only of parts of P-branches, for a wide range of vibrational bands. The centers of these bands are shifted towards longer wavelengths with increasing vibrational quantum number. Due to the wide J-manyfold of the lasing transitions, adjacent vibrational band strongly overlap. Thus a dense grid of fixed frequency laser lines covers the range from $2070 cm^{-1}$ ($v=2 \rightarrow 1$) down to $1200 cm^{-1}$ ($v=37 \rightarrow 36$). For one isotopic species, e.g. $^{12}C^{16}O$, we get more than 400 lines and the isotopomers $^{13}C^{16}O$ and $^{12}C^{18}O$ easily yield another 800 lines.

The pumping mechanism in the CO-plasma and the explanation for the "partial inversion" conditions is given in detail at a previous ASI [6]. Here we just recall that in this plasma the gain is not due to total inversion of the population for vibrational states, but due to a strong disequilibrium between the vibrational degree of freedom and the rotational degree of freedom:

$$T^*_{vib} \gg T_{rot} = T_{transl.} = T \tag{1}$$

T^* vib stands for a two level-temperature, for two adjacent vibrational states.

The gain is described by the Patel formula [7]. A shortened version of this formula, where only the essential factors are given explicitly, is the following:

$$\alpha_{v''J''}^{v'J'} = A \frac{g_{J'}}{Z(B_v, T)} \left| R_{v''}^{v'} \right|^2 \left\{ N_{v'} e^{-B_{v'} \frac{J'(J'+1)}{k_B T}} - N_{v''} e^{-B_{v''} \frac{J''(J''+1)}{k_B T}} \right\} \tag{2}$$

Here, g_J is the degeneracy factor, $Z(B_v T)$ the distribution function, $|R|^2$ the dipolar transition matrix element, N_v the total number of molecules in the vibrational state v and B_v the rotational constant, the factor a contains all those factors that are not taken into account explicitly.

One essential feature of this formula is the fact, that the gain originates from the pure Boltsmann part of the distribution rather than from the total vibrational population, $N_v(J,T)$. The result is demonstrated in Fig. 4.

This point is to be stressed since in spite of the fact, that Patel's formula gives a correct answer, its interpretation has often been rather misleading, in the past.

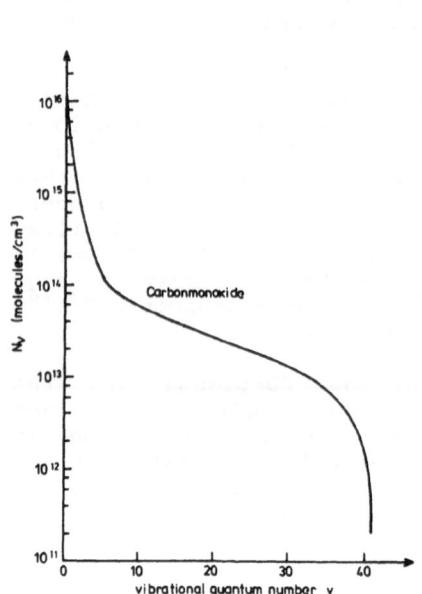

Fig. 3
Distribution of the vibrational population in a liquid N_2 cooled CO-plasma. The data are obtained for discharge conditions optimised for lasing at high vibrational transitions.
(Data from Ref. [5]).

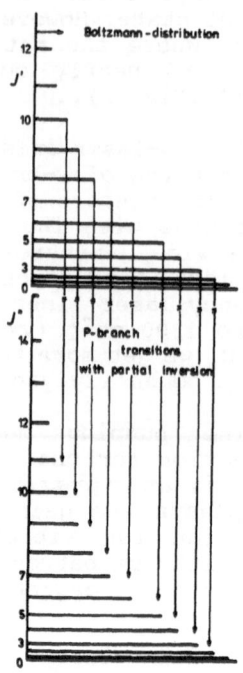

Fig. 4
Boltzmann distribution and partial inversion buildup for two adjacent vibrational states in CO. ($N_{v+1}/N_v \approx 0.95$,

$T_{rot} \approx 135K$, $B_v = 1,9cm^{-1}$).

In this formula another essential quantity is the dipolar transition moment, $|R|^2$. For the harmonic oscillator approach and $\Delta v=1$ it is directly proportional to the vibrational quantum number of the upper state v'. Thus this moment increases approximately proportional to the vibrational excitation level. This fact is very favorable for the CO-laser, since it partly compensates for the decreasing population in the higher vibrational states (Fig. 3).

3. CO-OVERTONE LASER $\Delta v=2$

3.1 The gain condition

So far we have discussed the conventional CO-laser since its understanding is prerequisite for the overtone laser.

There are two main differences for the latter.

(i) The transition moment $|R|^2$, that would be identically zero for v'-v"=2 in the harmonic case, increases proportional with the square of the quantum number of the upper vibrational level for an anharmonic potential.

(ii) The relative decrease of total population of the upper state for $\Delta v=2$ is twice as large as for normal $\Delta v=1$ transitions, if we refer to the plateau region (Fig.3).

This latter fact gives us a hint that the overtone laser will depend crucially on the slope of this plateau and thus it will respond most sensitive to the plasma conditions. The first point tells us, that it should be preferable to work at high vibrational states, due to the v^2 dependence of the transition moment that enters directly into the gain factor described by formula (2).

3.2 The Optimised Laser Plasma

Strong gain, particularly on the more elevated vibrational states can only be achived in a liquid nitrogen cooled plasma, and therefore only in a gas flow system. We need a plasma optimised for high vibrational states, which means a high concentration of CO gas. In a flowing gas plasma, the electron temperature can be reduced by a small amount of oxygen. With high CO concentrations this oxygen also serves to prevent dissociation of CO into C+O; however, much more oxygen is needed for this purpose. Oxygen also is a very fast VT-relaxer and thus too much oxygen is reducing the vibrational population. Therefore a most delicate equilibrium has to be established requiring just the right small amount of oxygen. The optimised plasma in the end needs the following composition (pressures given in m bar)

He: 7.5, N_2 : 0.7,   CO : 0.8 - 1.3,   air 0.05.

This is identical with a laser plasma optimised for high v-transitions in $\Delta v=1$.
Since the necessary amount of oxygen is so small its dosage is easier achieved by using air. In a flow system possible trace gas admixtures from the ambient air do not create a problem.

As for the optimised gas composition, one cannot use a premixed gas mixture, since the CO-concentration must be increased slowly to its final value. In order to get stable operation it may be better to start with twice the optimum air concentration and gradually reduce it to the final value.

These points may not be essential for the understanding of the physics going on in the CO-plasma, nevertheless it contains some information about the complex behaviour of a laser plasma.

3.3 The Laser Resonator

One essential aspect of a laser is its gain, another is the laser resonator. The main problem with the CO-overtone laser is not only the small gain for the overtone transitions, but the simultaneous high gain for $\Delta v=1$ transitions. Thus a very high selectivity for the tuning element is needed.

During recent years the technology of mechanically ruled blazed infrared reflection gratings has made considerable progress and reflection efficiencies of close to 99 % have been achieved. It is essential to have a ruling density so that only first order reflection is possible for the wavelength in question since the existence of a second order refraction automatically means further losses.

Our first overtone CO-laser was operating with 450 1/mm grating specially produced by B.Bach at Hyperfine Inc., Boulder, and the second laser ist equipped with a special grating ruled by B. Nelles at Zeiss, Oberkochen. These gratings have a wide range of high reflectivity and thus can operate the whole bunch of vibrational bands of the overtone laser.

In our setup the laser output is coupled out via the zeroth order of the grating in Littrow mount. The intensity distribution of the laser output must not necessarily reflect the characteristics of the gain curve, but is modified by the efficiency of the grating. If we assume that the efficiency is at maximum where the zero order output coupling is lowest, we understand the minimum laser intensity at the $v=34 \rightarrow 32$ band obtained with the "Hyperfine" grating. This minimum is shifted to $v=31 \rightarrow 29$ when using the "Zeiss" grating. (Fig. 5a and 5b). The fact that at these minimum intensity bands the J-manifold is highest is a proof for the statement of minimum loss at that region. A good measure for the intracavity intensity of the laser is the signal taken from the residual reflex of the Brewster windows. (Fig. 5c) The relative intensities at the output coupling via Littrow mount and the Brewster window reflection are plotted in Fig. 5b and 5c.

3.4 Power and Line Distribution

The very first observation of overtone transitions from a CO-laser plasma has been reported by Bergman and Rich [8]. They used a supersonic expansion gain medium capable to produce several hundred watts of $\Delta v=1$ CO-laser radiation. The spectral analysis showed several overtone bands obviously radiating in cascades from those bands that were not lasing on $\Delta v=1$.
Our CO-plasma is producing one to two watts single line power on $\Delta v=1$ and from the same plasma we recently have obtained up to 300 mW of overtone single line power [9].

The vibrational transitions observed from a 100 cm long liquid nitrogen cooled active medium starts with $v=15 \rightarrow 13$ and goes up to v=37→35. The J-manyfold is smaller for the lower states; some of the lines do not appear due to water

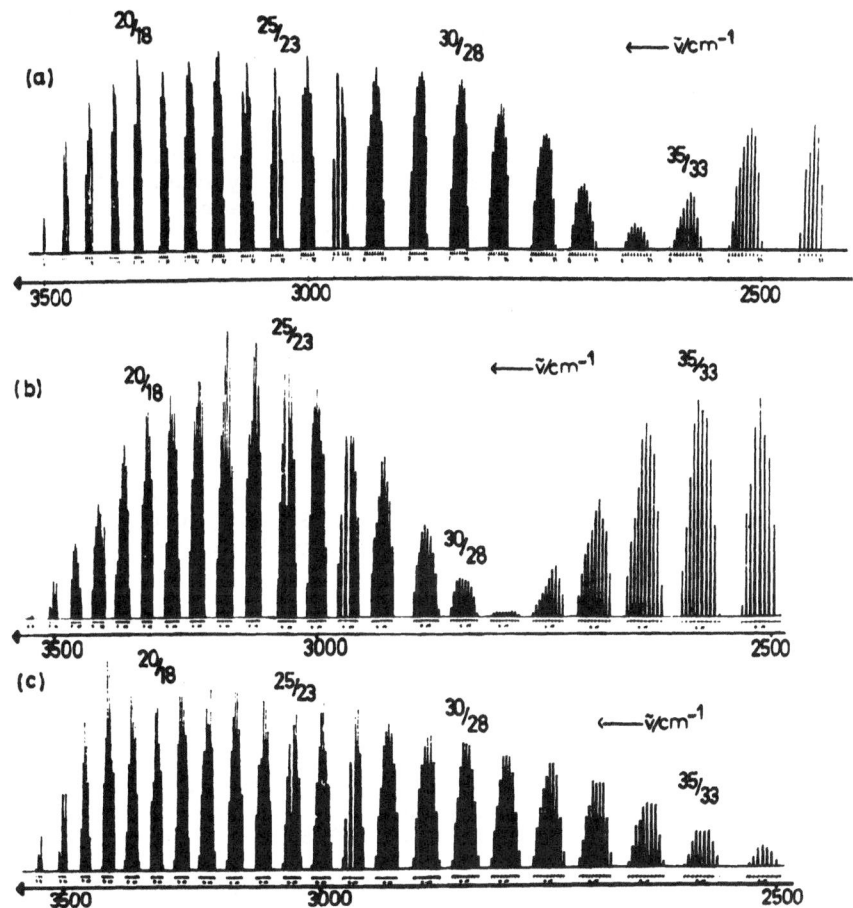

Fig. 5 Spectra of the CO-overtone laser taken under different resonantor conditions:
(a), (b): Gratings from two different manufacturers. These spectra are obtained by taking the zeroth order reflection from the grating. Both gratings: 450 lines/mm.
(c): Spectrum observed from residual reflection from one of the Brewster-windows sealing the laser plasma. Therefore this spectrum represents the intracavity power distribution for the various bands. It is obtained from the very same laser as Fig. (b).
The minima that become obvious in Fig. (a) and (b) are due to the location of maximum efficiency in the grating reflectivity. The intracavity intensity (c) does not show this behaviour. (Data taken from [3] and [10]).

absorption or due to coincidence with R-branch transitions of CO (Fig.5).

The overtone bands are shifted by twice the anharmionicity constant, therefore no overlap of the J-manyfold of adjacent vibrational bands occurs. In addition this manyfold is smaller for $\Delta v=2$ than for $\Delta v=1$ transitions of a liquid nitrogen cooled CO-laser.

4. CONCLUSIONS

As for the wavelength region, the CO-overtone laser covers a range, part of which so far has only been accessible by the difference frequency laser. Between 3 and 4 μm practically no other laser with more than 1mW power and high signal to noise ratio is available. Into this frequency range the CH, NH and also OH vibrations are falling, thus the overtone laser represents a source for very interesting spectroscopic applications, particularly in laser magnetic resonance (LMR) and photoacoustic spectroscopy.

ACKNOWLEDGEMENTS

The author expresses his thanks to Jörg Reuss for his critical comments on the manuscript. Part of the work that finally led to success with the overtone laser was supported by the Deutsche Forschungsgemeinschaft.

REFERENCES

[1] K.M.Evenson, H.P.Broida, J.S.Wells, H.J.Mahler, M.Mizushima, Electron Paramagnetic Resonance Absorption in Oxygen with the HCN-Laser, Phys.Rev.Letters, 21, 1038 (1968)

[2] W.Rohrbeck, A.Hinz, P.Nelle, M.A.Gondal, W.Urban Broadband Mid-IR-LMR-Spectrometer for the Range of 1200-2000 cm^{-1}, Appl.Phys. B 31, 13a (1983)

     A.Hinz, D.Zeitz, W.Bohle, W.Urban, A Faraday Laser Magnetic Resonance Spectrometer for Spectroscopy of Molecular Radical Ions, Appl.Phys. B 36, 1 (1985) (and others, e.g. references therein)

[3] M.Gromoll-Bohle, W.Bohle, W.Urban, Broadband CO-Laser Emission on Overtone Transition $\Delta v=2$, Optics Commun. 69, 409 (1989)

[4] C.E.Treanor, J.W.Rich, R.G.Rehm, Vibrational Relaxation of Anharmonic Oscillators with Exchange-Dominated Collision J.Chem.Phys. 48, 1789 (1968)

[5]   G.Guelachvili, D.DeVilleneuve, R.Farrenq, W.Urban,
        J.Verges, Dunham Coefficients for Seven Isotopic
        Species of CO, J.Mol.Spectros. $\underline{98}$, 64 (1983)

        R.Farrenq, C.Rossetti,
        Vibrational Distribution Functions in a Mixture of
        Excited Isotopic CO Molecules, Chem. Phys. $\underline{92}$, 401
        (1985)

[6]   W.Urban,
        Infrared Lasers for Spectroscopy in: Frontiers of
        Laser Spectroscopy of Gases, 4-42A.C.P.Alves,
        J.M.Brown, J.M.Hollas, edit. Cluwer Academic
        Publ.1988

[7]   C.K.N. Patel, Vibrational-Rotational Laser Action in
        Carbon Monoxide, Phys. Rev. $\underline{141}$, 71 (1966)

[8]   C.Bergman, J.W.Rich, Overtone Bands Lasing at 2.7-
        3.1μm in Electrically Excited CO, Appl.Phys.Lett. $\underline{9}$,
        597 (1977)

[9]   E.Bachem, A.Weidenfeller, M.Schneider,
        Private Communication, Bonn 1989

[10]   A.Weidenfeller, Diploma Thesis, Bonn 1990

TUNABLE SIDEBAND LASER SPECTROSCOPY OF ATOMS AND MOLECULES

Jean LEGRAND and Pierre GLORIEUX

Laboratoire de Spectroscopie Hertzienne, associé au C.N.R.S.
Université des Sciences et Techniques de Lille
F-59655 Villeneuve d'Ascq Cedex (France)

Lasers provide useful tools for spectroscopy as far as these systems have some reasonable tunability and their usefulness relies in large part on that characteristic of the lasers. In the medium infrared range (λ = 5-10 μm), molecular lasers have provided a large amount of spectroscopic datas but special techniques were required to overcome their very limited tunability. For instance the typical tuning range of a CO_2 laser line is 50-100 MHz while the separation between successive lines is of the order of 50 GHz and the available part of the spectrum is only 10^{-3} of the total range. One method that has been extensively used to take advantage of these lasers in spite of their rather restricted tunability is to tune the molecular (or ionic) absorption lines in resonance with the laser by Stark or Zeeman shifting the energy levels. This method is very helpful for the spectroscopic investigation of light molecules that must be paramagnetic or should have a large permenent electric dipole. Another way to compensate for the lack of tunability of molecular lasers is to add a tunable microwave (or radiofrequency) photon to the almost fixed frequency photon emitted by the laser by making the molecule undergo a two-photon process resonant with a molecular transition. This technique was introduced by Freund and Oka who could obtain high resolution spectroscopic datas on NH_3 for instance and later extended to other molecules[1]. However most of the spectroscopic applications of lasers require a tunable source and the tunable laser diodes appeared to be the adequate source in that spectral range. In practice, the fact that they emit simultaneously on many modes together with the very lòw power available in each mode usually does not allow to use them for saturation spectroscopy. Moreover their spectral purity in excess of 20 MHz or more makes them unsuitable for high resolution spectroscopy. An alternative to all these methods is provided by the tunable sideband technique. It is essentially based on the fact that when a coherent field at frequency ω_0 is modulated at a frequency ω_1, its power spectrum exhibits sidebands at frequencies $\omega_0 \pm \omega_1$ which benefit from the tunability of ω_1 (if a sufficiently wideband modulation is provided) while keeping the spectral purity of the original sources. To achieve a wideband modulation, sufficiently high modulation frequencies should be used. This implies practically the use of microwaves for the modulation.

In fact the posssibility of microwave modulation of CO_2 laser radiation has long been studied and to our knowledge the first such experiment is that of Corcoran *et al.* who could modulate 10.6 μm radiation by the electric field delivered by a klystron operated at 55 GHz [2]. These authors used that technique to

Applied Laser Spectroscopy, Edited by W. Demtröder and
M. Inguscio, Plenum Press, New York, 1990

generate infrared radiation at some specified wavelengths but their system was working only at fixed frequencies. However these authors could carry out point by point measurements of the gain and pressure broadening of laser lines. Their technique was later extended and developed in two directions, tending to increase both the efficiency and the tunability of the modulation. The efficiency was increased by using very high microwave modulation power. With 60 kW of 17 GHz radiation, Carter could reach an efficiency of 67 % [3]. This is of course limited to pulsed (<100 ns) experiments and has not been applied to spectroscopy. Most of the applications in that field are in fact related to the extension of the modulation bandwidth, a work which was initiated by Bonek and Magerl in

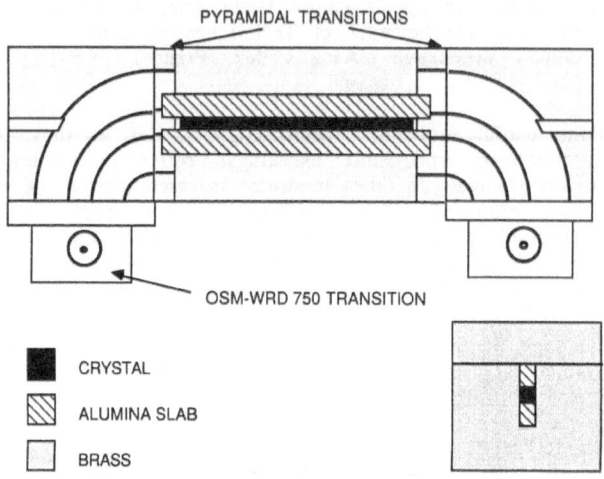

Figure 1. Design of the waveguide modulator.
The laser beam enters and goes out through the holes on the elbows.

1976 [4]. They first achieved a 2.75 GHz tunability around 53 GHz but the major step towards application came when they designed a modulator covering the whole 8-18 GHz range[5]. In the devices designed by these authors, the electrooptic crystal is installed inside a microwave guide and the infrared field is propagating freely in the space or inside the crystal which is typically a few millimeter thick. Unfortunately the efficiency of such devices is very limited and to improve it, P. K. Cheo proposed to realize an integrated optics version of the modulator in which the infrared field propagates inside a GaAs optical waveguide which also acts as the support of a stripline for microwave propagation[6]. Almost all spectroscopic investigations have used the Magerl-Bonek design of the bulk modulator. We are now going to discuss in more details the spectroscopic applications of these techniques in relation with the corresponding technological advances.

THE BULK MODULATOR

An overall view of the bulk modulator used in our experiments is given in Figure 1. The CdTe crystal is installed inside a microwave guide whose dimensions have been calculated so that the microwave and infrared phase velocities are matched at 13 GHz. The modulator also comprises two 90° elbows with a hole on the small side through which the infrared field is injected inside the modulator. Two tapered transitions match the crystal loaded waveguide to the double ridge elbows and alumina slabs on each side of the crystal are adjusted together with inner Teflon slabs to achieve wideband operation with as little ripple as possible. Extreme care is needed in the machining of the modulator housing in order to get a good velocity matching. Practically, before electronic leveling, the ripple on the frequency characteristic curve is 30 % typically. With 20 W microwave power and a 5 W infrared laser, the power available in each sideband is typically 0.5 mW.

SPECTROSCOPY WITH TUNABLE SIDEBAND SPECTROMETERS

In the first devices based on this design, the output power was very low and only linear spectroscopy experiments could be carried out. Magerl and collaborators first investigated the absorption spectrum of CH_3F and of several spherical top molecules such as SiH_4 , GeH_4 [7]. The design was optimized and could

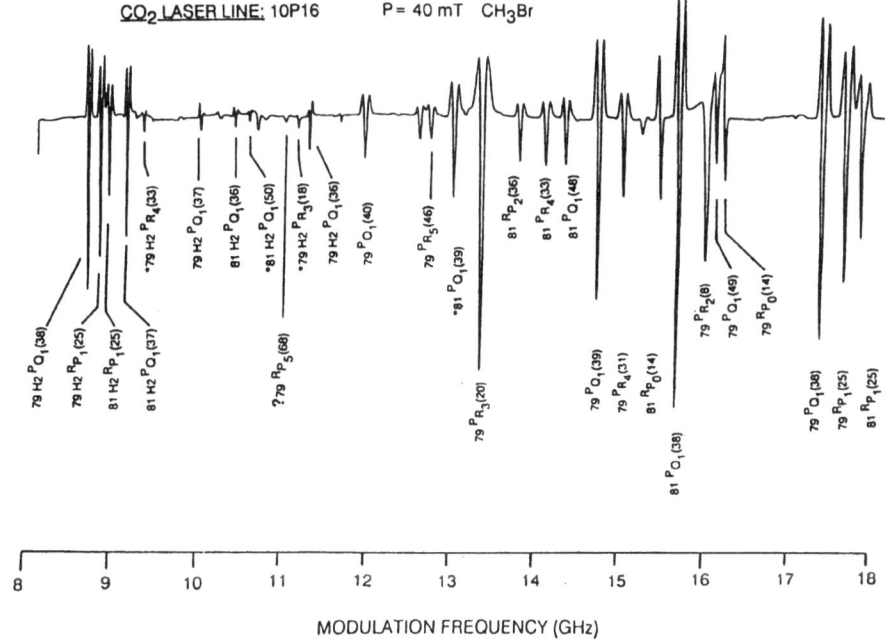

Figure 2. An example of linear sectroscopy with tunable sidebands :
the ν_6 band of CH_3Br near 948 cm^{-1}

generate sidebands with 0.5 mW power . In spite of its weakness, this is large enough to saturate with a beam diameter of a few millimeters, CH_3F, NH_3, SiH_4 and SiF_4 at a few milliTorr pressure, which was achieved by Magerl, Oka *et al.*. Infrared-radiofrequency double resonance which also requires saturation was

139

carried out at the same time by Scappini *et al.* on OsO$_4$ and by Jörissen *et al.* on SiF$_4$ [8,9]. To illustrate the flexibility of this method, we have reported on Figure 2 a part of the absorption spectrum of CH$_3$Br recorded by laser sideband spectroscopy with the 10 P16 CO$_2$ laser line[10]. Absorption lines with signal to noise ratio in excess of 10^3 are routinely recorded. Such spectra are helpful to improve the sets of molecular constants in particular high order centrifugal distortion constants since they provide high accuracy datas on high J and K lines. We have recently extended its use to unstable species namely excited atoms like Ar* and Ne* [11] but the method is sensitive enough to be applied also to free radicals and molecular ions.

The laser sideband technique has been extended to CO lasers by Magerl and Schwendeman who could also use it for double resonance experiments [12] but the major progress in that field should come from the work of P. K. Cheo who took advantages of the possibilities of integrated optics to increase by two orders of magnitude the efficiency without loosing the tunability. The last version of his device is described in details in [6]. It is essentially made of a 30 µm thick GaAs slab which acts as an optical waveguide and supports a microwave stripline which carries the modulating field. Such a device is able to deliver up to 100 mW with microwave and laser power equal to 30 W and 20 W respectively and according to Cheo[6] could still be increased by a factor of 5 . Up to now such devices have only served for feasability experiments but although they seem to be more delicate to use, they should be of larger use in the future.

This work was supported by DRET and la Région Nord-Pas de Calais. The expert technical assistance of B. Delacressonnière, J. M. Chevalier and C. Lizoret and discussions with G. Magerl are gratefully acknowledged.

REFERENCES

1.   see for instance T. OKA in "Frontiers in Laser Spectroscopy", vol. 2, 529-570, R. Balian, S. Haroche and S. Liberman eds., North-Holland, Amsterdam (1977).
2.   V. J. CORCORAN, R. E. CUPP, J. J. GALLAGHER and W. T. SMITH, Appl. Phys. Lett. 16, 316-318 (1970).
3.   G. M. CARTER and H. A. HAUS, I.E.E.E. J. Quant. Electron. QE-15, 217-224 (1979).
4.   G. MAGERL and E. BONEK, J. Appl. Phys. 47, 4901-4903 (1976).
5.   G. MAGERL, W. SCHUPITA and E. BONEK, I.E.E.E. J. Quant. Electron. QE-18, 1214-1222 (1982).
6.   P. K. CHEO, I.E.E.E. J. Quant. Electron. QE-20, 700-709 (1984).
7.   G. MAGERL, E. BONEK and W. A. KREINER, Chem. Phys. Lett. 52, 473-476 (1977) ; G. MAGERL, W. SCHUPITA, E. BONEK and W. A. KREINER, J. Chem. Phys. 72, 395-398 (1980).
8.   G. MAGERL, W. A. KREINER, J. M. FRYE and T. OKA, Appl. Phys. Lett. 42, 656-658 (1983) ; F. SCAPPINI, W. A. KREINER, J. M. FRYE and T. OKA, J. Mol. Spectrosc. 106, 436-440 (1984) ; G. MAGERL, W. SCHUPITA, J. M. FRYE, W. A. KREINER and T. OKA, J. Mol. Spectrosc. 107, 72-83 (1984).
9.   L. JORISSEN, W. A. KREINER, Y.T. CHEN and T. OKA, J. Mol. Spectrosc. 120, 233-235 (1986).
10.   J. LEGRAND, B. DELACRESSONNIERE and P. GLORIEUX, J. Opt. Soc. Am. B6, 283-286 (1989).
11.   J. M. CHEVALIER, J.LEGRAND, P. GLORIEUX, G. WLODARZAK and J. DEMAISON, J. Chem. Phys., 90, 6833-6839 (1989).
12.   S-C. HSU, R. H. SCHWENDEMAN and G. MAGERL, I.E.E.E. J. Quant. Electron., QE-24, 2294-2301 (1988).

HIGH RESOLUTION FAR INFRARED SPECTROSCOPY

L.R. Zink, M. Prevedelli, K.M. Evenson [*]
and M. Inguscio

European Laboratory for
Nonlinear Spectroscopy (LENS)
Florence, Italy

*National Institute of
Standards and Technology (NIST)
Boulder, Colorado, USA

INTRODUCTION

The far infrared (FIR) spectral region (which we will define as 0.3 - 10 THz [1mm to 30 μm]) plays an important role in molecular and atomic spectroscopy. The pure rotational transitions of light molecules, for example diatomic hydrides, occur in this region; FIR measurements provide the only accurate means of determining their rotational structure and associated characteristics such as line strengths, pressure broadening parameters, and permanent electric dipole moments. Heavier molecules can also be observed in the FIR. Their rotational transitions in this region involve high quantum numbers and therefore small but important centrifugal distortion effects can be measured. In atoms, transitions between fine structure levels as well as metastable levels occur at FIR frequencies.

Laboratory spectroscopic measurements in the far infrared are increasingly important for other branches of science. Radio astronomers are now making measurements in the FIR and rely heavily on the laboratory measured transition frequencies. The FIR molecular and atomic transitions are important in the characterization of interstellar molecular clouds; FIR measurements are useful for identifying species, measuring concentrations, and determining temperature profiles.[1] FIR spectroscopy is also important in the monitoring and modeling of the earth's atmosphere. For instance the concentration of the hydroxyl radical (an important molecule in the ozone production cycle) is monitored in balloon flights by observing its FIR spectrum.[2,3] Accurate determination of the concentration requires laboratory measurements of the pressure broadening coefficient at the appropriate frequency.

Applied Laser Spectroscopy, Edited by W. Demtröder and
M. Inguscio, Plenum Press, New York, 1990

Classical spectrometers in the FIR use interferometric techniques coupled with blackbody sources and high sensitivity bolometric detectors. At present, the best available Fourier transform spectrometers provide complete coverage of the FIR region with 20 MHz spectral resolution and accuracy approaching 1 MHz.[4,5] They also have the advantage of performing wide spectral scans in fairly short periods and have a frequency range extending into the visible.

FIR laser spectroscopy is limited by the fact that no broadly tunable, narrow-band lasers exist in this region. However, over 1000 optically pumped, fixed frequency FIR laser lines do exist, with about 1 kHz linewidth;[6] the only problem is lack of tunability between lines. To overcome this problem laser magnetic resonance (LMR) was developed.[7] LMR uses a magnetic field to Zeeman tune a molecule or atom into resonance with the fixed frequency laser. Intracavity FIR LMR is one of the most sensitive spectroscopic techniques (it exhibits a minimum detectability of 5×10^{-10} cm^{-1}) and has wide applications in spectroscopy and chemical kinetics. For detailed LMR reviews, the reader is referred to references 8 and 9, reference 10 provides a good overview of the use of LMR in atomic spectroscopy.

As powerful and useful as FTS and LMR are in the far infrared, there are limitations to both techniques. LMR is only applicable to species with open electronic shells, has an accuracy limited to about 1 MHz, and requires the unravelling of the Zeeman splittings . With FTS, the resolution is often insufficient to observe hyperfine structure and the maximum accuracy of several MHz is only available on a few of the best spectrometers. For these reasons, other coherent sources have been developed. These all take advantage of nonlinear mixing to generate tunable, narrow-band far infrared radiation.

Harmonic multiplication of microwave oscillators was developed in the 1960's and at present provides microwave resolution and accuracy at frequencies up to 1 THz.[11] Frequencies above one THz have been synthesized in several different ways: 1) CO_2 laser difference frequency generation in GaAs[12] and other crystals[13]; 2) Generation of microwave sidebands on FIR laser radiation[14-17]; and 3) CO_2 laser difference frequency generation in metal-insulator-metal (MIM) diodes.[18] The first technique has never been employed for spectroscopy and will not be discussed further.

The generation of tunable FIR laser sidebands in a Schottky diode is used for spectroscopy in several laboratories and has produced fine results; however there are several serious drawbacks to this method. Although over 1000 FIR laser lines exist, sideband generation requires the most powerful FIR laser lines. This limits the spectral range to frequencies below 3 THz and also limits the coverage below 3 THz. The molecular transitions which can be measured are limited to those in the vicinity of a strong FIR laser line. The typical absolute accuracy obtainable is $\pm$ 2 X 10^{-7},[19] ($\pm$ 200 kHz at 1 Thz) determined by the re-settability of the FIR laser. This value is for a system designed for high frequency reproducibility; the frequency uncertainty can increase for a system designed to yield high FIR power. This limitation can in principle be overcome by measuring the FIR laser frequency simultaneously with the FIR spectra. The separation of the sideband radiation from the carrier is a less serious problem . Although these systems generate tens of μw sideband power, the sensitivity is limited by the presence of the stronger carrier. Even after interferometric separation of the sidebands, the power at the carrier frequency still limits the sensitivity to about 2 X 10^{-6} cm^{-1}.

Generation of tunable FIR (TuFIR) radiation by the difference frequency between two CO_2 lasers in a MIM diode was first accomplished in 1984.[18] Later in the same laboratory microwave sidebands were generated on the difference frequency.[20] This method is superior in both frequency coverage and accuracy to the FIR laser sideband system and has comparable sensitivity. The frequency range extends from 0.3 to 6 THz, with greater than 95% coverage. The accuracy is limited by the accuracy of the CO_2 lasers, which are frequency stabilized to 25 kHz, yielding an FIR accuracy of 35 kHz. And although the FIR power generated is less than that using the FIR laser sideband method, the sensitivity is roughly the same because there is no strong carrier laser causing excess noise on the detector. The third order technique of CO_2 laser difference frequency plus microwave sidebands has been employed in a new spectrometer at LENS and will be described below.

THIRD ORDER SPECTROMETER AT LENS

Figure 1 is a schematic of the third order TuFIR spectrometer. The radiation from lasers I and II are combined on a beamsplitter and coupled onto the diode by a 25 mm focal length lens. The microwave radiation is coupled onto the diode by a bias tee connected to the diode junction. The generated FIR is radiated from the diode's whisker in a long wire antenna pattern.[21] This FIR is then collected and collimated by a 30 mm focal length off-axis section of a parabolic mirror. After passing through an absorption cell the FIR is detected on a liquid He cooled Si bolometer. The FIR radiation is frequency modulated (via frequency modulation of the CO_2 lasers) and phase sensitive detected in a lock-in amplifier. The microwave frequency is swept by a personal computer, which also collects the data from the lock-in. Thus we are able to average scans for increased sensitivity. This is essentially the same experimental arrangement as described in ref. 20 only without the acousto-optic modulators isolating the lasers from the diode.

Both CO_2 lasers are frequency stabilized to the 4.3 μm saturated fluorescence signal from low pressure CO_2 cells[22] (not shown in figure 1). These frequencies have been measured to an absolute frequency of better than 5 kHz,[23,24] but without the special locking techniques used in refs. 23 and 24 our measured stability is 25 kHz for each laser. The overall uncertainty of

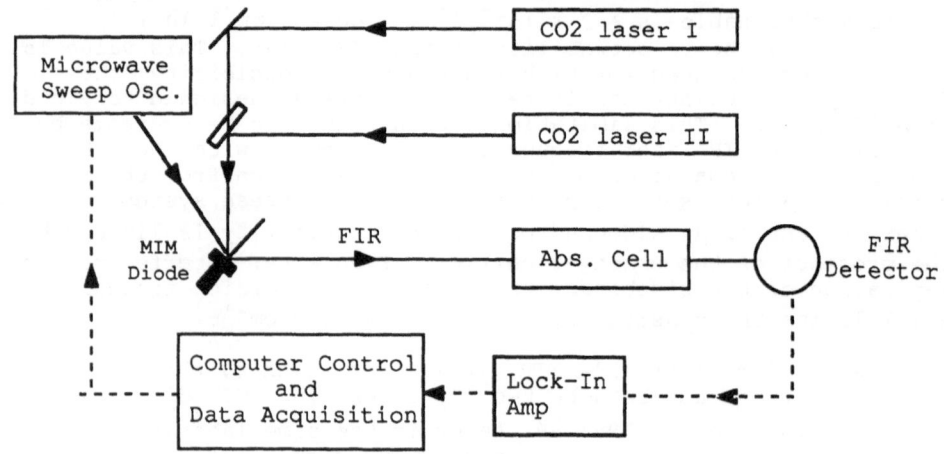

Fig. 1. Block diagram of the third order tunable FIR
spectrometer. $\nu_{FIR} = \nu_I - \nu_{II} \pm \nu_{\mu w}$.

our FIR is thus 2 X 25 kHz or 35 kHz. For further information on
frequency stabilization of CO_2 lasers, the reader is referred to
the article by L. Zink in this publication.

The frequency range of the spectrometer is from 0.3 to 6
THz. The lower limit is set by our bolometer and the upper limit
by the largest difference frequency between $^{12}C^{16}O_2$ lasers. Over
100 lines oscillate on each CO_2 laser plus ± 20 GHz tunability of
the microwave sweeper yield a 95% coverage of this region, the
only gaps occur above 4.5 THz. The range could possibly be
extended to 9.28 THz by substituting an ammonia laser for one of
the CO_2 lasers.

The spectrometer sensitivity is limited by the FIR power and
the sensitivity of the bolometer. 150 mW from each laser and 6-
10 dBm of microwave power are applied to the MIM diode,
generating about 10^{-7} W of FIR radiation. Although there are two
orders of magnitude less power than with the FIR laser sideband
technique, the noise sources are also less. Our typical minimum
detectable absorption is 10^{-4} in a 1 s time constant, which for a
meter long cell corresponds to 10^{-6} cm^{-1}. This is of comparable
sensitivity to the laser sideband technique.

MIM Diodes

The heart of the spectrometer is the metal-insulator-metal
diode. Used as a rectifier for many years, its possible use for
rectification of very high frequencies was noted as early as
1948.[25] The MIM diode has been used for laser frequency
measurements since 1969[26,27] and used to measure frequencies up
to 200 THz[28] and difference frequencies of visible light.[29] The
measurement at 200 THz enabled the most accurate determination of
the speed of light[30] and led to the re-definition of the meter.[31]

The diode consists of an electrochemically sharpened tungsten whisker (25 μm in diameter and typically 3 to 7 mm long) contacting a metal base. The metal base has a naturally occurring thin oxide insulating layer. In the generation of the difference frequency between two CO_2 lasers (second order) nickel is the best material for the base. In third order operation (difference frequency plus sidebands) cobalt appears to be superior. Good second order diodes generate very little third order FIR and vice versa. We also find that sharper tungsten whiskers produce more third order radiation. In general, third order diodes produce about 1/3 as much tunable FIR radiation as second order diodes and are more difficult to make, yet have the advantage of increased tunability. The physical properties of an ideal base material are not yet understood, hopefully in the future different materials or processing techniques will yield even greater TuFIR power.

Applications in Spectroscopy

The spectrometer at LENS has just been completed and the first results will be described below. However, a second order TuFIR spectrometer at the National Institute of Standards and Technology (NIST), in Boulder, Colorado has been in operation for several years and demonstrates the usefulness of this technique for high resolution spectroscopy. Spectra have been obtained for abundant, stable molecules such as CO,[20] providing accurate standards for Fourier transform spectrometers. Free radicals and transient species have been observed,[32] including the molecular ions H_2D^+ and H_3O^+. In addition to absorption spectroscopy the tunable FIR has been employed in an IR-FIR double resonance experiment;[33] for FIR Stark spectroscopy;[34] and in O_2 pressure broadening measurements.[35]

The first molecule observed with the third order system at LENS (during the first week of this school) was [13]CO. Traces of the J= 6 - 5 and the J= 26 - 25 lines are shown in figure 2. In all, seven transitions were measured, extending up to the J= 30 - 29 transition at 3.3 THz. New molecular parameters were determined with a 50 kHz (1σ) standard deviation of the fit.

Because of the high frequency accuracy of this technique, small pressure induced shifts of the transition frequency can be measured. Pressure broadening and shift measurements of the CH_3CN molecule have been performed on transitions ranging from J=31 - 30 at 570 GHz to the J= 81 - 80 transition at 1.5 Thz. Although this data is still being analyzed, it does serve to compare our instrument with one of the FIR laser sideband spectrometers. Figure 3 illustrates different K components of the J= 57 - 56 transition measured by the two techniques; eight components were observed with video detection with a FIR laser sideband technique[36] and 12 components were measured with phase sensitive detection using the LENS third order spectrometer (four of which are shown in figure 3). The FIR laser sideband experiment using phase sensitive detection yielded a signal to noise of 50,[17] slightly less than our S/N of 90. Not only were more k components measured with our system, but the accuracy of these measurements is about one order of magnitude higher. The

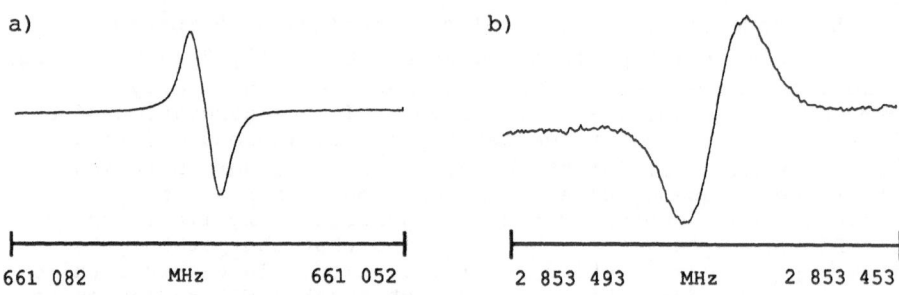

Figure 2. FIR rotational transitions of ^{13}CO observed with the TuFIR spectrometer at LENS. a) J=6-5 b) J=26-25.

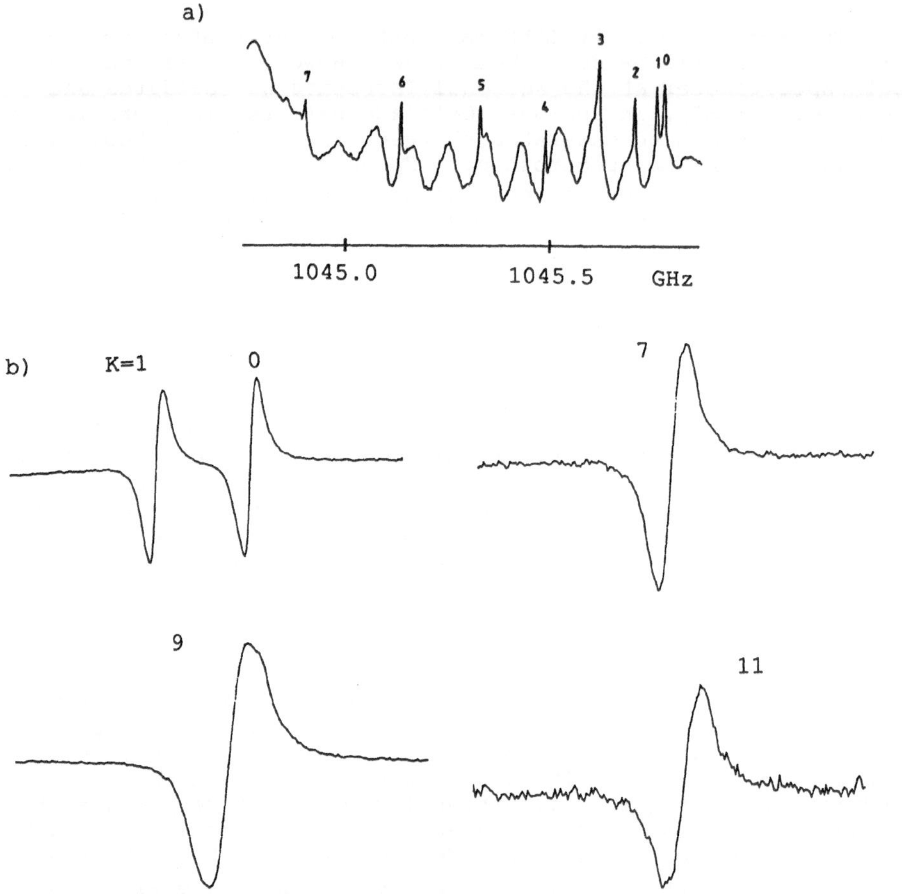

a)

1045.0   1045.5   GHz

b)   K=1   0   7

9   11

Figure 3. Observed transitions of the CH_3CN molecule; J=57-56, $\Delta K=0$. a) 8 K components seen in video detection with a FIR laser sideband spectrometer (from ref. 36). b) 4 of the 12 components observed in phase sensitive detection with the third order spectrometer at LENS.

146

accuracy in ref. 36 is 300 - 700 kHz (determined by the re-
settability of the FIR laser) compared to our accuracy of 55 -
100 kHz. Our uncertainty was limited by the S/N and the
transition linewidth.

SUMMARY

The new third order TuFIR spectrometer at the European
Laboratory for Nonlinear Spectroscopy has been described and the
first results presented. This spectrometer provides nearly
complete coverage from 0.3 to 6 THz, with 35 kHz accuracy and a
minimum detectable absorption of 0.01% in a 1 second time
constant. Plans for this system include the study of
astrophysically important molecules, the study of molecular ions,
and, taking advantage of the accuracy, measurements on trapped
atomic ions (see the article by G. Werth in this publication).

REFERENCES

1. D.M. Watson, R. Genzel, C.H. Townes, and J.W.V. Storey,
   Astrophys. J. 298, 316 (1985).
2a. B. Carli, F. Mencaraglia, A. Bonetti, M. Carlotti, and
   I. Nolt, Int. J. IR and MM waves 6, 149 (1985).
 b. W.A. Traub and K.V. Chance, Geophys. Res. Lett. 8, 1075
   (1981).
3. W.S. Heaps and T.J. McGee, J. Geophys. Res. 90, 7913 (1985).
4. J.W.C. Johns, J. Opt. Soc. Am. B 2, 1340 (1985).
5. P. DeNatale, L.R. Zink, F. Pavone, M. Inguscio, and
   K.M. Evenson, "Far Infrared Spectrum of 13CO", to be
   published in J. Mol. Spectrosc.
6. M. Inguscio, G. Moruzzi, K.M. Evenson, and D.A. Jennings,
   J. Appl. Phys. 60, R161 (1986).
7. K.M. Evenson, H.P. Broida, J.S. Wells, R.J. Mahler, and
   M. Mizushima, Phys. Rev. Lett. 21, 1083 (1968).
8. K.M. Evenson, R.J. Saykally, D.A. Jennings, R.F. Curl Jr.,
   and J.M. Brown, in Chemical and Biochemical Applications of
   Lasers: Vol. V edited by C.B. Moore (Academic Press, London,
   1980), pp 95-138.
9. K.M. Evenson, Farad. Disc. Roy. Soc. no. 71, (1981).
10. M. Inguscio, Physica Scripta 37, 699 (1988).
11. P. Helminger, J.K. Messer, and F.C. DeLucia, Appl. Phys.
   Lett. 42, 309 (1983).
12. R.L. Aggerwal, B. Lax, H.R. Fetterman, P.E. Tannenwald, and
   B.J. Clifton, J. Appl. Phys. 45, 3972 (1974).
13. R.L. Aggerwal and B. Lax, in Nonlinear Infrared Generation
   edited by Y.R. Shen (Springer-Verlag, Berlin, 1977),
   pp 19-80.
14. D.D. Bicanic, B.F.J. Zuidberg, and A. Dymanus, Appl. Phys.
   Lett. 32, 367 (1978).
15. W.A.M. Blumberg, D.D. Peck, and H.R. Fetterman, Appl. Phys.
   Lett. 39, 857 (1981).
16. J. Farhoomand, G.A. Blake, M.A. Frerking, and H.M. Pickett,
   J. Appl. Phys. 57, 1763 (1985).
17. G. Piau, F.X. Brown, D. Dangoisse, and P. Glorieux, IEEE J.
   Quant. Electron. QE-23, 1388 (1987).
18. K.M. Evenson, D.A. Jennings, and F.R. Petersen, Appl. Phys.
   Lett. 44, 576 (1984).

19. M. Inguscio, L.R. Zink, K.M. Evenson, and D.A. Jennings, "Accurate Frequency of the 119 μm Methanol Laser from Tunable Far Infrared Absorption Spectroscopy", submitted to IEEE J. Quant. Electron.

20. I.G. Nolt, J.V. Radostitz, G. DiLonardo, K.M. Evenson, D.A. Jennings, K.R. Leopold, M.D. Vanek, L.R. Zink, A. Hinz, and K.V. Chance, J. Mol. Spectrosc. 125, 274 (1987).

21. K.M. Evenson, M. Inguscio, and D.A. Jennings, J. Appl. Phys. 57, 956 (1985).

22. C. Freed and A. Javan, Appl. Phys. Lett. 17, 53 (1970).

23. F.R. Petersen, E.C. Beatty, and C.R. Pollock, J. Mol. Spectrosc. 102, 112 (1983).

24. L.C. Bradley, K.L. Soohoo, and C. Freed, IEEE J. Quant. Electron. QE-22, 234 (1986).

25. H.C. Torrey and C.A. Whitmer, Crystal Rectifiers (McGraw-Hill, New York, 1948), p. 7.

26. V. Daneu, D. Sokoloff, A. Sanchez, and A. Javan, Appl. Phys. Lett. 15, 398 (1969).

27. An excellent review of frequency measurements using MIM diodes as well as other devices is provided by: D.A. Jennings, K.M. Evenson, and D.J.E. Knight, Proc. IEEE 74, 168 (1986).

28. K.M. Evenson, D.A. Jennings, F.R. Petersen, and J.S. Wells, in Laser Spectroscopy III, edited by J.L. Hall and J.L. Carlsten (Springer-Verlag, Berlin, 1977), pp 56-58.

29. R.E. Drullinger, K.M. Evenson, D.A. Jennings, F.R. Petersen, J.C. Berquist, and L. Berkins, Appl. Phys. Lett. 42, 137 (1983).

30. D.A. Jennings, C.R. Pollock, F.R. Petersen, R.E. Drullinger, K.M. Evenson, J.S. Wells, J.L. Hall, and H.P. Layer, Opt. Lett. 8, 136 (1983).

31. Comptes Rendus des Seances de la 17e CGPM, BIPM, Sevres, France (1983).

32. For a more complete listing and references see: K.M. Evenson, D.A.Jennings, and M.D. Vanek, in Frontiers of Laser Spectroscopy of Gases (NATO ASI Series C; Vol. 234), edited by A.C.P. Alves, J.M. Brown, and J.M. Hollas (Kluwer Academic Publishers, Dordrecht, 1988).

33. M. Inguscio, L.R. Zink, K.M. Evenson, and D.A. Jennings, Opt. Lett. 12, 867 (1987).

34. L.R. Zink, D.A. Jennings, K.M. Evenson, A. Sasso, and M. Inguscio, J. Opt. Soc. Am. B 4, 1173 (1987).

35. D.A. Jennings, K.M. Evenson, M.D. Vanek, I.G. Nolt, J.V. Radostitz, and K.V. Chance, Geophys. Res. Lett. 14, 722 (1987).

36. F.X. Brown, D. Dangoisse, and J. Demaison, J. Mol. Spectrosc. 129, 483 (1988).

HIGH-RESOLUTION LASER SPECTROSCOPY

IN THE UV/VUV SPECTRAL REGION

Sune Svanberg

Department of Physics
Lund Institute of Technology
P.O. Box 118
S-221 00 Lund, Sweden

INTRODUCTION

In the present paper we will discuss high-resolution laser spectroscopy of free atoms and especially focus on the UV and VUV wavelength region. The demands for a high spectral resolution in this wavelength region are the same as in the more workable visible region, where a great deal of work has been performed. Frequency conversion to short wavelengths can be performed much more easily for pulsed lasers, which, however, necessarily have a much larger linewidth than the single-mode CW systems available at longer wavelengths. We will discuss recent UV/VUV work using pulsed lasers performed in our group to make high resolution readily available at "difficult" wavelengths, and will also illustrate the status of narrow-band CW work at short wavelengths. A wide variety of techniques is available and different techniques are discussed in monographs, such as those given in Refs. 1 and 2. High-resolution spectroscopy, free of Doppler broadening, is required for:

*   Accurate energy level position determination

*   Studies of fine- and hyperfine structures,
    isotopic shifts, Zeeman and Stark effects

*   Metrology   Frequency standards
                       Fundamental measurements

The oldest high-resolution techniques for atomic spectroscopy rely on the observation of rf resonances or atomic coherences. The Doppler broadening ($\Delta\nu/\nu = 10^{-6}$) is present but has a negligible effect because of the low frequency of the studied transitions (1-1000 MHz). Techniques can be listed as follows:

*   Atomic Beam Magnetic Resonance (ABMR)
*   Optical Pumping (OP)
*   Optical Double Resonance (ODR)
*   Level Crossing Spectroscopy (LC)
*   Quantum Beat Spectroscopy (QBS)

The last three techniques (ODR[3-5], LC[6-8,5], and QBS[9,10]) are very

Applied Laser Spectroscopy, Edited by W. Demtröder and
M. Inguscio, Plenum Press, New York, 1990

useful for studying short-lived excited states. The principles of the
techniques are schematically shown in Fig. 1. The methods were first
developed for use with classical (Doppler-broadened) light sources, but
their applicability has been greatly extended by using lasers as excitation sources. Since *the resolution is not limited by the linewidth of the
light source* but only by the natural radiative linewidth associated with
the finite lifetime τ, *the lasers can be broadbanded and pulsed.* Due to
the high intensity of the laser radiation, step-wise excitation can be
used to reach otherwise inaccessible states. In this way high-resolution
studies could be extended to numerous highly excited alkali atom states
employing CW multi-mode dye lasers. (For a review, see Ref. 11). As an
example ODR signals in the Paschen-Back region of the 7 and 8s $^2S_{1/2}$
states of ^{39}K are shown in Fig. 2[12]. Here, the number of resonances (4 =
2I+1) yields the nuclear spin I, the center of gravity the Landé g_J
factor, the signal spacing the magnetic dipole interaction constant and
the signal halfwidth $\Delta\nu$ the natural radiative lifetime $\tau = 1/\pi\Delta\nu$.

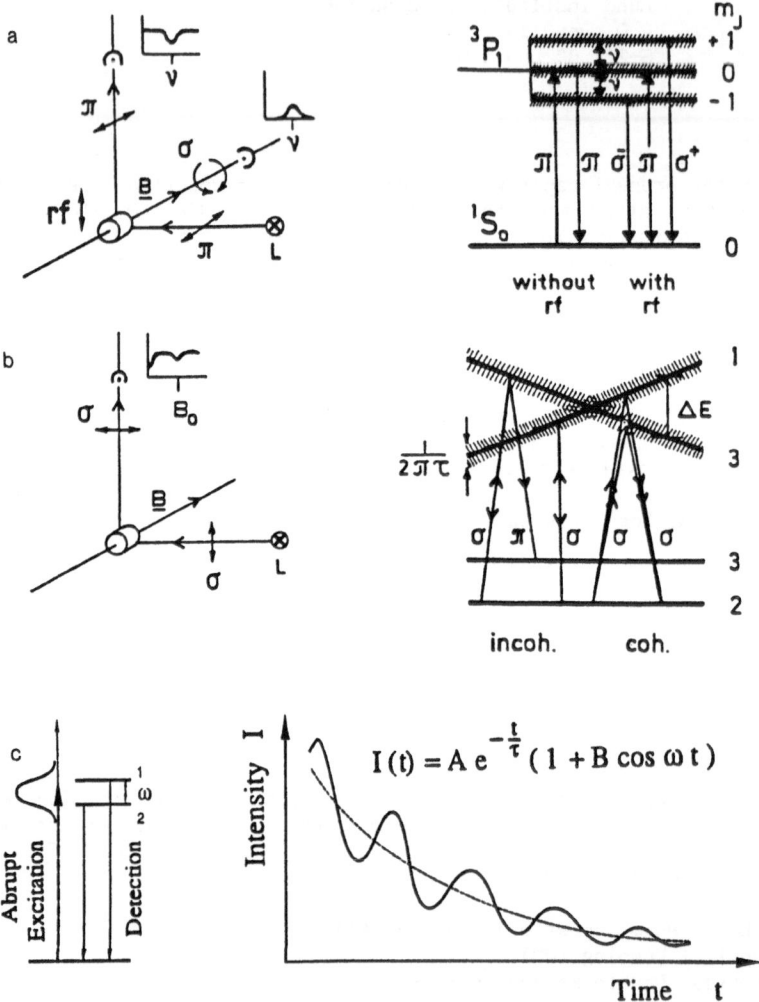

Fig. 1. Schematic principles of a) optical double resonance, b)
level-crossing spectroscopy and c) quantum beat spectroscopy.

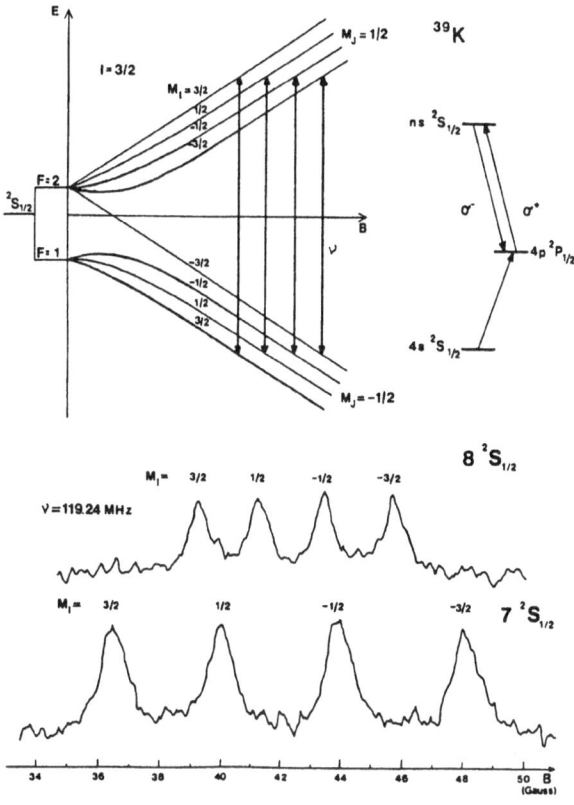

Fig. 2. Breit-Rabi diagram for an $^2S_{1/2}$ state and experimental ODR recordings for the 7 and 8s $^2S_{1/2}$ states of ^{39}K (I=3/2). (From Ref. 12.)

ODR and LC measurements can readily be performed using pulsed lasers, which also make it possible to achieve a resolution below the natural radiation width limit. This is obtained by using "conditional sampling", i.e. restricting the observation to "old" atoms[13-16]. An example showing narrowed level crossing signals in the 9 $^2D_{3/2}$ state of ^{133}Cs is shown in Fig. 3[17].

A different class of high-resolution techniques are the *Doppler-free laser spectroscopic techniques relying on single-mode lasers*, which should preferably have a linewidth small enough not to add to the natural radiative linewidth. A list of techniques is given below.

* Collimated atomic/ionic beams
         Perpendicular irradiation
         Collinear irradiation; kinematic compression
* Saturation spectroscopy
         Absorption detection (POLINEX, FM)
         Fluorescence detection
         Optogalvanic/opto-acoustic detection
* Two-photon absorption
* Trapped and cooled atoms/ions

Such techniques are schematically described in Fig. 4.

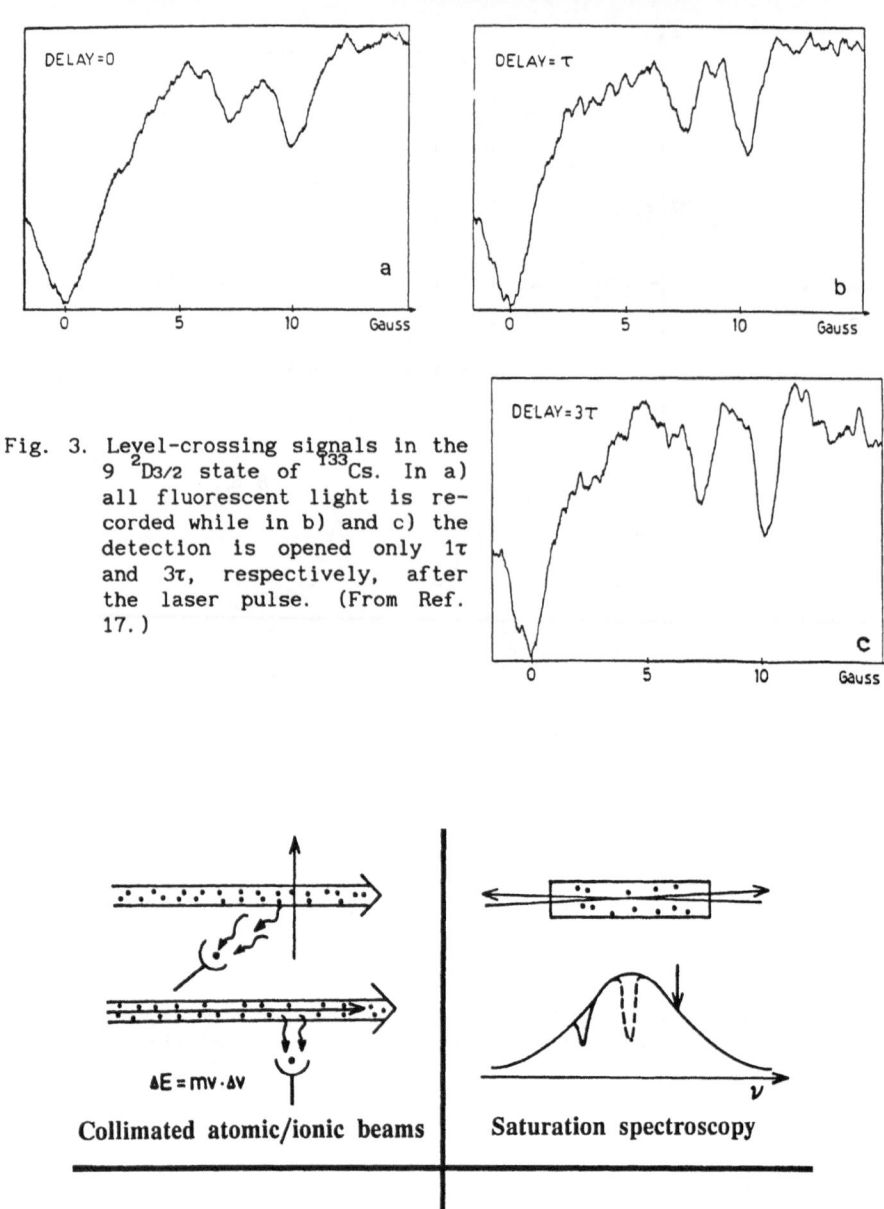

Fig. 3. Level-crossing signals in the $9\ ^2D_{3/2}$ state of ^{133}Cs. In a) all fluorescent light is recorded while in b) and c) the detection is opened only 1τ and 3τ, respectively, after the laser pulse. (From Ref. 17.)

Fig. 4. Schematic diagrams of high-resolution laser spectroscopic techniques.

A variety of methods is available for the generation of short-wavelength coherent radiation. Usually, tunable radiation in the UV or VUV spectral region is generated using nonlinear optics techniques.

Sum-frequency generation in crystals

Frequency doubling or mixing in nonlinear crystals is an efficient way of generating radiation in regions where the crystals are transparent and phase-matching conditions can be obtained. A review of modern materials and techniques can be found in Ref. 18. Of particular importance is β-barium borate, which provides frequency doubling down to 205 nm and tripling down to 197 nm. Frequency tripling in this material can be performed with high efficiency[19]. We have been able to generate almost 10 mJ of tunable radiation in the region around 200 nm using this technique applied to radiation from a Nd:YAG pumped dye laser.

Frequency conversion in gases

Gases are transparent at shorter wavelengths than crystals and different non-linear techniques can be used. A simple technique not requiring phase-matching is stimulated Raman scattering in high-pressure H_2 gas[20]. By generation of successively higher anti-Stokes components a photon energy increase of 4155 cm^{-1} or ≈ 0.5 eV can be gained in each step. This means, that using a primary laser wavelength of 200 nm the first anti-Stokes Raman component will be at 185 nm and the second anti-Stokes component at 170 nm, as indicated in Fig. 5. In this figure the on-set of absorption, due to the oxygen Schumann-Runge bands, requiring evacuation or nitrogen flushing of the optical paths, is also shown[21].

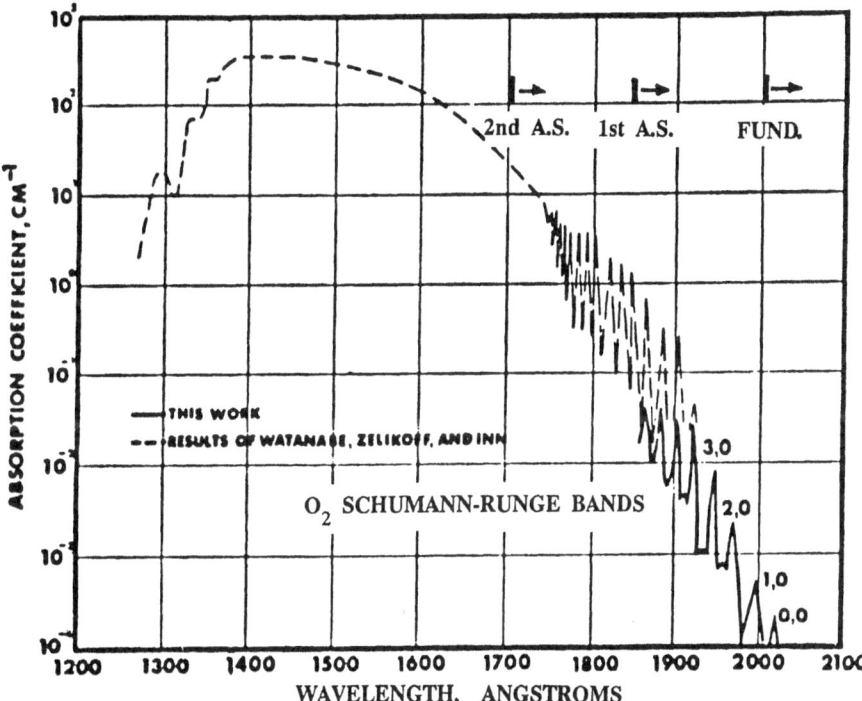

Fig. 5. Absorption of molecular oxygen with indications of anti-Stokes Raman-shifted laser frequencies. (From Ref. 21.)

Sum- and difference frequency mixing can also be performed in noble gases, such as Kr and Xe, and also in metal vapors such as Hg and Mg. The efficiency is greatly improved if resonance enhancement through two-photon resonances in the gas can be obtained. For sum frequency mixing phase-matching must be achieved. The index of refraction of the gas can be manipulated by adding noble gas to the converting metal vapor. Most of the range 100-200 nm can be covered in this way. Even shorter wavelengths can be achieved by high (odd) harmonic generation in the noble gases. The field of non-linear frequency conversion in gases is covered in Refs. 22,23.

Harmonic generation in an electron beam undulator

The generation of continuously tunable coherent VUV radiation is a long-term goal for free-electron laser (FEL) research. However, there are many obstacles on the way. One of the outstanding problems is to develop high-reflectivity mirrors which can withstand the high average light power from the undulator magnet. This has stimulated the investigation of a mirrorless device called an optical klystron. Odd harmonic generation of pulsed laser radiation focused onto the relativistic beam in the undulator magnet has been demonstrated[24,25]. The strong fields associated with the laser pulse cause a microbunching of the electrons which will start to radiate coherently. There is a potential for very high harmonic generation. We have recently observed third harmonic generation in a normal undulator magnet at the MAX synchrotron radiation facility in Lund, employing a Nd:YAG laser at 1.06 μm. Recordings are shown in Fig. 6 illustrating the increase in intensity from the electron bunch hit by the laser pulse[26]. The observation was performed through a narrow-band mono-chromator, and the coherent radiation at 355 nm was shown to be sharper than 0.15 Å.

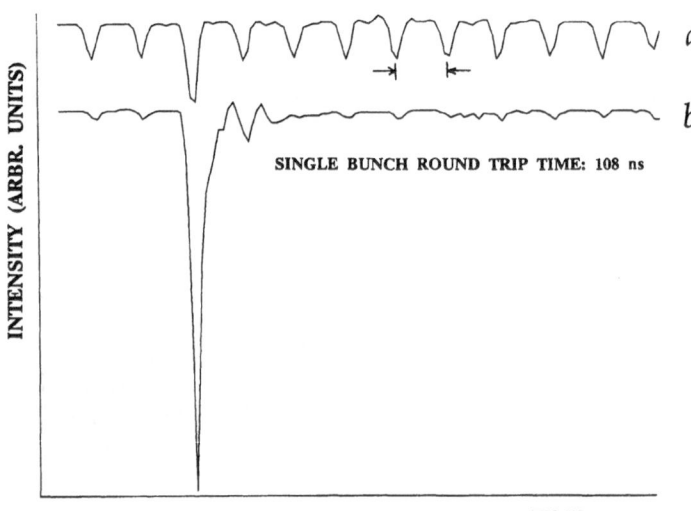

Fig. 6. Experimental observation of third harmonic generation from 1.06 μm radiation, irradiated into an undulator magnet. In a) the enhancement of the third bunch undulator signal is seen. In b) conditions are better optimized. (From Ref. 26.)

For some time, intra-cavity frequency doubling has been available for commercial ring dye lasers, making the high-resolution laser spectroscopic techniques readily applicable in the wavelength region from about 250 to 350 nm. Output powers of up to 10 mW are available. As an example of non-trivial collimated atomic beam spectroscopy using such equipment, a recording of the 276 nm line in copper, connecting the $3d^9 4s^2 \; ^2D_{3/2}$ metastable state with the $5p \; ^2P_{1/2}$ state is shown in Fig. 7[27]. The metastable atoms are produced in a hollow-cathode discharge from which an atomic beam effuses through a small hole to form a moderately collimated atomic beam.

The CW frequency doubling can also be performed in a separate build-up cavity, which can be a ring cavity of basically the same construction as the one used for the laser. The separation of the generation and conversion steps has advantages that are even more evident if non-linear CW frequency mixing is employed instead. As an example of the use of an external build-up mixing cavity, we choose the well-known experiment on the 1s - 2s two-photon transition in atomic hydrogen at 243 nm, performed by Hänsch and coworkers[28,29]. The setup and recordings are shown in Fig. 8. In an external, servo-controlled build-up cavity[30] 351 nm single-mode radiation from an Ar^+ laser is mixed in a KDP crystal with 790 nm radiation from a single-mode dye laser to produce several mW at 243 nm. The beam is back-reflected through the hydrogen measuring cell to achieve Doppler-free two-photon transitions at frequencies that can be related to a secondary tellurium standard. A laser beam from a tellurium saturation spectrometer operating at 486 nm is frequency doubled in a urea crystal

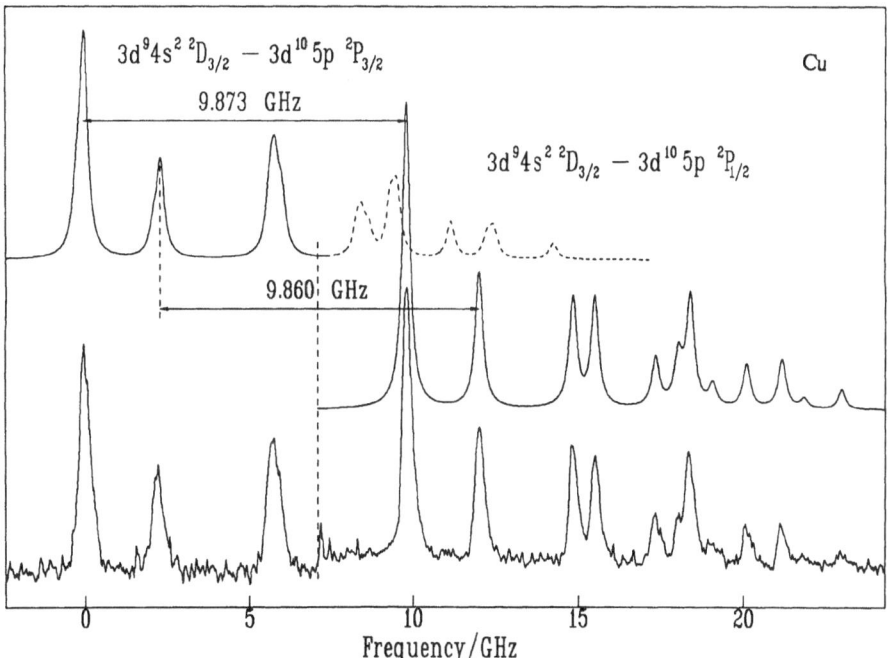

Fig. 7. Experimental recording (bottom) of the $3d^9 4s^2 \; ^2D_{3/2} \rightarrow 3d^{10} 5p \; ^2P$ transition. The upper spectra are theoretical simulations. (From Ref. 27.)

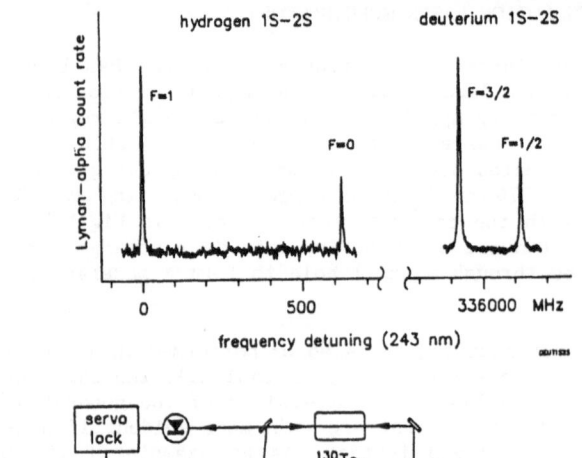

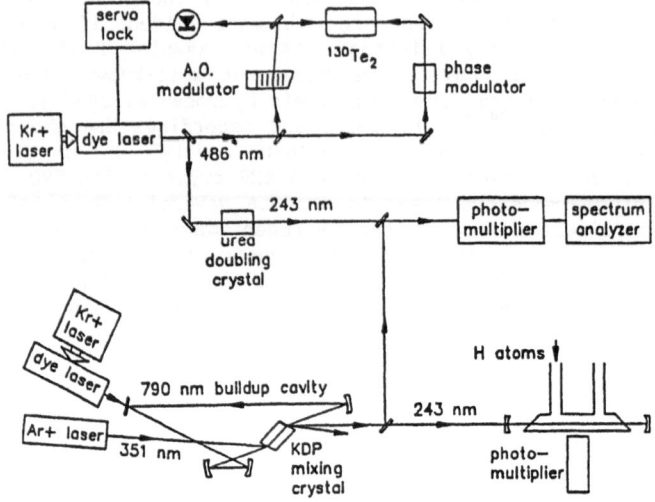

Fig. 8. Two-photon spectra of the 1s ⟶ 2s transition in hydrogen and deuterium (top). Experimental setup for hydrogen
1s ⟶ 2s experiments. (From Refs. 28 and 29.)

to 243 nm and is mixed with the radiation from the hydrogen setup to
produce an easily measurable beat note in the radio frequency regime. The
spectra for hydrogen and deuterium with proper frequency reference provide an accurate value for the ground state Lamb shift and can also be
combined with other data for an evaluation of the Rydberg constant.

As a further illustrative example of CW laser spectroscopy at even
shorter wavelengths, the frequency standard work on a single cooled and
trapped Hg ion, performed by Wineland and coworkers at NIST, Boulder, is
chosen. In Fig. 9 the relevant transitions in the Hg ion are shown,
including the 194 nm very strong cooling and detection transition and the
282 nm "clock" transition, which is extremely sharp due to the long lifetime of the upper $^2D_{5/2}$ state. The setup for generating the 194 nm radiation[31] is also included in the figure. First, a 515 nm single-frequency
Ar$^+$ ion laser is frequency doubled to 257 nm in a build-up cavity. It is
then mixed in a KB5 crystal placed in a build-up cavity for 792 nm
radiation, injected from a ring dye laser. Few µW of 194 nm radiation can

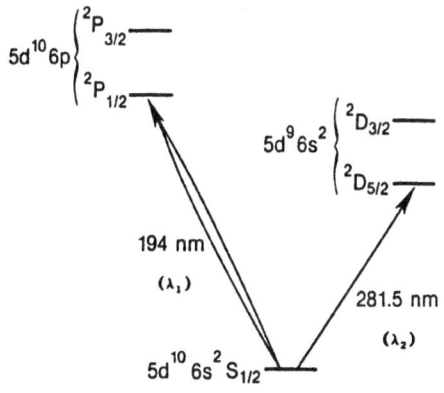

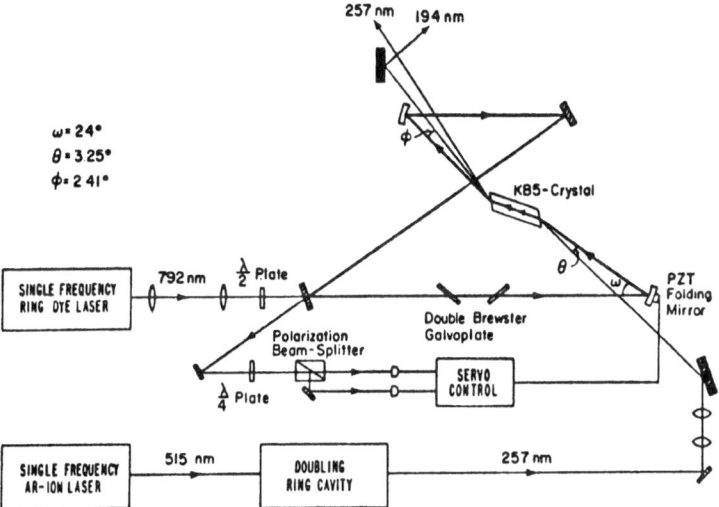

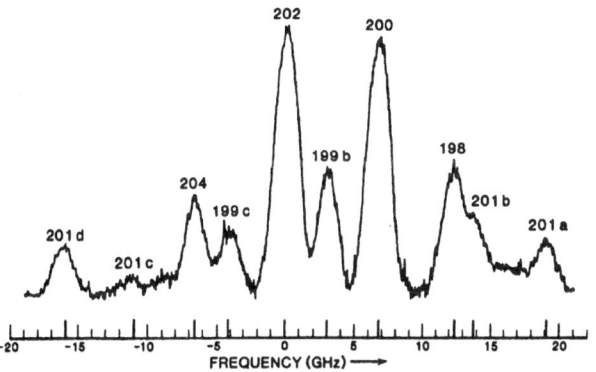

Fig. 9. Partial energy level scheme for Hg[+] (top). Laser setup for generating narrow-band CW radiation at 194 nm (middle). Experimental recording of the 6s $^2S_{1/2} \longrightarrow$ 6p $^2P_{1/2}$ line in Hg[+] (bottom). (From Refs 31 and 32.)

be obtained in this way to easily produce Doppler-broadened scans of the Hg ionic line, as shown in the figure.

When this radiation is applied to the single ion in the trap the induced fluorescence will show a typical "shelving" behavior if the clock transition is simultaneously driven with a highly stabilized frequency-doubled dye laser. The fluorescence intensity is either high, if the ion is available in the ground state, or zero, if it has been "shelved" in the long-lived $^2D_{5/2}$ state, as shown in Fig. 10. A high-resolution spectrum of the clock transition can be made by making a histogram of the shelving events as a function of laser detuning, as shown in the lower part of Fig. 10. A linewidth of 86 Hz has been achieved[32].

CW single-mode radiation at very low powers at even shorter wavelengths can be generated by frequency conversion in vapors. Thus, about 10^{-13} W at 144 nm was achieved already in 1983 by frequency tripling in Mg vapor[33]. The efficiency has now been greatly improved, as illustrated in the paper by R. Wallenstein in this volume.

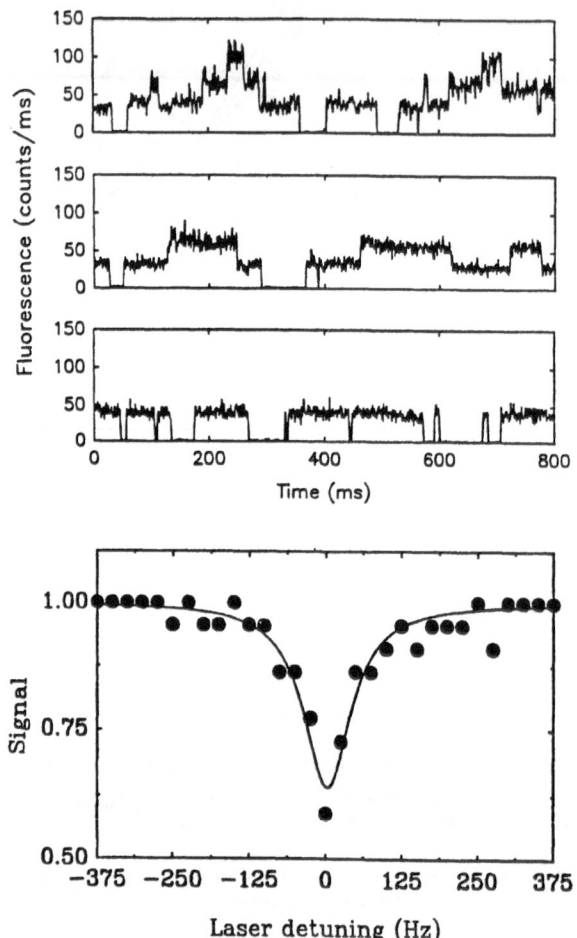

Fig. 10. Recordings of "quantum jumps" in 1, 2 and 3 ions (top) and tuning curve through the "clock" transition (bottom). (From Refs. 31 and 32.)

As we have seen, narrow-linewidth CW radiation at short UV and VUV wavelengths can be generated at low powers and with quite complex arrangements. As already pointed out, it is much easier to attain such difficult wavelengths using pulsed laser systems. A high resolution can be obtained with such broadbanded sources only if the excitation is combined with the use of a technique using resonances or atomic coherences, i.e. optical double resonance, level crossing or quantum beat spectroscopy. However, such a combination provides convenient general purpose spectroscopy in wide wavelength ranges, as recently demonstrated in our laboratory. However, the ODR, LC and QBS techniques only allow small energy intervals within the same atom to be determined (e.g. fine and hyperfine structures, Zeeman and Stark splittings). Isotopic shift or scalar Stark interactions must be determined by true wavelength scanning with a narrow-band laser. Also, for the purpose of cooling atoms or ions, sharp frequency lasers are mandatory.

In the 70's high-resolution laser spectroscopy of atomic splittings was expanded quite substantially by combining stepwise excitation with CW broadband and multimode sources with resonance techniques, providing a resolution down to the Heisenberg limit[34,35,11]. These same techniques, now adopted for pulsed excitation, can, in the same way, provide a convenient and general purpose gateway into UV/VUV high-resolution laser spectroscopy. Basically, the three techniques provide the same type of information and the choice of method will be determined by practical aspects. QBS requires good time resolution in both the laser and the detection electronics. In the ODR technique rf fields of suitable frequency must be available and at sufficient strength, which can be a limitation for short-lived states. In many respects, the level crossing method is the least demanding technique. We will now illustrate the use of the pulsed LC and the ODR methods in studies of Cu, Ag and Mg atoms at short excitation wavelengths.

Copper

The copper atom has a $4s\,^2S_{1/2}$ ground state, as has the potassium atom, but because of the easily perturbed $3d^{10}$ subshell the energy level system of copper is much more complicated than that of potassium. Levels belonging to the $3d^9 4s4p$ configuration strongly perturb the alkali-like doublet spectrum[36], leading to anomalous lifetime values in the ns and nd Rydberg sequences[37]. The fine structures in the np sequence are extremely perturbed and so are the lifetimes[36]. Clearly, hyperfine-structure studies in this sequence are of considerable interest. In 1966 Ney reported data from an accurate level-crossing investigation of the $4p\,^2P_{3/2}$ level, which was excited by a hollow-cathode lamp[38]. The pulsed level-crossing experiments now reported are the only ones with which it is believed that data will be yielded on the higher $^2P_{3/2}$ levels of Cu with a reasonable level of effort.

Resonance scattering of light by atoms is governed by the Breit formula[39]. The total intensity of scattered light with polarization vector **g** following excitation by light with another polarization vector **f** is given by the expression

$$ S = C \cdot \sum_{\substack{mm' \\ \mu\mu'}} \frac{f_{\mu m} \cdot f_{m\mu'} \cdot g_{m'\mu} \cdot g_{\mu'm'}}{\frac{1}{\tau} + \frac{1}{\hbar}(E_\mu - E_{\mu'})} $$

159

The symbols $f_{\mu m}$ denote matrix elements of the type $\langle\mu|\mathbf{f}\cdot\mathbf{r}|m\rangle$ where μ and m are wavefunctions for substates of an excited state and a ground state, respectively. The lifetime of the excited state is τ and the total energy of a substate is E_μ. The summation is performed over all ground and excited substates. From the Breit formula it follows that when two levels intersect, the denominator becomes resonant and a change in the intensity of the scattered light can be detected. When the detection and excitation polarizations are perpendicular to the magnetic field, such resonances can be experimentally seen only for intersecting levels whose magnetic quantum numbers differ by two units, as the four-fold product in the numerator is then non-zero. This is due to the fact that both states must be coherently excited by the same (σ) photon.

When performing an LC experiment, an external magnetic field is applied, the magnetic sublevels are scanned and the scattered light intensity is recorded as a function of the varying magnetic field.

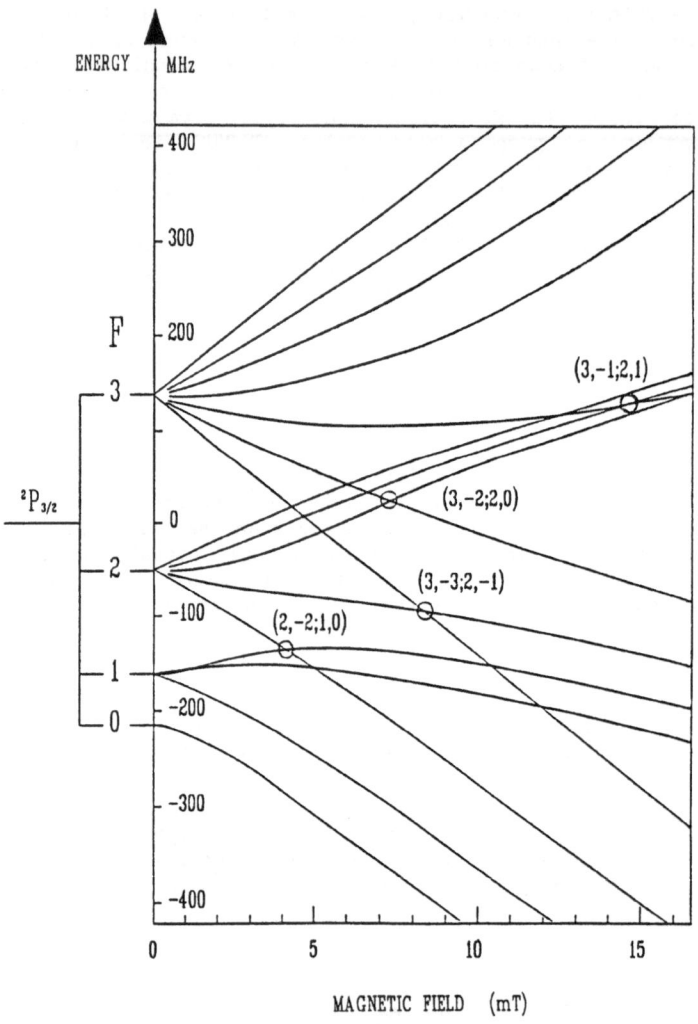

Fig. 11. Magnetic field dependence of the magnetic hyperfine substates for the 5p $^2P_{3/2}$ state of ^{63}Cu. $\Delta\mu=2$ crossings are indicated using the symbol $(F,\mu;F',\mu')$. (From Ref. 40.)

An energy level diagram for the 5p $^2P_{3/2}$ state of ^{63}Cu is given in Fig. 11. There are two stable isotopes of copper. Both have the nuclear spin I=3/2. The abundances of the isotopes are 69% for ^{63}Cu and 31% for ^{65}Cu. As can be seen from Fig. 11, in zero magnetic field, all magnetic sublevels of a hyperfine component have the same energy. A large zero-field or Hanle signal can thus be seen[6]. As shown in Fig. 11, four $\Delta\mu=2$ crossings occur in intermediate fields. The fourth of these is generally difficult to observe experimentally. The positions of the crossings fully determine both the magnetic dipole and electric quadrupole interaction constants, normalized to the g_j factor.

The experimental setup used in the pulsed LC experiments is illustrated in Fig. 12. The laser system used was a Quantel Datachrome 5000. It consists of a YG 581c Q-switched Nd:YAG laser and a TDL 50 dye laser, which was operated on a mixture of Rhodamine 640 and Rhodamine 610 in methanol. The dye laser output at 607.2 nm was frequency-tripled to 202.4 nm, which is the wavelength needed to reach the second excited $^2P_{3/2}$ state in copper. The atoms were produced as a beam from an oven in a vacuum system. Two pairs of Helmholtz coils produced a magnetic field. The intensity fluctuations of pulsed lasers due to pulse-to-pulse variations are normally of the order of 10%. This can be a problem when detecting signals smaller than these fluctuations. One way to compensate for these pulse-to-pulse variations is to monitor the laser intensity on a linear diode and normalize the detected signal. A better way is to use two fluorescent light detectors, one parallel and one perpendicular to the magnetic field, and divide the two signals. If the polarizations are chosen correctly this will double the signal amplitude and normalize the fluorescence intensity, which effectively takes care of the intensity fluctuations. The fact that normal photomultiplier tubes are slightly nonlinear will cause two detectors that are not perfectly matched, to

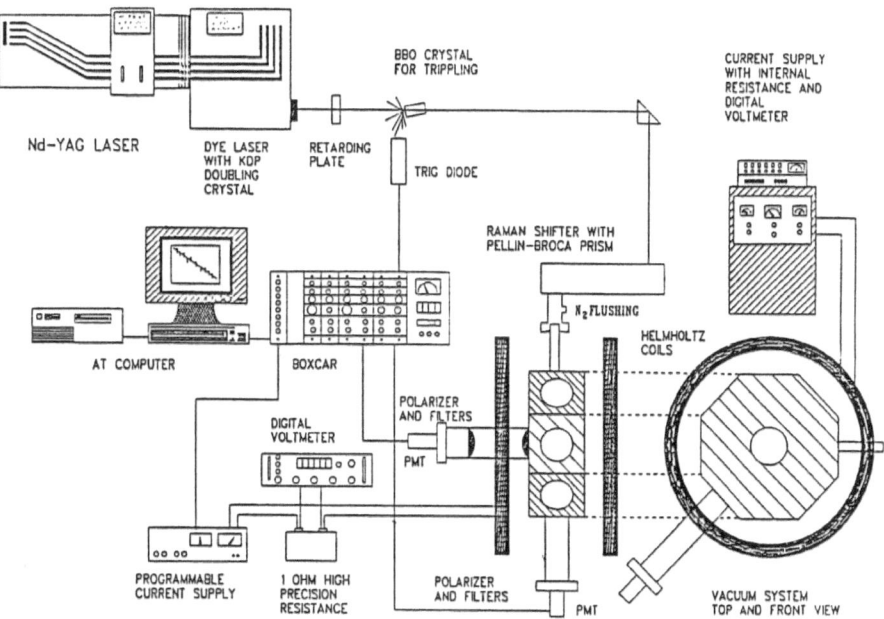

Fig. 12. Experimental setup for pulsed level-crossing measurements. (From Ref. 40.)

161

respond slightly differently to intensity fluctuations and the normalization will not be perfect. In order to compensate for this differential
nonlinearity we used a scheme proposed by Wolf & Tiemann.[41] The idea is
to record this differential nonlinearity and compensate for it. One way
to do this is to leave the experiment on stand-by and let both detectors
measure the fluorescence signal from the same sequence of laser pulses.
The response from one detector is plotted against that of the other and a
polynomial, for example, is fitted to this sequence of measured points.
This polynomial can be considered as a calibration curve which gives the
signal from one detector as it would have been if measured in the other.
When performing the actual experiment, the signal from one detector and
the recalculated signal from the other, will have the same response to
intensity fluctuations. A Stanford Research SR 265 boxcar integrator was
used to process the detected fluorescence. This was connected to an IBM-
compatible AT computer. The temporal gates on the boxcar integrator were
set so as to be long enough to record the whole fluorescence signal.

In Fig. 13 an experimental recording for the $5\ ^2P_{3/2}$ state is shown
together with a theoretical curve calculated with fitted magnetic dipole
(a) and electric quadrupole (b) hyperfine parameters. Final data for the
^{63}Cu isotope are included in the figure. The dipole interaction constant
is 4 times larger than the theoretical Hartree-Fock value, reflecting
strong interaction with the $3d^94s4p$ configuration.

Silver

Silver has a similar electronic configuration to copper. However,
the $4d^{10}np$ sequence is less perturbed since the $4d^95s5p$ configuration is
higher in energy relative to the P states of interest[42]. Since only the
lowest $^2P_{3/2}$ state had been investigated with non-laser techniques prev-

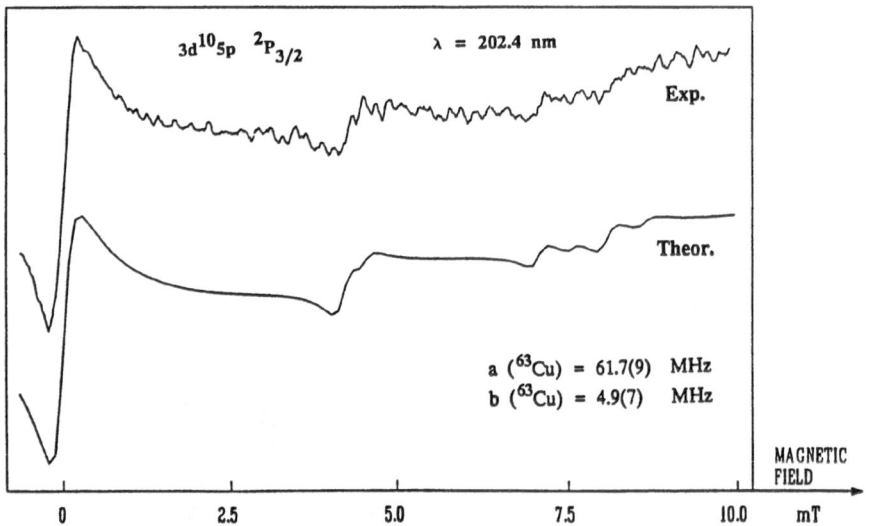

Fig. 13. A magnetic field scan over a large interval showing the
magnetic field dependence of the fluorescence light with
the zero-field Hanle effect and high-field crossing
signals for the $5p\ ^2P_{3/2}$ state. A theoretical curve
obtained from a computer calculation with the final
fitted parameters is included. (From Ref. 40.)

iously, we started our study with the second excited state (n=6) reachable with 206 nm radiation. An LC investigation revealed a structure that was only partially resolvable with crossing signals from both stable silver isotopes [107]Ag and [109]Ag. It turned out that a pulsed ODR experiment performed in the Paschen-Back region of the hyperfine structure yielded higher precision data than the LC measurements. Experimental recordings are shown in Fig. 14[43]. As in Fig. 2, 2I+1 signals are expected, and two signals, consistent with the known spin of 1/2 for both silver isotopes are observed. By restricting the observation to late times, a narrowing of the signals is observed providing a sub-natural resolution, as discussed in connection with Fig. 3. Data derived for the light silver isotope are given in the figure.

Magnesium

As a final example of a pulsed LC experiment we choose a lifetime determination for the 3s4p $^1P_1$ state of magnesium using the Hanle signal[44]. By observing the depolarization of the resonance radiation around zero magnetic field, an accurate lifetime can be determined for this short-lived state. In direct measurements of the fluorescence decay using time-resolved spectroscopy, a rather large deconvolution error occurs when excitation with a standard Q-switched pulse (5-10 ns long) is used. An experimental Hanle effect recording for the 203 nm line is shown in Fig. 15.

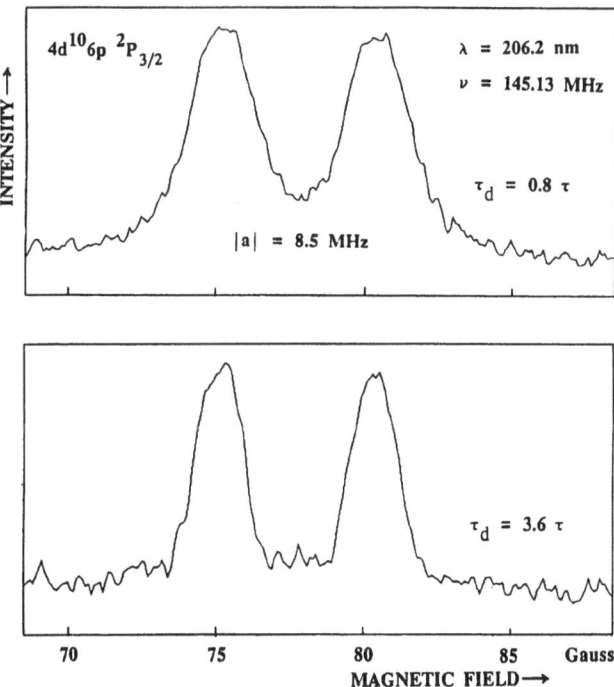

Fig. 14. Optical double resonance recordings at two different detection delays for the 6p $^2P_{3/2}$ state of silver. (From Ref. 43.)

DISCUSSION

The purpose of the present paper was to illustrate how the techniques of high-resolution laser spectroscopy which, in most cases started off with experiments on the yellow sodium line, can be put to work at short wavelengths, where there are many practical obstacles. It is illustrated how narrow-band CW sources through sophisticated techniques can be made useful for Doppler-free laser spectroscopy. However, since the techniques are presently elaborate and frequently specific to the particular wavelength needed, they have often been reserved for use by major groups working on fundamental atomic physics problems. Attention is drawn to the fact that while awaiting the commercial availability of advanced narrow-band sources at UV and VUV wavelengths, much simpler pulsed laser systems can be used in conjunction with "classical" resonance or coherence techniques. In this way, the same or a better spectral resolution than that

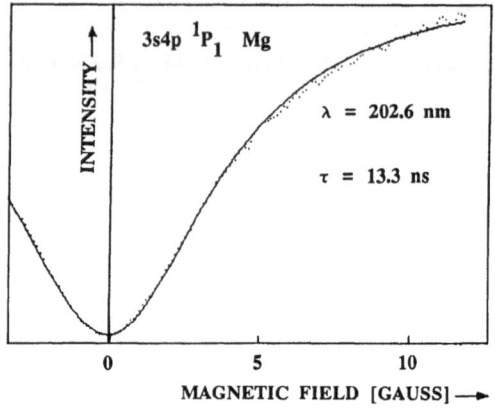

Fig. 15. Hanle effect recording for the 3s4p $^1P_1$ state of magnesium. (From Ref. 44.)

offered by the CW laser techniques can be attained in many spectroscopic applications. Also, for measurements of short radiative lifetimes the zero-field level-crossing (Hanle) technique is shown to be very competitive to time-resolved techniques requiring laser and detection equipment with a very high temporal resolution.

ACKNOWLEDGEMENTS

The author acknowledges very fruitful cooperation with a large number of past and present coworkers and graduate students in the field of basic atomic laser spectroscopy. This work was supported by the Swedish Natural Science Research Council.

REFERENCES

1. W. Demtröder, *Laser Spectroscopy*, 3rd corrected printing, Springer Series in Chemical Physics, Vol. 5 Springer, Heidelberg (1988).
2. S. Svanberg, *Atomic and Molecular Spectroscopy*, Springer Series in Atoms and Plasmas, Vol. 6, Springer, Heidelberg, in press.
3. A. Kastler, and J. Brossel, Comp. Rend. **229**:1213 (1949)
4. J. Brossel, and F. Bitter, Phys. Rev. **86**:368 (1952).
5. W. Happer, and R. Gupta, in *Progress in Atomic Spectroscopy, Part A*, W. Hanle and H. Kleinpoppen, eds., Plenum, New York (1979), p. 391.
6. W. Hanle, Z. Phys. **30**:93 (1924).
7. F.D. Colgrove, P.A. Franken, R.R. Lewis, and R.H. Sands, Phys. Rev. Lett. **3**:420 (1957).
8. P. Franken, Phys. Rev. **121**:508 (1961)
9. J.N. Dodd, and G.W Series, in *Progress in Atomic Spectroscopy, Part A*, W. Hanle and H. Kleinpoppen, eds., Plenum, New York (1978), p. 639.
10. S. Haroche, in *Topics in Applied Physics, Vol. 13*, K. Shimoda, ed., Springer, Heidelberg (1976), p. 253.
11. S. Svanberg, in *Laser Spectroscopy III*, J.L. Hall and J.L. Carlsten, eds., Springer Series in Optical Sciences, Vol. 7, Springer, Heidelberg (1977), p. 187.
12. G. Belin, I. Lindgren, L. Holmgren, and S. Svanberg, Phys. Scr. **12**: 287 (1975).
13. H. Figger, and H. Walther, Z. Phys. **267**:1 (1974).
14. J.S. Deech, P. Hannaford, and G.W. Series, J. Phys. **B7**:1311 (1975).
15. P. Schenk, R.C. Hilborn, and H. Metcalf, Phys. Rev. Lett. **31**:189 (1973).
16. P.D. O'Brien, P. Meystre, and H. Walther, in *Advances in Atomic and Molecular Physics*, **21**, D. Bates and B. Bederson, eds., Academic Press, Orlando, (1985)
17. J. Larsson, L. Sturesson, and S. Svanberg, Phys. Scr. **40**:165 (1989)
18. R.C. Eckardt, Y.X. Fan, M.M. Fejer, W.J. Kozlovsky, C.N. Nabors, R.L. Byer, R.K. Route, and R.S. Feigelson, in *Laser Spectroscopy VIII*, W. Persson and Sune Svanberg, eds., Springer, Heidelberg (1989).
19. W.L. Glab, and J.P. Hessler, Appl. Opt. **26**:3181 (1987).
20. A.P. Hickman, J.A. Paisner, and W.K. Bishel, Phys. Rev. **A33**:1788 (1986).
21. B.A. Thompson, P. Harteck, and R.R. Reeves, J. Geophys. Res. **68**:6431 (1963).
22. W. Jamroz, and B.P. Stoicheff, Progress in Optics **XX**:325 (1983).
23. R. Hilbig, G. Hilber, A. Lago, B. Wolf, and R. Wallenstein, Comments At. Mol. Phys. **18**:157 (1986).
24. J.M. Ortega, Y. Lapierre, B. Girard, M. Billardon, P. Elleaume, C. Bazin, M. Bergher, M. Velghe, and Y. Petroff, **QE-21**:909 (1985).
25. J.M. Ortega, Synch. Rad. News **2**:18 (1989).
26. S. Werin, M. Eriksson, J. Larsson, A. Persson, and S. Svanberg, to appear.
27. H. Bergström, W.X. Peng, and A. Persson, Z. Phys. **D13**:203 (1989).
28. R.G. Beausoleil, D.H. McIntyre, C.J. Foot, E.A. Hildum, B. Couillaud and T.W. Hänsch, Phys. Rev. **A35**, 4878 (1987).
29. T.W. Hänsch, R.G. Beausoleil, B. Couillaud, C.J. Foot, E.A. Hildum and D.H. McIntyre, in: *Laser Spectroscopy VIII*, W. Persson and S. Svanberg, eds., Springer, Heidelberg (1987).
30. B. Couillaud, L.A. Bloomfield, and T.W. Hänsch, Opt. Lett. **8**:259 (1983).

31. H. Hemmati, J.C. Bergquist, and W.M. Itano, Opt. Lett. **8**:73 (1983).
32. J.C. Bergquist, F. Diedrich, Wayne M. Itano and D.J. Wineland, in: *Laser Spectroscopy IX*, Academic Press, Orlando (1989).
33. A. Timmermann, and R. Wallenstein, Optics Lett. **8**:518 (1983).
34. S. Svanberg, P. Tsekeris, and W. Happer, Phys. Rev. Lett. **30**:817 (1973).
35 S. Svanberg, and P. Tsekeris, Phys. Rev. **A11**:1125 (1975).
36. J. Carlsson, Phys. Rev. **A38**:1702 (1988).
37. J. Carlsson, A. Dönszelmann, H. Lundberg, A. Persson, L. Sturesson, and S. Svanberg, Z. Physik **D6**:125 (1987).
38. J. Ney, Z. Physik **196**:53 (1966).
39. G. Breit, Rev. Mod. Phys. **5**:91 (1933).
40. J. Bengtsson, J. Larsson, S. Svanberg, and C.-G. Wahlström, Phys. Rev. **A41**, Jan. 1 issue (1990).
41. U. Wolf, and E. Tiemann, Appl. Phys. B **39**:35 (1986).
42. J. Carlsson, and P. Jönsson, to appear.
43. J. Bengtsson, J. Larsson, and S. Svanberg, to appear.
44. J. Larsson, and S. Svanberg, to appear.

HIGH RESOLUTION SPECTROSCOPY OF ATOMIC HYDROGEN :

MEASUREMENT OF THE RYDBERG CONSTANT

L.Julien, F.Biraben, J.C.Garreau and M.Allegrini[*]

Laboratoire de Spectroscopie Hertzienne de l'ENS[**]
Université Pierre et Marie Curie, 4 Place Jussieu
Tour 12 E01, 75252 Paris Cedex 05, France

INTRODUCTION

We have used Doppler-free two-photon spectroscopy of hydrogen Rydberg states to perform a measurement of the Rydberg constant which is the most precise one at the present time [1]. We briefly survey here the methods which have been chosen by various groups to measure this constant during the few past years, and we detail the experimental arrangement which has allowed us to obtain our result.

INTEREST OF THE RYDBERG CONSTANT

The Rydberg constant R_∞ is the scaling factor of energy levels in atomic hydrogen (more generally in atomic systems) and can be deduced from the wavelength of any transition between hydrogen levels having different principal quantum numbers n. It can be expressed as a function of the electron mass m, the electron charge e, and the Planck constant h :

$$R_\infty = \frac{m\,e^4}{8\varepsilon_0{}^2 h^3 c}$$

At the present time, this constant is known with a precision approaching 1 part in 10^{10}, but some groups are attempting to improve this precision to 1 part in 10^{11} or better. Indeed there are several motivations to measure the Rydberg constant with a very high accuracy :

• First R_∞ plays a key role in the least-squares adjustment of fundamental constants for which it is used as an auxiliary one (i.e. with a fixed value)[2]. The comparison of the values obtained for a less precisely known constant - such as the fine structure constant α - through different ways (involving R_∞ or not) then gives a test of the consistency between various domains of physics.

• A second motivation is that one needs a very precise value of R_∞ to test predictions of quantum electrodynamics on simple systems such as the positronium atom or the hydrogen atom itself : it allows one to deduce the 1S Lamb-shift from the 1S-2S measured frequency.

---

[*]Permanent address : Dipartimento di Fisica, Piazza Torricelli 2, 56100 Pisa, Italy.
[**]Associated with C.N.R.S.

Applied Laser Spectroscopy, Edited by W. Demtröder and
M. Inguscio, Plenum Press, New York, 1990

• Finally, since R_∞ can be deduced from transitions lying in a large range of wavelengths (from UV to microwaves), it gives a test of the 1/r dependence of the coulombian potential [3]. With this dependence, one can link optical and microwave frequencies through the hydrogen atom.

POSSIBLE WAYS OF MEASURING THE RYDBERG CONSTANT

During the last few years, three types of optical transitions have been successfully investigated to determine the Rydberg constant : the one-photon 2S-3P (H_α) or 2S-4P (H_β) Balmer transitions, the two-photon 1S-2S and 2S-nD transitions. These transitions differ in their wavelengths and in their natural widths but so far have given comparable precisions.

One can also investigate microwave transitions between two close circular Rydberg levels. Let us compare these various methods.

Transitions between circular Rydberg states

This method was proposed by Hulet and Kleppner [4]. Its main advantage is that these transitions lie in the microwave domain (448 GHz for n=24 ↔ n'=25) where frequency measurements are easy. Moreover, compared to other Rydberg states, circular states have the smallest radiative corrections and nuclear size effects, the smallest Stark effect and the longest radiative lifetimes, which makes them more convenient for a determination of the Rydberg constant. The lifetime of the n=25 circular state is 0.8 ms, which gives rise to a theoretical relative linewidth of 10^{-9} for the n=24 ↔ n'=25 transition. Two groups are currently investigating this way of measuring the Rydberg constant, one at the ENS in Paris [5] on the Li atom, the other at MIT [6] on hydrogen.

Balmer transitions

These transitions have been extensively studied by several groups since the advent of tunable dye lasers. The pioneering group was the Stanford group which used successively saturated absorption with a pulsed source [7] and polarization spectroscopy with a cw dye laser [8] to observe the Doppler-free H_α transition in a hydrogen discharge and deduce the Rydberg constant. The Yale group used a quite different method which consisted in exciting a 2S atomic beam with a crossed cw dye laser ; they recently performed very precise measurements of R_∞ on the H_α [9] and H_β [10] lines. The precision of this method seems to have reached its ultimate limit since the natural widths of the P levels involved are responsible for a relative linewidth of a few parts in 10^8.

1S-2S two-photon transition

This transition which lies in the UV range (excitation wavelength 243 nm) is the most promising one because of its extremely small natural width : 5 parts in 10^{16}! Two groups have recently succeeded in measuring its frequency using cw excitation [11][12], but because the 1S Lamb-shift is not precisely known, such a measurement by itself gives this Lamb-shift instead of the Rydberg constant.

2S-nD two-photon transitions

This is the method we use to measure the Rydberg constant. The studied transitions (n≥8) lie in the near infrared range. Their advantage is in the narrow natural width of the upper level (300 kHz for the 10D one) which allows us to observe linewidths as narrow as a few parts in 10^{10}. Another feature of this method is that, for the Rydberg constant measurement, we can use the experimental value of the 2S Lamb-shift which is known with a very high precision [13].

EXPERIMENTAL METHOD

Our aim is to obtain very narrow signals on the 2S-nD transitions in order to measure the Rydberg constant with very high precision. The basic idea of our method is to perform Doppler-free two-photon excitation of 2S metastable hydrogen atoms produced in an atomic beam. We use a geometry where the 2S metastable beam is irradiated by two counterpropagating laser beams collinear with it : the first-order Doppler effect is then eliminated and collisional and transit time broadenings are negligible.

The interaction region (1 m long) is carefully shielded against stray electric and magnetic fields. Metastable atoms are detected at the end of the beam : an applied electric field mixes the 2S and 2P states and the resulting fluorescence of the Lyman-α transition is measured.

The light source is an house-made ring LD700 dye laser working in the 730-780 nm range. The entire metastable beam apparatus is placed inside a Fabry-Perot cavity to enhance the two-photon transition probability. The power density in this cavity corresponds to about 50 W in each direction of propagation.

After being excited to the nD levels, atoms undergo radiative cascades towards the ground state so that optical excitation can be detected with the corresponding decrease of the metastable beam intensity. Our typical signal corresponds to a decrease up to 17% in this intensity. The observed linewidths are about 1 MHz depending on the transition and the light power. Relative linewiths as low as 6×10^{-10} have been observed and are the best resolution so far achieved in an optical transition of atomic hydrogen.

The signal frequency is measured by interferometric comparison between the wavelength of the dye laser and that of an I_2-stabilized etalon He-Ne laser. The key to this comparison is a high stability Fabry-Perot cavity built with two silver coated mirrors. Using this method, we obtained in 1986 a preliminary determination of the Rydberg constant with a precision of 5 parts in 10^{10} [14]. More recently we have achieved a new determination of R_∞ with a precision of 1.7 parts in 10^{10} [1].

RECENT IMPROVEMENTS TO OUR EXPERIMENT

During the last years several improvements to our set-up have been made to obtain this precision :

(i) We have built a new metastable beam apparatus which is evacuated by cryogenic pumps and can be heated. Stray electric fields seen by metastable atoms are then reduced to less than 2 mV/cm, and it is possible now to excite transitions towards high n levels (n≥20).

(ii) We have measured the velocity distribution of metastable atoms in the atomic beam. It has been deduced from the study of the Doppler shifted absorption profile of the 2S-3P line when the 2S beam is excited with a counterpropagating 656 nm laser beam. This velocity distribution is needed to evaluate the second-order Doppler shift in our lines and to perform calculations of the theoretical line profiles.

(iii) We have developed a new system for the control of our dye laser frequency using an acousto-optic device and a high finesse Fabry-Perot cavity locked to an I_2-stabilized He-Ne laser. This system is driven by a microcomputer and has a reproducibility of about 1 part in 10^{11}. It allows one to sweep the dye laser frequency over a range of 250 MHZ around any chosen frequency.

(iv) We have fitted the experimental lines to the calculated profiles ; the calculation takes into account the saturation of the transition probability and the inhomogeneous light shift experienced by the metastable atoms along each possible trajectory in the beam.

(v) The interferometric wavelength comparison has been performed very carefully. The reflective phase shifts due to the reflection off the metallic mirrors of the measurement FP cavity have been eliminated by the "virtual mirrors method" : we alternately used two cavity lengths 10 and 50 cm. Moreover, the Fresnel phase shift inside the cavity (due to the wavefront curvature) has been precisely measured [15].

RESULTS

We have measured the frequencies of six transitions : the $2S_{1/2}$-$nD_{5/2}$ transitions for $n = 8$, 10 and 12 in both hydrogen and deuterium. The six deduced independent determinations of the Rydberg constant are in very good agreement with one another. Our final result is :

$$R_\infty = 109\ 737.315\ 709\ (5)(18)\ cm^{-1}$$

The first quoted uncertainty does not include the uncertainty in the reference laser frequency. In fact the major part of our total uncertainty (1.7×10^{-10}) arises from this reference since the uncertainty relative to this reference is only 4.3×10^{-11}.

This result is in good agreement with our preceding result [14] and with those obtained by other groups from cw excitation of other transitions in hydrogen [9][10][11][12] even if it slightly differs from more ancient results [7][8][16].

CONCLUSION

The precision of our measurement of the Rydberg constant is mainly limited by that of the frequency reference used in the visible range. It is clear now that the frequency of the I_2-stabilized He-Ne laser at 633 nm needs to be remeasured, as is planned in Paris at the Laboratoire Primaire des Temps et Fréquences (LPTF). However, we think we have now reached the ultimate precision in interferometric wavelength comparisons, which is a few parts in 10^{11}. For this reason we are now preparing a direct frequency comparison between our transition frequencies and the frequency chain at the LPTF.

References

[1] F.Biraben, J.C.Garreau, L.Julien and M.Allegrini, Phys. Rev. Lett. 62, 621 (1989).

[2] E.R.Cohen and B.N.Taylor, Rev. Mod. Phys. 59, 1121 (1987).

[3] D.F.Bartlett and S.Lögl, Phys. Rev. Lett. 61, 2285 (1988).

[4] R.H.Hulet and D.Kleppner, Phys. Rev. Lett. 51, 1430 (1983).

[5] M.Gross, J.Hare, P.Goy and S.Haroche, in "The Hydrogen Atom" p.134, G.F.Bassini T.W.Hänsch and M.Inguscio eds (Springer-Verlag, Berlin, 1989).

[6] D.Kleppner, private communication.

[7] T.W.Hänsch, M.H.Nayfeh, S.A.Lee, S.M.Curry and I.S.Shahin, Phys. Rev. Lett. 32, 1336 (1974).

[8] J.E.M.Goldsmith, E.W.Weber and T.W.Hänsch, Phys. Rev. Lett. 41, 1525 (1978).

[9] P.Zhao, W.Lichten, H.P.Layer and J.C.Bergquist, Phys. Rev. $\underline{A34}$, 5138 (1986).

[10] P.Zhao, W.Lichten, H.P.Layer and J.C.Bergquist, Phys. Rev. Lett. $\underline{58}$,1293 (1987).

[11] R.G.Beausoleil, D.H.McIntyre, C.J.Foot, E.A.Hildum, B.Couillaud and T.W.Hänsch, Phys. Rev. $\underline{A35}$, 4878 (1987).

[12] M.G.Boshier, P.E.G.Baird, C.J.Foot, E.A.Hinds, M.D.Plimmer, D.N.Stacey, J.S.Swan, D.A.Tate, D.M.Warrington and G.K.Woodgate, Nature (London) $\underline{330}$, 463 (1987).

[13] V.G.Pal'chickov, Yu L.Sokolov and V.P.Yakovlev, JETP Lett. $\underline{38}$, 418 (1983).

[14] F.Biraben, J.C.Garreau and L.Julien, Europhys. Lett. $\underline{2}$, 925 (1986).

[15] M.Allegrini, F.Biraben, B.Cagnac, J.C.Garreau and L.Julien, in "The Hydrogen Atom" p.49, G.F.Bassini T.W.Hänsch and M.Inguscio eds (Springer-Verlag, Berlin, 1989).

[16] S.R.Amin, C.D.Caldwell and W.Lichten, Phys. Rev. Lett. $\underline{47}$, 1234 (1981).

LASER SPECTROSCOPY OF ATOMIC DISCHARGES

Massimo Inguscio

Dipartimento di Scienze Fisiche dell'Università
di Napoli and
European Laboratory for Nonlinear Spectroscopy
(LENS), Largo E.Fermi 2, I 50125 Firenze, Italy

WHY AN ATOMIC DISCHARGE

Discharges provide a powerful means for the production
of unstable molecular species to be used as samples for
spectroscopy [1]. However, they could seem not useful for the
investigation of atoms, for which more "clean" experimental
environments are provided by other configurations such as
traps or beams. On the contrary, discharges still play an
important role for atomic spectroscopy, whenever the final
goal is not necessarily an investigation with ultrahigh
precision. Discharges are the straightforward ways for the
production of atoms from molecular dissociation (H_2, O_2 ...),
they can be useful to produce refractory elements by
sputtering in hollow cathode configurations, or to produce
ionized species. It is worth noting that, thanks to the high
electron temperature in the discharge, also excited states
are populated and ready to be investigated by laser
techniques.

Interesting examples of the importance of discharges
for the production of the species can be found in the *far
infrared* spectroscopy of atoms[2]. Fine structure transitions
of important atoms occur in the far infrared, they are
magnetic dipole in nature and the high sensitivity necessary
for the detection is reached by producing the atom inside
the far infrared laser cavity itself (Laser Magnetic
Resonance)[3]. For instance, oxygen atoms were produced in
microwave O_2/noble gas discharges for the first far infrared
investigation of an atom [4]. As for refractory elements, in
cases where volatile compounds are available, suitable
discharges helped the chemical processes necessary to obtain
the atom. Important examples are given by carbon and silicon
atoms that could be investigated thanks to the availability
of CH_4 and SiH_4 respectively. In this case, fluorine atoms,
produced in a microwave discharge in a dilute mixture of F_2
in He, mixed with methane or silane and the reaction produced

Applied Laser Spectroscopy, Edited by W. Demtröder and
M. Inguscio, Plenum Press, New York, 1990

free carbon or silicon atoms. "Hollow-cathode" discharges
were used for the production of magnesium atoms, directly by
sputtering from the metal. In this case, the discharge also
allowed to populate the excited metastable states important
for far infrared investigations. On the contrary, a "hot
cathode" dc discharge was used to excite the ground state
atoms evaporated from an oven.

It is clear from this short overview, that discharges
are very useful in atomic spectroscopy and that they can be
designed and optimized for a given particular case. Laser
magnetic resonance usually provided the only precise
measurements available in the far infrared, with an accuracy
of the order of 1 MHz ($\Delta v/v = 10^{-6}$). In the particular case of
magnesium, the comparison with the more precise data existing
for one transition measured in an atomic beam, provided
evidence that the primary sources of uncertainty were indeed
due to errors in the magnetic fields and laser frequency
measurements and that the values obtained for the atomic
splittings were, to within the estimated uncertainty, free of
environmental effects.

DISCHARGES AND OPTOGALVANIC SPECTROSCOPY

Once the sample is produced in a discharge, one can
also take advantage of some detection schemes different from
the conventional absorption or fluorescence detections. Among
the "unconventional techniques"[5] ,optogalvanic spectroscopy
plays an important and well established role. It is based on
the change of the discharge impedance following the resonant
interaction between photons and atoms. As a first
approximation the effect can be explained by a simple model
which takes into account the difference between the
ionization cross-sections σ_i for the two states involved in
the optical transition. Actually, the complexity of all the
phenomena involved in the processes following the excitation
and finally leading to the ionization (for instance,
associative ionization, Penning ionization) prevents from
developing a general quantitative theory. Indeed several
parameters, strongly dependent on the particular types of
discharge have to be taken into account [6]. However a wide
success has been assured to this technique by the extreme
simplicity of the detection scheme. If the discharge is
direct current (DC), the signal can be measured as a change
of the voltage drop on the resistor R connected in series
with the discharge tube. For other types of discharge
slightly different technical solutions must be adopted, but
the logic scheme is the same. For instance in radiofrequency
dischargesthe change of impedance is detected as a feedback
from the RF oscillator to its stabilized power supply, as we
shall see below for the study of oxygen atoms.

As for applications to selective and sensitive
spectroscopy, it is very important to observe that all the
high resolution laser techniques[7] could be combined with
optogalvanic (OG) detection, including frequency
stabilization of lasers[8]. In 1978 Johnston[9] analysed one of
the Ne transitions using a single-mode dye laser and observed
the Lamb dip in the OG signal. Using intermodulated technique

Schawlow and co-workers [10] presented in 1979 experimental measurements of hyperfine splittings in ^{3}He, demonstrating the applicability of the OG technique to quantitative high-resolution spectroscopy. Figure 1 wants to be an illustration of the different levels of resolution that can be achieved by means of optogalvanic spectroscopy. An

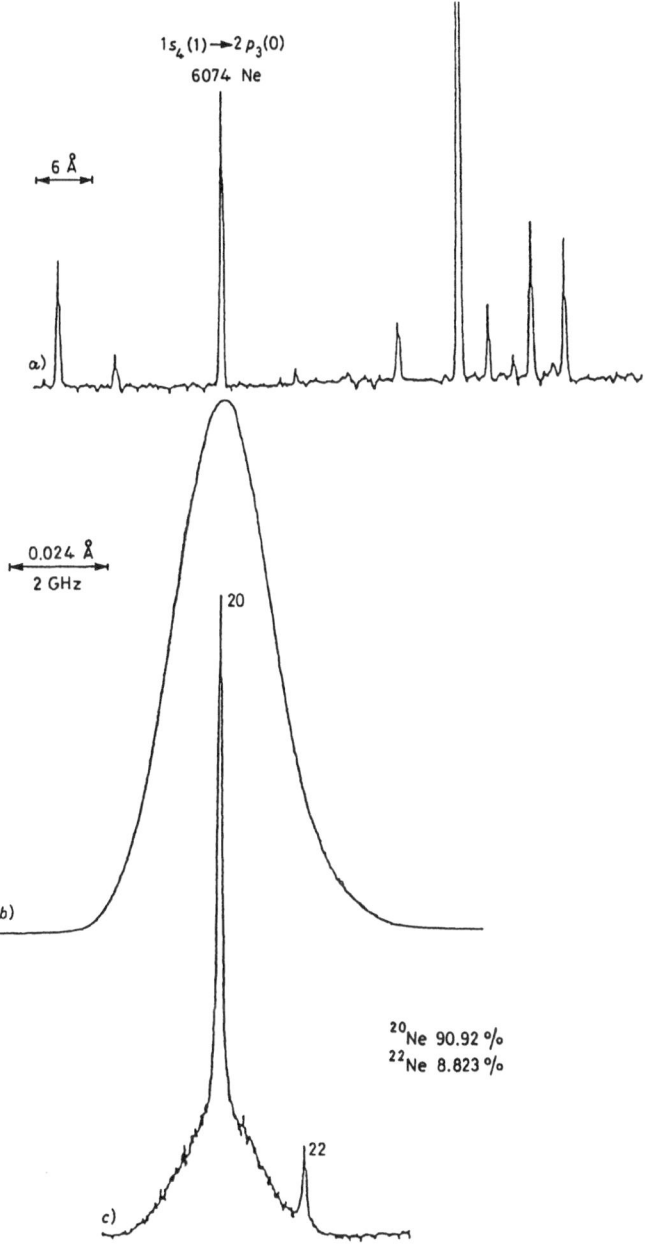

Figure 1 . Optogalvanic recordings in the same neon
discharge with different spectral resolution.

hollow-cathode configuration is used[5] for a 65mA discharge in 2 Torr of natural abundance neon. In Fig. 1a) a portion of the spectrum obtained with broadband scan around 607.4nm is shown. A narrow-band scan of the Doppler-broadened transition is recorded in b). Note the asymmetry caused in the line profile by the presence of two unresolved isotopes. The isotopic structure is resolved in c) by means of the *intermodulated optogalvanic* technique. This latter can be illustrated with the aid of Fig.2: the beam of a

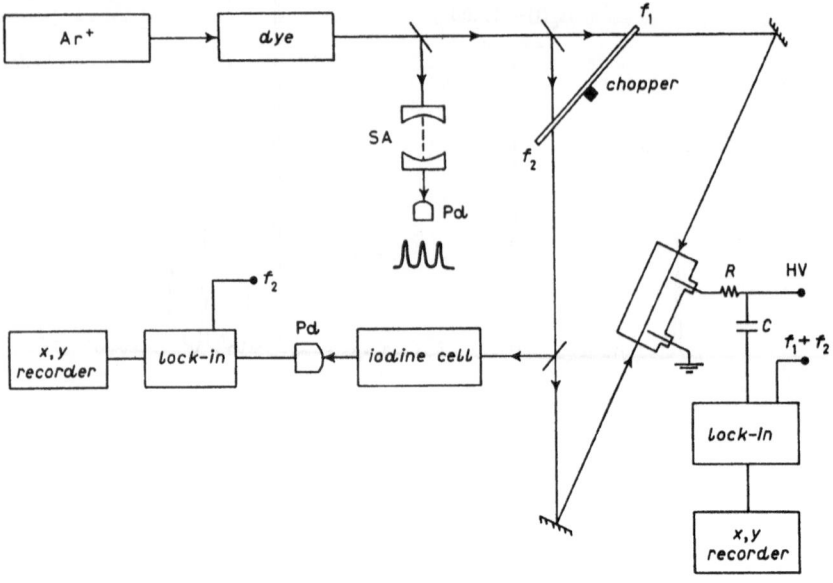

Figure 2. Experimental scheme for intermodulated optogalvanic spectroscopy.

single-frequency CW dye laser (frequency jitter about 1MHz) is split into two components of roughly equal intensity, which pass through the discharge in opposite directions. One beam is chopped at a frequency f_1 and the other at a frequency f_2 by the same mechanical chopper which also provides separate reference signals at each of the two frequencies f_1 and f_2 and at the sum frequency f_1+f_2. By using as reference for the lock-in detection f_1 or f_2, the Doppler broadened signal is recorded, while at the reference sum frequency f_1+f_2 only the Doppler-free signal should be detected since the signal is originated by atoms interacting with both the counterpropagating beams, i.e. by atoms with zero velocity component along the laser direction. In the actual experiment, however, velocity changing collisions (VCC) produce a saturation signal also when the laser frequency is off resonance, but still inside the Doppler profile of the line. Such collisions lead to a broad background superimposed on the narrow homogeneous resonance.

A quantitative analysis of the lineshape can be performed following the model introduced in ref.11 and then extended in ref.12 to take into account the case of weak collisions. In the hypothesis of a homogeneous width γ much smaller than the Doppler width $\Delta\nu_D$ (FWHM), the intermodulated lineshape can be assumed to a good approximation to be the superposition of a Lorentzian and a Gaussian pedestal:

$$Signal = A \left\{ \frac{\gamma/4}{\gamma/4 + (v-v_0)^2} + Cexp \left[\frac{-(v-v_0)^2}{\Delta_D^2} \right] \right\} \quad [1]$$

A is a normalization constant and C represents the weight of the collisional pedestal, whose dependence on the radiative and collisional parameters is

$$C = 2(\pi log2)^{1/2} \frac{\Gamma_\gamma}{\delta \Delta_D} \quad [2]$$

where δ is the decay rate of the longer effective lifetime level of the transition and Γ is its cross-relaxation rate.

Low C values, which are necessary for high-resolution spectroscopy, can be achieved by a proper choice of the discharge parameters. However the lifetimes of the levels involved in the transitions play an important role in determining the lineshape. For instance we show in Fig.3 intermodulated optogalvanic signals for three neon transitions in the same hollow-cathode discharge. Gas pressure and discharge current are similar for the three recordings, while the lifetime of the longer lived (lower) level is much different for each transition. In a), $1s_3 \rightarrow 2p_3$ at 616.3nm, lifetime is 10^{-4} s; in b), $1s_4 \rightarrow 2p_3$ at 607.4nm, it is 10^{-5} s; in c), $2p_4 \rightarrow 4s''_1$ at 590.2nm, lifetime is 10^{-8} s. Different lineshapes are related to these different lifetimes which for each transition determine the time interval in which atoms interacting with the two laser beams can undergo cross-relaxation processes. In fact the role played by velocity changing collisions is clear for transitions starting from metastable or quasi-metastable levels, where a more marked pedestal is present, while for the 590.2nm line the collisional pedestal is negligible. Correspondingly, by fitting the three curves to the expression [1], we obtain for the C factors a)5.3, b)0.9, and c)0.1. An important role is also played by collisions of metastable atoms with electrons: they decrease the effective lifetime and in fact the collisional pedestal is reduced when the discharge current is increased. A linear decrease of C as a function of the discharge current was for instance observed in the systematic investigation performed in ref.13. Some advantage can be provided by increasing the light chopper frequency to be comparable with the collisional frequency or by cooling the discharge temperature. In this latter case the improvement comes from the reduction in the doppler width and from a better stability of the discharge.

However, the real improvement is obtained by changing to a detection technique completely different and intrinsically much less affected by collisions. That is *laser polarization spectroscopy* [14] which is based on the atomic

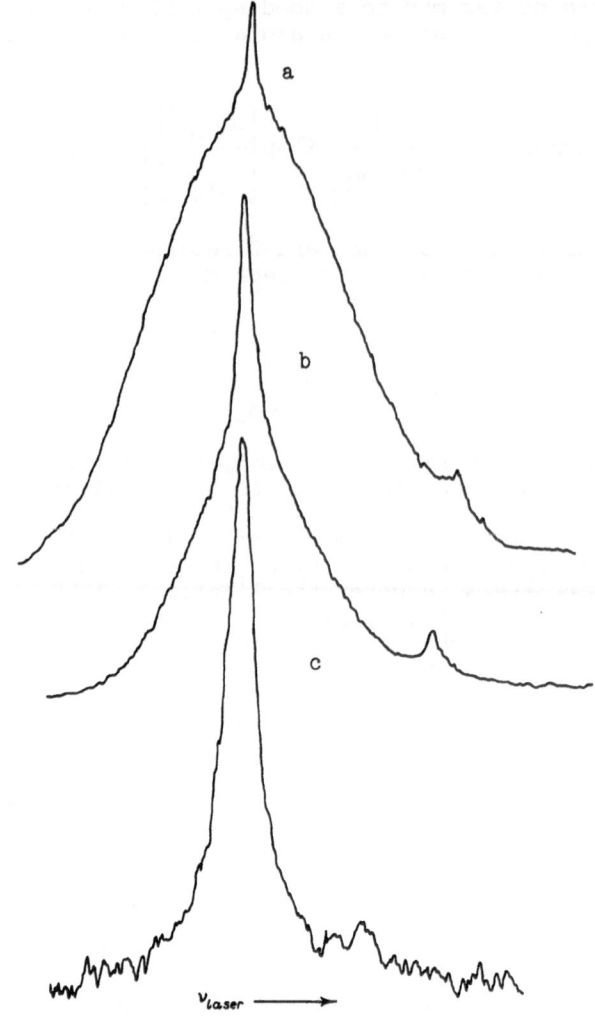

Figure 3. Intermodulated recording of three neon transitions involving levels with different lifetimes.

orientation and not only in the atomic population. The principle of the technique is illustrated with the help of the scheme in Fig.4. Radiation from a tunable laser is splitted into two beams of different intensities, counterpropagating through the sample. Now the stronger beam is circularly polarized in order to create the orientation of a velocity selected group of atoms. The induced birifringence is probed by the weaker linearly polarized beam. High sensitivity is achieved by detecting the transmitted intensity after a crossed linear polarizer with high extinction coefficient. The signal will be given by only those oriented atoms interacting with both the strong and weak beams, i.e. by atoms with zero velocity component, and will occurr when the laser frequency is tuned exactly at the center of the atomic transition. In addition, a small portion of the laser light is sent to a stable Fabry-Perot

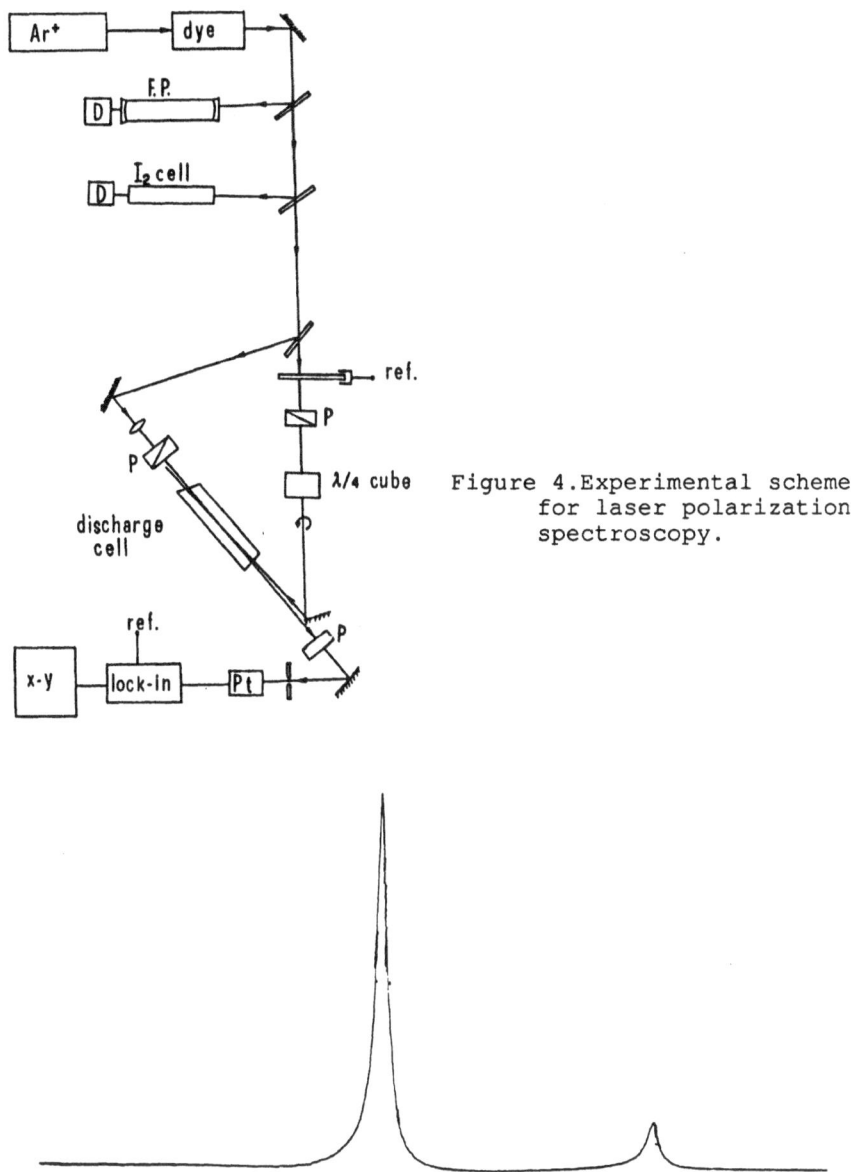

Figure 4.Experimental scheme for laser polarization spectroscopy.

Figure 5. Laser polarization recording of the same neon transition in Fig.1.

interferometer for calibration of the frequency scan, while the laser wavelength is determined by recording the absorption from a I_2 sample. Coming back to the polarization technique, the intensity of the signal can be evaluated accurately following the treatment in ref.15. In summary, the linearly polarized analysis beam can be regarded as the superposition of two circularly polarized σ^+ and σ^- beams. Correspondingly, the detected signal depends on the difference $(\alpha^+-\alpha^-)$ between the two absorption coefficients, which makes the probe to be elliptically polarized, and on the induced birifringence, which causes a rotation of the probe polarization. The effect can be computed depending on the change ΔJ in angular momentum and on the M sublevel. It turns out[7] that for a $\Delta J=\pm 1$ transition the dependence on M is much different for α^+ and α^-, while for $\Delta J=0$ it is nearly the same. As a consequence $\Delta J=\pm 1$ signals are much stronger than $\Delta J=0$ ones. This can be helpful in simplifying spectra, as we shall see for oxygen transitions with $\Delta J=\pm 1$ and $\Delta J=0$ overlapping components.

The power of polarization spectroscopy is illustrated by means of the comparison with intermodulated spectroscopy . In Fig.5 the same neon transition at 607.4nm, studied in Fig.1 is recorded by polarization spectroscopy. The much better signal to noise, and in particular the dramatic reduction of the doppler broadened pedestal with respect to intermodulated spectroscopy is evident.

ATOMIC OXYGEN: A CASE HISTORY

This final lecture will illustrate the power of high resolution atomic discharges laser spectroscopy by discussing a series accurate studies of oxygen. This atom plays a crucial role in many physical and chemical processes and it is the third abundant element after hydrogen and helium . High resolution investigations would be highly desirable for this light and theoretically still tractable atom, however this possibility was opened only very recently thanks to an intermodulated optogalvanic spectroscopy scheme[16]. Since then, it has been possible to measure fine structures, isotope shifs and to perform quantitative analysis, also thanks to the successful application of the polarization spectroscopy.

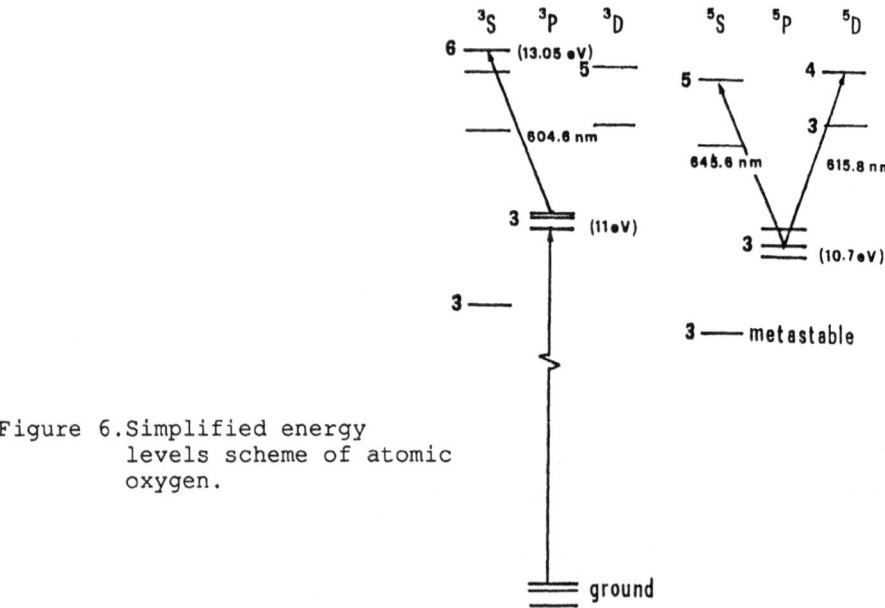

Figure 6.Simplified energy
         levels scheme of atomic
         oxygen.

The difficulties for the investigation of oxygen start with the production of the atom from the molecule, in environmental conditions suitable for high resolution spectroscopy. An efficient procedure is to use trace amounts of O_2 in a noble gas substained radiofrequency discharge. A detailed and quantitative analysis of the several different processes occurring in this particular plasma showing non-equilibrium thermal distributions,is presented in the contribution by A.Sasso and G.Tino[17], where a scheme for optogalvanic radiofrequency detection is also described.

Other difficulties are presented by the energy level scheme itself of the atom, as partially shown in Fig.6. The fundamental configuration $1s^2 2s^2 2p^4$ generates levels coupled by forbidden transitions in the visible, while the first level of the excited configuration lies at more than 9eV of energy, i.e., like for many other light atoms, the resonance lines are in the vacuum ultraviolet (VUV), where tunable radiation of high intensity and spectral purity are hard to produce. The use of pulsed sources is mandatory. That is well suited for lifetime measurements[18], however the pulsed linewidths limit the precision of the determination of spectral features[19,20]. Even if the radiation linewidths is reduced to the Fourier limit, as in the recent two-photon investigation[21] of the $2p^3P \rightarrow 3p^3P$ transition, the accuracy for the determination of the upper level fine structure is comparable with that of data from conventional optical spectra. It is important to stress, especially in this school on "applied" spectroscopy, the high interest of these pulsed techniques for the detection of oxygen in combustion processes, using either the ionization[19] or the fluorescence[20] following the two-photon excitation[22].

FINE STRUCTURE

Let us come back to the high resolution investigations. The ground-state fine structure splittings were the only high accuracy data available since they could be directly measured [4,23] at far infrared frequencies where Doppler broadening is negligible. On the other hand, as already stressed, optogalvanic spectroscopy is particularly convenient for studies of excited atoms and in fact an intermodulated optogalvanic recording of the $3p^5P_3 \rightarrow 5s^5S_2$ oxygen transition at 645.6nm is shown in Fig.7. The Doppler-free signal is superimposed onto the broader pedestal caused by velocity-changing collisions. By fitting the experimental curve to the sum of a Lorentzian and a Gaussian, as stated in Eq. 1, we obtain $C = 0.39$, $\gamma = 386$ MHz (FWHM), and $\Delta\nu_D = 2$ GHz (FWHM). The observed homogeneous linewidth is due primarily to pressure and power broadening. The forced use of high noble gas pressures implies a relatively large pedestal, as demonstrated by the large value of C. This feature can seriously affect the resolution of nearby components. This is for instance the case of the investigation of the transitions starting from the same lower $5P$ levels, but reaching the 4^5D_J multiplet. Here fine structure is of the same order of Doppler broadening and an intermodulated optogalvanic recordings is shown in Fig.8 for the $5P_2 \rightarrow 5D_{3,2,1}$: Doppler-free features do not present a very high contrast not only for the overlapping of the various

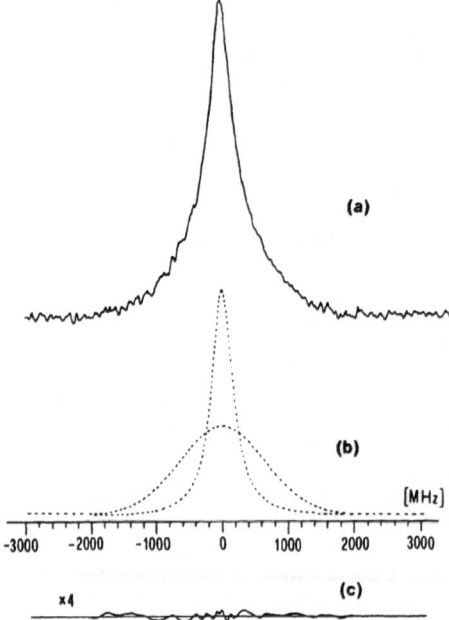

Figure 7. a) Intermodulated optogalvanic recording of the oxygen transition at 645.6nm transition as obtained in a O_2/Ar radiofrequency discharge (partial pressures of 0.04/2.3 Torr). b)The lineshape is given by the superposition of a Lorentzian and a Gaussian showing the separate contribution of the two curves. c) Difference between the experimental and fitted curve (note that the vertical scale is amplified by a factor of 4).

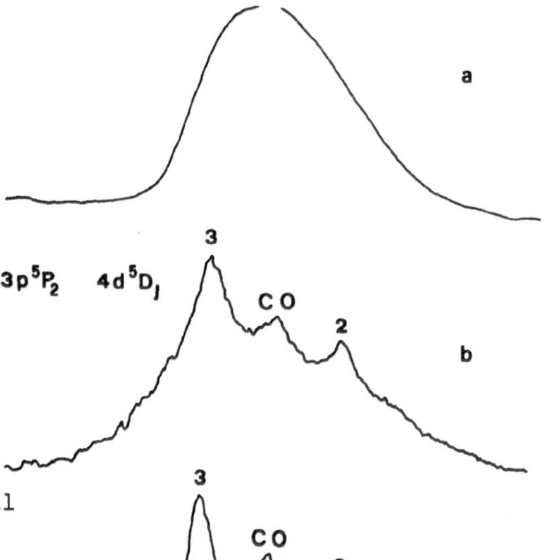

Figure 8. OG recordings of O transition at 615.8nm. The Doppler broadened signal is shown in a), and the intermodulated scheme is used in b) and c). Different gas mixtures were used:b,O-Ne; c, O-Ar.The total scan of the Doppler registration is 2.7 times broader than the Doppler-free recordings.

182

velocity changing pedestals, but also for the presence of
"cross over (CO)" signals originating in the middle of two
components sharing a common level. Also, it is worth noting
the importance of a proper choice of the noble gas
substaining the discharge. Because of the different types of
collisional processes involved[17], the contrast between
Doppler-free signals and Doppler broadened pedestals is
better with argon than with neon. These measurements
provided fine structure separations with a few percent
relative accuracy[24], i.e. one order of magnitude better than
for the values derived from conventional spectroscopy.

ISOTOPE SHIFT

As illustrated by the lecture of H.J.Kluge[25], the
relative displacements of spectral lines of different
isotopes provide valuable information on nuclear structure.
Isotope shifts are due to two distinct causes: the change
from one isotope to another in nuclear mass (mass shift) and
the distribution of nuclear charge over a finite region
(field or volume shift). The field shift involves the
electrostatic interaction between electrons and nucleus.
Within the nuclear region the electric field differs from
isotope to isotope, a larger charge volume (usually occurring
for heavier isotope) resulting in a smaller binding
potential. The volume difference has much more effect on s
electrons than on the others, since the s electron
wavefunction is more penetrating in the nucleus. Usually, in
light elements (Z<30) the field effect is very much smaller
than the mass-dependent effect and until recently it has been
possible to ignore it except as a factor in the
interpretation of hyperfine structure intervals. For instance
in the case of hydrogen, the correction in energy for 2s
levels due to the finite size of the nucleus is 0.12 MHz in H
and 0.73 MHz in D. However the precision of contemporary
laser spectroscopy has reached the point where it is becoming
necessary to be aware of the field effect in optical
spectra[26].The lighter the element, the more challenging is
the measurement, also because of the larger Doppler
broadening of spectral lines.
        Again, oxygen can play an interesting role also in this
kind of measurements. A small field effect, otherwise
nwgligible in such a light element, can originate from the
completely spherical symmetry of ^{16}O nucleus (doubly magic),
which is altered by the addition of nucleons outside the
closed shell. It is worth remembering that a light element
where deviations from the purely mass effect has been
unambiguously detected is ^{40}Ca, which in fact has a doubly
magic nucleus[27].
        A first investigation of the isotope shift has been
performed[28] using intermodulated optogalvanic spectroscopy
in a discharge containing an ^{18}O enriched sample. In Fig.9,
they are shown recordings for the (a) Doppler-limited, where
the Doppler width of 2 GHz is large enough to mask any
isotope effect, and (b) the Doppler-free $^{16,18}O$ $5P_3{\rightarrow}5S_2$
transition at 645.6 nm, the same as in Fig.7. Here the
isotope shift is well resolved, as in the recording (c) for
the weaker fine structure component. However the presence of
the strong and broad pedestal, due to the velocity-changing

collisions, reduces the accuracy and makes difficult the observation of the shift for [17]O.

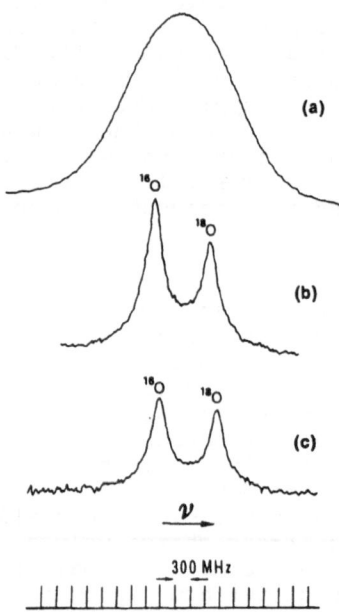

Figure 9. Optogalvanic measurements using an enriched O sample. a) The doppler limited and b) the Doppler-free transitions at 645.6 nm. c) The Doppler-free recording of the other fine structure $^5P_2-^5S_2$ component at 645.5 nm.

On the contrary measurements on all the three stable isotopes are necessary to distinguish the field effect from the predominant mass one. The requested improvement in sensitivity and resolution is achieved by means of polarization spectroscopy, according to the experimental scheme discussed in Fig.4. Typical recordings are shown in Fig.10. In a) the $3p^5P_3 \rightarrow 4d^5D_{4,3,2}$ transition at 615.8nm is recorded by means of the intermodulated OG spectroscopy, using a natural abundance sample. In b) the same transition is investigated using polarization technique, still in a natural abundance sample. Both the levels involved in the transition are radiative with a lifetime of the order of hundred nanoseconds, leading to a radiative broadening of tens of MHz. The observed width, essentially caused by pressure and power broadening, is 150MHz, lower than for the intermodulated optogalvanic recording one. This improvement is made possible by the increased sensitivity, which allows to operate at lower gas pressures. It must be noticed that only one of the three fine structure components, namely the $^5P_3 \rightarrow ^5D_4$, is actually detected. This is due to the different relative intensities of the fine structure components, but also to the fact that, when using circularly polarized pump beam, transitions corresponding to $\Delta J = 0$ have a smaller cross-section respect to transitions with $\Delta J = +1$, as discussed previously. This induces a simplification in the spectra avoiding overlapping of fine structure components which, in presence of different isotopes, would mask any structure. A typical recording, using a 50% ^{17}O-^{18}O enriched sample, is then reported in Fig.10c) for the $3p^5P_3 \to 4d^5D_4$ transition: isotope shift values can be obtained with an accuracy better than 1%. Unlike ^{16}O and ^{18}O, whose nuclear spins are zero, ^{17}O has a nuclear spin I = 5/2 and both a magnetic dipole and an electric quadrupole arise. The unresolved hyperfine structure is responsible for the broader linewidth of ^{17}O peak. Such measurements, repeated on different transitions and carefully analyzed in order to extract the predominant mass contribution, have evidenced[29] a field contribution to the isotope effect.

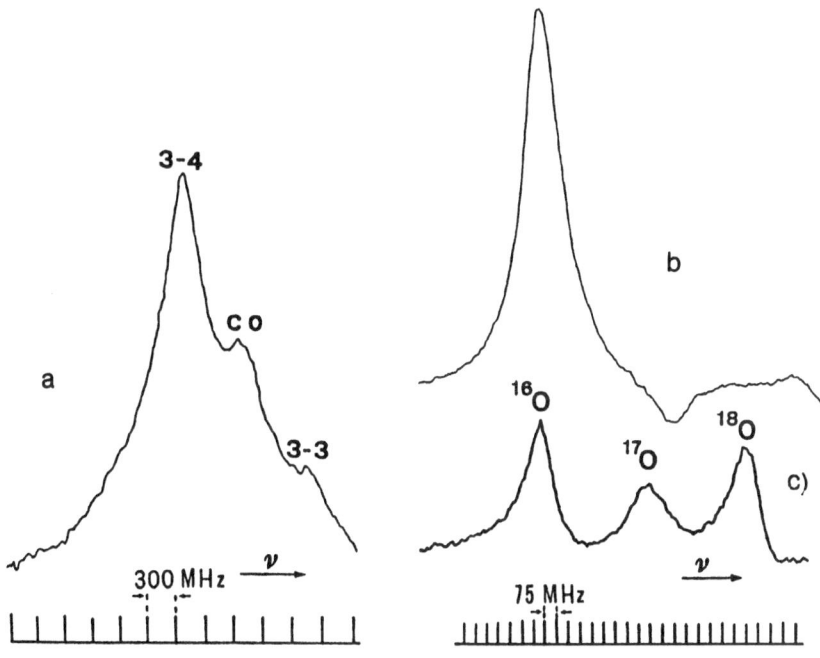

Figure 10. Comparison between intermodulated a) and polarization spectroscopy b) and c) on the same oxygen transition.

FUTURE DEVELOPMENTS

Among the future possibilities for high resolution spectroscopy of oxygen discharges, one should certainly consider the opportunity of using semiconductor diodes as laser sources, of the type illustrated in the lectures of L.Hollberg[30]. With reference to the scheme in FIG.6, one can see that transitions starting from both the lowest excited triplet and lowest quintet level are easily accessible to AlGaAs diode lasers (respectively at 850 and 780 nm). Since both the levels are the lowest S ones of the series, measurements of isotope shift could provide an unambiguous confirmation of the "field effect contribution" discussed in the previous section.*

More important, level $^5S_2$ is a metastable one, with a

lifetime of about 180 μs. The transitions at 777nm, accessible to diode lasers, are "closed" and can be possibly used to achieve *radiative cooling*[31] of atomic oxygen. The subsequent decay of the cooled metastable atoms into the ground state would produce cold oxygen in the ground state. Of course, this experiment at present represents an interesting challenge, because the lifetime of the $^5S_2$ level is comparable to the time required for the cooling of the atoms.

*Note added in proof: The weeks after the end of the "Applied laser spectroscopy" school, we decided to rest playing with diode lasers and oxygen atoms produced in a rf discharge. The result is a nice measurement of the hyperfine structure of the $3s^5S_2 \rightarrow 3p^5P_3$ transition: G.M.Tino, L.Hollberg, A.Sasso, M.Inguscio and M.Barsanti, to be published.

REFERENCES

1 See for instance: J.M.Brown: contribution to the present volume.
2 For a comprehensive review see: M.Inguscio: Physica Scripta **37** , 699 (1988).
3 K.M.Evenson: Farad.Disc.Roy.Soc.Chem. n.71 (1981).
4 P.B.Davies, B.J.Handy, E.K.Murray Lloyd, and D.R.Smith: Chem.Phys. **68**, 1135 (1978).
5 K.Ernst and M.Inguscio : Rivista Nuovo Cimento **11**, 1-66 (1988).
6 The reader can find a most recent and comprehensive review on the optogalvanic effect models and applications to spectroscopy in the work by N.Beverini, A.Sasso and B.Barbieri in press on the Review of Modern Physics.
7 W.Demtröder :"Laser Spectroscopy" Springer Series in Chemical Physics, vol. 5, Springer Verlag, Berlin, Heidelberg, New York, 1982.
8 See for instance L.R.Zink contribution to the present volume.
9 T.F.Johnston jr.: Laser Focus, March (1978), p.58.
10 J.E.Lawler, A.I.Ferguson, J.E.M.Goldsmith, D.J.Jackson and A.L.Schawlow: Phys.Rev.Lett. **42**, 2605 (1979).
11 P.W.Smith and T.Hänsch: Phys.Rev.Lett. **26**, 740 (1971).
12 C.Brechignac, R.Vetter, and P.R.Berman: Phys.Rev. A **17**, 1609 (1977).
13 A.Sasso, G.M.Tino, M.Inguscio N.Beverini and M.Francesconi: Nuovo Cim. **10** , 941 (1988).
14 C.Wieman and T.W.Hänsch: Phys.Rev.Lett.**36**, 1170 (1976).
15 R.E.Teets, F.V.Kowalski, W.T.Hill, N.Carlson and T.W.Hänsch: Proc.Soc.Phot.Opt.Instr.Eng.**113**, 80 (1977).
16 M.Inguscio, P.Minutolo, A.Sasso and G.M.Tino : Phys.Rev.A **37**, 4056 (1988).
17 A.Sasso and G.M.Tino: contribution to the present volume.
18 S.Kröll, H.Lundberg, A.Persson, and S.Svanberg: Phys.Rev.Lett. **55**, 284 (1985).
19 J.E.M.Goldsmith: J.Chem.Phys.**78**, 1610 (1983).

20   N.S.Nogar and G.L.Keaton: Chem.Phys.Lett. **120**, 327 (1985).

21   D.J.Bamford, M.J.Dyer, and W.K.Bishel: Phys.Rev.A **36**, 3497 (1987).

22   For an illustration of the application of optical techniques to the combustion diagnostic, see J.P.Taran and A. D'Alessio and F.Cavaliere contributions to the present volume.

23   R.J.Saykally and K.M.Evenson: J.Chem.Phys. **71**, 1564 (1979).

24   A.Sasso, P.Minutolo, M.I.Schisano, G.M.Tino, and M.Inguscio: J.Opt.Soc.Am. B **5**, 2417 (1988).

25   H.J.Kluge: contribution to the present volume.

26   For the recent progresses in atomic spectroscopy, see for instance: "The Hydrogen Atom", F.Bassani, M.Inguscio, T.W.Hänsch Eds, Springer Verlag, Berlin, Heidelberg, New York, 1989.

27   H.W.Brandt, K.Heilig, H.Knockel, and A.Steudel: Z.Phys. A **288**, 141 (1978).

28   K.Ernst, P.Minutolo, A.Sasso, G.M.Tino, and M.Inguscio: Opt.Lett. **14**, 554 (1989).

29   L.Gianfrani, A.Sasso, G.M.Tino and M.Inguscio : Opt.Commun.1990, in press.

30   L.Hollberg: contribution to the present volume.

31   For a reference to radiative cooling in this School, see: N.Beverini and F.Strumia: contribution to the present volume.

INFRARED LASER SPECTROSCOPY

John M. Brown

The Physical Chemistry Laboratory
South Parks Road
Oxford, OX1 3QZ
England

1. INTRODUCTION

Infrared spectroscopy (in the wavenumber range 300 to 4000 cm^{-1}) is a time honoured technique, having been in active use for more than 100 years[1].   In this period, an enormous number of molecules have been studied and characterized (it is estimated at more than 10^4) and low resolution spectra have been widely used by chemists for analytical purposes.   Until the mid 1970's, the highest resolution spectra were recorded with large grating spectrographs using a continuum source of radiation.   The linewidth was instrument limited at 0.01 cm^{-1} or 300 MHz (FWHM).   Sensitivity in these experiments was achieved by the use of multi–pass cells of considerable length[2].   Using these techniques, several attempts were made to record infrared (IR) spectra of transient molecules but none were successful. From about 1970 onwards, two separate developments have changed the face of IR spectroscopy almost beyond recognition.   As a result, the field has been re–invigorated and results are being obtained which could have been barely dreamed of 20 years earlier.

The first of these developments was that of Fourier Transform spectroscopy. In this technique, the continuum radiation is dispersed with a Michelson type interferometer, the spectrum being scanned by varying the pathlength in one of the optical arms.   By moving the reflecting mirror over a considerable distance, it is possible to obtain very high resolution in this way.   A state–of–the–art instrument has 5m travel which on Fourier transformation corresponds to a resolution of $1/(2 \times 500) = 0.001$ cm^{-1}, close to the Doppler linewidth.   Pierre Connes was a pioneer in this field, building instruments for astrophysical studies[3,4].   High resolution instruments are now commercially available, thanks in part to the developments in instrument making and in part to the availability of powerful microcomputers to store the interferograms and perform the Fourier transformation.

The second development is more relevant to this chapter, namely that of IR lasers. The first such laser was the CO_2 laser in 1964[5] but in succeeding years, several other systems became available. They are all very fine spectroscopic sources, producing highly collimated beams of high spectral brightness and very narrow linewidths (typically a few MHz, which is very much less than the Doppler linewidth). As a result, IR spectroscopy at the Doppler linewidth has become commonplace. Indeed, there have been several studies at sub–Doppler resolution using saturation effects. Even more excitingly, lasers have enabled spectroscopy to be performed with much greater sensitivity so that there are now many IR spectra recorded of transient species such as free radicals and van der Waals' complexes. Perhaps the major achievement in spectroscopy in the last decade has been the characterisation of a large number of molecular ions. The vast majority of these observations have been made in the IR region using laser spectroscopy.

Some landmarks in the field are recorded in this chapter. After a brief summary of the laser systems employed, the various experimental techniques used are described. The bulk of the chapter is given over to sketches of individual pieces of work in order to exemplify the range of problems which can now be tackled.

2. INFRARED LASER SYSTEMS

It is not my intention to give a detailed description of the various IR laser systems available. This has been done with much greater authority elsewhere[6,7,8]. However, it is worth appreciating the main features and differences of the various systems which are summarized in Table 1. There are two main types, *tunable* lasers whose frequency can be varied continuously over a certain range by application of a voltage ramp and *fixed frequency* lasers which oscillate only at particular frequencies (corresponding to spectroscopic transitions in the lasing molecules). Tunable lasers have a condensed phase active medium whereas fixed frequency lasers have a gaseous active medium. Reference to Table 1 shows that considerably more power is available from the gas lasers and it is this characteristic which makes them competitive with the less powerful, tunable lasers. The other important difference is that the various systems operate in different regions of the IR. Often the molecule which is to be studied will dictate which laser system is to be used. For tunable lasers, it is important to appreciate how far each device can be tuned in a *single scan*. Individual scans are usually very narrow (~ 1 cm^{-1}) but nowadays it is commonplace to control the laser with a computer. Under such control, it is possible to achieve broader scans by stitching individual scans together.

Table 1. Characteristics of Infrared Lasers

Laser	Active Medium	Wavenumber range/cm⁻¹	Typical output power (cw)	Range of tunability (single scan)	Linewidth	Disadvantages
Diode	p–n junction diode	380–3500	100μW	100 cm⁻¹ (1–5 cm⁻¹)	10^{-3}cm⁻¹	Multimode lasing, patchy coverage
Difference	dye laser + Ar⁺	2400–4500	10μW	100 cm⁻¹ (1 cm⁻¹)	2×10^{-4}cm⁻¹	No long λ coverage
F–centre	$F_{A(II)}$centre	2900–4000	10mW	500 cm⁻¹ (50 cm⁻¹)	10^{-4}cm⁻¹	No long λ coverage
CO_2	CO_2/N_2	850–1100	30W	line tunable	10^{-5}cm⁻¹	Not tunable 10μm only
CO	$CO/N_2/He$	1200–2050 (2450–3500)[a]	2W	line tunable	3×10^{-5}cm⁻¹	Not tunable

[a] Δv=2 transitions[9]

It is therefore this figure which is given in the fifth column of the table. Note the impressive scan range of the F–centre laser; this is the major advantage of this system. The linewidths are given in the sixth column. Although these show some variations, they are all considerably sub–Doppler (a typical Doppler linewidth in the IR is 100 MHz or 0.03 cm^{-1}, FWHM). Again the fixed frequency lasers have a considerable advantage of significantly narrower linewidths, which can be manifested in saturation experiments. The diode laser is the worst system in this respect. It is possible to reduce the linewidth by active stabilisation but the figure of 10^{-3} cm^{-1} is quite typical for the instabilities of a commercial diode laser.

Finally, it is worth highlighting a recent development of the CO laser. Professor W. Urban and his group at Bonn University have succeeded in building a system which will lase on $\Delta v=2$ transitions[9] (the normal 5μm transitions involve $\Delta v=1$). This is a result of great significance because it opens up the 3μm region (where most vibrations which involve stretching of the X–H bond occur) to the very sensitive laser magnetic resonance techniques. It is certain that many new observations will be made using this new laser system.

3. THE GENERATION OF UNSTABLE MOLECULES

Unstable or transient molecules by their very nature exist only fleetingly with typical laboratory lifetimes between 10^{-6} and 10^{-3} sec. Special methods are required to generate sufficient concentration for them to be detected. For most of these methods, a steady state concentration is built up by forming the species continuously.

(a) *Discharge/Flow methods*

Most methods of generating transient molecules involve flow systems in which streams of gas are pumped continuously through tubes. In the discharge/flow system, two such streams of gases are mixed in a reaction zone to generate the molecule of interest. Usually, one of the streams is rendered more reactive by first passing it through an electric discharge. A diagram of the production scheme is shown in Fig. 1. Although the discharge can be operated d.c. (or at low a.c. frequencies), it is preferable to use either an r.f. or microwave discharge because they do not require electrodes within the gaseous volume (electrodes tend to provide very efficient sites for the removal of transient molecules). An example of the use of this method is the production of the HCO radical[10]:

$$H_2CO + F \longrightarrow HCO + HF$$

where the F atoms are formed by discharging CF_4 or F_2.

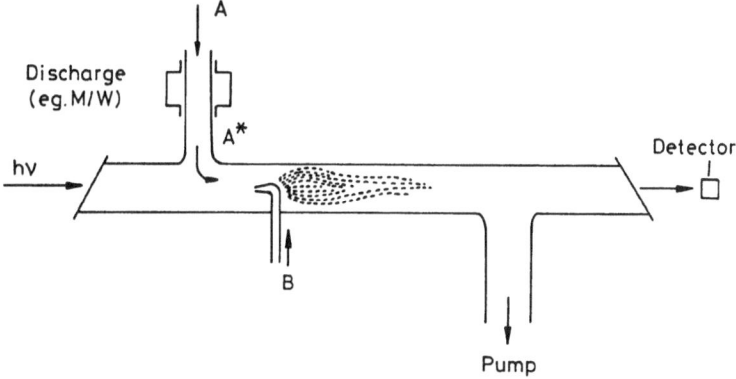

Fig. 1. A schematic diagram of the discharge/flow apparatus for the generation of short–lived molecules. The products of a discharge through flowing gas A are mixed with a continuous flow of a secondary gas B reacting downstream to generate the desired product.

(b) *Photolysis and ablation*

The photolytic generation of free radicals with U.V. lamps has been used for a long time. It provided a great flood of results for electronic spectroscopy in the 1950's, the technique being referred to as flash photolysis[11]. However, the concentrations of transient species generated in this way were not high enough to permit the detection of IR spectra. More recently, the availability of high power excimer lasers has changed the situation. It is usual to photolyse the stable parent molecule in the bulk phase. For example, the methyl radical can be generated efficiently by the photolysis of CH_3I[12]:

$$CH_3I + h\nu \longrightarrow CH_3{}^{\cdot} + I^{\cdot}$$

Radicals and even ions have also been generated by photolysis or photoionisation of a stable precursor seeded into a rare gas free–jet expansion[13]. The molecules are thereby generated cold, considerably simplifying the resultant spectrum.

(c) *Electric Discharges*

Much of the progress in spectroscopy of transient molecules over the last 20 years has depended on the use of electric discharges to generate the molecules concerned. This is particularly true for ionic species. The real advantage of an electric discharge is that it provides a method for generating transient molecules continuously over a reasonably large volume. Consequently, it is easier to achieve long absorption paths and hence high sensitivity; most other methods of generation tend to produce the species in a very confined volume.

Although gaseous electric discharges have been used for a variety of purposes for a very long time, the details of their operation are still only poorly understood. The last decade has seen some attempt at characterizing them from the point of view of the formation of both charged and neutral molecules. The main regions in a low frequency (or d.c.) electric discharge are shown in Fig. 2. The voltage drop between anode and cathode is not uniform; indeed most of it occurs close to the cathode in the cathode dark space. The concentration of positive ions is greatest in the negative glow region which is relatively short under normal conditions. Its length is equal to the distance travelled by the high energy electrons which have been accelerated across the cathode drop region and is proportional to the electron energy. It is typically a few cm at a pressure of 1 Torr. The ions are formed in this region by electron bombardment,

$$M + e \longrightarrow M^+ + 2e$$

The characteristics of a glow discharge are determined principally by cathode phenomena. In a normal discharge, the voltage across the cathode drop remains essentially constant as the current increases ($R \sim 0$), hence the need for a ballast resistor under these circumstances. At the same time, the cathode area covered by the discharge plasma increases. In the so–called "anomalous discharge", the plasma covers the entire cathode area and the voltage across the cathode drop rises

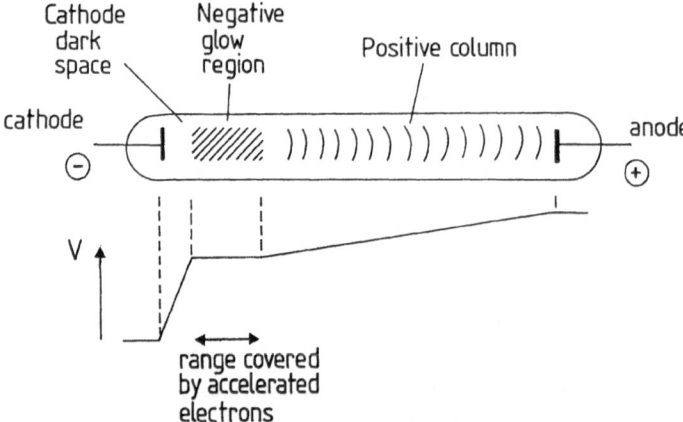

Fig. 2. A diagram showing the different regions in an electric discharge through a low pressure gas. The potential varies non–uniformly through the discharge as shown, the largest fall occurring in the cathode dark space. Positive ions are formed predominantly in the negative glow region.

roughly in proportion to the current.   As a result, the energy of the electrons and the distance they travel both increase with current.

Two main approaches have been used to the improvement of the discharge design for the production of ions.

(i) *Hollow cathode discharge cell*

The principle of the hollow cathode discharge is to wrap the cathode around as much of the discharge volume as possible, thereby increasing the extent of the negative glow region and hence the number of ions formed in the discharge.   Van den Heuvel and Dymanus published an effective design[14] which has since been copied by many other groups.   It is shown in Fig. 3.   The discharge is struck between the hollow cathode and an anode positioned at the end of a glass insert into the main tube.   Stable operation is possible over a large range of currents and pressures.   The positive column is confined to the glass insert between anode and cathode only and the negative glow is concentrated along the axis of the cathode tube.   At about 10 cm from the cathode surface, the glow changes abruptly into the cathode dark space.   The length of the glow is directly related to the voltage applied to the anode.   The cathode is cooled efficiently by a slow flow of liquid nitrogen through a helical copper tube soldered around the cathode tube.

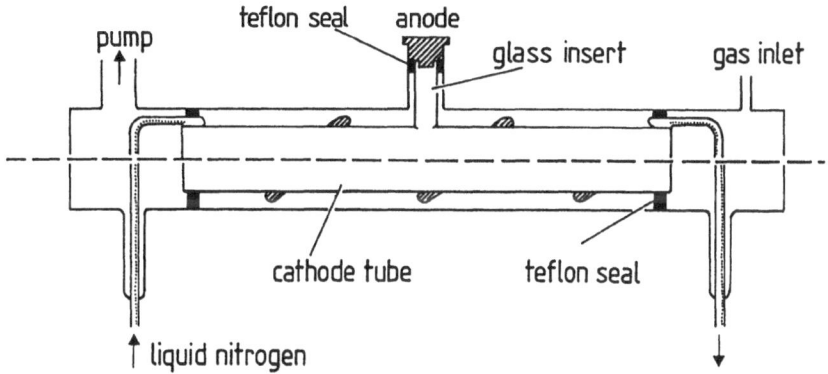

Fig. 3.   A hollow cathode discharge cell for the generation of positive ions, after the design of van den Heuvel and Dymanus[14]. The copper cathode fits inside a glass cell and is cooled by flowing liquid nitrogen through a spiral tube soldered to its outside surface.

(ii) *Magnetic Field Enhancement of the Negative Glow Region*

With a more conventional design of discharge tube (Fig. 2), the electrons which are accelerated through the cathode gap tend to spread out and consequently many are lost through collisions with the wall of the tube. The distance which they travel down the discharge tube is much less than it might otherwise be. It was observed many years ago[15] that the length of the negative glow could be extended by the application of a longitudinal magnetic field. De Lucia and his group at Duke University have exploited this observation to achieve dramatic increases in the signals recorded from ions formed in a discharge[16]. By passing a current through a solenoid wound on the outside of their discharge tube, magnetic field strengths of several tens of mT were achieved. Fields of this magnitude are sufficient to confine electrons with several hundreds of volts of transverse energy to a cyclotron radius of the order of 1 cm, roughly the cross section of the discharge probed by the laser beam. By restricting the size of the electrodes, it is possible to force the discharge to run in the anomalous mode. The increase in positive ion concentration achieved in this way is about 100 fold.

(d) *Thermal Generation*

On the face of it, the use of high temperatures to generate molecules for the study of their infrared spectra does not look a very promising approach, essentially because a furnace is a good source of infrared noise. However a laser beam is easy to collimate and so spatial separation of signal beam and furnace background can be achieved. Several groups have recorded infrared spectra of transient molecules by generating them in a furnace[17,18] but Jones and his group at Ulm University have really exploited this technique to the full in studying a whole series of diatomic hydride molecules[19].

Their apparatus is shown schematically in Fig. 4. The cell is made of an aluminium oxide ceramic and has two cylindrical water–cooled stainless steel electrodes fitted coaxially to the ends of the tube. The cell is placed inside a tube furnace capable of producing temperatures up to 1300^0C. A suitable metal M is placed in the centre of the cell in a ceramic boat and the cell is filled with a 20% mixture of hydrogen in helium at a total pressure of 5 mbar. The gas is discharged to generate H atoms which combine with the hot metal to produce MH. The technique works particularly well for the more volatile metals for which significant vapour pressure can be generated.

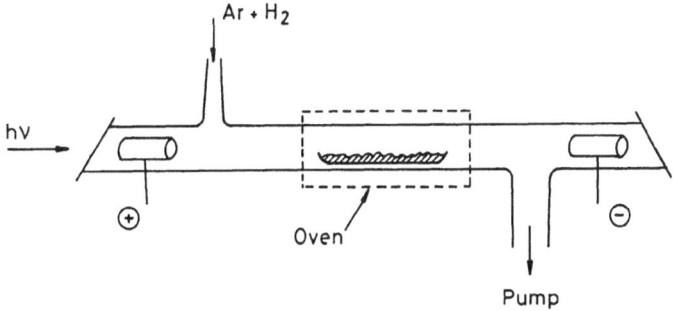

Fig. 4. Apparatus for the generation of metal hydrides by flowing (discharged) H_2 over metal heated in a furnace indicated by the dashed outline.

(e) *Supersonic Expansion*

When a gas is forced through a small hole from a high to a low pressure region, a diverging jet of molecules is formed for which equipartition of energy is no longer applicable. The translational velocity of the molecules is increased at the expense of the internal (vibrational and rotational) degrees of freedom[13]. With low vibrational "temperatures", it is possible to form and maintain weakly bound complexes such as van der Waals' molecules[20]. Much work has been done in the last five years on the infrared spectroscopy of such species[21,22], using an apparatus like that shown in Fig. 5. The compound which is to be involved in the complex is seeded in an inert gas carrier at moderately high pressures ($\sim$ 1 atm) and the whole is expanded through a pulsed nozzle into a high vacuum region. The region just in front of the nozzle (where the complexes are formed) is sampled with a beam from an infrared laser. The absorption of laser power by the sample is enhanced by the use of a multipass system; the plane mirrors are not parallel so that successive reflections of the beam "walk" in towards the nozzle and then out again. Greater path lengths and hence sensitivity can be achieved by using a slit nozzle[23].

4. SPECIAL EXPERIMENTAL TECHNIQUES

The advance of infrared spectroscopy of transient molecules has often depended on the ingenious exploitation of the special properties of the laser sources used and the molecules concerned. Two of these deserve special mention.

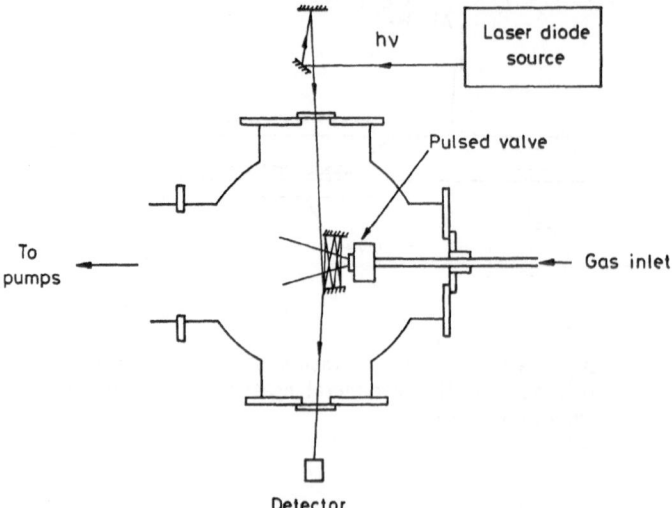

Fig. 5.
Apparatus for the study of infrared spectra of van der Waals'
molecules, formed by supersonic expansion of the component
gases through a pulsed nozzle. The system of tilted plane
mirrors causes the laser beam to make several passes through
the sample, thereby increasing the absorption.

(a) *Velocity modulation*

This is a method, attributed to Gudeman and Saykally[24], for increasing the
sensitivity for the detection of molecular ions in a discharge by absorption of
infrared radiation. Although quite high concentrations of positive ions can be
achieved in a discharge plasma, they are always much less (by about 3 orders of
magnitude) than the concentration of parent neutral molecules. The latter cause
infrared absorptions at very similar wavelengths which can easily swamp the signals
from the ions.

In a positive column discharge, the local electric field is typically 10 V cm^{-1}.
This causes a drift velocity v_d of the ions of about 500 m s^{-1}, in other words of the
same magnitude as the average random thermal velocity (at room temperature). In
the rest frame of the molecule, the radiation frequency ν_{ir} is Doppler shifted by

$$\Delta\nu_{Doppler} = (v_d/c)\,\nu_{ir}$$

where c is the speed of light. This shift is about the same size as the Doppler
linewidth. In the velocity modulation technique, a bipolar a.c. voltage (at a few
kHz) is applied to the discharge. This has the effect of shifting the absorption line
in and out of resonance with the tunable infrared laser at the modulation frequency.
Neutral species are not affected by the alternation of the discharge polarity since

they have purely random motion. The ion signal is thus readily separated from that of the parent molecule with a phase sensitive detector tuned to the modulation frequency. The extent to which the neutral molecule signals are suppressed depends on the ability to balance the two halves of the discharge cycle. Any mismatch gives rise to a difference in concentrations which in turn causes a signal.

(b) *Doppler Tuning*

Much spectroscopy is performed with fixed frequency lasers. Doppler tuning is a technique borrowed from atomic physicists for scanning regions of molecular ion spectra with a fixed frequency laser. For a molecule moving towards a radiation source of frequency ν_0 with a velocity of magnitude v, the frequency perceived in the molecule's frame of reference is

$$\nu = \nu_0 \left(1 + \text{v}/\text{c}\right).$$

Thus by changing the velocity with which the molecule moves relative to the source, it is possible to vary the frequency of the radiation in accordance with this relationship. The velocity of charged molecules can be controlled quite easily by accelerating them to a set potential V in an ion beam. This method of tuning was first applied to molecular species by Wing et al.[25]. It has the additional advantage from the spectroscopic point of view that the velocity spread of the ions is compressed by their acceleration to the potential V, essentially because the slower moving ions spend longer in the accelerating field and so catch up the faster ones. The spectroscopic linewidth which depends on this velocity spread is correspondingly reduced.

5. EXAMPLES

The remainder of this chapter is given over to brief descriptions of individual molecules which have been studied in the infrared by laser spectroscopy. The examples, all of which are transient species, have been selected because they are in some way important and because they demonstrate in full the various methods described earlier.

(a) *Tunable Lasers*

We deal first with some examples of spectra recorded with tunable lasers, i.e. simply as frequency scans.

(i) The Hydroperoxyl Radical, HO_2

HO_2 is a short–lived free radical which plays an important part in combustion processes and also in atmospheric chemistry. It has been studied spectroscopically from every point of view and its major structural properties are well characterized. Its ground electronic state is antisymmetric with respect to reflection in the plane of the molecule $(\tilde{X}^2 A'')$ and its geometry is well determined:

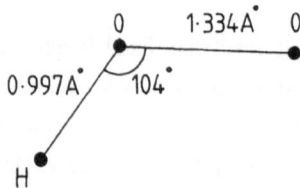

HO_2 has 3 vibrational modes, all of a' symmetry and all of which have been studied in the infrared by laser spectroscopy. The wavenumber of the vibrations are:

$$\tilde{\nu}_1 = 3436.20 \text{ cm}^{-1} \qquad \text{(H–O stretching vibration)}[26]$$
$$\tilde{\nu}_2 = 1391.75 \text{ cm}^{-1} \qquad \text{(bending vibration)}[27]$$
$$\tilde{\nu}_3 = 1097.63 \text{ cm}^{-1} \qquad \text{(O–O stretching vibration)}[28]$$

Note that the O–O stretching vibration has a lower frequency than the bending vibration, an unusual occurrence which is consistent with the anomalous length (and weakness) of the O–O band. In the laboratory, the molecule is usually formed in a discharge flow system by the reaction of F atoms with H_2O_2[29] or of O atoms with allyl alcohol[28]. Part of the ν_2 band recorded by diode laser spectroscopy is shown in Fig. 6. All of the observed transitions[27] are induced by the a component of the dipole moment and hence obey the selection with $\Delta K_a = 0$. From the inertial point of view, the molecule is a near–prolate symmetric top. Each rotational level $N_{K_a K_c}$ is split into a doublet by spin–rotation coupling

$$J = N + S.$$

The two Q–branches ($\Delta J = 0$) which arise in this way for $K_a = 3$ are shown in Fig. 6. The spin splitting is seen to be quite large showing that the spin–rotationinteraction changes with vibrational excitation. Each line in the spectrum appears as a 2nd derivative of an absorption lineshape. This is because the signals were recorded with Zeeman modulation and detected at twice the modulation frequency.

(ii) H_3^+

The detection of the infrared spectrum of H_3^+ by Takeshi Oka in 1980[30] was a

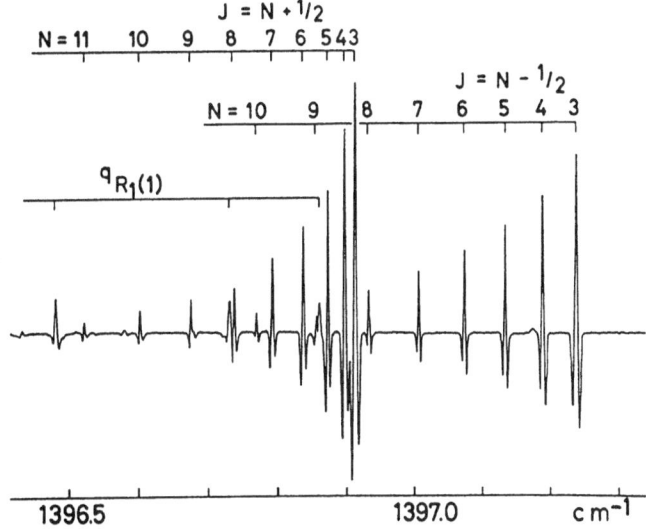

Fig. 6.   Part of the ν_2 band of the HO_2 radical recorded with a diode laser[27].   The lineshape arises from the use of Zeeman modulation.   The two Q–branches shown arise from the effects of spin doubling.

landmark in molecular spectroscopy, marking as it did the beginning of a decade of intensive study of molecular ions.   The species itself is of fundamental importance. H_3^+ is the simplest possible polyatomic molecule, it is a very stable component of ion–molecule reactions and acts as a protonating source in many situations; it is important in both terrestrial and astrophysical environments, largely because of the high abundance of hydrogen in the universe.   The molecule has the symmetrical shape of an equilateral triangle in its ground electronic state $(\tilde{X}^1A_1{}')$. However it has no bound excited electronic states of the same geometry so that it is not possible to detect and study the molecule through its electronic spectrum.   This is one reason why H_3^+ is such a latecomer onto the spectroscopic scene.   It can be and has been detected through infrared spectroscopy, a demonstration of a general maxim that all polyatomic molecules have infrared spectra but they may not have electronic spectra.   Thus, though infrared spectroscopy may be a less sensitive technique, it is more reliable.

The H–H bond length of H_3^+ in its ground state is now known to be 0.8763Å. The molecule has 2 vibrational modes which can be classified in its point group (D_{3h}).   The totally symmetric stretching vibration ν_1 (breathing mode) is of $a_1{}'$ symmetry and the 2–fold degenerate ring distortion mode ν_2 is of e' symmetry. The former is not infrared active and has still not been measured experimentally. Oka detected transitions in the ν_2 band induced by the perpendicular components of

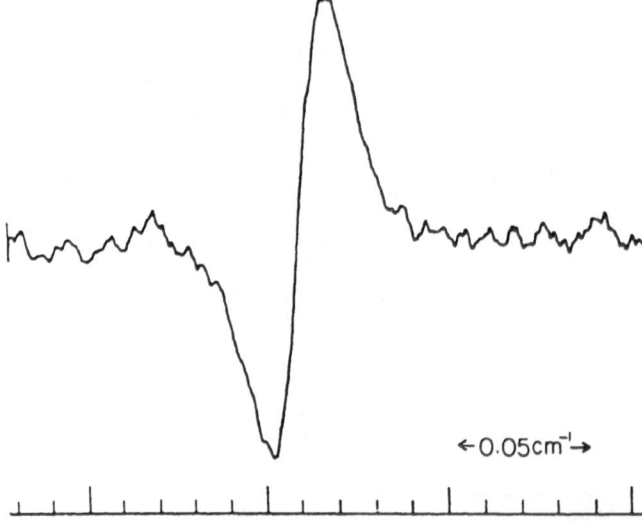

$\leftarrow 0.05 \text{cm}^{-1} \rightarrow$

Fig. 7.   The observed $R_0(1)$ transition of H_3^+ at 2725.885 cm^{-1} recorded by Oka with a difference frequency spectrometer[30]. The first derivative lineshape is caused by the use of frequency modulation.

the dipole moment and obeying the selection rule $\Delta K = \pm 1$. His observations andsubsequent work[31,32] have established the wavenumber for this vibration to be 2521.31 cm^{-1}. An example from Oka's original spectrum recorded with a difference frequency laser is given in Fig. 7. This shows the $J=2 \leftarrow 1$, $K = 1 \leftarrow 0$ line, the first derivative line–shape arising from the use of frequency modulation in recording the spectrum. The H_3^+ was generated by a discharge through H_2 in a long tube, the walls of which were cooled to liquid nitrogen temperatures. It has been possible to improve the signal–to–noise ratio greatly in later experiments, some of which have involved the use of diode lasers.

(*iii*) H_2F^+

This molecule has been selected as a representative example of the work of R.J. Saykally and his group at the University of California, Berkeley. A very large number of infrared studies of both positive and negative ions have come from this laboratory where the velocity modulation method was developed[24]. H_2F^+ is isoelectronic with water and therefore has a closed shell $^1A_1$ ground state. Its geometry is very similar to that of water

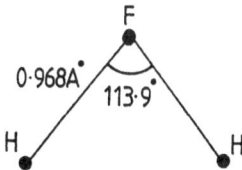

and it has 3 vibrational modes:

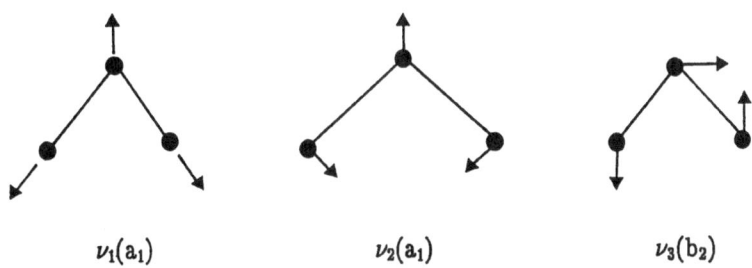

$\nu_1(a_1)$ $\qquad\qquad$ $\nu_2(a_1)$ $\qquad\qquad$ $\nu_3(b_2)$

All three vibrations are infrared active.   The ν_1 and ν_3 fundamental bands at
3348.71 cm^{-1} and 3334.67 cm^{-1} respectively have been studied using a colour–centre
laser and velocity modulation[33]. H_2F^+ was formed in an a.c. discharge through
150 Pa H_2 and 25 Pa HF.   Part of the rotational structure of the ν_3 band is shown
in Fig. 8.   Like water, H_2F^+ is a strongly asymmetric rotor and the rotational
structure is rather irregular.   The two equivalent H nuclei give rise to ortho and
para spin states with weights of 3 and 1 respectively.   The two K–doubling
components of the $J = 4 \leftarrow 5,\ K_a = 3$ transition which can be seen in the centre of

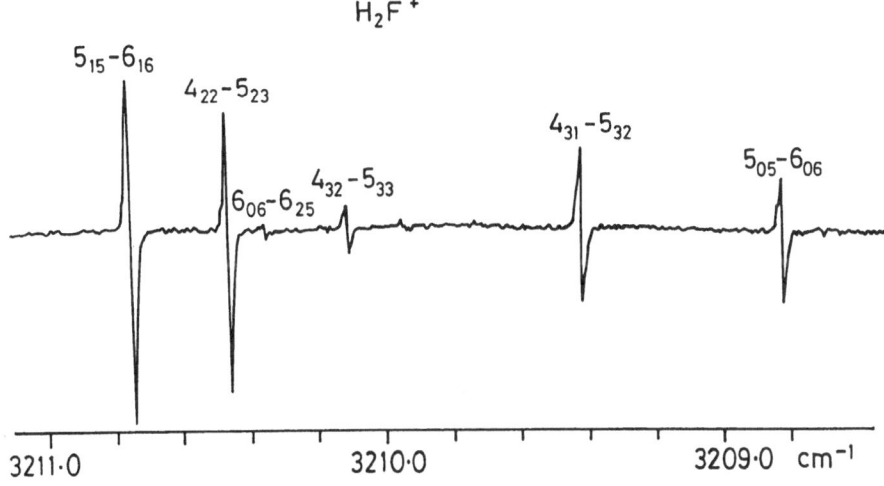

Fig. 8.   A portion of the P–branch of the ν_3 band of H_2F^+, recorded
with a colour–centre laser[33].   Note the 3:1 relative intensity
of the 4_3–5_3 asymmetry doublet.

the scan in Fig. 8 show these relative intensities very clearly.

(iv) Ar...CO$_2$

This is an example of the class of compound called van der Waals' complexes, in which the Ar atom is loosely bonded to the CO$_2$ to form a T–shaped molecule. Many such species have been studied in absorption by infrared laser spectroscopy by exciting a vibration in the parent molecule, in this case the antisymmetric stretching vibration of CO$_2$ at 2349 cm^{-1}. The energy in each infrared photon is much larger than the dissociation energy of the van der Waals' bond ($\sim$ 100 cm^{-1})[34]. Nevertheless, sharp lines can be seen in such spectra because it takes a long time for the vibrational excitation energy in the CO$_2$ molecule to transfer into the van der Waals' bond. The molecule is T–shaped with a long bond (3.29Å) between the Ar and C atoms[34]. It is therefore a near–prolate symmetric top with the a inertial axis lying along this bond. The transition moment is perpendicular to this direction, resulting in the selection rule $\Delta K_a = \pm 1$. Furthermore, the ^{16}O nuclei have zero nuclear spin so that all levels with odd values for K_a are missing by the exclusion principle. Part of the absorption spectrum of ArCO$_2$ recorded with a diode laser is shown in Fig. 9. The complex is formed by expanding a mixture of 0.5% CO$_2$ in 2 atm of argon through a 350 μm pulsed nozzle into a high vacuum region[35]. Information on the geometric structure of the complex can be obtained from such a spectrum. In addition, Randall et al.[35] showed that the antisymmetric

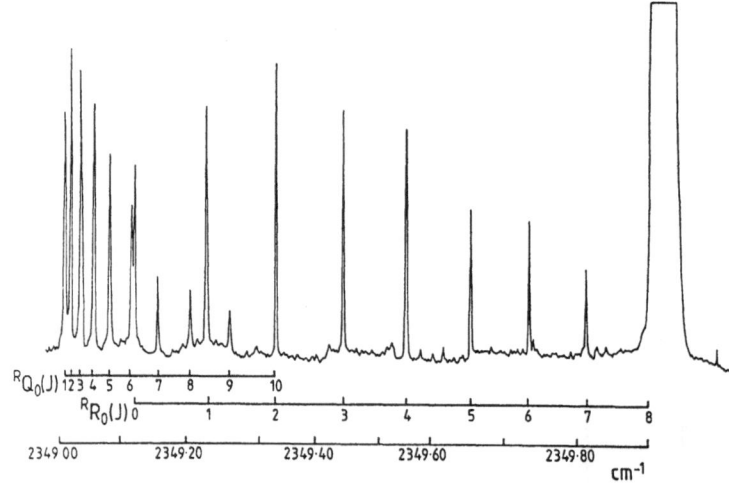

Fig. 9.   The $^IQ_0(J)$ and $^IR_0(J)$ branches in the diode laser spectrum of ArCO$_2$. Note the strong line arising from the parent molecule, CO$_2$. Note (i) the low rotational temperature in the spectrum (2.5K) which results from the isentropic expansion and (ii) the immensely strong line due to the parent molecule CO$_2$.

stretching vibration has a slightly lower wavenumber (by 0.470 cm^{-1}) in the complex. The smallness of this shift confirms the assumption that the vibrational motion is localised on the CO_2 moiety. The shift also implies that the binding energy of the complex *increases* on excitation of the antisymmetric stretching vibration in CO_2, a rather surprizing result.

(b) *Fixed Frequency Lasers*

Table 1 shows that fixed frequency lasers have some advantages over their tunable counterparts, chiefly those of higher power and greater spectral purity. There is therefore some incentive to use these lasers to record molecular spectra. This is done by tuning the molecular transitions into coincidence with the laser rather than the other way round. Molecular tuning can be achieved in one of three ways: Stark tuning, Zeeman tuning and Doppler tuning. Stark tuning is accomplished by applying a variable electric field to the samples. The electric field E interacts with the dipole moment μ to cause a shift in the energy of an individual M state:

$$\Delta E = - < \mu \, . \, E >$$

The frequency of a transition between a given pair of M states thus varies as a function of the applied electric field. With a modicum of luck, a field exists for which the transition frequency exactly equals the laser frequency and energy is absorbed from the radiation field by the molecule. The spectrum is therefore recorded as a series of resonances as the electric field is increased. The experiment can be performed on any molecule which possesses a permanent electric dipole moment.

Zeeman tuning is the analogous experiment performed with a magnetic field instead of the electric field. In this case, the M states are shifted through the interaction of the magnetic dipole m with the applied magnetic flux density B:

$$\Delta E = - < m \, . \, B >$$

The experiment in this case is called laser magnetic resonance (LMR) and requires the molecule to have a sizeable magnetic moment. It is therefore restricted to the study of open shell molecules (free radicals) where the electron orbital or spin angular motion produces a large magnetic moment.

The third method is Doppler tuning which has been discussed in some detail in an earlier section. In practice, this procedure, which involves scanning the velocity with which a molecule moves, is restricted to molecular ions.

(v) The formyl radical, HCO

The HCO radical plays an important part in combustion processes and, like HO_2, has been studied in almost every conceivable spectroscopic experiment. It is non–linear in its ground $^2A'$ state with the following geometry

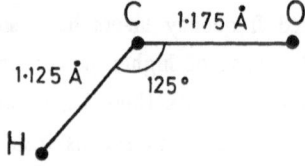

The 3 vibrational modes (all of a′ symmetry) are well–established:

$$\tilde{\nu}_1 = 2434.48 \text{ cm}^{-1} \qquad \text{C--H stretching vibration}[26]$$
$$\tilde{\nu}_2 = 1080.76 \text{ cm}^{-1} \qquad \text{bending vibration}[37,38]$$
$$\tilde{\nu}_3 = 1868.17 \text{ cm}^{-1} \qquad \text{C--O stretching vibration}[39-41]$$

The most common method for generating HCO is the reaction of F atoms with H_2CO in a discharge flow system[10]. The radical can also be formed quite efficiently by the photolysis of CH_3CHO with an excimer laser[36]. HCO is a good candidate for study by fixed frequency laser spectroscopy because the ν_2 and ν_3 vibrations fall in the 10 μm (CO_2) and 5 μm (CO) regions respectively. It has a sizeable magnetic moment from the spin of the unpaired electron and it has been studied extensively by LMR spectroscopy. However, it is almost the only transient molecule to have been studied by laser Stark spectroscopy[37]. This technique requires the sample to be located between two closely spaced metal plates across which the electric field is generated. In practice, it is very difficult to form reactive molecules in such an environment because they tend to recombine on the metal surface. Largely for this reason, laser Stark spectroscopy has been used to study stable molecules.

Since laser Stark spectroscopy depends on the interaction between the electric dipole moment and the applied electric field, it also provides a measurement of the dipole moment. In the case of HCO, the dipole moments are determined to be

$$\mu_a' = 1.3474(48) \text{ Debye,}$$
$$\mu_a'' = 1.3626(39) \text{ Debye.}$$

The electric dipole moment does not lie along the a inertial axis but only its a component can be determined.

(vi) The Methylene Radical, CH_2

The methylene radical is a fundamental hydrocarbon fragment which has proved to be a very elusive spectroscopic target. It took a long time before its electronic spectrum was firmly identified[42]. More recently, it has succumbed to attack in the far– and mid–infrared which has resulted in most of its major properties being characterized. The ground state has an open shell electronic structure $(\tilde{X}^3B_1)$ with a very obtuse bond angle[43]:

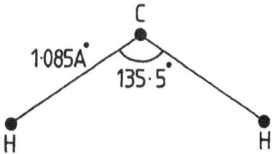

There is also a low–lying singlet state $(\tilde{a}^1A_1)$, only 3165 cm^{-1} above the triplet state[44] in which the bond angle is more nearly acute at 104°. Of its 3 vibrational modes, only the bending vibration ν_2 has been measured: $\tilde{\nu}_2 = 963.0995$ cm^{-1} [45]. The two C–H stretching vibrations are thought to be only weakly infrared active and have not yet been detected. Their calculated wavenumbers are 2992 cm^{-1} $(\tilde{\nu}_1)$ and 3213 cm^{-1} $(\tilde{\nu}_3)$.

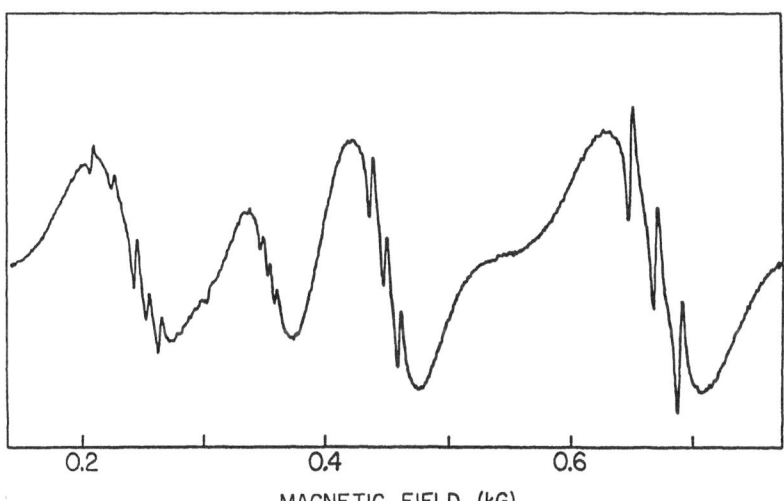

MAGNETIC FIELD (kG)

Fig. 10   Part of the mid–infrared LMR spectrum of CH_2 observed with the $^{12}C^{16}O_2$ P(34) laser line at 931.001 cm^{-1}. The transition is $2_{20} - 1_{11}$ of the ν_2 band. Note the saturation dips with the triplet hyperfine structure expected for ortho levels.

The bending vibration has been widely studied by LMR (the transitions fall conveniently in the range of the CO_2 laser) and also by diode laser spectroscopy[46]. The LMR spectrum of the ν_2 fundamental showed characteristic Zeeman patterns which provided the key to the analysis of both this spectrum and also the far infaredspectrum which had been obtained previously[43,47]. Fig. 10 shows part of the mid–IR LMR spectrum. The first derivative line shapes are caused by the use of Zeeman modulation. Notice also the saturation dips on each M component. The triplet structure arises from the proton hyperfine interaction and confirms that ortho levels are involved in this transition.

(vii) DCl^+

In addition to the large number of neutral free radicals which have been studied in the infrared by LMR, a few positive ions have also been detected. DCl^+ in its $X^2\Pi$ state is a good representative of this class of molecules. It was the first ion to be detected by LMR, by Saykally and Evenson in the far–infrared[48]. It has also been studied in the mid–IR (in its deuterated form because the vibrational frequency of HCl^+ lies outside the range of the CO laser) by Prof. W. Urban's group at Bonn University[49,50] using a LMR spectrometer based on the Faraday rotation effect. This system offers some advantages of increased sensitivity and is ideally suited to the study of long discharge plasmas since the magnetic field is provided by a longitudinal, super–conducting solenoid.

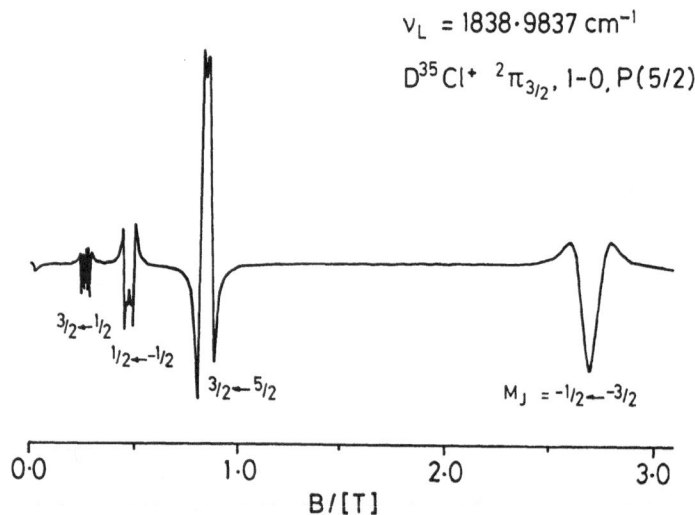

$$\nu_L = 1838\cdot9837 \text{ cm}^{-1}$$

$$D^{35}Cl^+ \ {}^2\pi_{3/2}, 1\text{-}0, P(5/2)$$

$3/2 \leftarrow 1/2$

$1/2 \leftarrow -1/2$

$3/2 \leftarrow 5/2$

$M_J = -1/2 \leftarrow -3/2$

0·0   1·0   2·0   3·0

B/[T]

Fig. 11.   The LMR spectrum of DCl^+ in an anomalous discharge. The transition involved is $P(5/2)$ in the fundamental band. The spectrometer works on the Faraday rotation effect which causes a 2nd-derivative type of lineshape. The sign of the signal gives information on the sign of the tuning rate and whether ΔM is $+1$ or -1. Note the Cl hyperfine structure on the lowest field resonance.

DCl$^+$ was formed in an electric discharge through a few mTorr of DCl in 2 torr of He. The discharge was operated in the anomalous mode so that the negative glow extended over the whole absorption region (30 cm in length). The magnetic field also confined the ions to a region near the axis, thereby increasing the concentration achieved. The discharge was cooled down to about 150K with liquid nitrogen, causing an increase in the population of the lower rotational levels. Under these conditions it was easy to resolve the Cl nuclear hyperfine splitting (see Fig. 11). On the other hand, hot bands up to $v = 7 - 6$ were observed suggesting extensive population of excited vibrational levels. It is postulated that this vibrational distribution may be the result of Penning ionisation by metastable He atoms.

(viii) The N_3 radical

Several polyatomic molecules have been studied by mid–IR LMR. The latest example is N_3, a linear symmetrical molecule with a $^2\Pi_g$ ground state. The electronic spectrum of this molecule was first studied some time ago[51] but gave no information as to the ground state vibrational intervals. Very recently, vibration–rotational transitions involving the antisymmetric stretching vibration ν_3 have been detected, first by FTIR[52] and then by LMR with a CO laser[53]. The vibrational wavenumber is determined from these studies to be 1644.679 cm^{-1}. The

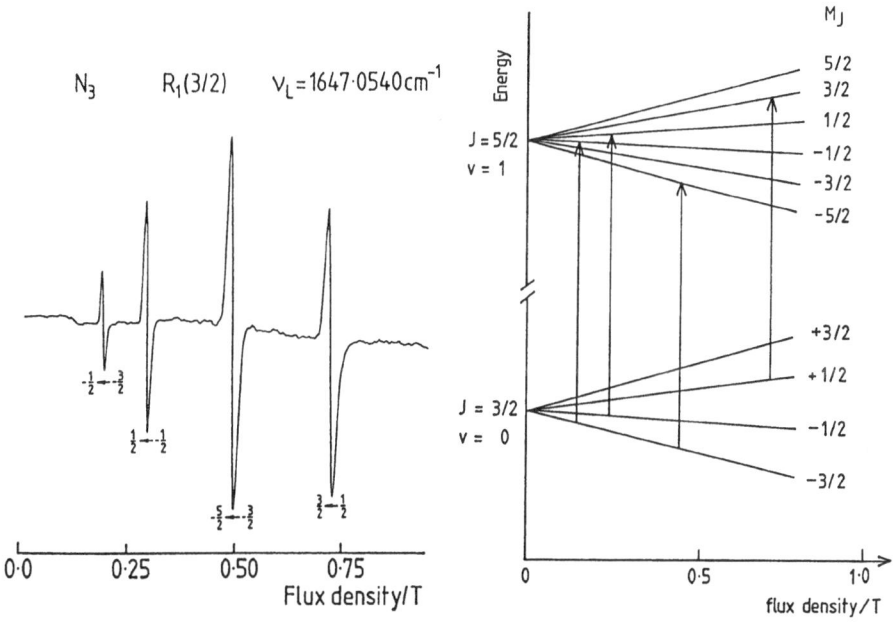

Fig. 12.   The LMR spectrum associated with the R(3/2) transition in the ν_3 fundamental band of N_3 and the energy level diagram which shows how the magnetic resonance spectrum arises.

LMR experiment suffers from being a very poor search technique but once transitions are detected, it offers much greater sensitivity. It has therefore proved possible to detect several hot bands and also to resolve ^{14}N hyperfine structures with Lamb dips. The N_3 radical is generated in a discharge flow system by the reaction of F atoms with hydrazoic acid, HN_3. Fig. 12 shows the LMR spectrum associated with the R(3/2) transition of the fundamental (ν_3) band of N_3, together with the energy level diagram which shows how these transitions arise.

(ix) The Hydrogen Molecular Ion, HD^+

H_2^+ is a molecule of fundamental importance for much the same reasons as H_3^+. It is the simplest possible molecule with only one electron and can be subjected to very accurate *ab initio* calculations. Despite this, there have been very few spectroscopic studies of this molecule until quite recently. Of course, H_2^+ is a symmetrical, non–polar molecule and therefore has no vibration–rotation spectrum. For this reason, HD^+ (in its ground $^2\Sigma^+$ state) has been intensively studied instead, using the technique of Doppler tuning of ion beams.

This work was started (as indeed was all molecular ion beam spectroscopy) by Wing et al.[25] in their study of vibration–rotation transitions in HD^+. Because HD^+ is a one electron system, its bond is comparatively weak and falls within the compass of the CO laser (around 1915 cm^{-1}). A diagram of the layout of the

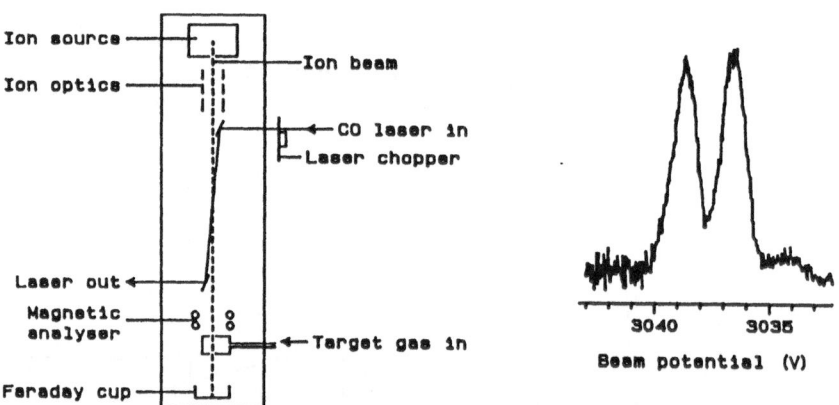

Fig. 13.   The apparatus used to study vibration-rotation transitions in HD$^+$ by Doppler tuning. The absorption of photons from the CO laser beam is detected by a change in the ion beam intensity at the Faraday cup as a result of differential charge transfer with a target gas (H_2). A typical signal is shown on the right.

apparatus used is shown in Fig. 13. A beam of HD^+ is formed by electron bombardment of HD. The ions are accelerated to an energy of several keV into a region of constant electrostatic potential where the ion beam crosses the CO laser beam at a small angle. The accelerating potential is adjusted to shift an ion transition into resonance with a nearby laser line through the Doppler effect. The ions then pass through a gas target where they are partially neutralized by charge transfer

$$HD^+ + H_2 \rightarrow HD + H_2^+$$

and are collected on the Faraday cup. The efficiency of charge transfer changes a few percent between adjacent vibrational levels so that when HD^+ is pumped from one vibrational to another by the laser, the intensity of the beam at the detector changes. In this way, the absorption of the laser photon can be detected. Wing et al. were able to detect vibration–rotation transitions in this way in the (1,0), (2,1) and (3,2) bands. Furthermore, the compression of the velocity distribution on acceleration of the ions gave narrow enough linewidths that proton hyperfine structure could be resolved.

The ideas sown by Wing in his initial experiment have borne fruit in the hands of other workers. In particular, Carrington and his co–workers have made extensive studies of vibration–rotation transitions in HD^+ in high–lying vibrational levels near the dissociation limit using the Doppler tuning method[54-56]. They detect the absorption of an IR photon by a second and subsequent photodissociation

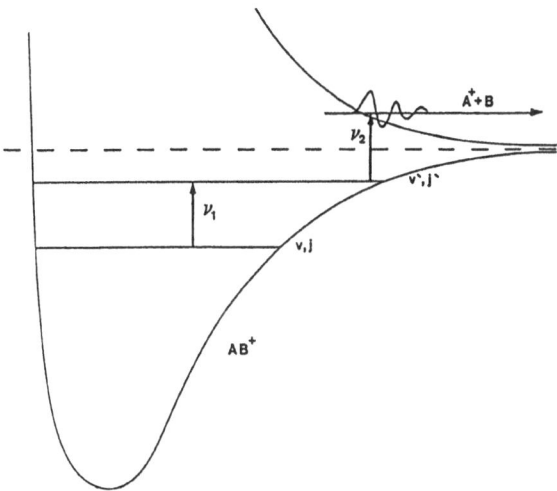

Fig. 14.   A diagram to show the principle of the detection of vibration–rotation transitions in HD^+ by laser photo–dissociation from the upper vibrational level.

step usually excited by the same laser source:

$$HD^+(v'') \xrightarrow{h\nu_L} HD^+(v') \xrightarrow{h\nu_L} H + D^+$$

The appearance of D^+ in the ion beam thus signals the absorption of photons in the initial step. The principle of their experiment is shown in Fig. 14. It can be seen that only levels within 2 quanta of the dissociation limit can be studied in this way. For HD^+, this means levels with v=14 to v=21 (the highest vibrational level). Experiments have been performed with both CO_2 and CO lasers. Transitions which involve changes in v by 3 or 4 are seen relatively easily because the anharmonic nature of the potential energy curve in this region makes the harmonic oscillator selection rule ($\Delta v = \pm 1$) inapplicable.

Very recently, Carrington, McNab and Montgomerie have shown that HD^+ in levels very close to dissociation (v=21) can be dissociated simply by applying an

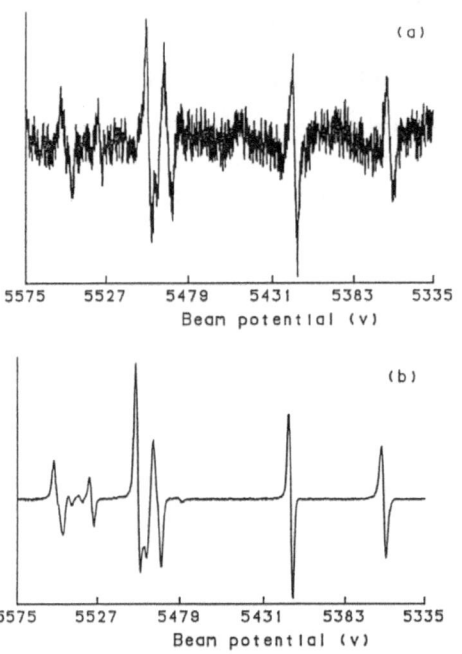

Fig. 15.   Recordings of the v = 21 − 17, N = 2 ← 3 transition in HD^+ obtained by monitoring the resonant increase in D^+ ions. The signals are produced by photodissociation with 40W of CO_2 laser power in (a) and by electric field dissociation in (b). The improvement in sensitivity is evident. The lines are different hyperfine components of the rotational transition.

electric field[53].   Indeed, for these levels, this a much more efficient method than photodissociation and gives a dramatic improvement in signal–to–noise.   An example is shown in Fig. 15.

ACKNOWLEDGEMENTS

I am particularly grateful to Iain McNab for copies of Figs. 14 and 15 and to Jonathan Towle and Valerie Heinrich for their help with the preparation of this manuscript.

REFERENCES

1.   R. Robertson, *Faraday Soc. Trans.* **25**, 899 (1929).
2.   J.U. White, *J. Opt. Soc. Amer.* **32**, 285 (1942); H.J. Bernstein and G. Herzberg, *J. Chem. Phys.* **16**, 30 (1948).
3.   J. Connes and P. Connes, *J. Opt. Soc. Amer.* **56**, 896 (1966).
4.   J. Pinard, *Ann. Phys. (Paris)* **4**, 147 (1969).
5.   C.K.N. Patel, *Phys. Rev. A* **136**, 1187 (1964).
6.   J. Hecht, *"The Laser Guidebook"*, McGraw Hill (1987).
7.   L.F. Mollenauer and J.C. White (eds.) *"Tunable Lasers"*, Topics in Applied Physics, Vol. 59, Springer (1987).
8.   W. Urban in *"Frontiers of Laser Spectroscopy of Gases"*, p9 (A.C.P. Alves, J.M. Brown and J.M. Hollas, eds.) Kluwer Academic (1988).
9.   M. Gromoll–Bohle, W. Bohle and W. Urban, *Optics Comm.*, **69**, 409 (1989).
10.   I.C. Bowater, J.M. Brown and A. Carrington, *J. Chem. Phys.*, **54**, 4957 (1971).
11.   G. Porter, *Proc. Roy. Soc.* **200A**, 284 (1950).
12.   G.E. Hall, T.J. Sears and J.M. Frye, *J. Chem. Phys.* **90**, 6234 (1989).
13.   S.C. Foster, R.A. Kennedy and T.A. Miller in *"Frontiers of Laser Spectroscopy of Gases"*, p421 (A.C.P. Alves, J.M. Brown and J.M. Hollas, eds.) Kluwer Academic (1988).
14.   F.C. van den Heuvel and A. Dymanus, *Chem. Phys. Letts.*, **92**, 219 (1982).
15.   J.J. Thomson and G.P. Thomson, *"Conductivity of Electricity through Gases"*, Cambridge University Press (1928).
16.   F.C. De Lucia, E. Herbst, G.M. Plummer and G.A. Blake, *J. Chem. Phys.* **78**, 2312 (1983).
17.   N.N. Haese, D.J. Liu and R.S. Altman, *J. Chem. Phys.* **81**, 3766 (1984).
18.   B. Lemoine, C. Demuynck, J.–L. Destombes and P.B. Davies, *J. Chem. Phys.* **89**, 673 (1988).
19.   U. Magg and H. Jones, *Chem. Phys. Letts.* **146**, 415 (1988) and subsequent papers.
20.   D.H. Levy, *Ann. Rev. Phys. Chem.* **31**, 197 (1980).
21.   G.D. Hayman, J. Hodge, B.J. Howard, J.S. Muenter and T.R. Dyke, *J. Chem. Phys.*, **86**, 1670 (1987).
22.   C.M. Lovejoy, M.D. Schuder and D.J. Nesbitt, *Chem. Phys. Letters*, **127**, 374 (1986).
23.   C.M. Lovejoy and D.J. Nesbitt, *Rev. Sci. Instrum.* **58**, 807 (1987).
24.   C.S. Gudeman and R.J. Saykally, *Ann. Rev. Phys. Chem.* **35**, 387 (1984).
25.   W.H. Wing, G.A. Ruff, W.E. Lamb and J.J. Spezeski, *Phys. Rev. Lett.* **36**, 1488 (1976).
26.   C. Yamada, Y. Endo and E. Hirota, *J. Chem. Phys.* **78**, 4379 (1983).
27.   K. Nagai, Y. Endo and E. Hirota, *J. Mol. Spec.* **89**, 520 (1981).

28. J.W.C. Johns, A.R.W. McKellar and M. Riggin, *J. Chem. Phys.* **68**, 3957 (1978); A.R.W. McKellar, *Disc. Faraday Soc.* **71**, 63 (1981).
29. H.E. Radford, K.M. Evenson and C.J. Howard, *J. Chem. Phys.* **60**, 3178 (1974).
30. T. Oka, *Phys. Rev. Lett.*, **45**, 531 (1980).
31. J.K.G. Watson, *J. Mol. Spec.*, **103**, 350 (1984).
32. J.K.G. Watson, S.C. Foster, A.R.W. McKellar, P. Bernath, T. Amano, F.S. Pan, M.W. Crofton, R.S. Altman and T. Oka, *Can. J. Phys.*, **62**, 1875 (1984).
33. E. Schäfer and R.J. Saykally, *J. Chem. Phys.*, **80**, 2973 (1984).
34. J.M. Steed, T.A. Dixon and W. Klemperer, *J. Chem. Phys.*, **70**, 4095 (1979).
35. R.W. Randall, M.A. Walsh and B.J. Howard, *Disc. Faraday Soc.*, **85**, 13 (1988).
36. C.B. Dane, D.R. Lander, R.F. Curl, F.K. Tittel, Y. Guo, M.I.F. Ochsner and C.B. Moore, *J. Chem. Phys.*, **88**, 2121 (1988).
37. B.M. Landsberg, A.J. Merer and T. Oka, *J. Mol. Spec.*, **67**, 459 (1977).
38. J.W.C. Johns, A.R.W. McKellar and M. Riggin, *J. Chem. Phys.*, **67**, 2427 (1977).
39. J.M. Brown, J. Buttenshaw, A. Carrington, K. Dumper and C.R. Parent, *J. Mol. Spec.* **79**, 47 (1980).
40. J.M. Brown, K. Dumper and R.S. Lowe, *J. Mol. Spec.*, **97**, 441 (1983).
41. A.R.W. McKellar, J.B. Burkholder, J.J. Orlando and C.J. Howard, *J. Mol. Spec.*, **130**, 445 (1988).
42. G. Herzberg, *Proc. Roy. Soc.*, **262A**, 291 (1961).
43. T.J. Sears, P.R. Bunker, A.R.W. McKellar, K.M. Evenson, D.A. Jennings and J.M. Brown, *J. Chem. Phys.*, **77**, 5348 (1982).
44. A.R.W. McKellar, P.R. Bunker, T.J. Sears, K.M. Evenson, R.J. Saykally and S.R. Langhoff, *J. Chem. Phys.*, **79**, 5251 (1983).
45. T.J. Sears, P.R. Bunker and A.R.W. McKellar, *J. Chem. Phys.*, **77**, 5363 (1982).
46. M.D. Marshall and A.R.W. McKellar, *J. Chem. Phys.*, **85**, 3716 (1986).
47. J.A. Mucha, K.M. Evenson, D.A. Jennings and C.J. Howard, *Chem. Phys. Letts.*, **66**, 244 (1979).
48. R.J. Saykally and K.M. Evenson, *Phys. Rev. Letters*, **43**, 515 (1979).
49. A. Hinz, W. Bohle, D. Zeitz, J. Werner, W. Seebass and W. Urban, *Mol. Phys.*, **53**, 1017 (1984).
50. W. Bohle, J. Werner, D. Zeitz, A. Hinz and W. Urban, *Mol. Phys.*, **58**, 85 (1986).
51. A.E. Douglas and W.E. Jones, *Can. J. Phys.*, **43**, 2216 (1965).
52. C.R. Brazier, P.F. Bernath, J.B. Burkholder and C.J. Howard, *J. Chem. Phys.*, **89**, 1762 (1988).
53. R. Pahnke, S.H. Ashworth and J.M. Brown, *Chem. Phys. Letts.*, **147**, 179 (1988).
54. A. Carrington and J. Buttenshaw, *Mol. Phys.*, **44**, 267 (1981).
55. A. Carrington, J. Buttenshaw and R.A. Kennedy, *Mol. Phys.*, **48**, 775 (1983).
56. A. Carrington and R.A. Kennedy, *Mol. Phys.*, **56**, 935 (1985).
57. A. Carrington, I.R. McNab and C.A. Montgomerie, *Chem. Phys. Lett.*, **151**, 258 (1988).

OPTO-THERMAL SPECTROSCOPY

Davide Bassi, Andrea Boschetti[s] and Mario Scotoni

Dipartimento di Fisica
Università degli Studi di Trento
I-38050 Povo (TN), Italy

[s]Centro CNR di Fisica degli Stati Aggregati, Trento

INTRODUCTION

About a decade ago, the first opto-thermal spectrometer was developed by Gough, Miller and Scoles [1] at the
Waterloo University. The basic idea of this technique is to
use a cryogenic thermal detector (bolometer) for measuring
the internal energy of a supersonic molecular beam. When the
molecular beam is illuminated by means of resonant radiation,
the internal state of molecules may change and the corresponding energy variation is detected by means of the bolometer. Up to now opto-thermal methods have found application
in the following fields:

- molecular beam diagnostics and internal state preparation;
- sub-Doppler infrared spectroscopy;
- photodissociation spectroscopy of dimers and clusters;
- spectroscopy of highly excited vibrational states (multiphoton excitations, overtone and combination bands).

In this paper we briefly recall the basic operating
principles of opto-thermal spectrometers. The use of pulsed
laser sources in combination with fast superconducting bolometers is discussed. Experimental results on multiphoton and
overtone spectroscopy of polyatomic molecules are presented
and discussed.

THE BASIC EXPERIMENTAL SETUP

Supersonic beams are, in principle, ideal spectroscopic tools. In fact they are characterized by a low number
density (typical values are summarized in Table 1) and low
temperatures (typically below ≈ 30 K, at least for rotational
and translational degrees of freedom). Under these conditions
pressure broadening and other collision effects are
undoubtedly negligible, while both Doppler broadening and
spectrum congestion are dramatically reduced. Moreover the
lack of collisions makes possible to study highly reactive

Applied Laser Spectroscopy, Edited by W. Demtröder and
M. Inguscio, Plenum Press, New York, 1990

215

(free radicals) or weakly bound species (Van der Waals
clusters) which are not stable in a room temperature cell.
Unfortunately the low number density and the small volume of
the illuminated region make actually impossible to use molecular beams in conventional absorption measurements.

A possible approach which overcomes the above problem
is based on the opto-thermal method. In a typical experiment
(see Fig.1) a skimmed supersonic beam [2] is crossed by a
tunable laser beam which can induce :

a) a change of internal energy (positive in case of
excitation or negative in case of stimulated emission) or,
alternatively, b) the loss of particles leaving the molecular
beam axis after a photodissociation process [3].

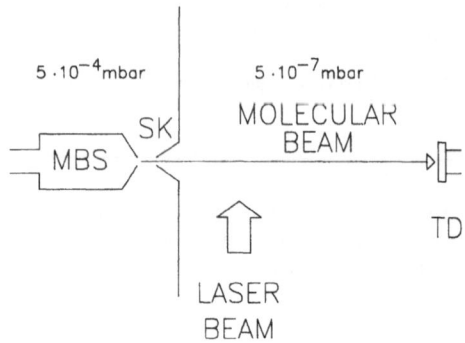

Fig.1   Schematic view of an opto-thermal spectrometer. The
vacuum system consists of two differentially pumped chambers.
The first vacuum chamber contains the molecular beam source
(MBS). The vacuum chambers are connected through a conical
skimmer (SK) which samples the core of the jet expansion. The
molecular beam is illuminated by the laser beam in a
collisionless region. Typical values of the pressure in the
two vacuum chambers are shown. The molecular beam energy flux
is measured by means of a thermal detector (TD), typically a
cryogenic bolometer.

Internal state excitations can be opto-thermally
detected if their natural lifetime Γ is longer than the time-
of-flight τ of the molecular beam from the illuminated region
to the thermal detector. In fact, if $\Gamma < \tau$ molecules lose
internal energy by fluorescence emission before sticking on
the detector surface. Typical values of τ range from 0.050 to
0.5 ms and the condition $\Gamma > \tau$ is generally satisfied in
the case of ro-vibrational excitations. A new interesting
application of opto-thermal methods is related to the study
of long-lived electronically excited metastable states, with
excitation energy below a few eV. The standard detection
method of metastable species is based on their conversion to
secondary electrons by means of a metal plate, followed by an
electron multiplier. This method does not work when the
energy of the metastable state is lower than the work
function of the conversion plate (typically a few eV) [5]. In
this case the metastable energy can be efficiently detected
by a cryogenic bolometer [4].

TABLE 1

Typical opto-thermal experimental parameters.

Supersonic beam intensity (see note 1)	$10^{16} \div 10^{19}$ mol./(ster.·s)
Molecular beam velocity	$2 \cdot 10^{4} \div 2 \cdot 10^{5}$ cm/s
Number density of a molecular beam (see note 2)	$10^{9} \div 10^{12}$ mol./cm^3
Volume of the laser/molecular beam interaction region	$\approx 10^{-3}$ cm^3
Detection solid angle	$10^{-3} \div 10^{-5}$ ster.
Energy per molecule	$10^{-19} \div 10^{-20}$ J
Power at the detector (see note 1)	$10^{-5} \div 10^{-8}$ W
Detection limit of a cryogenic bolometer (see note 3)	$10^{-12} \div 10^{-13}$ W

Notes

1) Centerline intensity of a full beam.
2) Calculated at a distance of ≈ 10 cm, downstream from the nozzle.
3) Beam fluctuations and gas condensation noise limit the best achievable sensitivity at a few part per million of the total power impinging on the bolometer.

Cryogenic bolometers [5] are the most popular detectors for opto-thermal experiments. Room temperature pyro-electric detectors have found limited applications [6] when sensitivity is not a must. Cryogenic bolometers can be grouped in two main categories : i) semiconducting and ii) superconducting bolometers. Their main properties are reported in Table 2.

Semiconducting bolometers are currently commercially available and are widely used as infrared radiation (IR) detectors. They can be applied to molecular beam detection with minor modifications. In particular it is necessary to remove the liquid helium cooled optical window which is used in most IR applications for limiting background radiation noise. In molecular beam applications, spike noise is often observed after a few hours of operation. It is probably originated by fast rearrangements of the cryodeposit growed on the detector surface. This effect can be particularly important when the bolometer is operating below ≈ 2 K, i.e. in the temperature range where semiconducting detectors have their maximum sensitivity. It has been empirically found that spike noise is much less critical when He-seeded super-sonic beams are used. Helium atoms do not contribute to the opto-thermal signal and probably promote a better energy exchange between molecules absorbed on the bolometric surface, reducing surface instabilities.

The response time of a bolometer depends on the ratio between the thermal capacity (C) of the sensitive element and the thermal conductivity (G) of the link which connects the sensitive element to the cryostat. This response time can be reduced increasing the conductivity of the thermal link, but this produces also a decrease of sensitivity, which is proportional to 1/G. In order to produce fast and sensitive bolometers, C must be reduced to the lower possible limit. C depends on both the volume of the sensitive element and the operating temperature. Practical considerations limit the response time of "fast" semiconducting bolometers to about 0.1 ms.

Bolometers with response time in the ms range are not suitable for carrying out opto-thermal experiments in combination with pulsed laser sources. In this case two additional noise sources must be considered : i) electro-magnetic inter-ferences (EMI) and ii) radiation noise. The first one is originated by fast high voltage discharges which are currently used in pulsed laser sources. EMI effects can be reduced by adopting a proper electrical installation [7], but cannot be completely eliminated. The second noise source is due to the light scattering in the beams interaction region. The use of optical baffles, shields and anti-reflecting coatings can reduce unwanted light diffusion. On the other hand the peak power of a laser pulse is often greater than 1 MW and it is straightforward to observe that even a very small fraction of this power can produce a large signal on the bolometer. Fortunately both EMI and radiation signals are well separated in time with respect to the true opto-thermal signal. In fact both spurious signals are produced immediate-ly after or during the laser firing, while the opto-thermal signal is produced after a delay time of the order of τ. It can be separated from spurious signals only if the response time of the bolometer is much lower than τ.

TABLE 2

Characteristic parameters of cryogenic bolometers used in opto-thermal experiments.

	Semiconducting	Superconducting
Response time	0.1÷10 ms	1÷5 μs
Working temperature	1.2÷4.2 K	3÷10 K
Responsivity	10^5 V/W	10^3 V/W
Noise Equivalent Power (N.E.P.)	10^{-13} W/√Hz	10^{-12} W/√Hz

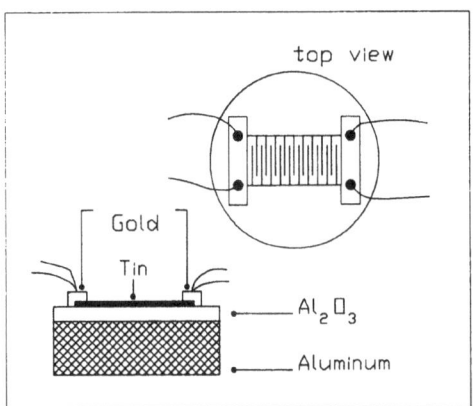

Fig.2   Schematic view of a tin film superconducting bolometer. The aluminum body temperature is stabilized at ≈3.7 K, around the center of the tin superconducting transition.

Bolometers with fast response time can be built using a superconducting film as sensitive element [8/10]. In Fig.2 we show a schematic view of a superconducting bolometer based on a tin thin-film. The typical electric resistivity of the film is about 10 Ω/square, at room temperature. Two gold

conctacts are used for connecting the bolometer to a
polarization circuit (bias current ≈0.1 mA) and to a fast
ultra-low-noise amplifier (input noise ≈0.1 nV/√Hz, at 1 kHz
[11]), using a four wires configuration. The Sn film is
deposited on a Al_2O_3 substrate which ensures both electrical
and thermal insulation. The aluminum body temperature is
stabilized around the center of the tin superconducting
transition (≈3.7 K), using the low-frequency component of the
bolometric signal as a feed-back for the temperature control
circuit [10].

Bolometers built using different superconducting
materials and different insulating substrates have been
tested in our laboratory. The choice of different superconducting materials enabled us to develop detectors having
operating temperatures from ≈3 K up to ≈10 K. When the
operating temperature is increased a decrease of the N.E.P.
is generally observed. On the other hand at higher temperatures condensation problems may be reduced (see next page).

We found that both structural and thermal properties
of the insulating substrate, strongly affect the bolometer
responsivity. In Fig.3 we show the superconducting transition
of two Sn films used for bolometric detection. Both films are
deposited on electrochemically formed Al_2O_3 substrates. The
two samples have Al_2O_3 substrates with different thickness :
1 μm for full circles and 4 μm for open circles data. In the
latter case the transition is too broad and the bolometer has
unsatisfactory characteristics.

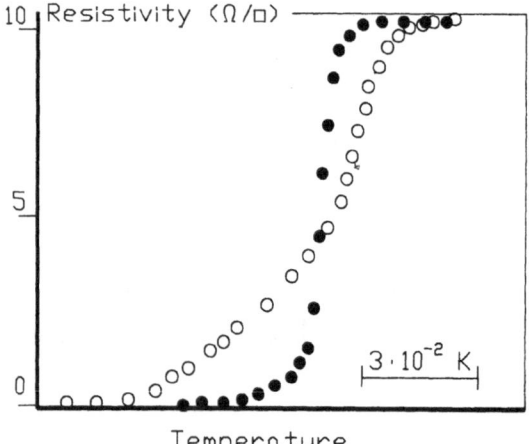

Fig.3 Electrical resistivity of two Sn films, measured as a
function of temperature, around the superconducting
transition. See text for details.

Superconducting bolometers with response time below 1 µs are easy to build. However such fast detectors are not really necessary for pulsed opto-thermal experiments : for practical purposes a response time in the range of 1÷5 µs must be choosen. This is the so called "initial" response time, which characterizes the bolometer when its surface is not appreciably covered by a cryodeposit. We should consider however that the residence time of most atomic and molecular species on the bolometric surface (at temperatures below 10 K) is virtually infinite. After long exposures to intense supersonic beams, the thermal capacity of a superconducting bolometer may be appreciably increased and a corresponding increase of the response time is observed. This effect is clearly shown in Fig.4.

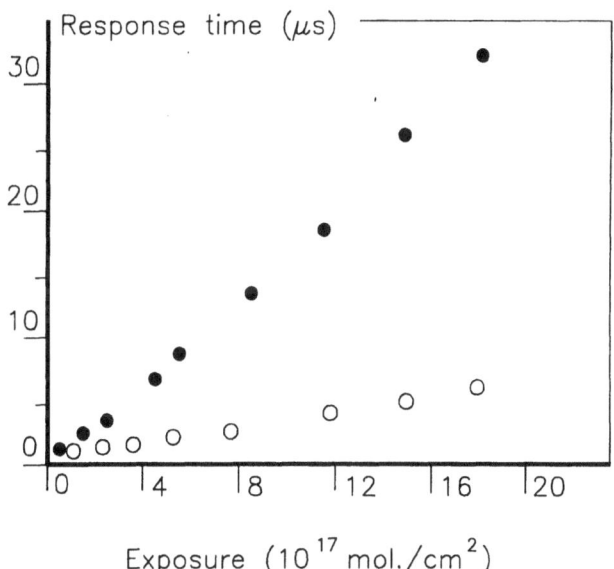

Fig.4   Response time of a superconducting bolometer measured as a function of the exposure to different supersonic beams (open circles = Ar; full circles = SF_6).

Data shown in Fig.4 have been obtained measuring the bolometric signal generated by means of a high Mach number He beam, modulated by a fast mechanical chopper. If the same measurement is carried out by means of a fast infrared emetting diode, the observed response time does not increase appreciably, even after long exposure to the molecular flux. This fact can be understood taking into account that the bolometric detection of radiation involves a mechanisms different with respect to molecular beam detection. In the last case the energy transfer process occurs on the top of the bolometric surface, producing a temperature variation of both the cryodeposit and the superconducting film. On the contrary non-resonant infrared radiation is not absorbed by the cryodeposit and reaches directly the superconducting film. In this last case the growing of a cryodeposit on the bolometric surface produces only second-order effects.

In the particular case of weakly bound adsorbates (typically H_2) condensation problems may be solved by choosing a superconducting film with a sufficiently high transition temperature. In fact an increase of the bolometric surface temperature produces a reduction of the residence time of molecules. The residence time on a surface is proportional to a factor $\exp(\epsilon/kT)$ where ϵ is the adsorption energy, k is the Boltzmann constant and T is the surface temperature. In the case of H_2 an operating temperature of about 10 K is sufficient to avoid any appreciable increase of the bolometer response time even after long exposures : bolometers operating around this temperature have been built in our laboratory by means of Nb films.

An alternative obvious approach consists in limiting the exposure of the bolometer to the molecular beam. In our opto-thermal spectrometer the laser repetion rate is 10 Hz and a c.w. supersonic beam is modulated by a mechanical chopper having a duty-cycle of 2:100. The laser firing system is synchronized with the mechanical chopper and the detector is exposed to the molecular beam only during a time interval of about 2 ms, immediately after each laser shot. This system works well with c.w. supersonic beams and no appreciable increase of the response time is observed during a typical run time (about five hours). Of course we cannot exclude that condensation effects could limit the response of fast superconducting bolometers if they are used in connection with very high intensity pulsed molecular beam sources [12].

EXPERIMENTAL RESULTS

Multiphoton spectroscopy

The multiphoton excitation of several polyatomic molecules has been investigated in our laboratory by means of the opto-thermal method. The experimental apparatus is described in detail elsewhere [13/15]. The molecular beam is produced by a variable temperature supersonic source (nozzle diameter 75 µm) and is irradiated by means of a Lumonics TEA 820 CO_2 laser. A tin superconducting bolometer is used for detecting the opto-thermal signal. The temporal behavior of the signal gives information about the transverse energy distribution of the molecular beam. The integral of the signal is proportional to the average energy absorbed by each molecule. Following the calibration procedure described in ref. [13] it is possible to obtain the average number of photons $\langle N \rangle$ absorbed by each molecule. An estimation of the initial rotational temperature of the molecular beam is obtained by means of the enthalpy balance [14].

In Fig.s 5-6 the multiphoton excitation spectra of two freon molecules are shown. CF_3I spectra (Fig. 5) have been measured using a pure CF_3I beam with two different source temperatures (T_s): 248 and 348 K respectively. The source pressure is 0.2 bar and the laser fluence is 0.33 J/cm². Both spectra show a main peak centered at ≈ 1072 cm^{-1} which is related to the presence of a Fermi resonance between the ν_1 mode at 1075 cm^{-1} and the overtone $2 \cdot \nu_5$ at 1080.8 cm^{-1} [16].

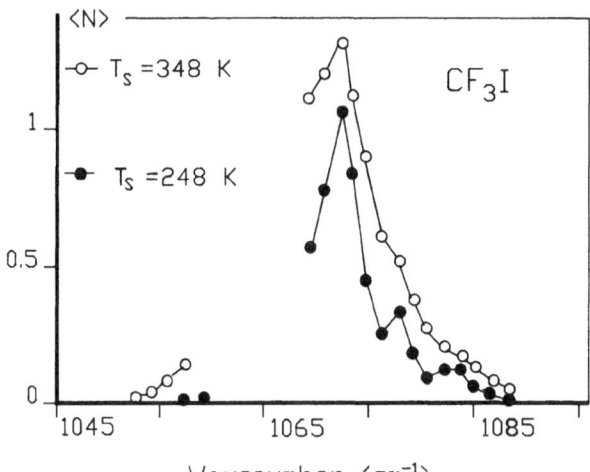

Fig.5   Multiphoton absorption spectra of CF$_3$I : $\langle N \rangle$ is the average number of photons absorbed by each molecule. The two spectra have been taken using different molecular beam source temperatures (T_s). See text for details.

The spectrum taken at the lower source temperature shows also a structure in the wavenumber range from ≈ 1075 to 1085 cm^{-1}. This structure is strongly attenuated when T_s is increased from 248 to 348 K. This source temperature variation produces a change of the initial rotational temperature from ≈ 25 to ≈ 40 K. The initial vibrational temperature is always very close to T_s. The modification of the excitation spectrum as a function of T_s may be partially attributed to different contributions of rotational broadening and hot-bands. Recent work by Liedenbaum [17] have demonstrated the presence of strong photodissociation peaks for CF$_3$I clusters in the region from ≈ 1065 to ≈ 1085 cm^{-1}, just in coincidence with the structure observed in our low temperature spectrum. In our experiment we checked for the presence of clusters tuning the laser outside the multiphoton absorption region [13]. However Liedenbaum's measurements show that this region is almost coincident with the cluster photodissociation spectrum. In conclusion we cannot exclude that the low T_s data reported in Fig.5 are partially affected by systematic errors due to the interference between the positive multiphoton signal with the negative signal originated by spurious cluster dissociation.

The multiphoton spectrum of C_2F_5Cl is shown if Fig.6. The experimental conditions are the same of the CF_3I spectra. When the source temperature is set at 223 K the rotational temperature of the molecular beam is ≈50 K. A single absorption peak is observed which is related to the presence of the C-Cl stretch ν_4 mode at 982 cm^{-1}. When T_S is increased up to about 400 K (rotational temperature ≈140 K) the absorption spectrum becomes much more intense and a second absorption peak is found red shifted of about 5 cm^{-1} with respect to the first one.

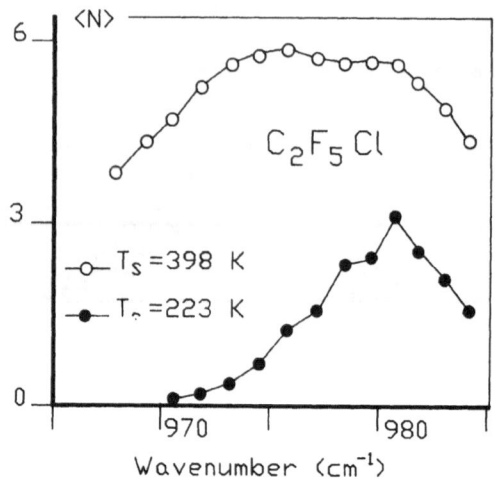

Fig.6   Multiphoton absorption spectra of C_2F_5Cl : $\langle N \rangle$ is the average number of photons absorbed by each molecule. The two spectra have been taken using different molecular beam source temperatures (T_S). See text for details.

Overtone spectroscopy

Vibrational overtone and combination band spectroscopy of polyatomic molecules have been extensively investigated in recent years. These studies have demonstrated that highly vibrationally excited states are not adequately described by the conventional normal mode model. When the excitation level increases a transition from normal to local mode [18] behavior is observed. This is particularly evident in the case of highly anharmonic oscillators, weakly coupled with the rest of the molecule. X-H bond stretches (with X=C,O,..) are well known examples of such a local mode vibrations.

The C-H stretch of benzene is probably the most studied example of a local mode system. Room temperature cell absortion experiments have shown broad spectra which have been explained assuming that intramolecular energy redistribution is responsible for strong homogeneos broadening [19]. Molecular beam experiments may provide important information about overtone spectroscopy because the dramatic reduction of spectrum congenstion makes possible a detailed investigation of broadening effects. Unfortunately the cross section of overtone transitions is quite small and decreases by about a factor 10 for every increase of the overtone order. For example in the case of benzene the absorption cross section for the 0-3 transition (second order overtone) is ≈640 barn and decreases to ≈23 barn [19] when the third overtone is considered.

Recently the C-H benzene overtones have been studied by the Lee's group [20]. In this experiment a pulsed beam of Ar seeded benzene has been analyzed using a two laser, state selective multilevel saturation spectroscopy. Lee's experiments have demonstrated the existence of absorption lines much narrower than expected from room temperature data.

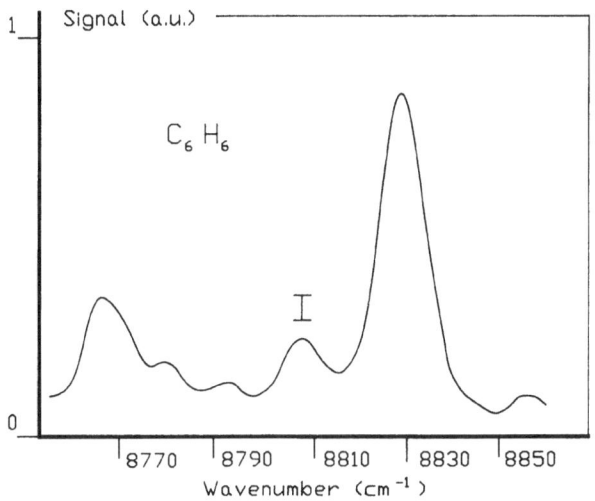

Fig.7   Low resolution opto-thermal absorption spectrum for the second C-H overtone of benzene. The error bar indicates ± 3 standard deviations.

In Fig.7 we show preliminary results of an opto-thermal study carried out in our laboratory for investigating the second overtone transition of benzene. The experimental apparatus has been described elsewhere [20]. The molecular beam source has a nozzle diameter of 0.2 mm and is filled with a 10% benzene/helium mixture. The source pressure is 300 mbar and the temperature is 300 K.

Our experimental results confirm the findings of ref. [20]. The main absorption peak is found at 8827 cm^{-1} with a

FWHM of about 10 cm^{-1}. Two other peaks are clearly identified at about 8767 and 8807 cm^{-1}. The analysis of the spectrum and the extension of measurements toward higher wavenumbers are presently in progress.

REFERENCES

[1]   T.E.Gough, R.E.Miller and G.Scoles;
      Appl.Phys.Letters 30 (1977) 338.
[2]   For a recent review about molecular beam methods see :
      "Atomic and Molecular Beam Methods"; edited by G.Scoles;
      D.Bassi, U.Buck and D.Laine' Ass.Eds.; Oxford University
      Press (New York) 1988, Vol.1.
[3]   T.E.Gough, R.E.Miller and G.Scoles;
      J.Chem.Phys. 69 (1978) 1588.
[4]   C.Douketis; Ph.D. thesis, Indiana Univ.(1989).
[5]   M.Zen; in ref.[2], Ch.10.
[6]   R.E.Miller; Rev.Sci.Instrum. 53 (1982) 1719.
[7]   D.Bassi; in Ref.[2], Ch.6.
[8]   G.Gallinaro and R.Varone; Cryogenics 15 (1975) 292.
[9]   G.Gallinaro, G.Roba and R.Tatarek;
      J.Phys.E 11 (1978) 628.
[10] D.Bassi, A.Boschetti, M.Scotoni and M.Zen;
      Appl.Phys. B26 (1981) 99.
[11] G.Fontana; "Amplificatore a Si-BJT, operante a 77 K" (in
      italian), Trento Univ., Internal Report (1987).
[12] W.R.Gentry; in Ref.[2], Ch.3
[13] D.Bassi, A.Boschetti, G.Scoles, M.Scotoni and M.Zen;
      Chem.Phys. 71 (1982) 239.
[14] A.Boschetti, M.Zen, D.Bassi and M.Scotoni;
      Chem.Phys. 87 (1984) 131.
[15] D.Bassi, A.Boschetti, S.Iannotta, M.Scotoni and M.Zen;
      Laser Chem. 5 (1985) 143.
[16] E.Borsella, R.Fantoni, A.Giardini-Guidoni, D.R.Adams and
      C.D.Cantrell; Chem.Phys.Letters 101 (1983) 86.
[17] C.Liedenbaum; Ph.D. thesis, Nijmegen Univ. (1989).
[18] B.R.Henry; Acc.Chem.Res. 10 (1977) 207.
[19] K.V.Reddy, D.F.Heller and M.J.Berry;
      J.Chem.Phys. 76 (1982) 2814.
[20] R.H.Page, Y.R.Shen and Y.T.Lee;
      J.Chem.Phys. 88 (1988) 4621.
[21] M.Scotoni, M.Zen, D.Bassi, A.Boschetti and M.Ebben
      Chem.Phys.Letters 155 (1989) 233.

MULTIPHOTON IR SPECTROSCOPY

J. Reuss and N. Dam

Fysisch Laboratorium, Katholieke Universiteit Nijmegen
Toernooiveld, 6525 ED Nijmegen, The Netherlands

1. Two-level case

Introduction

One can really consider multiphoton effects in a two-level system. First one can consider two (or more) laser fields and follow the absorption and eventual re-emission of photons of (slightly) different frequencies. But, second, even if only one photon mode is excited (i.e. one kind of photon is present) consecutive absorption and emission processes can lead to multiple exchanges between a molecule and a laser field, i.e. to multiphoton effects. Besides this, the two level system serves as introduction to multilevel systems.

1.1. Schrödinger equation for a two-frequency field

We assume here that both frequencies are similar and also not too far from resonance

$$\omega_1 \approx \omega_2 \approx \omega_0 = \frac{(E_2 - E_1)}{\hbar} \tag{1}$$

Further, electric dipole interaction couples the e.m. fields to the molecule, $H_{int} = -E_1\mu_{12}\cos\omega_1 t - E_2\mu_{12}\cos\omega_2 t$. Then the Schrödinger equation yields

$$\frac{d}{dt}\begin{pmatrix} a_1 \\ a_2 \end{pmatrix} = \begin{pmatrix} 0 & ie^{-i\omega_0 t}\sum_j \Omega_j \cos\omega_j t \\ ie^{i\omega_0 t}\sum_j \Omega_j \cos\omega_j t & 0 \end{pmatrix}\begin{pmatrix} a_1 \\ a_2 \end{pmatrix}, \quad j = 1,2 \tag{2}$$

The Rabi-frequency is given by $\Omega_j = \mu_{12}E_j/\hbar$; the amplitudes of the state populations are described by a_1 and a_2.

1.2. The rotating wave approximation

On the right hand side of Eq. 2 one finds weakly time-dependent terms, proportional to $\cos(\omega_j - \omega_0)t$, together with fast oscillating ones, proportional to $\cos(\omega_j + \omega_0)t$. The rotating wave approximation consists of neglecting the latter ones,

$$\frac{d}{dt}\begin{pmatrix} a_1 \\ a_2 \end{pmatrix} = \begin{pmatrix} 0 & (i/2)\sum_j \Omega_j e^{-i(\omega_0 - \omega_j)t} \\ (i/2)\sum_j \Omega_j e^{i(\omega_0 - \omega_j)t} & 0 \end{pmatrix}\begin{pmatrix} a_1 \\ a_2 \end{pmatrix}. \tag{3}$$

Applied Laser Spectroscopy, Edited by W. Demtröder and
M. Inguscio, Plenum Press, New York, 1990

For $|a_1|^2=1$ at $t=0$ and $\Omega_2=0$, the general solution of Eq. 3 yields

$$|a_2|^2 = \frac{\Omega_1^2}{(\omega_0-\omega_1)^2+\Omega_1^2} \cdot \sin^2\left\{[(\omega_0-\omega_1)^2+\Omega_1^2]^{1/2}\cdot\frac{t}{2}\right\}, \qquad (4)$$

with $|a_1|^2+|a_2|^2=1$.

Eq. 4 describes the Rabi-oscillations of the molecular system, which - on resonance, $\omega_0=\omega_1$ - yields an inversion of the population at $t=\Omega_1^{-1}\cdot\pi,\ \Omega_1^{-1}\cdot3\pi,\ \Omega_1^{-1}\cdot5\pi,\ \ldots$.

On the other hand, the $\sin^2$ factor can be replaced by $1/2$ if dephasing effects are present. The absorption line shape then becomes Lorentzian with a FWHM$=2\Omega_1$. This width is due to power broadening.

1.3. The density matrix

The density matrix elements are defined by

$$\rho = \begin{pmatrix} |a_1|^2 & a_1^* a_2 e^{i\omega_0 t} \\ a_1 a_2^* e^{-i\omega_0 t} & |a_2|^2 \end{pmatrix}. \qquad (5)$$

With Eq. 5 the Schrödinger equation yields

$$\frac{d\rho}{dt} = -\frac{i}{\hbar}[H,\rho], \qquad (6)$$

where

$$H = \begin{pmatrix} E_1 & H_{12} \\ H_{21} & E_2 \end{pmatrix} \qquad (7)$$

and, if the molecule considered is a symmetric top,

$$H_{12} = H_{21}^* = \sum_{p,q,j}(-)^p \epsilon_j(1,-p)E_j \cos\omega_j t[(2J'+1)(2J''+1)]^{1/2}\cdot$$

$$(-)^{J'-M'}\begin{pmatrix} J'' & 1 & J' \\ M'' & p & -M' \end{pmatrix}(-)^{J'-K'}\begin{pmatrix} J'' & 1 & J' \\ K'' & q & -K' \end{pmatrix}\cdot$$

$$< v'||\mu(1,q)||v'' > . \qquad (8)$$

It may seem strange to introduce M-degeneracy through Eq. 8 into the two-level problem of Eq. 7. One could imagine the application of a d.c. electric field, strong enough to lift the M-degeneracy and weak enough to produce no significant J-mixing yet.

J' (J''), K' (K'') and M' (M'') characterize the final (initial) state. The polarization of the E_1 and E_2 fields are described by the spherical tensor components ϵ_j of the unit vector along the polarization direction. The last factor on the right hand side of Eq. 8 stands for the reduced matrix element of the dipole transition operator, sandwiched between initial and final vibrational states. This factor is zero for IR-forbidden transitions. The selection rules are contained in Eq. 8.

Although Eq. 7 is a real matrix, the equation of motion 6 is essentially a complex equation which can be reduced to a real form only in case of resonance, i.e. $\omega_1=\omega_2=\omega_0$.

1.4. Optical Bloch equations

The next and final step is now to construct, from the density matrix elements, a vector which allows a simple discussion of coherence effects, Rabi-oscillations and rapid adiabatic passage (RAP) effects. To this end one defines $\bar{R} = (\rho_{12} + \rho_{21}, \frac{1}{i}(\rho_{12} - \rho_{21}), \rho_{22} - \rho_{11})$, whose components are all real. Note that $|R|^2 = 1$, i.e. R is a unit vector. The equation of motion then assumes the form

$$
\dot{\bar{R}} = (\omega_0 - \omega_r) \begin{pmatrix} 0 & -1 & 0 \\ 1 & 0 & 0 \\ 0 & 0 & 0 \end{pmatrix} \bar{R}
$$

$$
+ \begin{pmatrix} 0 & 0 & 2\sin\omega_r t \sum_j \Omega_j \cos\omega_j t \\ 0 & 0 & 2\cos\omega_r t \sum_j \Omega_j \cos\omega_j t \\ -2\sin\omega_r t \sum_j \Omega_j \cos\omega_j t & -2\cos\omega_r t \sum_j \Omega_j \cos\omega_j t & 0 \end{pmatrix} \bar{R}
$$

(9)

For $\omega_r = 0$, Eq. 9 reduces to the form valid for a non-rotating $\bar{R}$-vector, $\bar{R}_{non-rotating}$, whereas with $\omega_r \neq 0$ a rotation around the 3-axis is introduced,

$$
\bar{R}_{rotating} = \begin{pmatrix} \cos\omega_r t & \sin\omega_r t & 0 \\ -\sin\omega_r t & \cos\omega_r t & 0 \\ 0 & 0 & 1 \end{pmatrix} \bar{R}_{non-rotating}
$$

(10)

This rotation is assumed to obey $\omega_0 \approx \omega_1 \approx \omega_2$; the rotating wave approximation then yields

$$
\dot{\bar{R}} = (\omega_0 - \omega_r) \begin{pmatrix} 0 & -1 & 0 \\ 1 & 0 & 0 \\ 0 & 0 & 0 \end{pmatrix} \bar{R}
$$

$$
+ \begin{pmatrix} 0 & 0 & +\sum_j \Omega_j \sin(\omega_r - \omega_j)t \\ 0 & 0 & +\sum_j \Omega_j \cos(\omega_r - \omega_j)t \\ -\sum_j \Omega_j \sin(\omega_r - \omega_j)t & -\sum_j \Omega_j \cos(\omega_r - \omega_j)t & 0 \end{pmatrix} \bar{R}
$$

(11)

On the right hand side one finds two asymmetric matrices; their action can be transformed into a vector product

$$
\dot{\bar{R}} = \bar{D} \times \bar{R} + \bar{F} \times \bar{R}
$$

(12)

with $\bar{D} = (0, 0, \omega_0 - \omega_r)$ and $\bar{F} = (-\sum_j \Omega_j \cos(\omega_r - \omega_j)t, -\sum_j \Omega_j \sin(\omega_r - \omega_j)t, 0)$.

Before we start to discuss practical cases, we want to stress that all relaxation effects (due to collisions or to spontaneous decay) are left out, up to this point. We are going to include them below.

1.5. Rabi oscillations

First we consider a single frequency field, i.e. $\Omega_2 = 0$. In this case we put $\omega_r = \omega_1$. For very small Ω_1 values we find a precession of the R-vector around the 3-axis, with an angular velocity of $\omega_0 - \omega_1$.

On resonance, $\omega_1 = \omega_0$, we have $\bar{D} \equiv 0$ and $\bar{F} = (-\Omega_1, 0, 0)$. Consequently, a precession around the 1-axis produces an oscillation of $\rho_{22} - \rho_{11}$ between -1 and $+1$, inversion occurring with a period of $\tau = 2\pi \cdot \Omega_1^{-1}$.

Off resonance a precession is observed around the stationary $\bar{D}$ vector, yielding a motion in agreement with Eq. 4.

Experimentally, these Rabi-oscillations are exploited to determine transition dipole moments [1,2,3,4].

1.6. Rapid adiabatic passage

If, during a laser pulse, the ω_1 laser frequency is linearly swept through the resonance value $\omega_1=\omega_0+\Delta t$, Rabi-oscillations can be suppressed. Condition is that the sweep rate is sufficiently small; however, the laser field must also be sufficiently strong at detunings of the order of the Rabi-frequency. The phenomenon observed in this case is called RAP. This means that the $\bar{R}$-vector follows the $\bar{D} + \bar{F}$-vector which moves from a direction anti-parallel to the 3-axis ($t \ll 0$) to a direction along the 1-axis ($t = 0$), to point finally parallel to the 3-axis ($t \gg 0$). The solution obtained this way is $\bar{R} = \pm(\bar{F}+\bar{D})/|\bar{F}+\bar{D}|$. Obviously the vector product remains zero all the time, whereas $d\bar{R}/dt$ assumes very small values only if the sweep is slow enough.

Experimentally this means that inversion can be obtained and Rabi-oscillations are avoided altogether. Realization of the proper sweep conditions can be achieved by molecular beams passing through a laser beam near to the minimum waist. The curvature of wave fronts produces the desired frequency sweep, since the molecules experience an appropriate Doppler shift [5].

With a second laser crossing the once inverted population can be de-excited coherently (by stimulated emission). Actually, this de-excitation serves as an indicator for an achieved inversion. De-excitation can be observed readily by optothermal detection [6,7].

RAP-processes are not accompanied by power broadening. The physical picture is that Rabi-oscillations, which otherwise would lead to residence time shortening and thus broadening effects, are avoided [7].

1.7. Intracavity crossing

In this case one clearly has $\Omega_1 = \Omega_2$, $\omega_1 = \omega_0 + \Delta t$, $\omega_2 = \omega_0 - \Delta t$ in Eq. 12. Thus, with $\omega_r = \omega_0$ one finds $\bar{D} \equiv 0$ and $\bar{F}=(2\Omega_1\cos(\Delta \cdot t^2),0,0)$.

One obtains a torque around the 1-axis, oscillating with a frequency $\Delta \cdot |t|$. For large $|t|$ values the torque becomes zero due to the amplitude factor $\Omega_1(t)$. Only for $t \approx 0$ a stationary torque occurs, producing a rotation of $\bar{R}$ around the 1-axis. Superimposed on this motion fast oscillations are calculated. At the end of the laser pulse the rotation has produced R_3 values which signify inversion (Fig. 1), or for larger Ω_1 values return to the initial state, or, for still larger Ω_1 values, inversion again, etc. [8].

If the sweep is zero, $\Delta=0$, pure Rabi-oscillations occur with a doubled frequency, $2\Omega_1$. Experimentally this condition prevails if the molecular beam passes through the minimum waist.

2. Three-level case

Introduction

Three-level cases can be distinguished in those where two available lasers pump two consecutive transitions (both leading up the energy ladder or one up and the other one down) and those where they pump two transitions starting from the same initial level. In the following we shall restrict ourselves to the first case. The sequence of events as experienced by the molecules will be of primary importance and interest.

In contrast to chapter 1, relaxation processes will be incorporated in the following discussion.

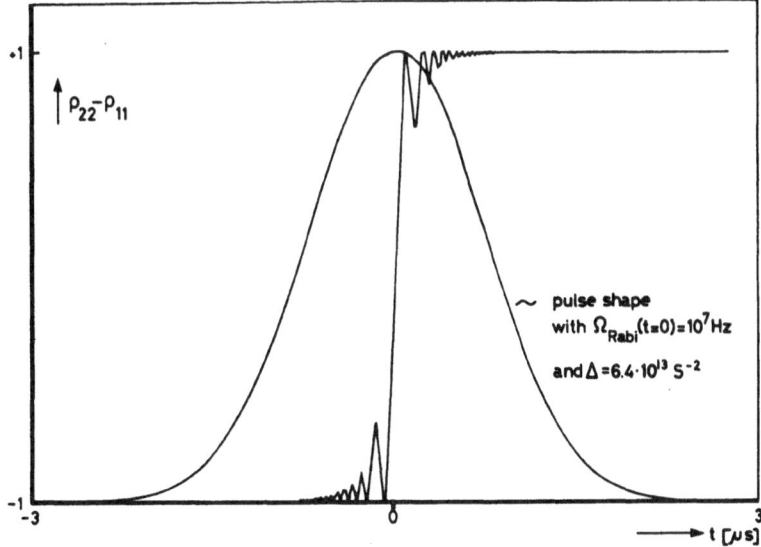

Figure 1. *The time dependence of population for a resonant two-level system passing through an intracavity focus. Besides the global effect (i.e. inversion for the chosen parameters) fast oscillations are calculated. For increasing laser intensities the inversion disappears like if one goes from a π to a 2π pulse, to show up again for still larger intensities, etc..*

2.1. The density matrix formulation

In the case of three coupled non-degenerate levels the Schrödinger equation yields [9]

$$\dot{\rho}_{11} = -\gamma_1(\rho_{11} - \rho_{11}^{eq}) - \frac{i}{2}\Omega_1(\tilde{\rho}_{12}^* - \tilde{\rho}_{12})$$

$$\dot{\rho}_{22} = -\gamma_2(\rho_{22} - \rho_{22}^{eq}) + \frac{i}{2}\Omega_1(\tilde{\rho}_{12}^* - \tilde{\rho}_{12}) - \frac{i}{2}\Omega_2(\tilde{\rho}_{23}^* - \tilde{\rho}_{23})$$

$$\dot{\rho}_{33} = -\gamma_3(\rho_{33} - \rho_{33}^{eq}) + \frac{i}{2}\Omega_2(\tilde{\rho}_{23}^* - \tilde{\rho}_{23})$$

$$\dot{\tilde{\rho}}_{12} = -\gamma_{12}(\tilde{\rho}_{12} - \tilde{\rho}_{12}^{eq}) - i\Delta_{12}\tilde{\rho}_{12} + \frac{i}{2}[\Omega_1(\rho_{11} - \rho_{22}) + \Omega_2\tilde{\rho}_{13}]$$

$$\dot{\tilde{\rho}}_{13} = -\gamma_{13}(\tilde{\rho}_{13} - \tilde{\rho}_{13}^{eq}) - i\Delta_{13}\tilde{\rho}_{13} + \frac{i}{2}[\Omega_2\tilde{\rho}_{12} - \Omega_1\tilde{\rho}_{23}]$$

$$\dot{\tilde{\rho}}_{23} = -\gamma_{23}(\tilde{\rho}_{23} - \tilde{\rho}_{23}^{eq}) - i\Delta_{23}\tilde{\rho}_{23} + \frac{i}{2}[\Omega_2(\rho_{22} - \rho_{33}) - \Omega_1\tilde{\rho}_{13}] \qquad (13)$$

with

$$\gamma_{ij} = \frac{1}{2}(\gamma_i + \gamma_j) + \frac{1}{2}(\gamma_i^{deph} + \gamma_j^{deph})$$

$$\rho_{22}^{eq} = \rho_{33}^{eq} = \tilde{\rho}_{ij}^{eq} = 0 \quad i \neq j$$

$$\Delta_{12} = \frac{(E_1 - E_2)}{\hbar} + \omega_1$$

231

$$\Delta_{23} = \frac{(E_2 - E_3)}{\hbar} + \omega_2$$

$$\Delta_{13} = \frac{(E_1 - E_3)}{\hbar} + \omega_1 + \omega_2$$

$$\tilde{\rho}_{jk} = \rho_{jk} \exp\left\{-\frac{i}{\hbar}(E_k - E_j)t\right\}$$

Eq. 13 contains relaxation terms. Here, γ_i stands for inelastic collision rates, changing the population of a level and γ_i^{deph} describes the rate of dephasing collisions, yielding a decay of coherence within a mixture of state amplitudes. So-called strong-collision conditions prevail if all collisions lead to a change of state and all γ_i^{deph} can be neglected.

With this system of equations we can simulate two laser pulses (or the passage of a molecular beam through two possibly overlapping laser beams), including a time dependent tuning of the frequencies ω_1 and ω_2. The interesting and new part of the story is that, even for time-independent resonance conditions, $\hbar\omega_1 = E_2 - E_1$, $\hbar\omega_2 = E_3 - E_2$, RAP processes can be produced by chosing an appropriate (anti-intuitive) sequence of events.

In Fig. 2 experimental results are compared to calculations using Eq. 13 with $\gamma_1=15$ MHz, $\gamma_2=20$ MHz, $\gamma_3=0$, $\gamma_{ik}=1/2(\gamma_i + \gamma_k)$, $\Omega_1(t = 0)=1366$ MHz, $\Omega_2(t = 0)=28$ MHz. The experiments were performed on a C_2H_4 molecular jet. In Table 1 the pump and probe laser parameters are collected. The time scale is converted to a Δz scale by $v \cdot t = \Delta z$, $v=896$ m/s. The local density corresponds to 0.6 Torr* (1 Torr* $\sim 3.27 \cdot 10^{22}$ molecules/m^3).

Fig. 2 shows qualitative agreement between the measured points and the calculated curves. The $\Delta z > 0$ peak is what one expects from a traditional pump and probe experiment with both lasers tuned on resonance. The minimum at $\Delta z = 0$ occurs as a consequence of the ac-Stark effect, which dynamically detunes the $(E_3\text{-}E_2)$ distance with respect to $\hbar\omega_2$, by action of the strong pump laser (frequency ω_1). For increasing Δz values the γ_2 relaxation depopulates the 2-level resulting in a decrease of the probe laser absorption. For a fixed $\Delta z > 0$ value a variation of power of laser 1 leads to the observation of Rabi-oscillations between levels 1 and 2.

For $\Delta z < 0$ values, the probe laser partially preceeds the pump laser. As a consequence RAP processes occur which will be discussed in the next section.

2.3. The Schrödinger equation for three-level systems

Actually we go back to a simpler problem, neglecting relaxation phenomena. However, the issue of three-level adiabatic following will become clearer, in the Schrödinger formulation of the dressed atom picture. The corresponding Hamiltonian takes the form (note that the off-diagonal elements are taken in the semi-classical limit)

$$H = \begin{pmatrix} E_1 + n_1\hbar\omega_1 + n_2\hbar\omega_2 & \frac{\Omega_1}{2} & 0 \\ \frac{\Omega_1}{2} & E_2 + (n_1 - 1)\hbar\omega_1 + n_2\hbar\omega_2 & \frac{\Omega_2}{2} \\ 0 & \frac{\Omega_2}{2} & E_3 + (n_1 - 1)\hbar\omega_1 + (n_2 - 1)\hbar\omega_2 \end{pmatrix} \quad (14)$$

First we treat the problem "on exact resonance", i.e. the three-diagonal terms can be replaced by zero. Second, although $\Omega_1(t)$ and $\Omega_2(t)$ are time-dependent quantities, describing the momentanous amplitudes of the laser pulses, we consider H to be piecewise time-independent and follow, if possible, adiabatically the time evolution of the system through eigenstates.

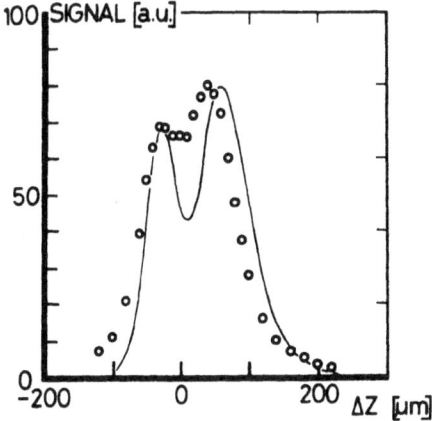

Figure 2. *Double resonance signal as a function of overlap of the two laser foci.* Δz *is the distance of the probe laser downstream of the pump laser. Data points are obtained from a jet of pure ethylene (400 Torr through a 0.5 mm circular nozzle at room temperature). The solid curve is a best fit, assuming strong collisions and neglecting population relaxation out of the uppermost level. For the corresponding level scheme see Fig. 3.*

The three eigenvalues are

$$\lambda_I = \frac{1}{2}\sqrt{\Omega_1^2 + \Omega_2^2}$$
$$\lambda_{II} = 0$$
$$\lambda_{III} = -\frac{1}{2}\sqrt{\Omega_1^2 + \Omega_2^2} \tag{15}$$

and the eigenvectors are

$$
\begin{pmatrix}
a_{I,1} = \dfrac{-i\Omega_1}{\sqrt{2}\sqrt{\Omega_1^2+\Omega_2^2}} \\[2mm]
a_{I,2} = \dfrac{1}{\sqrt{2}} \\[2mm]
a_{I,3} = \dfrac{i\Omega_2}{\sqrt{2}\sqrt{\Omega_1^2+\Omega_2^2}}
\end{pmatrix},
\begin{pmatrix}
a_{II,1} = \dfrac{\Omega_2}{\sqrt{\Omega_1^2+\Omega_2^2}} \\[2mm]
a_{II,2} = 0 \\[2mm]
a_{II,3} = \dfrac{\Omega_1}{\sqrt{\Omega_1^2+\Omega_2^2}}
\end{pmatrix}
\text{ and }
\begin{pmatrix}
a_{III,1} = \dfrac{-i\Omega_1}{\sqrt{2}\sqrt{\Omega_1^2+\Omega_2^2}} \\[2mm]
a_{III,2} = \dfrac{1}{\sqrt{2}} \\[2mm]
a_{III,3} = \dfrac{-i\Omega_2}{\sqrt{2}\sqrt{\Omega_1^2+\Omega_2^2}}
\end{pmatrix}.
\tag{16}
$$

The interesting solution belongs to $\lambda_{II}=0$. If (anti-intuitively) $\Omega_1=0$, $\Omega_2 \neq 0$ for $t \ll 0$, one has $a_1=1$, $a_2=a_3=0$ as initial situation. We follow this solution, initially an eigenstate of the dressed molecule, (adiabatically) through $t=0$, $\Omega_1(t=0)=\Omega_2(t=0)$, $a_1=a_3=1/\sqrt{2}$, $a_2=0$ to $t \gg 0$, $\Omega_1=0$, $\Omega_1 \neq 0$, $a_1=a_2=0$, $a_3=1$. Here we see precisely what adiabatic passage means: a molecule follows an eigenfunction which slowly changes its character. During this process the degeneracy of states I, II, III is lifted, to become restored at the end of the interaction. Incidentally (and without special consequences), λ_{II} remains constant all the time [10].

233

Another important point is that for this *II*-solution of the Schrödinger equation a_2 remains zero, i.e no population can be found on level 2 at any time though the full population is inverted and transferred from level 1 to level 3.

Note that no frequency sweep is needed to produce RAP in the three-level case. Further, the ratio between Ω_1 and Ω_2 (i.e. the relative strength of the two lasers) is arbitrary. Third, for very weak lasers the RAP scheme breaks down since the level splitting becomes comparable to frequencies present in the Fourier-spectrum of the time-dependent laser pulse.

2.4. Relaxation measurements, C_2H_4

In a real system collisional effects often compete with coherent changes of population. The spectacular total absence of population in the intermediate level is hereby lifted. Still, the maximum occupation of level 2 can be kept low by applying sufficiently strong lasers, such that relaxation from this level is rendered negligible for $\Delta z < 0$. On the other hand, for $\Delta z > 0$, the tail for large Δz values depends entirely on γ_2 as it becomes for a conventional pump and probe experiment (see Fig. 3) [10].

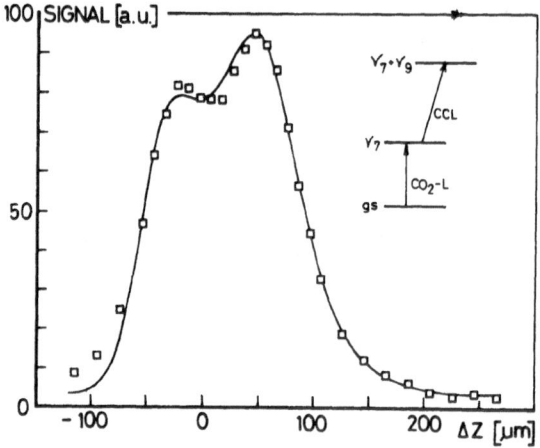

Figure 3. *As Fig. 2, but with the solid curve now obtained from a fit including population relaxation, involving all levels and not adhering to the strong collision assumption. The corresponding relaxation parameters are collected in Table 2.*

For comparison with a real system, C_2H_4 was chosen, excited by a CO_2 laser (ω_1) and a CCL laser (ω_2) to couple the levels indicated in Fig. 3.

The two lasers have mutually perpendicular polarisations. They cross the molecular jet axis at distances z_E (CO_2 laser) and z_p (CCL laser) from the nozzle, with $\Delta z = z_p - z_E$. The laser and jet parameters are displayed in Table 1.

Table 1. *Pump and probe laser parameters for the double resonance experiments on a free jet of ethylene (C_2H_4); μ is the transition dipole moment, θ^{max} the Rabi frequency in the center of each laser focus (FWHM as indicated).*

	pump	probe
laser	CO_2 waveguide	CCL
transition	$(4,1,3) \rightarrow \nu_7(5,0,5)$	$\nu_7(5,0,5) \rightarrow \nu_{7+9}(5,1,4)$
μ (Debye)	0.16	0.054
FWHM (μm)	60	30
P (Watt)	4.8 (max.)	$2.4 \cdot 10^{-3}$
θ^{max} (MHz)	250 (max.)	4

Table 2. *Relaxation parameters (in MHz) of Eq. 13 as obtained from the fit of Fig. 3. The fit is insensitive to the value of γ_2^{deph}, which therefore has been fixed at the value indicated.*

i	γ_i	γ_i^{deph}
1	15.9	40
2	5.0	12
3	18.0	47

In Fig. 3 a perfect fit is achieved with the relaxation parameters collected in Table 2. Actually, for this quantitative comparison Eqs. 13 were replaced by a similar set, taking into acount the m-degeneracy of the participating levels [10]. Dephasing collisions only influence the regime of overlapping laser foci, especially $\Delta z < 0$, where RAP processes occur. There, however, the intermediate level is hardly populated, that is, γ_2^{deph} escapes from observation. The positive consequence of this outcome is that one has to determine one parameter less; there remain γ_1 (determined by an independent pump and probe experiment), γ_1^{deph}, γ_2, γ_3 and γ_3^{deph} (through $\gamma_{ik} = \frac{1}{2}(\gamma_i + \gamma_k) + \frac{1}{2}(\gamma_i^{deph} + \gamma_k^{deph})$). Since $\gamma_3 \approx \gamma_1$, we assume $\gamma_1^{deph}/\gamma_1 = \gamma_3^{deph}/\gamma_3$. In Table 2 one sees that γ_2 is significantly smaller than $\gamma_1 \approx \gamma_3$. The intermediate level with $K_{-1}=0$ lacks the $|K_{-1}| \neq 0$ double-degeneracy. This restricts possible pathways for R-R relaxation severely and leads to approximately three times lower relaxation rates. The number of dephasing collisions is larger than the rate of inelastic collisions, especially at low temperatures, i.e. *the strong collision model does not work well for this system.*

2.5. Two-photon excitation, SF$_6$

SF_6 molecules have often been studied in experiments of IR-multiphoton excitation. Even multiphoton dissociation can easily be achieved applying conventional pulsed CO_2 lasers with laser fluences ≥ 1 J/cm^2.

Here we discuss the use of cw CO_2 lasers crossing a molecular beam of typically 1% SF_6 in He. The excitation of SF_6 molecules is observed by optothermal detection [6]. The in first instance unexpected and nasty consequence of the molecular beam expansion is that nearly none of the otherwise plenty levels remains sufficiently populated to be tuned easily into resonance with the radiation from a CO_2 waveguide laser [11,12]. In order to remedy this difficulty we extended the tuning range of our 8 Watt waveguide laser from ± 115 MHz to ± 300 MHz by utilizing an opto-acoustic modulator, operated at a fixed frequency of 90 MHz. The laser beam passes twice through this device so that in first order two times 90 MHz can be added to or subtracted from the laser frequency.

In Fig. 4 the P4 quadruplet of one-photon transitions is displayed for two different laser intensities. In both cases the transitions are saturated completely. In Fig. 4a we still are in the RAP regime, observing inversion as tested by a two-laser experiment. Accordingly, no power broadening is found. Increasing the laser intensity by a factor of about 5000, Fig. 4b, power broadening is seen together with a typical signal decrease (factor of 2) showing that we no longer have an inversion but a final 50% excitation as belongs to normal saturation behaviour.

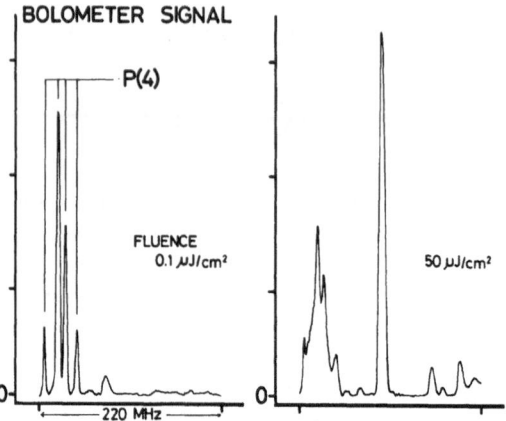

Figure 4. SF_6 bolometer signal vs waveguide laser frequency for two very different laser intensities. Actually the fluence obtained by multiplying the intensity by the beam inter-action time is indicated. For low fluence the P(4) quadruplet of one-photon transitions is clearly recognized as possessing a theoretical strength of 3:4:3:1. For strong fluence, power broadening and loss of inversion is observed, together with a sharp and large two-photon transition, which has not been seen before. This new peak has, in this low intensity regime, a quadratic intensity dependence and can be saturated at 50 $\mu J/cm^2$.

In Fig. 4b a narrow strong peak belonging to a two-photon transition (presumably $J = 7 \rightarrow J = 6 \rightarrow J = 5$) is shown near to the P(4) quadruplet. This transition is not saturated. Two-laser experiments reveal that we deal with an inverted population. By further increasing the laser power the peak height was found not to increase any more. The level scheme belonging to this two-photon transition is displayed in Fig. 5. Δ indicates the detuning of the intermediate level, estimated to amount to a couple of GHz. As it should, the bolometer signal shows a quadratic dependence on laser power for small intensities.

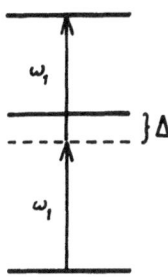

Figure 5. *Three level system, excited by a two-photon process;* Δ *stands for the detuning of the intermediate level.*

3. The multi-level case

Introduction

As discussed in section 2.5., instead of increasing the number of lasers to N-1, if N is the number of participating levels in an excitation process, we use multiphoton excitation through nearly resonant real intermediate levels by a single strong laser. In contrast to many previous measurements, we employ a cw single mode, single frequency CO_2 laser to produce coherent excitation of a multilevel system, such as is represented by the SF_6 molecule. There are two important features which make such an investigation really worthwhile. First, the SF_6 level system is bottlenecked, i.e. there is a highest level that can be reached by excitation in a strong single frequency laser field. Second, by again exploiting Doppler sweeps due to the curvature of wavefronts, an inverted population can be produced in this highest state. Inversion persists at least for about 1 ms in a collisionfree molecular beam. These are the aspects which will be discussed in the following sections.

3.1. Inversion, calculated

Since under molecular beam conditions relaxation processes can be neglected, it is rather straightforward to let the computer construct e.g. the Schrödinger equation for a (non-degenerate) eight-level system in a pulsed radiation field, the frequency of which is linearly swept through resonance due to the passage of molecules through a laser beam near to the point of minimum waist. The result is displayed in Fig. 6. The population of the nth level is indicated by a sequence of numbers n. At the beginning of the interaction (the shape of the laser pulse is displayed, too) the system is in level 1. The population "1" decays, showing some resonant Rabi-oscillations. Other levels show transient populations with Rabi-oscillations more strongly damped the higher the level. The population of level "8" shows a small rise up to an entirely inverted situation.

3.2. Inversion, observed

The case just sketched can be found in the SF_6 molecule where the anharmonicity is stronger than the symmetry splitting for increasing excitation of the ν_3-mode. At $n_3=5$, the exciting laser is effectively detuned so that no further excitation can be realized,

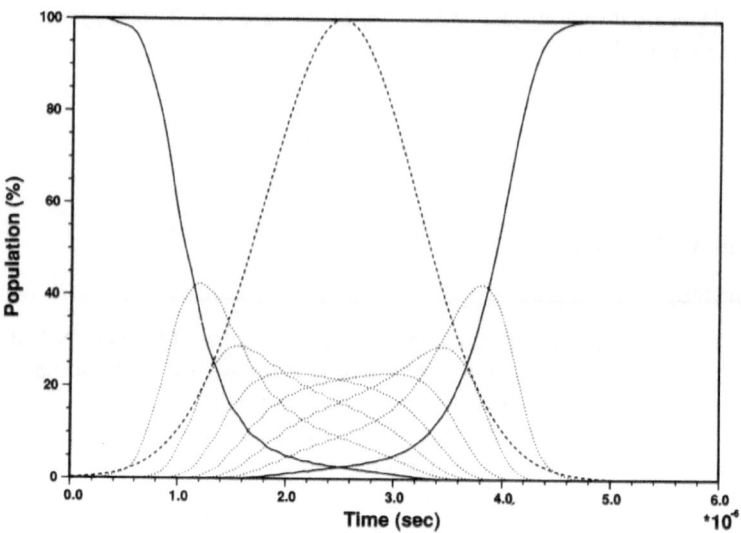

Figure 6. *For an equidisant ystem, the population of the different levels is displayed as a function of time if a laser pulse (intensity indicated by − − − − −) is applied. The laser frequency $\omega = \omega_0$ is exactly on resonance at 2.5 µs (moment of maximum laser intensity, $\Omega_{Rabi} = 150$ MHz). The laser frequency is swept linearly, with $\frac{d\omega}{dt} = 8 \cdot 10^{11}$ $1/s^{-2}$, i.e. from − 2 MHz to +2 MHz in 5 µs. The resonance energy is given by $\hbar \omega_0 = 952$ cm^{-1}. The descending full curve corrsponds to the population of the initial (ground) state "1", the rising one to the final state "7". The intermediate levels "2" to "6" show populations in the sequence of increasing energy. The total process leads to a completely inverted popuated by merit of a multilevel RAP process. The figure is neary symetrical in time; consequently, the inversion can be undone by a separatd second indentical laser pulse.*

with a single single-frequency laser - a theoretical prediction [13] which is borne out by experiments, at least for the RAP case where power broadening is avoided [14].

Moreover, with two-laser experiments it has been demonstrated that indeed strong inversion has been obtained which is undone (by stimulated de-excitation) by the second laser. For observing this stimulated effect the two lasers have to possess the same frequency within 0.5 MHz.

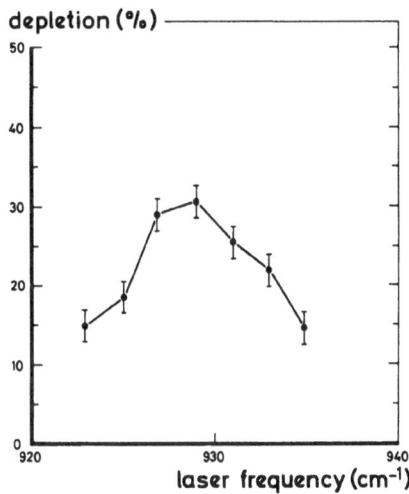

Figure 7. *The bolometer signal for pre-excited SF_6 (pre-excitation at 936.8 cm^{-1}, $I=10$ kW/cm^2). Broad hole burning (depletion) is observed around 929 cm^{-1}, where a probe laser (I=2 kW/cm^2) excites the pre-excited SF_6, as a function of a pump laser (I=10 kW/cm^2) with variable frequency (abcissa).*

The second laser can be delayed up to 100 ns without there being found a change in the magnitude of the stimulated de-excitation. This means, on the other hand, that within the 0.5 MHz of instrumental resolution, every excited SF_6 molecule finds only one excited final level. Otherwise dephasing effects would lead to an attenuated de-excitation. To translate this into a statement on the density of vibrational states, we conclude that per wavenumber less than 60,000 rovibrational infrared-active levels are present, at an energy which corresponds to the observed and predicted bottleneck level, i.e. at $n_3=5$. (Actually, at $n_3=10$ the density of vibrational states approaches this value of 60,000 for SF_6).

3.3. Hole burning for higher excitations of SF_6

Once given the inversion of SF_6 molecules at $n_3=5$ one can perform a two-laser experiment again, where the first laser pumps the molecules up the vibrational ladder and the second (in frequency coincident) laser probes either the depletion of the lower state or the possibly inverted population of the upper state. The experimental outcome is twofold; a) no sharp spectral dip has been observed with FWHM near to 1 MHz; b) a very broad dip has been found with a FWHM of about 10 cm^{-1} [7], Fig. 7 The absence of the sharp dip, a), may be explained by the increased density of states at the final level of excitation, large enough so that the first laser excites more than one single level coherently. Before the second laser interacts, however, dephasing takes place (within about 200 ns) so that the stimulated de-excitation passage (identical to the way up) becomes at least partially blocked. Observation b) indicates that - again due to the increased density of involved states - a single (pre-excited) initial level has various ways of becoming (further) excited with a bandwidth of about 10 cm^{-1}.

References

[1] A.G. Adam, T.E. Gough, N.R. Isenor and G. Scoles, Phys. Rev. A32 (1985) 1451

[2] B. Zhang, X.J. Gu, N.R. Isenor and G. Scoles, Chem. Phys. 126 (1988) 151

[3] W.Q. Cai, T.E. Gough, X.J. Gu, N.R. Isenor and G. Scoles, J. Mol. Spectrosc. 120 (1986) 374

[4] T.E. Gough, X.J. Gu, N.R. Isenor and G. Scoles, Int. J. Infrared Millim. Waves 7 (1986) 1893

[5] W. Demtröder, in: Laser Spectroscopy, Springer Series in Chemical Physics 5 (Springer Verlag Berlin, 1981) p. 102

[6] T.E. Gough, R.E. Miller and G. Scoles, Appl. Phys. Lett. 30 (1977) 338

[7] C. Liedenbaum, S. Stolte and J. Reuss, Chem. Phys. 122 (19088) 443

[8] B. Wichman, C. Liedenbaum and J. Reuss, submitted to Chem. Phys.

[9] N. Dam, S. Stolte and J. Reuss, Chem. Phys. (in press)

[10] N. Dam, L. Oudejans and J. Reuss, submitted to Chem. Phys.

[11] R.S. McDowell, H.W. Galbraith, C.D. Cantrell, N.G. Nereson and E.D. Hinkley, Opt. Comm. 17 (1976) 178

[12] Ch.J. Bordé, M. Ouhayoun, A. van Leberghe, C. Salomon, S. Avrillier, C.D. Cantrell and J. Bordé, Laser Spectroscopy IV, Eds. K. Walther and K.W. Rothe, (Springer Verlag New York/Berlin 1979) p. 142-153

[13] M.S. Child and L. Halonen, in: Advances in Chemical Physics, Vol LVII (Wiley, New York)

[14] C. Liedenbaum, S. Stolte and J. Reuss, Phys. Reports 178 (1989) 1

LIGHT INDUCED KINETIC EFFECTS IN ALKALI VAPORS

L. Moi

Istituto di Fisica Atomica e Molecolare

CNR - via del Giardino 7 - 56127   Pisa - Italy

INTRODUCTION

During the last years light has been successfully used to modify atomic motion in beams or in gases. In atomic beams, where the collision rate is negligible, resonance radiation pressure (RRP) is very effective. RRP is due to the transfer of the photon momentum to the atom in the absorption-emission processes. Deceleration of the atoms as well as compression of their velocity distribution can be achieved by using a counterpropagating laser beam. In this case the Doppler shift of the absorption line must be compensated in order to avoid the interruption of the cooling cycle. This compensation can be obtained either by chirping the laser frequency[1] or by Zeeman tuning of the atomic resonance[2]. Other schemes can be adopted in which a broad band laser source interact instantaneously with the whole Doppler profile[3]. This particularly favorable situation is no more valid when the RRP has to be exploited in a gas. In this case, in fact, the collision rate is in general high and the collisions with the gas container inevitable. These two circumstances imply a continuous interruption of the cooling cycle and a thermalization of the atomic velocity. Therefore in an unidimensional geometry (i.e. a configuration in which the transversal dimensions of the vapor sample are much smaller than the longitudinal ones), no cooling can be obtained, but only a modification of the gas diffusion inside the cell. Cooling of the vapor can be obtained if a three dimensional geometry is assumed, as suggested by Hansch and Schawlow[4]. Let us, for the moment, consider a vapor confined in a capillary cell having the radius r≪ L, where L is the cell length. A laser beam propagates along its major axis z. The radiation force pushes the vapor along the capillary until an equilibrium between the RRP and the diffusion pressure is reached. The force is proportional to the fraction of excited atoms and it is maximum when the laser is on resonance. In fig. 1a the macroscopic result of this effect is sketched. RRP has been observed in 1933 by Frish[5], who was able to deflect a sodium beam by using a resonance lamp. RRP on a vapor has been, instead, put in evidence only recently by Xu and Moi[6]

Applied Laser Spectroscopy, Edited by W. Demtröder and
M. Inguscio, Plenum Press, New York, 1990

241

They observed sodium vapor drift in a capillary by using a broad band dye laser with a special cavity configuration ("lamp laser"[7]). When a gas or a vapor are considered a new mechanical effect induced by light can be observed, as proposed by Gel'mukhanov and Shalagin[8]. This effect is known as Light Induced Drift (LID) and it is due to the combined action of both the laser excitation and the collisions with a perturbing gas. The role played by the collisions is to break the symmetry of the Maxwellian distribution of the atomic velocities. Laser excitation must be velocity selective in order to create two fluxes of atoms, one of excited and the other of ground state atoms, propagating along opposite directions. As the diffusion coefficients of these two fluxes are in general different, the net result is a macroscopic drift of the gas. The direction of this drift depends on the sign of the laser detuning inside the Doppler profile. LID has, therefore, a dispersive behaviour and it vanishes when the laser is on resonance (see Fig. 1b,c). In general, LID is observable, with atomic or molecular gases, every time the environment is sensitive to the velocity selective excitation induced by a laser. For example "surface" LID originates from velocity selective excitation followed by state dependent molecule-surface interaction[9]. The first experimental evidence of LID has been obtained with sodium vapor by Antsygin et al.[10]. Few years later Woerdman and coworkers observed LID at optically thick regime ("optical piston"[11]) and wall frictionless LID experiments have been performed by using paraffin[12] and silane[13,14] coated cells. LID depends on the kind of buffer gas, on the laser detuning and intensity and, also, on the laser bandwidth. Drift velocities of the order of 30 m/s have been measured by excitation with a "lamp laser"[15]. LID can be effective in isotopic separation process and it has been demonstrated with Rb atoms[16].

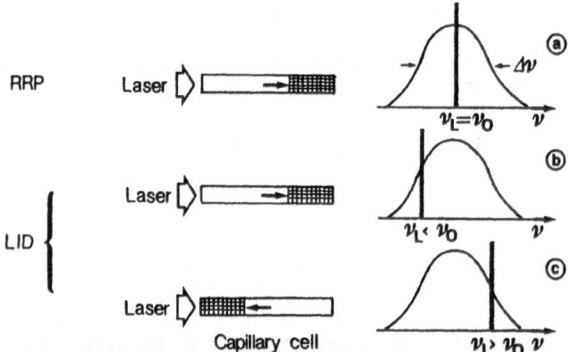

Fig. 1 - Schematic representation of unidimensional RRP and LID effects. The curves to the right represent the Doppler broaden absorption line profile. The bold line represents the laser excitation.

Few remarks must be done at this point. The atom-wall interaction plays a determinant role on the evolution of both RRP and LID effects. This interaction can be characterized by the adsorption energy parameter E_a and by the adsorption time

T_a which are dependent through the relation

$$T_a = T_o \ e^{Ea/kT} \tag{1}$$

where $T_o = h/kT$ is of the order of 10^{-13} s. In the case of alkali-pyrex system Ea is about 1 eV and $T_a \cong 10^{-3}$ s. This very long adsorption time causes both a very slow diffusion and the formation of an alkali layer at the cell walls. The number of atoms at the surface may be order of magnitude larger than those in the vapor phase. This gives a huge impedance as these atoms too have to be removed in order to see the RRP or LID effects. To get negligible this problem a wall coating must be used which, by lowering E_a, makes very short T_a. For example the silane coating presents an adsorption energy of about 0.1 eV and a $T_a \cong 10^{-11}$ s^{13}. Therefore, except when explicitly indicated, all the experiments described below make use of coated cells, prepared by starting from an heater solution of dymethylpolysiloxane. This coating, which can be heated up to 200 °C, permits to work with saturated vapor up to relatively high densities. In an unidimensional geometry, anyway, the direct interaction of the vapor with the cell walls cannot be avoided. The observation of three dimensional LID geometry would permit to switch off this interaction and open new possible applications of LID. Very recently LID has been observed in a spherical cell[17]. Another important remark concerns the laser spectral characteristics. A single mode laser permits a careful tuning of its wavelength and, as a consequence, a precise knowledge of its detuning. Drawbacks are represented by the poor coupling with the vapor and by the hyperfine optical pumping. Both of these phenomena drastically reduce the light induced kinetic effects. To improve the laser induced force a broad band laser, having special characteristics, can be used. It must have both a very narrow mode pattern, in order to instantaneously excite all the atoms regardless their speed, and a bandwidth comparable to the Doppler linewidth, in order to maintain the velocity selectivity. This last condition is determinant in LID experiments. This laser has been obtained by prolonging the resonant cavity. For example in order to get a mode spacing comparable to the 10 MHz linewidth of sodium absorbing line, a 15 m long cavity has been realized[7] In the experiments described below laser cavities of 5, 10 and 15 m respectively have been used depending on the effect to be studied.

RESONANCE RADIATION PRESSURE

When a vapor is illuminated by a resonant laser it is submitted to a force F which is proportional to the fraction of excited atoms per unit of time. This force, for a two level atom and for a single mode laser, is given by[6].

$$F = (h/\lambda \tau) \ n_o/n = \frac{h}{\lambda \tau} \cdot \frac{I_L / I_s}{(1 + I_L / I_s)} \cdot \frac{\Delta \nu_{eff}}{\sqrt{\Pi} \ \Delta \nu_D} \tag{2}$$

where λ = laser wavelength, $\Delta \nu_D$ = Doppler linewidth, τ =

lifetime of the excited state, I_L= laser power density, I_s= saturation power density, $\Delta\nu_{eff}= \Delta\nu_n\sqrt{1 + I_L / I_s}$ effective linewidth of the absorption line, $\Delta\nu_n$= natural atomic linewidth. This force can be increased by using a multimode dye laser with the characteristics outlined before:

$$F = \sum_n \frac{h}{\lambda\tau} \frac{I_n / I_s}{(1 + I_n / I_s)} \cdot \frac{\Delta\nu_{eff} \; e^{-\left(\frac{n\,\Delta\nu_n}{\Delta\nu_D}\right)^2}}{\sqrt{\pi}\;\Delta\nu_D} \qquad (3)$$

where the sum extends over all the laser modes, and I_n is the power density of each mode. F_{max}, that can be in this last case one or two order of magnitude larger than that induced by the single mode laser, is obtained when all the atoms in the vapor are instantaneously excited. To a given RRP force corresponds a drift velocity of the vapor given by

$$v = \mu F = \frac{FD}{kT} \qquad (4)$$

where D is the diffusion coefficient, μ is the mobility coefficient. The diffusion of the vapor inside the capillary is described by two coupled differential equations, one describing the vapor diffusion and the other one the laser intensity dependence on the position. Under optically thin regime, I_L and v are constant and the RRP effect can be described only by

$$\frac{\partial n(z,t)}{\partial t} = D \frac{\partial^2 n(z,t)}{\partial z^2} - v \frac{\partial n(z,t)}{\partial z} \qquad (5)$$

When the boundary conditions corresponding to a capillary cell with closed ends and with the metal reservoir to one of its ends are imposed, the following stationary solution is found

$$n(z) = n_o \cdot e^{\frac{F(z-L)}{kT}} \qquad (6)$$

where the laser crosses the cell from the side opposite to the reservoir. The density follows an exponential profile whose maximum is determined by the density in the reservoir. When the laser comes from the opposite side n exponentially increases in the capillary because the atoms are pumped from the reservoir. The general solution n(z,t) is given by a sum of exponential terms having different time constants, corresponding to different propagation modes. In first approximation and for a relatively long delay the fluorescence follows an exponential decay whose time constant is proportional to v^2/D. The experiments have been performed with a capillary 50 cm long and with 0.2 cm diameter, permanently connected to an ion pump, in order to make negligible the background pressure. This condition is important to avoid LID contributions. The metallic sodium is contained in a reservoir

placed to one end of the cell. Therefore the cell is
asymmetrical and this asymmetry is reflected in the different
behaviour of the RRP effect when the laser beam crosses the
cell from the capillary side or from the reservoir side. In
the first case a decreasing of the vapor density is expected
(eq. 6), while in the other case an increasing must be
observed. The effect is detected through the fluorescence
collected by an optical fiber placed along the capillary. In
Fig. 2 the temporal evolution of fluorescence signals are
reported for the two laser directions. In Fig. 2a the
fluorescence decreases. A fit of this curve with the general
solution of the diffusion equation gives a drift velocity of
the order of 20 m/s. This velocity is in agreement with the
value derived from eq. (4). In Fig. 2b the fluorescence signal
increases. In this case the time constant is longer than
before due to the different boundary conditions.

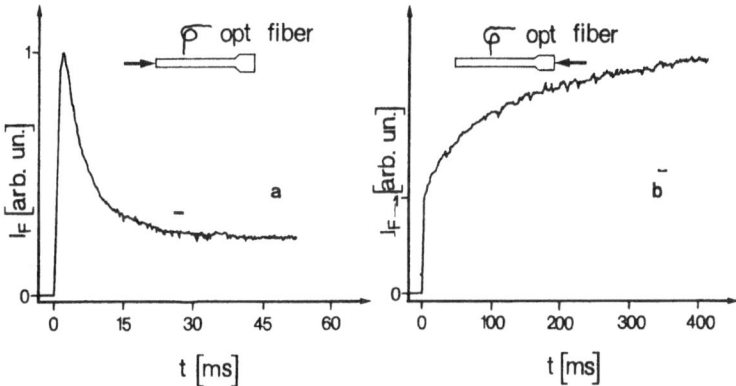

Fig. 2 - Fluorescence intensity versus time; laser
beam direction as shown in the insert.
Signals obtained with a 15m laser cavity.

LIGHT INDUCED DRIFT

Also for LID the time evolution of the vapor density is
described by two coupled differential equations[20] similar to
those describing RRP effect. The situation simplifies
significantly in the optically thin regime. In this case the
laser is assumed to be only slightly attenuated by the vapor
and the drift velocity can be assumed uniform along the cell.
The diffusion equation is similar to eq. (5) with the
differences that, in this case, D is the diffusion coefficient
of the ground state atoms in the buffer gas and v is the drift
velocity of the vapor immersed in the buffer gas. The
solution, for a semifinite cell (z>0), with the initial
condition of a uniform distribution and with the boundary
condition of a vanishing flux at the cell entrance becomes,
by defining $x=(v/D)z$ and $\tau=(v^2/D)t$

$$n(x,\tau) = \left[1 - \int_{\frac{(x-\tau)}{2\sqrt{\tau}}}^{\infty} e^{-s^2} \frac{ds}{\sqrt{\pi}}\right] + A\,(x,\tau) \qquad (7)$$

where $A(x,\tau)$ is a term rapidly vanishing by going away from the cell ends[20]. The drift velocity can be derived by solving the set of rate equations describing the excited and ground-state velocity distributions after a suitable collision model has been adopted. An heuristic approach gives

$$v = \frac{D_g - D_e}{D_g}\, v_L \eta \qquad (8)$$

where $D_{g,e}$ represent the diffusion coefficients of ground and excited state respectively, η the fraction of excited atoms and v_L the velocity selected by the laser. A detailed calculation for two level atoms has been done by Haverkort[19] and its solutions applied to single mode and broad band laser excitations by Gozzini et al.[20]. These calculations show that, under the same conditions, the broad band (BB) laser impresses a drift velocity about twice that induced by the single mode (SM) one. The experiments have been runned by using coated capillary cells (internal diameter = 0.2 cm, length = 15 cm) filled with a few Torr of neon or krypton as a buffer gas. In Fig.3 a typical signal with the fitting curve obtained from eq. 7 is reported. In this case the drift velocity results v = 8±1 m/s. In Fig. 4 a comparison between the v induced by the SM and the BB dye lasers is shown. As predicted the v induced by the BB laser is always about twice that induced by the SM one. The BB laser is very effective and the maximum speed obtained in a coated cell with 6 Torr of krypton and 6 W/cm^2 has been v = 30±3 m/s.

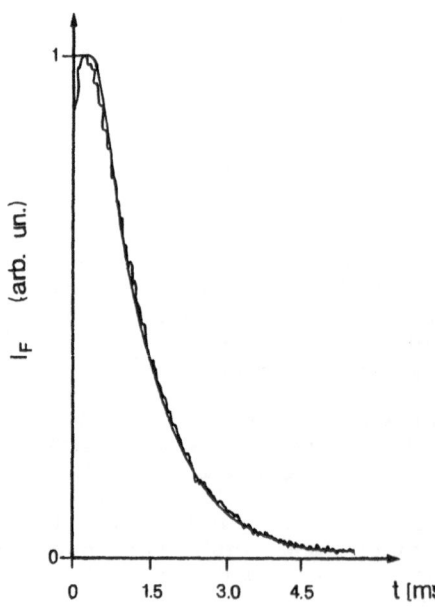

Fig. 3
Fluorescence signal showing the LID effect in a coated cell as induced by broad band dye laser. The continuous curve is the best fit of eq.7.

Quite recently LID has been observed in a three dimensional geometry. The capillary cell has been changed with a spherical cell (diameter = 1.5 cm) connected through a capillary to the Na reservoir. The cell is filled with a few Torr of krypton as a buffer gas and silane coated. The broad band laser beam is coupled to a six optical fiber bunch. Then the fibers are splitted apart and positioned around the cell along the three orthogonal directions. Each fiber has an output coupler which produces a parallel light beam whose diameter is comparable to that of the cell.

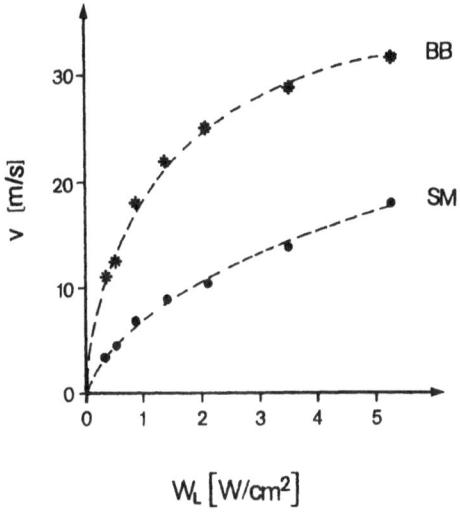

Fig.4 - Measurements of v plotted as a function of the laser-power density W of SM and BB dye lasers. The detunings of the two lasers have been adjusted to get the maximum speed.

The detection is accomplished by two optical fibers monitoring the fluorescence coming from the center and the outer part of the cell. In Fig. 5 the fluorescence variations induced by the simultaneous presence of the six laser beams are reported. A decreasing of the fluorescence in correspondence of the external part of the cell and a larger increasing in its center are detected. These curves demonstrate a compression of the vapor in the center of the cell. By switching off only one of the laser beams, a completely different behaviour is obtained.

CONCLUSIONS

The experiments, recently performed on alkali vapors, clearly demonstrate that it is possible to apply strong light induced mechanical actions on gases. RRP and LID can be conveniently used to study the transport properties of gases as well as their interactions with surfaces. Interesting applications are represented also by isotope separation and chemical reaction dynamics studies. Experiments are in progress to cool down a vapor through RRP effect and to realize wall free supersaturated vapors through LID.

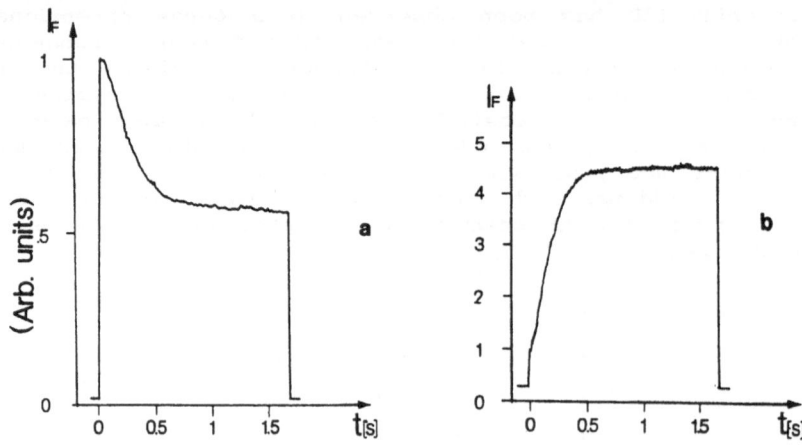

Fig.5 - Fluorescence signals induced by the six laser
beams as a function of time; Fig.5a: signal
collected from the other part of the spherical
cell; Fig.5b: signal collected from the center.

REFERENCES

1. V.S.Letokhov and V.G.Minogin, Phys.Rep. 73,1 (1981)
2. W.D.Phillips and H.Metcalf, Phys. Rev. Lett. 48, 596 (1982)
3. L.Moi, Optics Commun. 50, 349 (1984)
4. T.W.Hansch and A.L.Schawlow, Optics Commun. 13, 68 (1975)
5. O.R.Frisch, Z.Phys. 86, 42 (1933)
6. J.H.Xu and L.Moi, Optics Commun. 67, 282 (1988)
7. J.Liang, L.Moi and C.Fabre, Optics Commun. 52, 131 (1984)
8. F.Kh. Gel'mukhanov and A.M.Shalagin, Zh.Eksp.Teor.Fiz.
    78, 1674 (1980) (Sov.Phys.-JEPT 51, 839 (1980))
9. R.W.M.Hoogeveen, R.J.C.Spreeuw and L.J.F.Hermans, Phys.
    Rev. Lett. 59, 447 (1987)
10. V.D.Antsygin, S.N.Atutov, F.Kh.Gel'mukhanov, G.G.Telegin
    and A.M.Shalagin, Sov.Phys.-JEPT Lett. 30,243 (1980)
11. H.G.Werij, J.P.Woerdman, J.J.M.Beenakker and I.Kusher,
    Phys.Rev. Lett. 52, 2237 (1984)
12. S.N.Atutov, St.Lesjak, S.P.Podjachev, and A.M.Shalagin,
    Optics Commun. 60, 41 (1986)
13. J.H.Xu, M.Allegrini, S.Gozzini, E.Mariotti and L.Moi,
    Optics Commun. 63, 43 (1987)
14. E.Mariotti, J.H.Xu, M.Allegrini, G.Alzetta, S.Gozzini and
    L.Moi, Phys. Rev. A38, 1327 (1988)
15. C.Gabbanini, J.H.Xu, S.Gozzini and L.Moi, Europhysics
    Lett. 7, 505 (1988)
16. A.Streater, J.Mooibroek and J.P.Woerdman, Optics Commun.
    64, 137 (1987)
17. S.Gozzini, D.Zuppini, C.Gabbanini and L.Moi, Europhysics
    Lett. submitted
18. G. Nienhuis, Phys. Rev. A31, 1636 (1985)
19. J.E.M.Haverkort, Ph.D.thesis, Leiden University, 1987
20. S.Gozzini, J.H.Xu, C.Gabbanini, G.Paffuti and L.Moi,
    Phys.Rev. A40, XX (1989)

ANOMALOUS DOPPLER BROADENING

IN O_2-NOBLE GASES RADIO-FREQUENCY DISCHARGES

A. Sasso and G.M. Tino

Dipartimento di Scienze Fisiche, Universitá di Napoli
Mostra d'Oltremare Pad.20, 80125 Napoli, Italy

I. INTRODUCTION

The difficulties for spectroscopic investigation of atomic oxygen start with the production of the excited atoms since no allowed optical transitions are available from the 2^3P ground state. Recently we made atomic oxygen accessible to cw high resolution investigation by means of several sub Doppler techniques [1-4]. Oxygen atoms are produced from trace amounts of O_2 in a noble gas substained radiofrequency discharge. The investigation of oxygen plasma is difficult as it is a complex medium composed of electrons, positive and negative ions, excited atoms and molecules. The purpose of this work is to summarize the collisional physics leading to the formation of atomic oxygen. We also suggest a simple model to explain anomalous Doppler broadening originated by non thermal equilibrium of the produced atomic species.

II. EXPERIMENT

Oxygen atom was produced in a radio-frequency (r.f.) discharge. The sample cell was a pyrex tube of 7 mm internal diameter containing an oxygen-noble gas mixture. Typical O_2 pressure was varied in the range 0.05-0.3 Torr while noble gas pressure was changed between 0.8 and 10 Torr.

A scheme of the experimental apparatus is shown in Fig.1. The r.f. discharge of moderate power (50W at 60 MHz) was maintained by an oscillator fed by a current stabilized power supply and coupled to the cell by two sets of coils. For high resolution spectroscopy, tunable radiation was provided by an actively stabilized single-mode ($\Delta \nu_L = 1$ MHz) ring dye laser (Coherent 699-21); a multimode ($\Delta \nu_L = 20$ GHz) dye laser (Coherent 599-03) was used for quick and broad scans to detect various atomic species and transitions present in the discharge.

When the laser radiation is resonant with the optical transition of a species in the discharge region the resulting absorption induces changes in the impedance of the discharge (optogalvanic effect). These variations reflect on the feedback loop between the oscillator and the current stabilized power supply. Signals can be detected with high sensitivity by chopping the laser beam and using phase sensitive detection. More details can be found in ref.s [5,6]. The laser wavelength was determined by recording the absorption spectra from a I_2 cell, and the frequency scan was calibrated by means of the markers from a 300 MHz and a 75 MHz free spectral range confocal Fabry-Perot interferometers. Light from the discharge was dispersed by a Littrow mounted grating monochromator with about 1 Å resolution.

Applied Laser Spectroscopy, Edited by W. Demtröder and
M. Inguscio, Plenum Press, New York, 1990

249

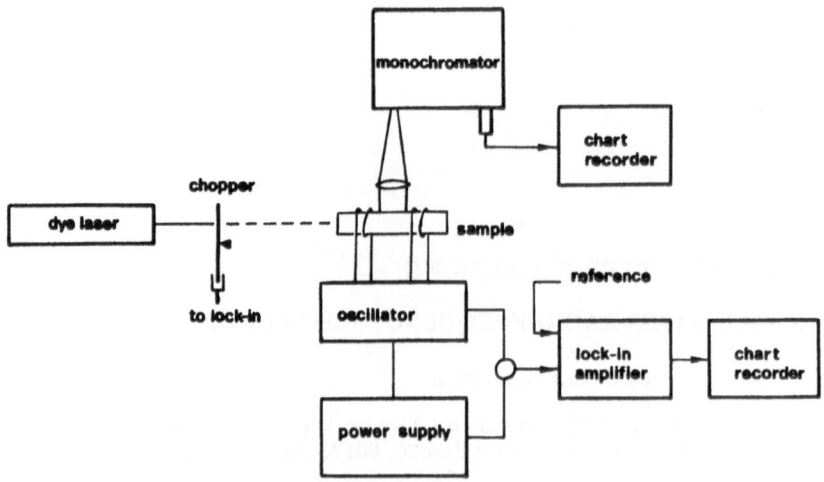

Fig.1 Schematic of the experimental apparatus. The atomic oxygen sample is produced in a O_2-noble gas radio-frequency discharge. Spectroscopic analysis is performed either detecting fluorescence or monitoring the optogalvanic effect.

III. O_2-METASTABLE NOBLE ATOMS COLLISIONS

The dissociation of molecular oxygen by electron impact is an important channel for the production of atomic oxygen in pure O_2 glow discharges [7]. Neverthless we have observed that oxygen atoms production is largely enhanced in presence of noble gases. Metastable levels of noble gas atoms are efficiently populated by means of electron-impact excitation and represent a reservoir of energy in the discharge sample. Thermal collisions of a rare gas metastable atom X^* with an oxygen molecule O_2 can give place to Penning ionization

$$X^* + O_2 \rightarrow X + O_2^+ + e^- \tag{1}$$

or to quasi-molecule formation

$$X^* + O_2 \rightarrow [XO_2] \tag{2}$$

It is known [8,9] that process (1) dominates in presence of He$^*(2^1S, 2^3S)$ and Ne$^*(^3P_{0,2})$ while for Ar*, Kr*, and Xe* Penning ionization is ineffective. As a consequence, different mechanisms are expected to contribute to oxygen atom formation depending on the specific metastable noble gas atom involved in the dissociative collision. We have used helium, krypton, neon and argon as buffer gas. Systematic investigations have been performed only with the last two gases, which also seem to act in different ways in the formation of the oxygen atoms. The different behaviours can be motivated by the different position of the metastable levels relative to the O_2 molecule energy levels, and can be understood with the help of Fig.2. Both processes (1) and (2) are followed by dissociation. The production of molecular ions O_2^+ in the excited metastable $a^4\Pi_{ui}$ state is quenched by fast dissociative recombination

$$O_2^+(a^4\Pi_{ui}) + e^- \rightarrow O(2^3P) + O^*. \tag{3}$$

Excited atomic oxygen O^* is produced in highly excited 3^5P and 3^3P levels.

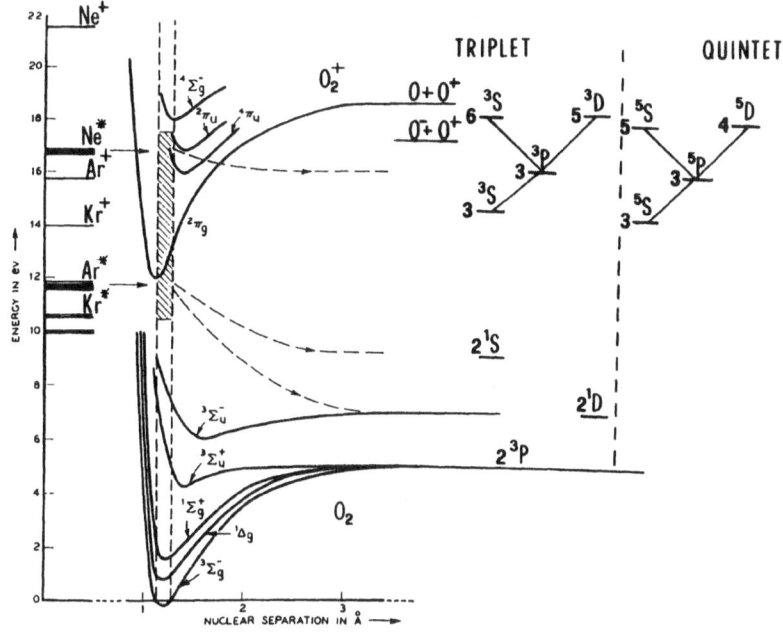

Fig.2 Excited O atoms are produced by means of collisions involving O_2, O_2^+, and a noble gas in a metastable state. The internal energy levels of the colliding partners are shown together with the schematic illustration of the O transitions investigated.

In $Ar - O_2$ discharges the production of O atoms seems to take place prevalently through the dissociation of the quasi-molecules indicated in Eq.(2):

$$[ArO_2] \rightarrow Ar + O^* + O(2^3P) + \Delta E \qquad (4)$$

where O^* represents now O atoms in 2^1S_0 or 2^1D_2 metastable level. The energy defect ΔE, of several eV, is released to the produced oxygen atoms as kinetic energy. As a result, both metastable and ground state O atoms can have a velocity considerably larger than that of the discharge-temperature gas. However, while most of the atoms in the ground state are essentially "cold", due to thermalizing collisions with the cell walls or with the buffer gas, atoms in the metastable levels can keep the anomalous velocity distribution; in fact, thermalizing collisions for this atoms have also the effect of quenching them into the ground state; metastable levels are then essentially populated by collisions-free, non thermalized atoms. Such kinetic energy distribution can be transferred to higher excited oxygen atoms through electron impact excitation.

IV. LASER DIAGNOSTIC OF OXYGEN PLASMA

The diagnostic of the oxygen atom formation is performed by monitoring the population of several excited atomic levels by means of optogalvanic (OG) and/or fluorescence detection. OG signal amplitude for the oxygen $3p^5P_2 - 4d^5D_j(\lambda=615.7$ nm) and neon $1s_5 - 2p_6(\lambda=614.3$ nm) transitions obtained in a $O_2 - Ne$ mixture is reported in Fig.3 as a function of O_2 pressure. It is evident how critical the production of atomic oxygen (circles) is with the oxygen-neon pressure. Indeed, in the state of equilibrium distribution of the O atoms there is a balance between dissociation of the O_2 molecules and recombination of the O atoms to re-form the O_2 molecules. On the other hand, the strong decreasing of the OG signals relative to neon transitions (triangles) is an evidence of the reduction of metastable neon atoms due to resonant collisions with O_2 molecules (process (1)). Similar behaviours were observed analysing the fluorescence emission.

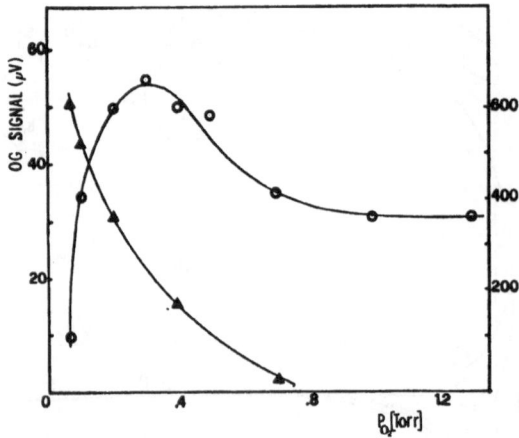

Fig.3 OG signal amplitude for atomic oxygen and neon transitions as a function of O_2 pressure $(P_{Ne} = 10 Torr)$. Left scale is relative to oxygen signal$(\odot)$, right scale refers to neon signal$(\blacktriangle)$.

V. LINESHAPES ANALYSIS OF ATOMIC OXYGEN TRANSITIONS

Optical transitions of atomic oxygen are essentially Doppler broadened. In our experimental conditions, in fact, homogeneous broadening is mainly due to collisions and it is much smaller than the Doppler broadening.

Measurement of the Doppler width gives informations on the value of the temperature of the gas in the discharge or, equivalently, on the average kinetic energy of the atoms. We have investigated lineshapes of several optical transitions of oxygen involving both triplet and quintet levels by varying the type of buffer gas in the discharge. For example, in Fig. 4 typical Doppler limited lineshapes for quintet $3^5P_3 - 5s^5S_2$ transition in $Ne - O_2$ and $Ar - O_2$ mixtures are shown; Doppler width in both the cases is 2 GHz corresponding to a temperature of 570 K.

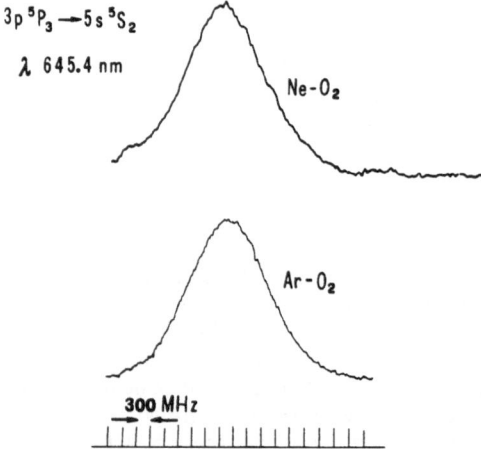

Fig.4 Recording of the Doppler broadened lineshape allows a determination of the kinetic energy of the atoms. In case of transitions between quintet levels this is nearly the same for Ar and Ne buffer gas $(\Delta\nu_D \approx 2$ GHz, T$\approx$ 570 K in the figure).

In Fig. 5 the entire multiplet $3p^3P_{1,2,0} - 6s^3S_1$ at 604 nm is shown, recorded in different mixtures with He, Ne, Ar, and Kr. By using He, Kr, or Ne as buffer gas (Fig.5 a,b,c) each component of the triplet is resolved and well fitted by a Gaussian whose FWHM is about 2 GHz. A result at first surprising is given by the observation of an extremely large Doppler effect when argon (Fig.5d) is used to substain the discharge. The large width prevents the observation of three separated components in spite of fine structure splittings in the $^3P_{1,2,0}$ level of 4.7 (0-2) and 16.8 (2-1) GHz respectively. Furthermore, in this case each component cannot be accurately fitted by a Gaussian curve nor by a Voigt profile.

The isolated fine structure component $3^3P_1 - 6^3S_1$ has been analysed in more detail by fitting the experimental line profile with the superposition of two Gaussian curves of different amplitude and width (Fig.6). As can be noted in Fig.6, the experimental lineshape is slightly asymmetric thus leading to an unexplainable shift of about 300 MHz between the two Gaussians. The width of the smaller Gaussian is of the same order of magnitude of those obtained with other buffer gases or for quintet transitions, i.e. $\Delta\nu_D=2$ GHz, while for the larger Gaussian we obtain a width of 5.7 GHz corresponding to a temperature of about 4500 K, i.e. nearly one order of magnitude larger than that "usually" observed.

Similar results were found investigating the triplet $3p^3P_{1,2,0} - 5d^3D_{3,2,1}$ transition at 595.9 nm.

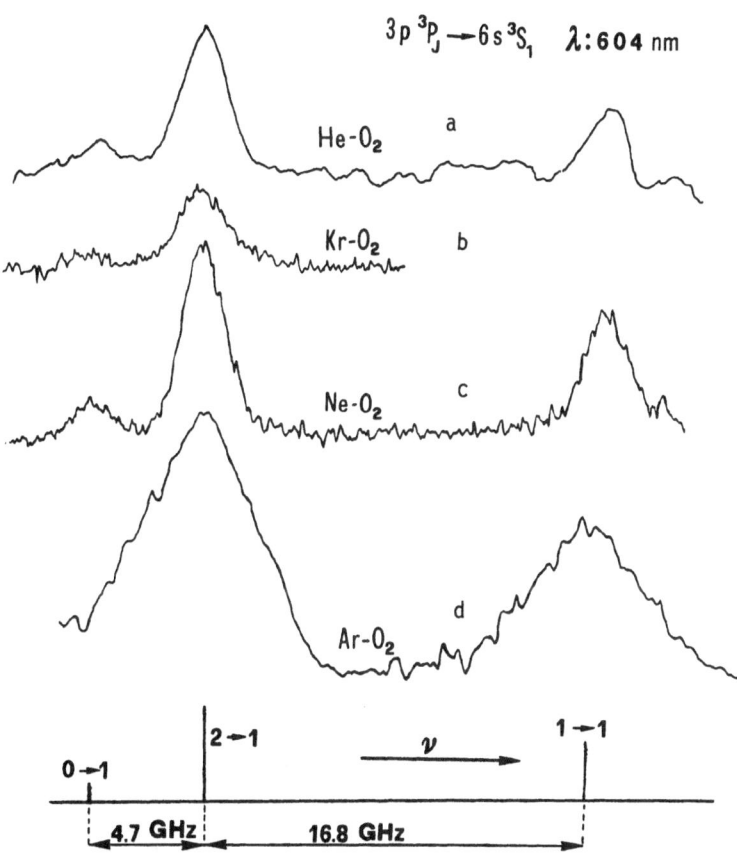

Fig.5 In case of O transitions between triplet states the "temperature" value is anomalous when argon is used as buffer gas. The recordings in the figure show for Ar a Doppler width about 3 times larger than for He, Ne, Kr.

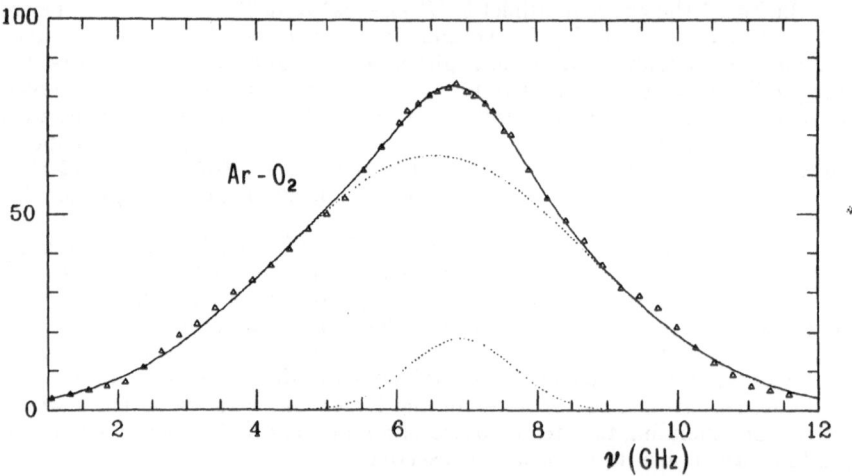

Fig.6 The anomalously broad lineshape in case of Ar can be fitted with the sum of two Gaussian profiles. The narrower corresponds to the same 570 K temperature observed for other noble gases or quintet transitions. The wider Gaussian corresponds to a temperature of about 4500 K. All the recordings have been obtained using the same experimental conditions (partial pressure, discharge current, etc.).

To check if the discussed anomalous broadening is caused by reduction of the lifetime of the involved oxygen levels we have analyzed the homogeneous lineshape. For this purpose we have developed a high resolution laser spectroscopy technique which allows us to remove the Doppler broadening. In particular by using polarization spectroscopy, described in detail in ref.[4], we obtain a homogeneous linewidth of 160 MHz (Fig.7) ascribed to pressure and saturation broadening.

The anomalous temperature value could be understood according to the model discussed in Sec. III; it was first suggested by Feld et al. [10] to explain the peculiar effects observed in the O laser emission at 844.6 nm sharing the 3^3P excited level with the transition at 604.6 nm (see Fig.2). In that model only metastable $O(2^1S, 2^1D)$ atoms have a wider velocity distribution ("warm" atoms) while ground-state atoms are essentially in thermal equilibrium ("cold" atoms). The electron-impact excitation of higher excited quintet and triplet state are regulated by selection rules similar to those for optical excitation. In particular, triplet states can be populated through electron collisions with ground state ($\Delta s = 0$) and metastable singlet state ($\Delta s = \pm 1$). As a result, the velocity distribution of such atoms and the resulting Doppler profiles of the relative transitions, are given by the superposition of a broad ("warm" excitation channel) and narrow ("cold" excitation channel) contribution. On the contrary, quintet atoms are produced essentially by excitation of "cold" ground-state atoms and, as a consequence, for optical transitions involving such levels a normal Doppler width is found.

While some aspects of the phenomenon have been, at least phenomenologically, understood some open questions remain.
One of these concerns the collisional mechanisms leading to O atoms formation in $Kr - O_2$ discharges; it should be very similar to that involving argon. On the contrary, the experimental results (Fig. 5b) do not show any anomalous broadening.
A more complete and deeper understanding of the basic mechanisms leading to the production of oxygen atoms in non thermal equilibrium with the discharge sample requires further investigations which are in progress in our laboratory.

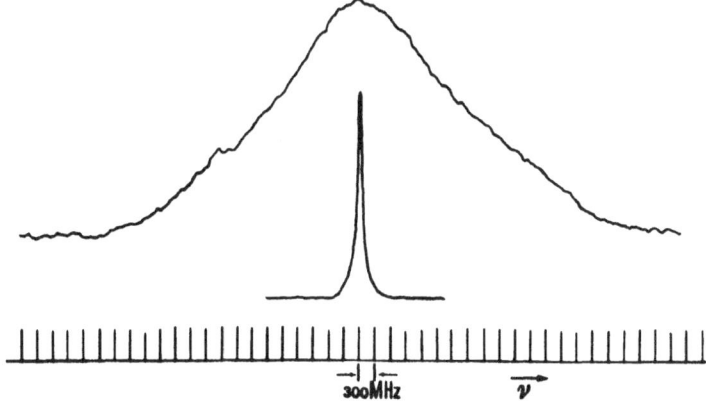

Fig.7 A comparison of the broad Doppler limited profile (upper trace) with the much narrower sub Doppler profile (lower trace) demonstrates the kinetic origin of the anomalous broadening. Sub Doppler lineshape has been resolved by polarization spectroscopy.

REFERENCES

1. M.Inguscio, P.Minutolo, A.Sasso, and G.M.Tino: Phys.Rev. A $\underline{37}$, 4056 (1988).
2. A.Sasso, P.Minutolo, M.I.Schisano, G.M.Tino, and M.Inguscio: J. Opt. Soc. Am. B $\underline{5}$, 2417 (1988).
3. K.Ernst, P.Minutolo, A.Sasso, G.M.Tino, and M.Inguscio: Opt. Lett., $\underline{14}$, 554 (1989).
4. M.Inguscio, L. Gianfrani, A.Sasso, and G.M. Tino: Laser Spectroscopy IX, M.S.Feld, A. Mooradian, and J.E.Thomas Eds., Academic Press (1989).
5. A.Sasso, G.M.Tino, M.Inguscio, N.Beverini, and M.Francesconi: Nuovo Cim. D $\underline{10}$, 941 (1988).
6. A.Sasso, M.Inguscio, G.M.Tino, and L.R.Zink : in "Non-Equilibrium Processes in Partially Ionized Gases", M.Capitelli, and T.H.Bardsley Eds., NATO ASI Series, Plenum, New York (1989).
7. M.Touzeau, G.Gousset, J.Jolly, D.Pagnon, M.Vialle, C.M.Ferreira, J. Loureiro, M.Pinheiro, and P.A.Sá, M.Capitelli and T.H.Bardsley Eds., NATO ASI Series, Plenum, New York (1989).
8. L.Apolloni, B.Brunetti, F.Vecchiocattivi, and G.G.Volpi: J. Phys. Chem. $\underline{92}$, 918 (1988).
9. W.P. West, T.B.Cook, F.B.Dunning, R.D.Rundel, and R.F.Stebbings: J. Chem.Phys $\underline{63}$, 1237 (1975).
10. M.S.Feld, B.J.Feldman, and A.Javan: Phys. Rev. A $\underline{7}$, 257 (1973).

THE N_2^*-OCS SYSTEM — MICROWAVE SPECTROSCOPY MEASUREMENTS OF

VIBRATIONAL POPULATIONS:THE PERTURBED STATIONARY STATE

P.G.Favero[1], M.C.Righetti[1] and L.B. Favero[2]

[1]Centro di Studio di Spettroscopia a Microonde
dell'Universita' di Bologna-Bologna Italy
[2]Istituto di Spettroscopia Molecolare del C.N.R.
Bologna Italy

INTRODUCTION

The determination of rate constants and mechanisms for collision-induced vibrational energy flow in poliatomic molecules is of basic and practical interest. Such informations are used as a guide and test for approximate teoretical models in the prediction of potential surfaces for collisional interactions. Moreover,they can be used to evaluate the possibility of influencing the course of a chemical reaction or developing new lasers.

In the two past decades great advances in the understanding of energy transfer phenomena have been obtained mainly due to the development of spectroscopic "state-to-state" methods for studying energy flow, especially in the case when the time behaviour of specific quantum states is given.

In these type of experiments the population of a particular and known state is consistently enhanced during a short time by a suitable means and immediately after the population of the same state and of other states is monitored by time-resolved measurements. Whenever possible, the use of a pulsed laser and a time-resolved fluorescence technique has proved to be the most powerful method for this type of studies. The overpopulation is obtained through an allowed transition (pumping transition) to the state of interest from a lower state (usually the fundamental state). As vibrational energy transfer studies are predominantly made in the gas phase and the pumping transition frequency falls in the IR , the laser method is somewhat limited by the lack of high power tunable laser in this region. As a matter of fact up to now only those molecular systems have been analyzed for which the pumping transition falls in the CO_2 laser frequency range (or in its second harmonics),the limited tunability of the CO_2 laser being compensated for by the generally resolved rotational structure of a vibrational band.

In other cases the overpopulation of a certain state can be obtained by mixing the gaseous molecules under study with another type of molecules that are separately excited by some means.Mainly those diatomics are used that have a vibrational state whose energy is in the neighborhood of the energy of this state. An intermolecular energy transfer process takes then place which does not require as stringent conditions of frequency match as

Applied Laser Spectroscopy, Edited by W. Demtröder and
M. Inguscio, Plenum Press, New York, 1990

in the case of the laser method. To monitor the population of the states any suitable time-resolved technique can be used.

In particular experimental conditions the whole system can reach a stationary state in which the population deviations from the Boltzmann distribution are stationary. In this case time indipendent population measurements can be made and informations similar to those obtained with the time-resolved experiments are obtained.

In what follows it will be shown how rotational microwave spectroscopy may be used in this type of problems taking as an example the N_2^*-OCS system and reporting some preliminary results obtained by the authors[1].

A SHORT SUMMARY OF BASIC CONCEPTS

Vibrational levels and rotational spectrum of a linear triatomic molecule

For a linear triatomic molecule there are two stretching and a doubly degenerate bending normal mode of vibration and therefore three fundamental frequencies. Four quantum numbers classify the vibrational states and define their energy. Different symbols are used throughout the literature to indicate such states; here the following symbol will be used: (v_1, v_2^l, v_3) where v_1, v_3 and v_2 are the quantum numbers relative to the lower energy and the higher energy stretching and the bending mode respectively. The l quantum number is relative to the vibrational angular momentum wich arises in degenerate modes and takes on the values $-v_2, -v_2+2 \ldots, v_2-2, v_2$. An alternative symbol used here is $nv_i^{\pm l}$ where n is the number of quanta excited in the ith mode (for the stretching modes l=0 and it is not specified). While in first approximation the vibrational energy does not depend on l, a more accurate treatment shows that the degeneracy of the bending mode levels is partially removed giving rise to sublevels. In fig.1a is shown a semplified scheme of the vibrational energy levels of OCS while fig.1b shows a detailed view for the bending manifold where the residual degeneracies are indicated by double orizontal lines.

The rotational energy E_R of a linear triatomic molecule is,to a good approximation for our purposes, a function of the total angular momentum number J, the vibrational quantum number l,the rotational constant B_v and the distortion constant D_v (much smaller than B_v). The index V means that the values of these constants are different (at least in principle) for different vibrational states. It can be seen from theory that the residual degeneracies in the bending states (other than the accidental degeneracies) are completely removed by rotation-vibration interaction. The rotational spectrum is made up by nearly equidistant lines starting at a frequency equal to $2B_v$ and due to the selection rule J-->J+1. The intensity of each line is a function of several parameters among which the fraction of the total number of molecules that are in a certain rotational (F_J) and vibrational(F_v) state are here of major interest. The rotational line intensities I are in fact directly proportional to the product of such two factors. In fig.2 the logarithm of the ratio I/I_0 (I_0 is the intensity of the line relative to the fundamental vibrational state) for the rotational J=8-->9 transition of OCS in several vibrational states is shown. The solid bars give the situation in condition of Boltzmann equilibrium at room temperature.

It is seen that the high resolution of microwave spectroscopy allows to observe the rotational line relative to practically all vibrational

258

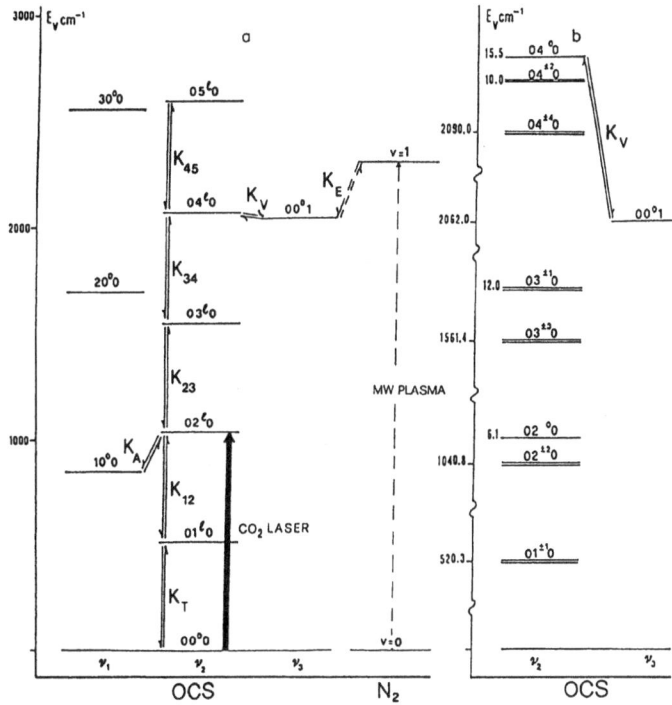

Fig. 1. a) Semplified scheme of the vibrational energy levels of OCS.
Solid arrows show the predominant energy transfer mechanism as
established with laser experiments. Dotted arrows indicate the
path of the energy transfer from Nitrogen to OCS.
b) Expanded view of the bending manifold.

states considered giving therefore through the intensities the needed
information on the vibrational populations. It may be useful pointing out
that while F_J changes with J, the ratio I/I_0 will be the same, at the same
temperature, for all rotational transitions and therefore the same type of
information will be obtained from any one of them.

It must be pointed out however that in microwave spectroscopy only
relative line intensities are feasible with sufficient precision. In the
present type of experiments the intensities (populations) are measured
relatively to the intensities (populations) in conditions of Boltzmann
equilibrium and to know the absolute population the temperature of the
system is required.

Kinetics and stationary states

When one has to deal with a collision induced energy transfer process
for bimolecular collisions the following formal relation is used:

$$A+B \underset{K_b}{\overset{K_f}{\longleftrightarrow}} C+D+\Delta E \tag{1}$$

259

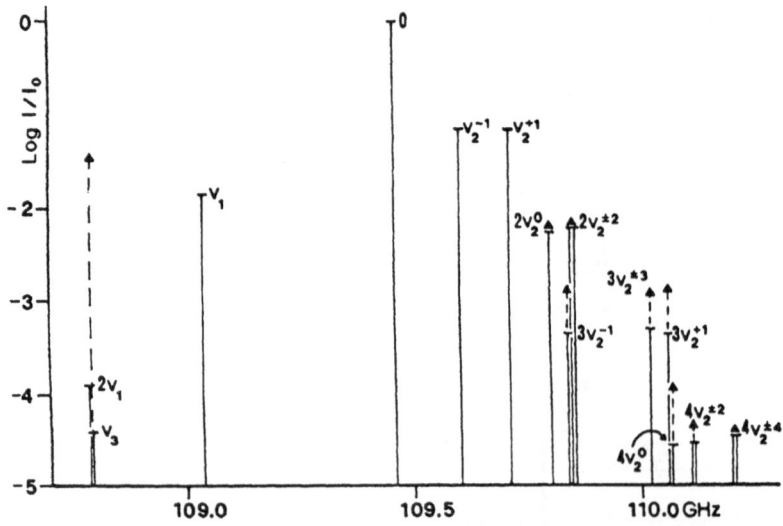

Fig. 2. The rotational transition J=8->9 of OCS in various vibrational
states. The solid vertical bars give the logarithm of the ratio
I/I_0 with I the intensity of a generic line and I_0 the intensity of
the transition in the fundamental vibrational state when the
system is at room temperature. The dotted arrows give the same
quantity in the presence of excited Nitrogen.

where A an B indicate the initial state of the molecule and its colliding
partner and C and D the relative final states. ΔE is the energy balance of
the whole process. If the state of the colliding partner is not specified
the letter M is used. In a gas mixture the letter M will mean a undefined
state of all components. If N_i is the number density of the ith species and
it is supposed thatthe process is evolving towards some equilibrium
condition, one can write down the velocity with which the species A changes
to species C as follows:

$$\dot{N}_A = -K_f N_A N_B + K_b N_C N_D \qquad (2)$$

where K_f and K_b (appearing in the formal expression (1)) are the so called
forward and backward rate constants of the process and are generally
different; the number densities N_i are now functions of time. When the
system is in Boltzmann equilibrium $\dot{N}_A = 0$ and therefore the ratio $K=K_f/K_b$
is known and has the value $N_C^B N_D^B / N_A^B N_B^B$. If one assumes that $K=K_f/K_b$
is true also for non equilibrium situations (detailed balance assumption)
only one constant needs to be determined.

As it will be seen shortly, an energy transfer process is in general
more complicated than shown by (1) and one will have to deal with a system
of coupled non linear differential equations. Whenever possible a
linearization procedure is performed so that such system appears to have,
in matrix formulation, the following form:

$$\underline{\dot{n}} = \underline{A}\,\underline{n} + \underline{V} \qquad (3)$$

where $\underline{n}$ is the vector of the displacements of N_i from an equilibrium or
stationary situation, A is a matrix whose elements contain the rate

constants and the vector $\tilde{V}$ is related to the initial conditions. The solution of (3) is of the type $n_1 = \Sigma_m c_{1,m} \exp(-k_m t)$, where the k_m are in general combinations of rate constants. A general energy transfer process implies a substantial complication even for very simple molecules like triatomics and even if an upper value of the vibrational energy is taken into consideration. For instance, if one includes 10 levels in the analysis, 45 rate constants are needed to describe the behaviour of the system. Fortunately, in many cases few rate constants are predominantly important in determining the energy flow map while the others are practically unimportant.

In laser experiments the system is perturbed during a very short time and then it evolves (see below) to a steady state in which all the $\dot{n}$ in (3) are now zero: this is the Boltzmann steady state. As it has been pointed out earlier, in flow experiments a perturbed steady state can be reached. In this case the matrix elements of $\underline{A}$ contain also terms related to the flow velocities of the various components. Again the $\dot{n}$ are all zero and equations (3) become a system of linear algebraic equations that can be solved for the rate constants once the population (number density) of the various states included in the energy transfer model are known from experiment.

From the rather large number of experimental results it is possible to infer some general rules on how to predict the relative values of the rate constants in a complex energy transfer process. The most favored processes will be those in which the number of exchanged quanta and the energy balance are minimum (the first condition may be not necessary if between two levels Coriolis or Fermi perturbation is operating which mixes the corresponding states). A vibration-translation (V-T) process will be less and less probable as the vibrational energy increases and therefore of all V-T processes, that originating from the lowest level will be the most probable. For more details see the reviews listed under reference [2].

THE VIBRATIONAL ENERGY TRANSFER IN OCS

Laser and time-resolved fluorescence experiments

OCS is well known as a FIR[3] and IR[4] optically pumped laser medium. The map for its vibrational energy transfer has been established by Mandich and Flynn[5] (see reference [5] for previous attempts) by using laser excitation and time-resolved fluorescence technique. After a laser pulse the fluorescence from the system is collected and suitably filtered to select the part coming from a predetermined vibrational level. The fluorescence signal is proportional to the population of that level. In fig.1 the solid arrows show the path of the excitation: once the $2v_2$ level population is enhanced by the laser pulse, the perturbation spreads rapidly into the bending manifold and into the v_3 and v_1 levels going ultimately to zero through the V-T transfer ($v_2 \longrightarrow 0$).

The predominant energy transfer process can therefore be defined by the following relations:

$$OCS(0) \longrightarrow OCS(2v_2) - 1042 \ cm^{-1} \qquad\qquad a)$$

$$OCS(2v_2) + M \quad \xleftrightarrow{K_A} \quad OCS(v_1) + M \ \ +118 \ cm^{-1} \qquad b)$$

$$OCS(v_2) + OCS(v_2) \xleftrightarrow{K_{12}} OCS(2v_2) + OCS(0) - 7 \ cm^{-1} \qquad c)$$

$$OCS(2v_2) + OCS(v_2) \xleftrightarrow{K_{23}} OCS(3v_2) + OCS(0) + 6 \ cm^{-1} \qquad d)$$

$$OCS(3v_2) + OCS(v_2) \xleftrightarrow{K_{34}} OCS(4v_2) + OCS(0) + 12 \ cm^{-1} \qquad e)$$

$$OCS(4v_2) + M \xleftrightarrow{K_v} OCS(v_3) + M + 43 \ cm^{-1} \qquad f) \qquad\qquad (4)$$

$$OCS(v_2) + M \xleftrightarrow{K_T} OCS(0) + M + 520 \ cm^{-1} \qquad g)$$

where, for semplicity, the backward rate constants have been omitted and the intra bending mode relations have been written only up to the $(4v_2)$ level.

The population of various levels has been monitored at various values of the OCS pressure in the range of 1 to 10 torr. A general behavior of the time-dependent fluorescence is shown in fig.3a. It is of the form to be generally espected for the solution of (3) with the Boltzmann number densities as initial conditions. The linearization is here possible due to the smallness of the overpopulation . From the dependence of the k_m from the OCS pressure the rate constants K_T, K_{12}, K_A and K_v (see Fig.1a) have been obtained[5].

Intermolecular transfer with N_2^* and microwave spectroscopy

The first experiment on the energy transfer from N_2^* to OCS studied by microwave spectroscopy has been performed by Bogey and Bauer who measured the rate constant K_E [6](see Fig.1). The method is based on the assumption that the rotational relaxation time is much shorter than any vibrational relaxation time of interest. The intensity of a rotational

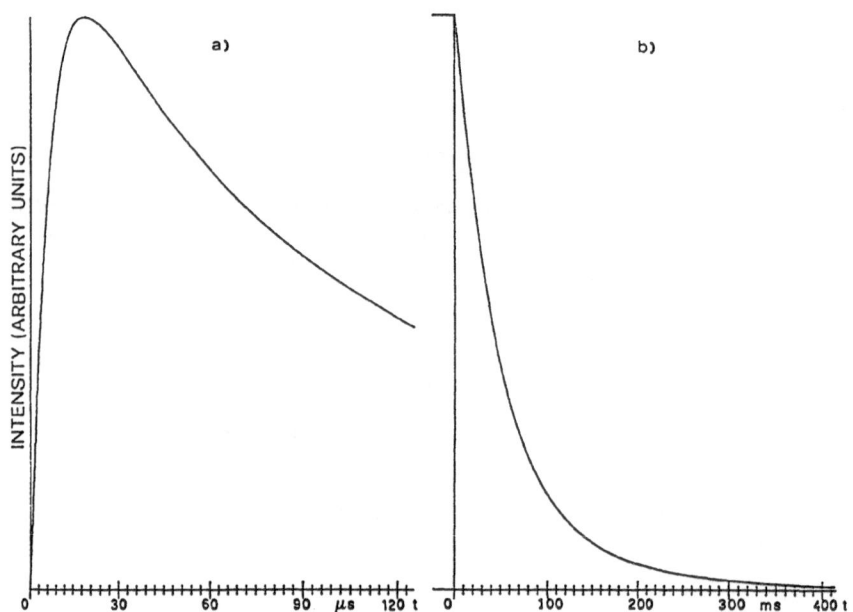

Fig. 3. a) Temporal behaviour of the fluorescence relative to an excited vibrational level after a laser pulse. b) Temporal behaviour of the intensity of a rotational line relative to an excited vibrational level after switching off the microwave plasma.

262

line is than proportional to the vibrational population of the state in which the rotational transition is observed even during a kinetic event.

In this laboratory the energy transfer from active nitrogen has been used to sufficiently enhance the population of the first excited vibrational state of the high frequency stretching mode of some triatomic linear molecules so that the rotational spectrum in that state becomes observable[6].

Active nitrogen is obtained in sizable amount by flowing pure nitrogen through the plasma produced by microwave power and contains,among other species,nitrogen molecules in excited vibrational states.These species are responsible for the vibrational energy transfer to OCS.

Active nitrogen and OCS are mixed in an absorption cell which is part of a flow system operated in controlled conditions of pressure and flow rates.Our cell was a short (14cm) and small volume (700cm³) interferometric absorption cell. At frequencies around 110 GHz the equivalent absorption path length of the cell is about 6 m and enough sensitivity to observe the rotational lines of OCS in excited vibrational states at Boltzmann equilibrium as high in energy as 2300cm^{-1} is secured.

The essential lay-out of the apparatus is shown in Fig.4. The microwave millimetre spectrometer is based on a highly frequency stabilized klystron operating around 37 GHz followed by a frequency tripler whose power is fed into the absorption cell. The millimetre power is frequency modulated and the output signal from the detector is demodulated by a lock-in amplifier whose reference frequency is twice the modulation frequency so that the absorption signal appears as the second derivative

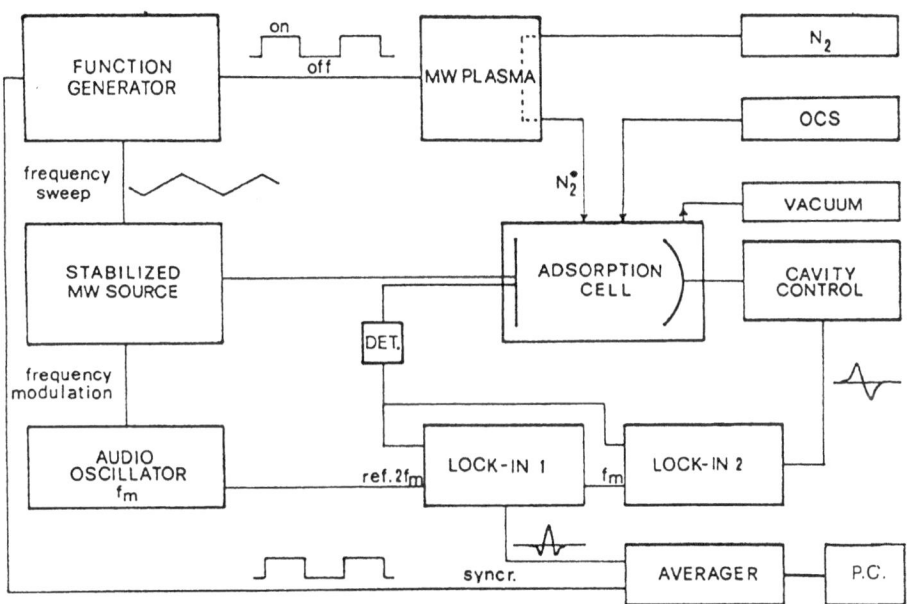

Fig. 4. Block diagram of the apparatus to monitor the population of the vibrational levels by microwave spectroscopy.

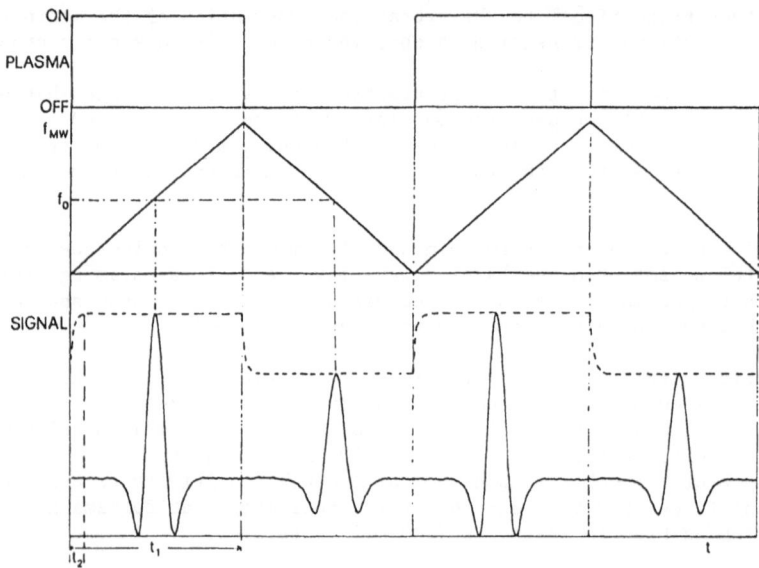

Fig. 5. Temporal sequence of the various operations for the determination of the R value.

of the line shape. The absorption signal is accumulated in a multichannel analyzer to enhance the signal to noise ratio and the data elaboration is made with a personal computer. If the source is kept at the frequency f_o of the maximum of the line, time-resolved measurements of intensities are possible (see Fig.3b).

To measure the intensity ratio R between a line in the perturbed and Boltzmann stationary state,the microwave plasma is switched on and off by a multifunction generator that produces also a synchronous wave for the millimetre frequency sweep (see Fig.5). The dotted lines show the time dependence of the line intensity, t_2 being the time at which the kinetics is practically complete (t_1 is of the order of one second).

Well resolved rotational spectra are obtained at rather low pressure. The measurements have been made in the range from 20 to 50 mtorr. In these conditions the characteristic kinetic times are around three order of magnitude lower than those observed in IR experiments as it shown in Fig.3.

In Fig.1 the dotted arrows show how energy is poured from excited nitrogen into the OCS vibrational energy system. On the basis of their experimental results Bogey an Bauer [7] have confirmed that the predominant mechanism is that proposed by Mandich and Flynn even in presence of nitrogen and up to now our results are in accord with this conclusion.

However, for the present experimental conditions, the energy transfer process is somewhat more complicated then shown by the relations (4) in that, once (4a) is eliminated, the following relations are to be added:

$$OCS(0)+N_2(v=1) \; <\xrightarrow{K_E}> \; OCS(v_3) + N_2(v=0) + 269 \; cm^{-1} \quad h) \qquad (4)$$

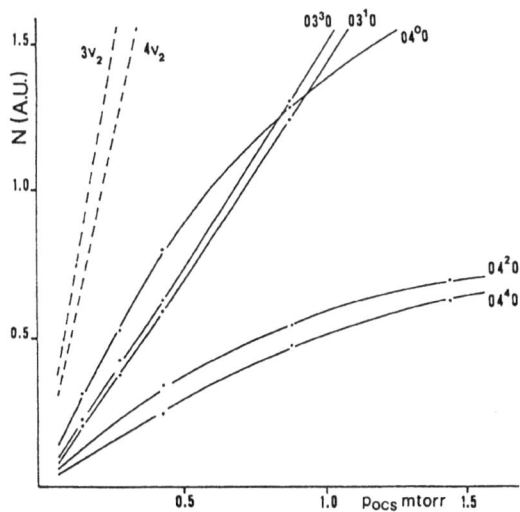

Fig. 6. Number densities N (populations) of the *single* sublevels of $3v_2$ and $4v_2$ (sublevels labeled by +1 or -1 have been found to have the same population). The dotted curves show the behaviour of the *total* population of the $3v_2$ and $4v_2$ levels.

$$N_2(v=1) \longrightarrow N_2(v=0) + 2330 \text{ cm}^{-1} \qquad \qquad \text{i)}$$

$$OCS(v_3) \longrightarrow OCS(0) + 2062 \text{ cm}^{-1} \qquad \qquad \text{l)}$$

where h) corresponds to the energy transfer from Nitrogen to OCS, i) is the process of deactivation of the excited Nitrogen at the walls and l) is the process of deactivation of $OCS(v_3)$ by spontaneous emission which must be taken into account due to the low experimental pressure.

From Fig.2 one sees that, in presence of excited Nitrogen, the v_3 level population is enhanced by around three orders of magnitude and that a consistent enhancement is experienced by the population of nearly all the other levels (all populations are in fact perturbed, but for some of them the variation does not show up with the scale used). In such situation it is clear that the system of differential equations relative to the new set of relations (4) is not linearizable and that the system of algebraic equations obtained by setting all $\dot{N}$ equal to zero is also non linear.

The inspection of Fig.2 and 6 reveals a rather peculiar population distribution of the $3v_2$ and $4v_2$ sublevels. With reference to Fig.1b, it is seen that the overpopulation is higher for higher sublevels. Due to the smallness of the energy differences a Boltzmann like distribution could be expected. In Fig.6 the population of the *single* sublevels of $3v_2$ and $4v_2$ as function of the partial pressure of OCS is shown: another interesting feature is that below .9 mtorr the population of the (04⁰0) sublevel is higher then the population of the sublevels (03¹0) and (03³0). This population inversion is not seen if one plots the *total* population of the two levels (dotted curves).

Table 1. Population ratios for $p_{ocs}=1.46$mtorr[a]

state	F_B	F_{C1}	F_{C2}	F_{EX}
(04⁰0)	.45	4.00	10.59	1.30
(04²0)	.92	.89	2.43	1.10
(04⁴0)	1.00	1.00	1.00	1.00
(03¹0)	.94	.92	2.11	1.08
(03³0)	1.00	1.00	1.00	1.00
(02⁰0)	.49	.48	.47	.48
(02²0)	1.00	1.00	1.00	1.00

[a] The degeneracy for $l \neq 0$ has been taken into account.

All these features are evidently observable only because of the highly detailed informations given by microwave spectroscopy. As a matter of fact analogous experiments performed with a fluorescence technique would give average population over all sublevels of a given vibrational level. IR diode laser spectroscopy can give similar detailed informations, but to our knowledge this technique has been used to give qualitative informations only [9].

A comparison with the results obtained by Mandich and Flynn [5] can be made inserting into the algebraic equation system for the perturbed stationary state the experimental total populations and solving it for the rate constants. The value $(320 \pm 140) \cdot 10^3$ torr^{-1}s^{-1} for K_{12} compares reasonably well with the previous value $(480 \pm 240) \cdot 10^3$ torr^{-1}s^{-1} [5].

To interpret the finer details it is necessary to substitute the relations (4c,d,e,g) with others that allow transfer processes between any sublevel. Furthermore relation (4f) is to be substituted by the one that represents the transfer between the (00°1) and the (04°0) levels(see Fig.1b and [5]):

$$OCS(04°0) + M \xrightarrow{K_v} OCS(00°1) + M + 53.5 \text{ cm}^{-1} \qquad f') \qquad (4)$$

Although the tranfer process is now rather complex, with few resonable assumptions and the use of known rate constants it is possible to calculate the populations and compare them with the experimental ones. In the table the results of these calculations are given through the ratios F_{C1} of the sublevels populations to the population of the lowest sublevel of a given vibrational level. F_B and F_{EX} are the ratios for a Boltzmann like and the observed distributions respectively.

It is seen that a consistent enhancement of the (04°0) population is predicted, but that the other sublevels show a Boltzmann like distribution. It may be that in the energy transfer process of the bending manifold some kind of propensity rule is operating. Among others, the assumption that the total vibrational angular momentum is conserved during the collision (dipole-dipole interaction) has been made. This implies that the condition $l_A+l_B=l_C+l_D$ must now be inserted into the modified relations (4). Again it is possible to calculate the populations ratios and they are shown in the Table as F_{C2}. The enhancements of the (04²0) and (03¹0) are now appearing, but no reasonable change of the rate constants seems to make the F_{C2} more similar to the F_{EX}.

Finally both models predict the previously mentioned inversion among the populations of the (04°0) and the $3v_2$ sublevels only if the assuption

(4f') is made and therefore the experimental results should prove its validity.

CONCLUSIONS

The preliminary results presented here on the N_2^*-OCS system show that Microwave Spectroscopy may contribute in the field of the collision-assisted transfer of vibrational energy by giving more detailed informations than the fluorescence technique. It seems to produce at present more questions than answers so that more experimental and theoretical work should be in order.

REFERENCES

1. P.G.Favero, M.C.Righetti and L.B.Favero, XI-th Colloquium on High Resolution Spectroscopy, Giessen (WD), September 1989.
2. J.D.Lambert, "Vibrational and rotational relaxation in gases", University Press, Oxford (1977).
    J.T.Yardley, "Introduction to molecular energy transfer", Academic Press, New York (1980).
    B.J.Orr and I.W.M.Smith, J.Chem.Phis. 91, 6106 (1987).
3. T.F.Deutsch, Appl.Phys.Letters 8, 334 (1966).
    J.C.Hassler and P.D.Coleman, Appl.Phys.Letters 14, 135 (1969).
4. H.R.Schlossberg and H.R.Fetterman, App.Phys.Letters 26, 316 (1975).
5. M.L.Mandich and G.W.Flynn, J.Chem.Phys. 73, 3679 (1980).
6. M.Bogey and A.Bauer, Chem. Phys. 46, 393 (1980).
7. G.Cazzoli, P.G. Favero and C.Degli Esposti, Chem.Phys.Lett. 50,336 (1977).
    G.Cazzoli, C.Degli Esposti and P.G.Favero, J.Mol.Struct. 48,1(1978).
    C.Degli Esposti, P.G.Favero, S.Serenellini and G.Cazzoli,J.Mol. Struct. 82, 221 (1982).
8. M.Bogey and A.Bauer, J.Mol.Spectrosc. 84, 170 (1980).
9. K.M.T.Yamada and M. Klebsch,J.Mol.Spectrosc. 125, 380 (1987).

PHONON LIFETIMES IN MOLECULAR CRYSTALS

Salvatore Califano

European Laboratory for non Linear Spectroscopy
Largo Enrico Fermi 2, 50125 Florence, Italy

1. INTRODUCTION

The subject of phonon relaxation is the study of the processes which contribute to the finite lifetime of phonon states in a crystal. Harmonic phonons are non-interacting, independent crystal excitations with infinite lifetime. Real phonons, however, are not independent, since they interact through anharmonic terms of the crystal hamiltonian. These anharmonic phonon-phonon interactions are responsible for the finite phonon lifetime[1]. In the specific case of the crystals discussed in this lecture, i.e. molecular crystals, phonons have relatively short lifetimes, even at very low temperatures. Typically phonon lifetimes in molecular crystals range from few picoseconds to some hundred picoseconds at the liquid He temperature. In some special cases, for very isolated phonon levels, these can even reach higher values, up to several nanoseconds. As the temperature increases the lifetime decreases, since the number of decay channels increases with temperature.

In this text the word phonon is used to denote any kind of collective vibrational excitation in a molecular crystal. The excitation may either involve translational or librational motions of the molecules as a whole (external or lattice phonons) or internal molecular vibrations (internal phonons or vibrons). For many crystals the internal phonons occur at higher frequency than the lattice phonons. It may happen, however, that some low frequency internal modes fall in the frequency range of the lattice vibrations. in these cases the formal separation between lattice and internal modes is lost and the crystal motions assume a mixed character[2].

The anharmonic phonon-phonon coupling terms of the crystal hamiltonian give rise to two basically different relaxation processes[1]. The first involves the simple loss of phase of the collective excitation due to scattering processes with thermal bath phonons. By analogy to spin relaxation theory, these pure dephasing processes are associated to a T_2^* relaxation time. The second process is due to the decay of the phonon energy into the thermal bath, i.e. to a depopulation of the phonon state.

Applied Laser Spectroscopy, Edited by W. Demtröder and
M. Inguscio, Plenum Press, New York, 1990

These energy decay processes are associated with a T_1 relaxation time. The total relaxation time T_2 of a phonon state is then given by [1]

$$\frac{1}{T_2} = \frac{1}{2T_1} + \frac{1}{T_2^*} \tag{1}$$

2. EXPERIMENTAL DETERMINATION OF PHONON LIFETIMES

The relaxation of phonon states can be studied either in the time or in the frequency domain[3,4,5]. In the time domain one measures directly the phonon lifetime by time-resolved spectroscopic techniques. In the frequency domain one measures instead the profile of the band associated to the absorption or scattering of light by the phonons. The two types of experiments are equivalent in principle, since the Fourier transform of the time domain relaxation curves gives the band profile in the frequency domain. When the relaxation time T_2 is truly exponential, the band profile is a Lorentzian, with full width γ at half maximum (FWHM in units of cm^{-1}) given by[1]

$$\gamma = \frac{1}{\pi c T_2} \tag{2}$$

Owing to the fact that all spectroscopic techniques are limited to phonons with zero wavevector, we shall limit our discussion to these optically active phonons.

In the frequency domain the most used experimental techniques are infrared absorption and Raman scattering spectroscopy. These techniques are well-known and do not need to be discussed here, except for the fact that in most of the cases of interest, especially at low temperatures, high resolution is needed for accurate band profile measurements.

Commercially available infrared interferometers have today resolution up to 0.001 cm^{-1}, largely sufficient for the determination of accurate band profiles in solids. Also in the case of Raman spectroscopy there exist today grating spectrometers with sufficiently high resolution (0.01 cm^{-1}) for most practical purposes. For higher resolution, necessary for very long lived phonons, it is convenient to use the kind of set-up, shown schematically in Fig.1, which consists of a Fabry-Perot interferometer coupled to a double monocromator used as pass band filter [6].

A typical output of this type of tandem system, which can reach a resolution of 0.001 cm^{-1}, is shown in Fig.2.

Another convenient technique for the determination of the band profiles of Raman active phonons is that of CARS (Coherent Antistokes Raman Spectroscopy)[3,4,5]. In a CARS experiment two laser beams of frequency ω_1 and ω_2, choosen so that $\omega_1 - \omega_2 = \omega_{phonon}$, are sent on a sample to coherently populate a phonon level. The same laser beam of frequency ω_1 (in some cases a third laser beam of different frequency ω_3 is used), stimulates the emission at the antistokes frequency $\omega_{as} = 2\omega_1 - \omega_2$, as shown schematically in the diagram of Fig. 3 .The process is a parametric one, in

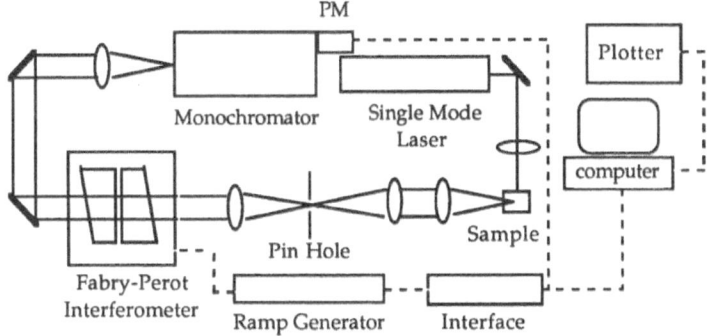

Figure 1. Fabry- Perot and Raman Spectrometer Tandem instrument for high resolution Raman Spectroscopy.

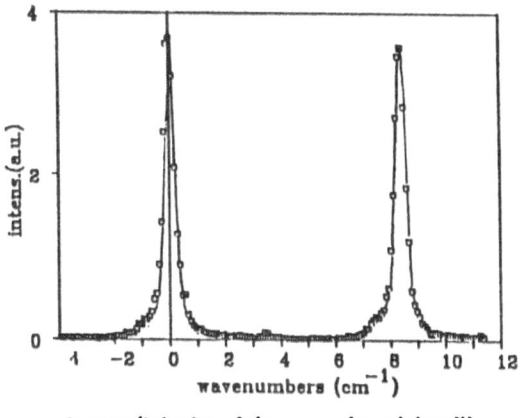

• Raman/Interfer. data ——— lorentzian fit

Figure 2. High resolution interferometric spectrum of a benzene phonon at 62 cm^{-1} measured at 10 K.

which the sample plays the role of transferring photons between different modes of the field with energy and momentum conservation. The momentum conservation relation imposes that, if the beams ω_1 and ω_2, are incident on the sample with an angle Θ, the new beam ω_{as} will be emitted in the direction defined by the triangle shown in the diagram.

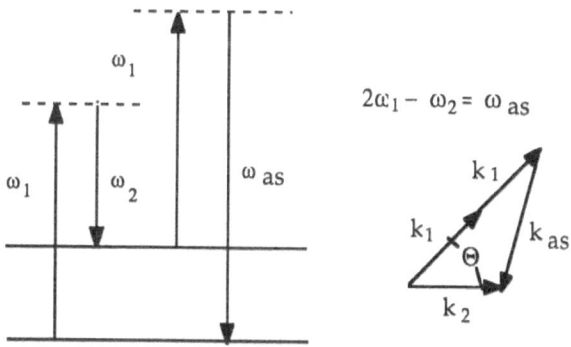

$$2\omega_1 - \omega_2 = \omega_{as}$$

Figure 3. Scheme of a CARS process.

271

A frequency resolved CARS system is shown in Fig. 4. A mode locked Nd Yag laser with second harmonic generation system pumps two dye lasers. The two beams, or alternatively one beam and the second harmonic of the pump laser, are focused on the sample. Delay lines are used in order that the two pulses arrive on the sample at the same time.

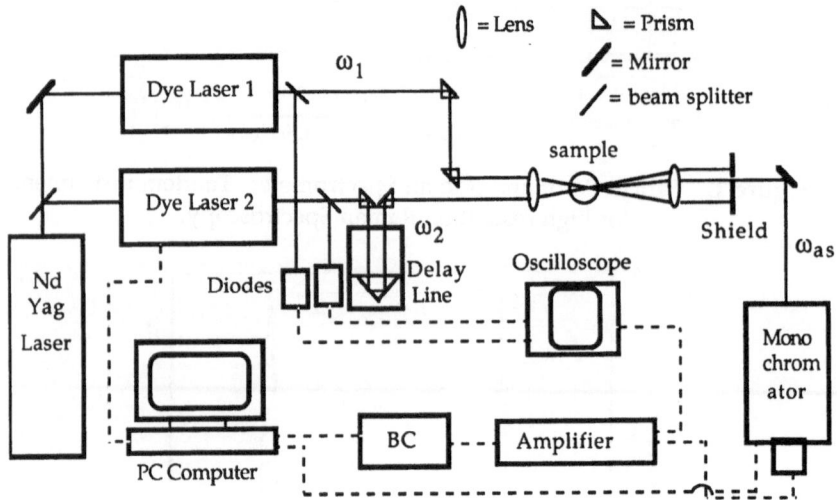

Figure 4. Schematic drawing of a frequency domain CARS instrument.

The antistokes beam emitted by the sample is spatially filtered, sent into a double monochromator and detected by a photon counting system..

The CARS technique is ideally suited for time resolved spectroscopic measurements of phonon lifetimes. In a time resolved experiment the ω_1 beam is split in two and one of the two beams is passed through a variable delay line.

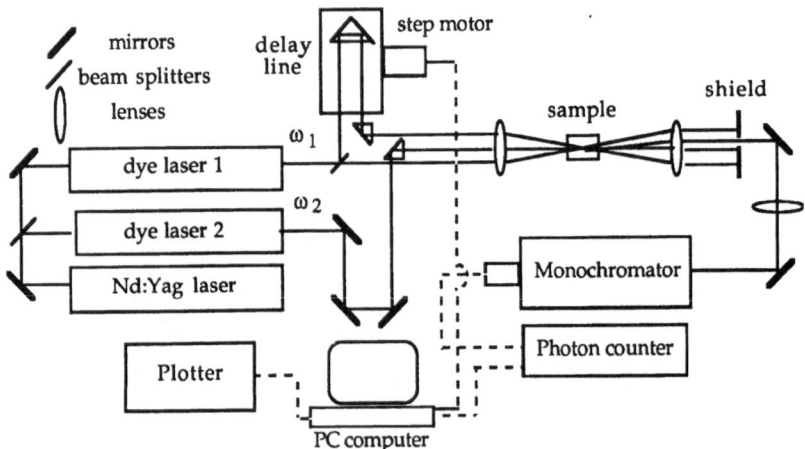

Figure 5. System for time resolved CARS Spectroscopy with ps pulses.

Two systems will be described here for time resolved spectroscopy. The first, shown schematically in Fig.5, is made of a mode locked, frequency doubled Nd. Yag laser synchronously pumping two dye lasers and has a time resolution of about 3 ps.. The output of the pump laser is divided into two beams by means of a beam splitter. One beam is used to pump the first dye laser, tuned at the frequency ω_1, whereas the other beam pumps the second dye laser, tuned at the frequency ω_2. The output of the first laser is again divided into two, by means of another beam splitter. One of the two ω_1 beams is used, together with the beam ω_2 to coherently populate the phonon state. The second beam of frequency ω_1 is passed through a motorized delay line so that the pulses reach the sample with a variable delay, controlled by a PC computer. The CARS beam at frequency $\omega_{as} = 2\omega_1 - \omega_2$ is spatially filtered, collimated and sent to a double monochromator. The signal from the cooled photomultiplier is then processed by a photon counting system. By retarding the delayed ω_1 pulse with respect to the other two, one observes the decay of the CARS signal over several decades. A typical output is shown in Fig. 6.

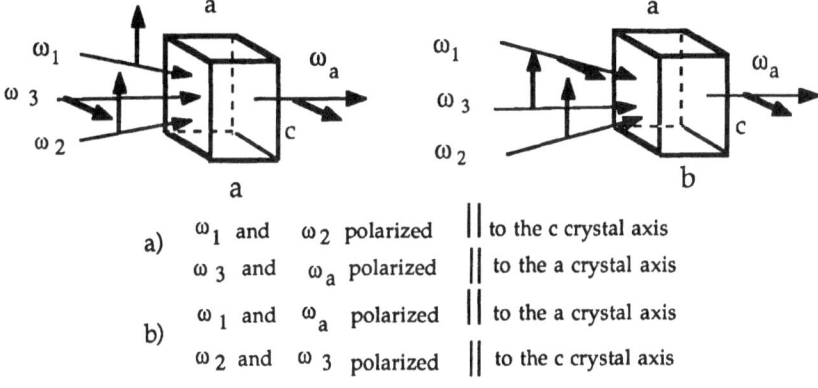

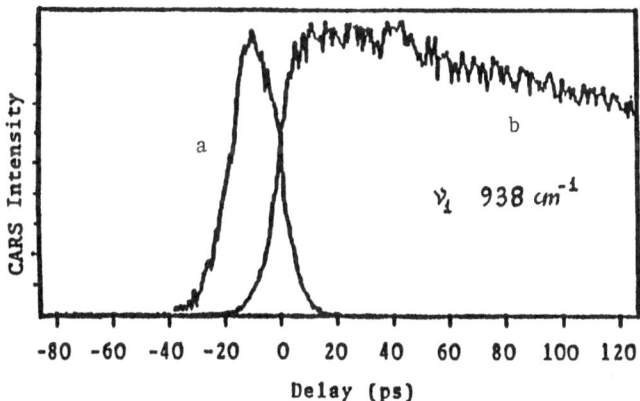

Figure 6. Time variation of the CARS signal for the vibron v_1 of the KClO$_4$ crystal.

The second system is designed for shorter pulses (from 250 to 100 femtoseconds) and is shown in Fig 7.

The infrared pulses of a mode locked Nd. Yag laser of about 80 ps duration (80 MHz repetition rate) are compressed in an optical fiber and through four passes on a grating, to 4 ps. The pulses are then frequency doubled and used to pump a dye laser which emits pulses of 250 fs. These are amplified by a three stage amplifier pumped by a second Nd. Yag laser (10 Hz repetition rate) and the output beam is devided into two parts by a beam splitter. One beam is directly used whereas the second is sent into a water cell to generate a 250 fs continuum. A frequency component of the continuum is then selected and amplified three times by a second amplifier, pumped by the same Nd. Yag laser. If pulses shorter than 250 fs are needed, the output of the first dye laser is again fiber compressed before being sent to the amplification system.

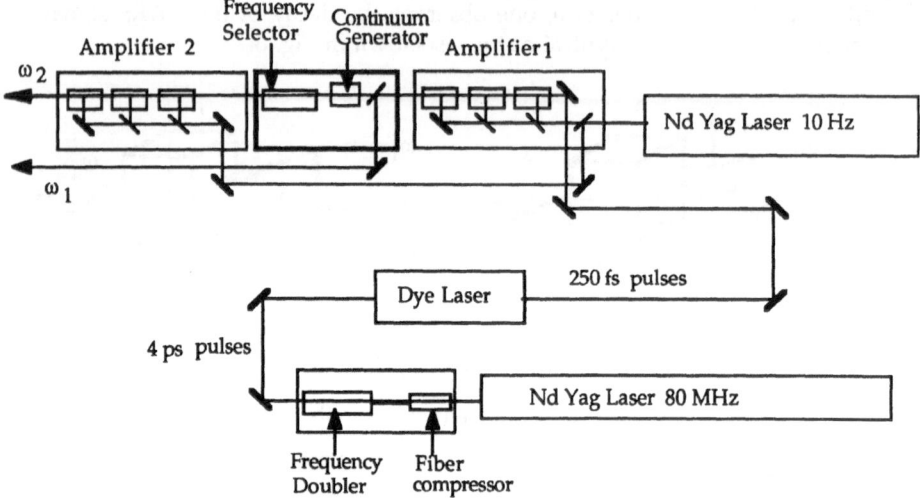

Figure 7. System for time resolved CARS Spectroscopy with fs pulses.

3. THEORY OF PHONON RELAXATION

As discussed before phonon relaxation processes are due to anharmonic phonon-phonon coupling terms of the crystal hamiltonian. For molecular solids the most convenient approach to the dynamics of anharmonic crystal vibrations is the perturbative expansion of the hamiltonian in terms of phonon normal coordinates. Many-body perturbation techniques are then used to evaluate anharmonic phonon shifts and bandwidths. The perturbation expansion can be carried out in principle to any order, although in practice the inclusion of fourth order terms is already a major task. The theory is discussed in details in many books and review articles [1,2]. Here we present a summary of the theory, limiting the exposition to the basic parts, necessary for the interpretation of the experimental results.

First we expand the crystal potential V in powers of the crystal normal coordinates Q

$$V = V_2 + V_3 + V_4 + \ldots \tag{3}$$

$$V = \frac{1}{2}\sum_{lm}C_{lm}Q_lQ_m + \frac{1}{3!}\sum_{lmn}C_{lmn}Q_lQ_mQ_n + \frac{1}{4!}\sum_{lmnp}C_{lmnp}Q_lQ_mQ_nQ_p + \ldots \tag{4}$$

where l,m,n,p are composite indices, comprehensive of both the phonon branch label j and of the phonon wavevector label k. Whenever necessary we shall specify these two labels separately. The coefficients C are derivatives at equilibrium of the potential with respect to normal coordinates [1,2]

$$C_{lm} = \left(\frac{\partial^2 V}{\partial Q_l \partial Q_m}\right)_0 \delta(\mathbf{k}_l + \mathbf{k}_m) \tag{5}$$

$$C_{lmn} = \left(\frac{\partial^3 V}{\partial Q_l \partial Q_m \partial Q_n}\right)_0 \delta(\mathbf{k}_l + \mathbf{k}_m + \mathbf{k}_n)$$

$$C_{lmnp} = \left(\frac{\partial^4 V}{\partial Q_l \partial Q_m \partial Q_n \partial Q_p}\right)_0 \delta(\mathbf{k}_l + \mathbf{k}_m + \mathbf{k}_n + \mathbf{k}_p)$$

where the δ functions assure the momentum conservation condition.

The crystal hamiltonian can be then written in the form

$$H = H_0 + H_1 + H_2 + \ldots$$

where

$$H_0 = T + V_2 \tag{6}$$
$$H_1 = V_3$$
$$H_2 = V_4$$
$$\ldots\ldots\ldots\ldots$$

For the perturbation approach the crystal normal coordinates are conveniently expressed in terms of phonon creation Q_l^+ and annihilation Q_l^- operators

$$Q_l = \left[\frac{\hbar}{2\omega_l}\right]^{\frac{1}{2}}(Q_l^- + Q_{-l}^+) = \left[\frac{\hbar}{2\omega_l}\right]^{\frac{1}{2}}A_l \tag{7}$$

where the index -l stands for -kj, i.e. for reversal of the wavevector direction. In terms of the sum operators $A_l = (Q_l^- + Q_{-l}^+)$ the various terms of the hamiltonian can be rewritten in the form

$$H_0 = \sum_l \hbar\omega_l^0(Q_l^+Q_l^- + \tfrac{1}{2}) = \sum_l \hbar\omega_l^0(n_l + \tfrac{1}{2}) \tag{8}$$

$$H_1 = \sum_{lmn} B_{lmn} A_l A_m A_n$$

$$H_2 = \sum_{lmnp} B_{lmnp} A_l A_m A_n A_p$$

In these expressions ω_l^0 is the harmonic frequency of the l-th phonon, n_l is the corresponding phonon occupation number

$$n_l = \left[\exp\left(\frac{\hbar\omega_l}{kT}\right) - 1 \right]^{-1} \tag{9}$$

and the B coefficients are defined as

$$B_{lmn} = \frac{1}{3!} \left[\frac{\hbar^3}{2^3 \omega_l \omega_m \omega_n} \right]^{\frac{1}{2}} C_{lmn} \tag{10}$$

$$B_{lmnp} = \frac{1}{4!} \left[\frac{\hbar^4}{2^4 \omega_l \omega_m \omega_n \omega_p} \right]^{\frac{1}{2}} C_{lmnp}$$

The calculation of phonon anharmonic shifts and bandwidths is normally made using the Green's function method. The Green's functions are appropriate generalization of time-dependent correlation functions, which are statistical averages of products of time-dependent operators in the Heisenberg representation. A one-phonon Green's function is called a "phonon propagator". The Green's function method is well-known in solid state physics and extensively discussed in many books and review articles [1,2,7]. Without presenting the method in detail we recall that it leads to the Dyson equation for the anharmonic phonon propagator

$$G_l(i\omega_m) = G_l^0(i\omega_m) + G_l^0(i\omega_m) \sum_{ll'}(i\omega_m) G_l(i\omega_m) \tag{11}$$

where $\sum_{ll'}(i\omega_m)$ is called the crystal self-energy, and $G_l(i\omega_m)$, $G_l^0(i\omega_m)$ are anharmonic and harmonic phonon propagators, respectively. The calculation of the self energy is a complex task and a full account of it is outside the scope of this lecture. We notice, however, that the self-energy is the sum of a series of elementary contributions which are conveniently expressed in terms of diagrams. In these diagrams phonon propagators are represented by arrows and phonon-phonon coupling terms by circles. The order of the coupling terms is given by the number of phonon arrows arriving at or leaving the circles. Arrows leaving a circle represent created phonons whereas arrows arriving at a circle represent annihilated phonons. The diagrams of interest for our problem are those which describe the relaxation of optical phonons which have wavevector k=0, i.e. $Q_l = Q_{j0}$. The lowest order diagrams, representing the lowest order processes, are shown below and involve two cubic coupling terms

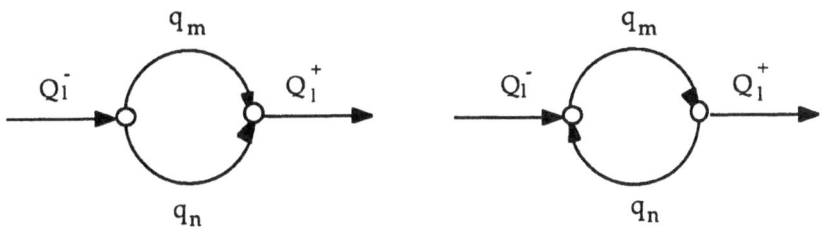

Diagram (a)	Diagram (b)
Three phonon down	**Three phonon up**
conversion process	**conversion process**
$\omega_l = \omega_m + \omega_n$	$\omega_l = \omega_m - \omega_n$

Diagram (a) represents an energy decay process in which the phonon ω_l is annihilated, giving rise to two lower energy phonons ω_m and ω_n, or is created by the fusion of two phonons of the thermal bath. For momentum conservation the ω_m phonon has wavevector $\mathbf{k}$ and the ω_n phonon has $-\mathbf{k}$ wavevector. The process is called **down** since the phonon energy is sent down in the phonon manifold. In the same way diagram (b) represents a process in which the ω_l phonon is fused with a thermal bath phonon ω_n to produce a higher energy phonon ω_m (**up** conversion) or is created from a decay process involving two bath phonons.

The next higher order diagrams involve either two quartic, or one quartic and two cubic, or four cubic coupling terms. Those with two quartic terms are [1,2]

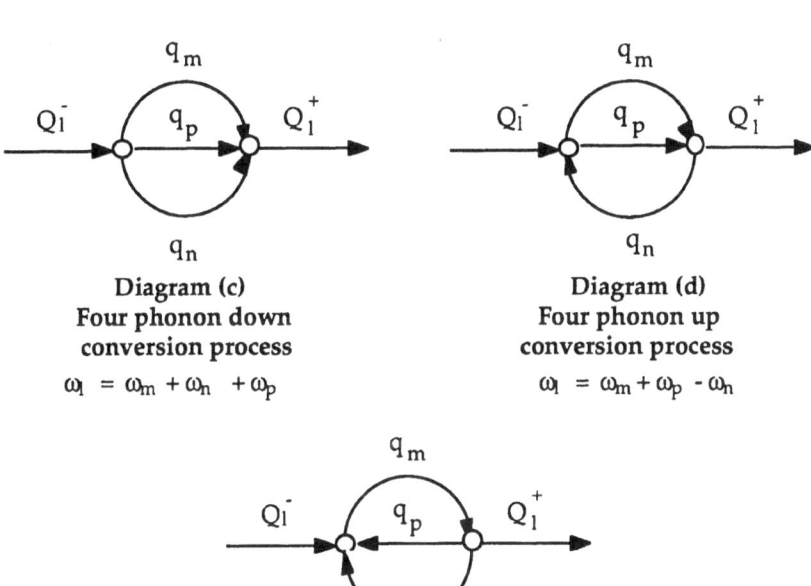

Diagram (c)	Diagram (d)
Four phonon down	**Four phonon up**
conversion process	**conversion process**
$\omega_l = \omega_m + \omega_n + \omega_p$	$\omega_l = \omega_m + \omega_p - \omega_n$

Diagram (e)
Four phonon up
conversion process
$$\omega_l = \omega_m - \omega_n - \omega_p$$

Diagrams of the same order in the perturbation expansion, representing processes with one quartic and two cubic or with four cubic terms [1,2], are shown below

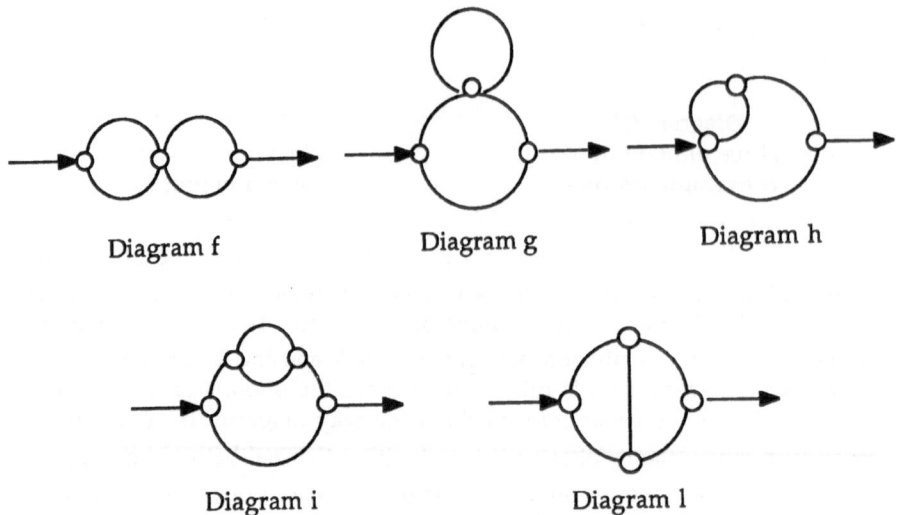

Diagram f   Diagram g   Diagram h

Diagram i   Diagram l

Each of the diagrams from f to l is representative of a whole series of diagrams with all possible combinations of creation and annihilation of phonons.

The calculation of the contribution of all these processes to the phonon bandwidth is extremely complex[1,2,8]. We shall limit ourselves, as an example, to the calculation of the contribution to the self-energy of the down- conversion diagram (a). From this we shall evaluate the contribution to the bandwidth and to the phonon anharmonic shift.

Using retarded Green's functions, we write the one-phonon Green's function for the ω_1 phonon (phonon propagator) [5]

$$G_1(i\,\omega_1) = \ll Q_1^-(t)\,;\,Q_1^+(0)\gg_a = \int_{-\infty}^{+\infty} dt\ e^{i\omega t}\{-i\theta(t)<Q_1^-(t)\,;\,Q_1^+(0)>_a\} \quad (12)$$

whose equation of motion is

$$\hbar\,\omega \ll Q_1^-\,;\,Q_1^+\gg_a = \hbar\,<[\,Q_1^-\,,\,Q_1^+\,]> + \ll[\,Q_1^-\,,H\,]\,;\,Q_1^+\gg_a \quad (13)$$

$$= \hbar\,<[\,Q_1^-\,,\,Q_1^+\,]> + \ll[\,Q_1^-\,,H_0\,]\,;\,Q_1^+\gg + \ll[\,Q_1^-\,,H_1\,]\,;\,Q_1^+\gg$$

where the label a means that it refers to diagram (a), [A,B] means the commutator of the operators in brackets and where the time dependence of the operators has been omitted. Since ω_1 is an optically active phonon, $l= j0$, i.e. the phonon belongs to the j branch with zero wavevector.

We notice that $\ll[\,Q_1^-\,,H\,]\,;\,Q_1^+\gg_a$ is a higher order Green's function, this being a general feature of the method, in the sense that the equation of motion of one Green's function always generates a function of higher order[7]. Continuing the process one obtains a chain of coupled equations with a hierarchy of Green's functions. In order to solve the system of coupled equations, one must uncouple the chain at a given order, introducing specific approximations in the higher order functions.

We consider first the commutator with the harmonic part of the hamiltonian, using normal rules for commutators of phonon operators

$$[Q_1^- , Q_m^-] = [Q_1^+, Q_m^+] = 0 \tag{14}$$

$$[Q_1^-, Q_m^+] = \delta_{lm}$$

According to eqs. (8) and (14), the only non-zero commutator is

$$[Q_1^-, H_0] = [Q_1^-, \hbar \omega_1 (Q_1^+ Q_1^-, + 1/2)] = \tag{15}$$

$$= \hbar \omega_1 \{Q_1^+[Q_1^-, Q_1^-] + [Q_1^-, Q_1^+]Q_1^-\} = \hbar \omega_1 Q_1^-$$

and therefore eq. (13) becomes

$$\hbar (\omega - \omega_1)\ll Q_1^- ; Q_1^+ \gg_a = \hbar + \ll [Q_1^-, H_1] ; Q_1^+ \gg_a \tag{16}$$

For the calculation of the commutator $[Q_1^-, H_1]$, we notice that in (8) there are three possibilities to couple Q_1^- with $(Q_1^- + Q_1^+)$ and that $B_{lmn} = B_{mln} = B_{mnl}$ by symmetry. We obtain then

$$[Q_1^-, H_1] = 3\Sigma_{mn} B_{lm-n}[Q_1^-, (Q_1^- + Q_1^+)(q_m^- + q_{-m}^+)(q_{-n}^- + q_n^+)] \tag{17}$$

$$[Q_1^-, H_1] = 3\Sigma_{mn} B_{lm-n}[Q_1^-, Q_1^+](q_m^- + q_{-m}^+)(q_{-n}^- + q_n^+)$$

$$[Q_1^-, H_1] = 3\Sigma_{mn} B_{lm-n}(q_m^- + q_{-m}^+)(q_{-n}^- + q_n^+)$$

$$[Q_1^-, H_1] = 3\Sigma_{mn} B_{lm-n}(q_m^- q_{-n}^- + q_{-m}^+ q_{-n}^- + q_m^- q_n^+ + q_{-m}^+ q_n^+)$$

The only operator of interest for diagram (a) is $q_{-m}^+ q_n^+$. Therefore

$$[Q_1^-, H_1]_a = 3\Sigma_{mn} B_{lm-n} q_{-m}^+ q_n^+ \tag{18}$$

and thus eq. (16) becomes

$$\hbar (\omega - \omega_1)\ll Q_1^- ; Q_1^+ \gg_a = \hbar + 3\Sigma_{mn}B_{lm-n}\ll q_{-m}^+ q_n^+;Q_1^+\gg_a \tag{19}$$

This equation involves the higher order Green's function $\ll q_{-m}^+ q_n^+;Q_1^+\gg_a$. Its equation of motion is

$$\hbar \omega \ll q_{-m}^+ q_n^+;Q_1^+\gg_a = \hbar <[q_{-m}^+ q_n^+;Q_1^+] > + \ll [q_{-m}^+ q_n^+,H];Q_1^+\gg_a \tag{20}$$

$$\hbar \omega \ll q_{-m}^+ q_n^+;Q_1^+\gg_a = \ll[q_{-m}^+ q_n^+,H]; Q_1^+\gg_a = \ll[q_{-m}^+ q_n^+,H_0 + H_1] ; Q_1^+\gg_a$$

since $[q_{-m}^+ q_n^+;Q_1^+] = 0$.

For the calculation of the commutator with H_0 we recall that the only non-zero terms are those involving the ω_m and ω_n phonons. We have then

$$[q_{-m}^+ q_n^+, H_0] = [q_{-m}^+ q_n^+, \hbar \omega_m(q_{-m}^+ q_{-m}^- + 1/2) + \hbar \omega_n(q_n^+ q_n^- + 1/2)] \tag{21}$$

279

$$= \hbar\,\omega_m[q\text{-}m^+,q\text{-}m^+q\text{-}m^-]\,q\;n^+ + \hbar\,\omega_n\,q\text{-}m^+[q\;n^+,qn^+qn^-]$$

$$= \hbar\,\omega_m q\text{-}m^+q\;n^+ + \hbar\,\omega_n\;q\text{-}m^+q\;n^+ = \hbar\,(\omega_m + \omega_n)q\text{-}m^+q\;n^+$$

By substitution of (21) in (20) we have

$$\hbar\,(\omega - \omega_m - \omega_n)\,\ll q\text{-}m^+q\;n^+; Q_1^+\gg_a = \ll [q\text{-}m^+q\;n^+, H_1]\,; Q_1^+\gg_a \tag{22}$$

For the calculation of the commutator in (22) we notice that in H_1 there are three equivalent choices for the ω_m and two for the ω_n phonon [7], thus giving a factor 6. Therefore

$$[q\text{-}m^+q\;n^+, H_1] = 6\sum_{mn} B_{l\text{-}mn}[\,q\text{-}m^+q\;n^+,(Q_1^- + Q_1^+)(q\text{-}m^- + q\;m^+)(q\;n^- + q\text{-}n^+)] \tag{23}$$

Using standard commutation properties we obtain

$$[q\text{-}m^+q\;n^+, H_1] = 6\sum_{mn} B_{l\text{-}mn}Q_1^-\{q\text{-}m^+q\text{-}m^-[q\;n^+,q\;n^-] + [q\text{-}m^+,q\text{-}m^-]\,q\;n^-q\;n^+\}$$

$$= -6\sum_{mn} B_{l\text{-}mn}\,Q_1^-\,(q\text{-}m^+q\text{-}m^- + q\;n^-q\;n^+) \tag{24}$$

If we substitute eq. (24) in (22), we obtain again a higher order Green's function whose equation of motion leads to a still higher function and so on. In order to break the chain of Green's functions, we use a decoupling approximation [7], using the well-known properties of phonon operators

$$
\begin{aligned}
q\text{-}m^+q\text{-}m^- &= n_m \\
q\;n^-q\;n^+ &= n_n + 1
\end{aligned}
\tag{25}
$$

Therefore

$$[q\text{-}m^+q\;n^+, H_1] = -6\sum_{mn} B_{l\text{-}mn}\,(n_m + n_n + 1)Q_1^- \tag{26}$$

By substitution of (26) in (22) we have

$$\hbar\,(\omega - \omega_m - \omega_n)\ll q\text{-}m^+q\;n^+;Q_1^+\gg_a = -6\sum_{mn} B_{l\text{-}mn}(n_m + n_n + 1)\ll Q_1^-\,; Q_1^+\gg_a \tag{27}$$

If we substitute the expression of $\ll q\text{-}m^+q\;n^+; Q_1^+\gg_a$ in eq. (19), we obtain

$$(\omega - \omega_1)\ll Q_1^-; Q_1^+\gg_a = 1 - 18\hbar^{-2}\sum_{lmn}|B_{lm\text{-}n}|^2\frac{(n_m+n_n+1)}{(\omega - \omega_m - \omega_n)}\ll Q_1^-; Q_1^+\gg_a \tag{28}$$

and by comparison with eq. (11) we obtain for the self-energy

$$\Sigma_1(i\omega_1) = -18\hbar^{-2}\sum_{lmn}|B_{lm\text{-}n}|^2\frac{(n_m+n_n+1)}{(\omega - \omega_m - \omega_n)} \tag{29}$$

This expression has a resonant denominator for $\omega = \omega_m + \omega_n$. We use therefore the standard procedure of replacing ω with $\omega + i\varepsilon$ and of using the relation [1]

$$\lim_{\varepsilon\to 0}\sum_\kappa \frac{f(\kappa)}{(A_\kappa + i\varepsilon)} = \sum_\kappa \frac{f(\kappa)}{(A_\kappa)_p} - i\pi\sum_\kappa f(\kappa)\delta(A_\kappa) = \Delta_\kappa - i\pi\Gamma_\kappa \tag{30}$$

where p means principal part of the sum. The real part of eq. (30) represents the anharmonic shift and the imaginary part the half band width.

We obtain in this way for the anharmonic shift

$$\Delta_{1(a)} = -18\hbar^{-2}\sum_{mn}|B_{lmn}|^2[\frac{(n_l+n_m+n_n)}{(\omega - \omega_m - \omega_n)_p}]$$ (31)

and for the full bandwidth

$$\gamma_{1(a)} = 32\pi\hbar^{-2}\sum_{mn}|B_{lmn}|^2(n_m+n_n+1)\delta(\omega - \omega_m - \omega_n)$$ (32)

where the label (a) means that it refers to diagram (a). The factor 32 originates from the fact that the full bandwidth is $\gamma = 2\Gamma$.

Using the same technique we can compute the anharmonic shift and bandwidth contribution of all diagrams shown above. The anharmonic shifts are, however, not spectroscopic observables and therefore we shall not consider them anymore.
The contribution of diagram (b) to the bandwidth is [1]

$$\gamma_{1(b)} = 72\pi\hbar^{-2}\sum_{mn}|B_{lmn}|^2(n_m - n_n)\delta(\omega + \omega_m - \omega_n)$$ (33)

The calculation of the bandwidth of fourth order diagrams is very complex[8,9]. We report here only those of diagrams (c), (d) and (e) which are[1,9]

$$\gamma_{1(c)} = 192\pi\hbar^{-2}\sum_{mn}|B_{lmnp}|^2[(n_m+1)(n_n+1)(n_p+1) - n_m n_n n_p]\delta(\omega - \omega_m - \omega_n - \omega_p)$$ (34)

$$\gamma_{1(d)} = 576\pi\hbar^{-2}\sum_{mn}|B_{lmnp}|^2[n_n(n_m+1)(n_p+1) - n_m(n_n+1)n_p]\delta(\omega - \omega_m + \omega_n - \omega_p)$$ (35)

$$\gamma_{1(e)} = 576\pi\hbar^{-2}\sum_{mn}|B_{lmnp}|^2[(n_m+1)n_n n_p - n_m(n_n+1)(n_p+1)]\delta(\omega - \omega_m + \omega_n + \omega_p)$$ (36)

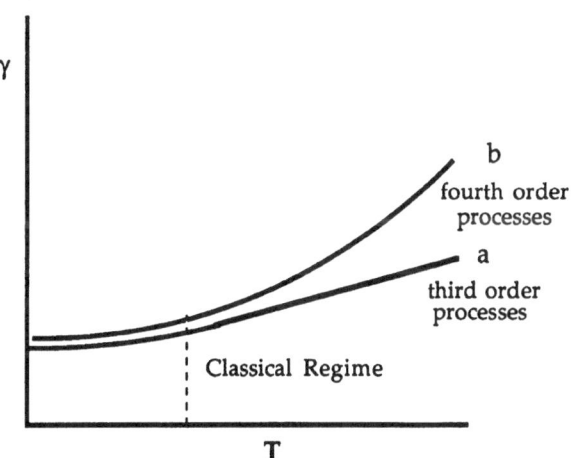

Figure 8. Schematic drawing of the variation of the bandwidth with T for third and fourth order processes.

There is an important difference between third- and fourth-order contributions to the bandwidth. Those of third-order (diagrams a and b) depend on single occupation numbers (see eqs. 32 and 33). In the classical regime, when $\hbar\omega < kT$, the occupation numbers are proportional to T, as can be seen from the expansion of eq. (9) as a function of T [1,2]

$$n = \left(\frac{k}{\hbar\omega}\right)T - \frac{1}{2} + \frac{1}{12}\left(\frac{\hbar\omega}{k}\right)\frac{1}{T} - \frac{1}{720}\left(\frac{\hbar\omega}{k}\right)^3\frac{1}{T^3} + \dots \qquad (37)$$

The fourth order contributions depend instead on products of two occupation numbers and thus are proportional to T^2 in the classical regime. Plots of the phonon bandwidth as a function of T, of the type of Fig. 8, allow us therefore to decide whether or not fourth-order contributions are important.

In addition to the energy decay processes discussed above there are pure dephasing processes which also contribute to the phonon bandwidth. These processes randomize the phase of the phonons, without changing their occupation number. The simplest of these processes[1] is represented by diagram (m) below

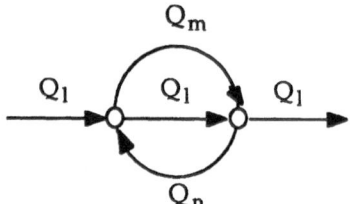

Diagram (m)
Energy exchange
dephasing process

which is a special case of diagram (d), with $Q_p = Q_l$. In this process two phonons Q_m and Q_n of the thermal bath exchange energy and this affects the width of the Q_l phonon. The contribution[1] to the bandwidth of diagram (m) is easily obtained from that of diagram (d) and is

$$\gamma_{l(m)} = 576\pi\hbar^{-2}\sum_{mn}|B_{lmnp}|^2[n_n(n_m+1)]\delta(\omega_n - \omega_m) \qquad (38)$$

4. INTERPRETATION OF THE EXPERIMENTAL RESULTS

In this section we present the interpretation of experimental lifetime measurements in terms of the theory discussed before. We shall consider two different problems: a) relaxation of internal phonons (vibrons) and b) relaxation of lattice phonons. The interpretation of relaxation processes for vibrons is simpler for two reasons. The first is that these modes occur normally at relatively high frequencies, above 400 cm^{-1}. Their occupation numbers are then vanishingly small in the temperature range from liquid He to room temperature and play therefore no role in all equations from (32) to (38). The second reason is that the number of decay channels is small for high frequency vibrations. We shall illustrate these ideas on the basis of few selected examples taken from our recent works.

4a. Benzene

Benzene crystallizes in the orthorombic system, space group D_{2h}^{15} with four molecules per unit cell on C_i sites. Each molecule has 30 internal modes, which classify in the symmetry species of the D_{6h} molecular group as 10 Raman active modes ($2a_g$, $1a_{2g}$, $2b_{2g}$, $2e_{1g}$, $3e_{2g}$) and 10 infrared active modes ($1a_{2u}$, $2b_{1u}$, $2b_{2u}$, $3e_{1u}$, $2e_{2u}$), all e modes being doubly degenerate[10]. Each non-degenerate Raman active internal mode splits in the crystal into four components and each degenerate mode into eight components according to the scheme

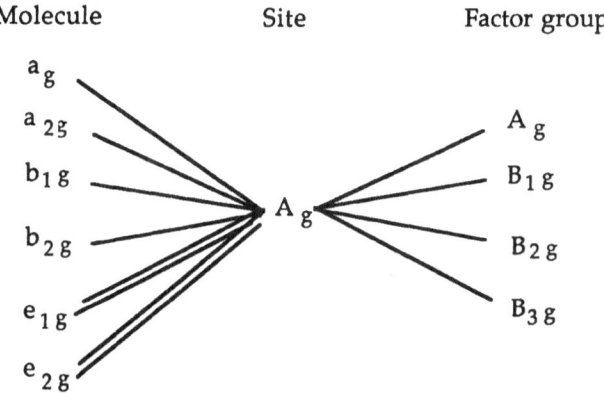

The same occurs for the infrared active modes and the corresponding correlation diagram can be obtained from the previous one changing the labels from g to u.

Here we consider the relaxation processes of the Ag crystal components of the four Raman-active vibrons v_1(a_g) at 991 cm^{-1}, v_6(e_{2g}) at 606 cm^{-1}, v_9(e_{2g}) at 1174 cm^{-1} and v_{10}(e_{1g}) at 854 cm^{-1}. The evolution of the inverse lifetime (bandwidth) with temperature in the range 10 - 180 K of these four vibrons is shown in Fig.9. The experimental data[11] are represented by circles. The curves in the Figures are calculated according to the discussion that follows.

Inspection of Fig. 9 shows that the temperature variation of the bandwidth of v_1, v_6 and v_{10} is linear in the classical regime, whereas that of v_9 is clearly quadratic. This means that for the first three vibrons only third order processes are important whereas for the v_9 vibron also fourth order processes must be considered.

In order to interpret the data, we consider the scheme of vibron levels of Fig.10. The figure is devided in four parts, one for each of the vibrons of interest. The vertical arrows in the figure have a lenght of 135 cm⁻¹, corresponding to the frequency range of the lattice phonons. Each internal level has a width corresponding to its dispersion in the crystal.

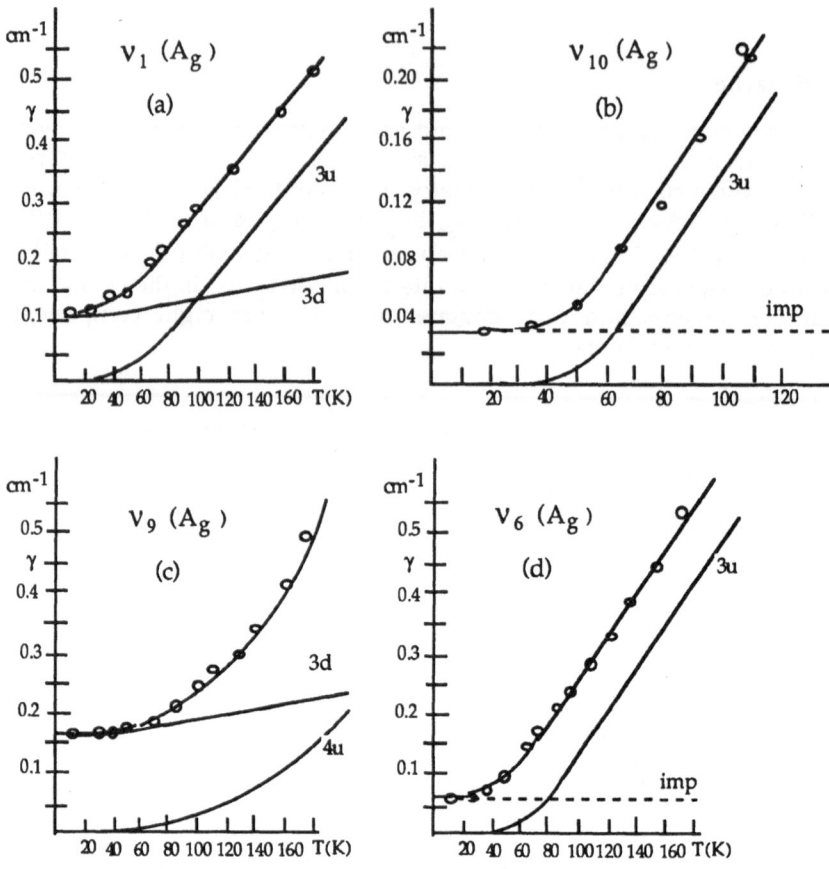

Figure 9. Experimental and calculated vibron bandwidths of benzene

We consider first the relaxation of V_1 which is linear with T in the classical regime. Fig. 10 shows that V_1 can decay, by creation of a lattice phonon, in the levels V_{10} and V_{17} or, by fusion with a lattice phonon, in the levels V_{18}, V_{12} and V_5. If we consider a single average process as representative of all possible processes taking place, we have for the two down-processes the equations

$$V_1[\,991\ cm^{-1}\,] = V_{10}\,(\,k)[\,860\ cm^{-1}\,] + \omega_{lattice}(\,-k)[\,130\ cm^{-1}\,] \qquad (39a)$$
$$V_1[\,991\ cm^{-1}\,] = V_{17}\,(\,k)[\,980\ cm^{-1}\,] + \omega_{lattice}(\,-k)[\,11\ cm^{-1}\,] \qquad (39b)$$

and for the three up-processes

$$V_1[\,991\ cm^{-1}\,] + \omega_{lattice}(\,-k)[\,18\ cm^{-1}\,] = V_{12}\,(\,k)[\,1009\ cm^{-1}\,] \qquad (40a)$$
$$V_1[\,991\ cm^{-1}\,] + \omega_{lattice}(\,-k)[\,21\ cm^{-1}\,] = V_5\,(\,k)[\,1012\ cm^{-1}\,] \qquad (40b)$$

$$\nu_1 [\ 991\ cm^{-1}\ ] + \omega_{lattice}(\ -k)[\ 45\ cm^{-1}\ ] = \nu_{18}(\ k)[\ 1036\ cm^{-1}\ ] \qquad (40c)$$

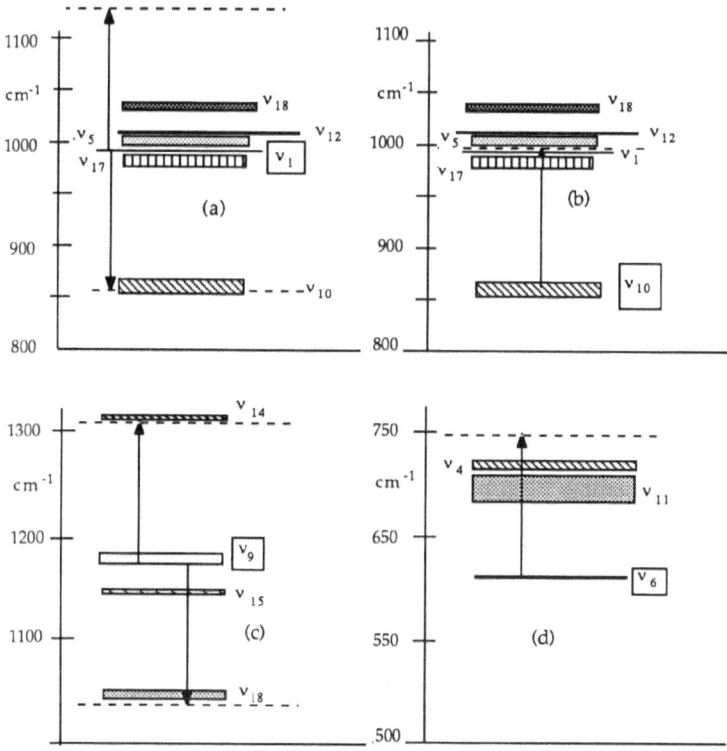

Figure 10. Schematic representation of the vibron levels of Benzene

Since the occupation numbers of the vibrons are vanishingly small, eqs. (32) and (33) can be rewritten in the simpler form

$$\gamma_{l(a)} = D_{l(a)} [\ n_{l(a)} + 1] \qquad (41)$$

$$\gamma_{l(b)} = M_{l(b)} [\ n_{l(b)}\ ] \qquad (42)$$

Using eqs. (41) and (42), the experimental data can be perfectly reproduced (full line curve in Fig. 10a) using the equations

$$\gamma_{down}(\nu_1) = D_{(130)} [\ n_{(130)} + 1] + D_{(11)} [\ n_{(11)} + 1] \qquad (43a)$$

$$\gamma_{up}(\nu_1) = U_{(18)}[\ n_{(18)}] + U_{(21)}[\ n_{(21)}] + U_{(4\,5)}[\ n_{(4\,5)}\ ] \qquad (43b)$$

where $D_{(abc)}$ represents the third order coupling coefficient for a down-process, $U_{(ab)}$ the third order coefficient for an up-process and $n_{(abc)}$ the occupation number of the involved lattice phonon. The curve 3d of Fig. 10a is calculated from eq. (43a) with $D_{(130)} = 0.125\ cm^{-1}$ and $D_{(11)} = 0$. The curve 3u is calculated from eq. (43b) with $U_{(4\,5)} = 0.132\ cm^{-1}$ and $U_{(21)} = U_{(4\,5)} = 0$. The full line curve of Fig. 10a is the sum of the two calculated curves[11].

285

The variation with temperature of the width of the v_{10} vibron is shown in Fig. 10b and is also linear in the classical regime. The v_{10} vibron cannot decay, through a three-phonon down-process, into lower vibron levels by creation of one lattice phonon, since the next vibron level occurs at 707 cm^{-1}, i.e. outside the range of 135 cm^{-1}, which is the upper limit of the lattice phonon frequencies. From Figure 10b it can be seen, however, that it can decay, through up processes, into v_1 and v_{17}, according to the equations

$$v_{10}[\ 854\ cm^{-1}\ ] + \omega_{lattice}(-k)[\ 121\ cm^{-1}\ ] = v_{17}(\ k)[\ 975\ cm^{-1}\ ] \tag{44a}$$

$$v_{10}[\ 854\ cm^{-1}\ ] + \omega_{lattice}(-k)[\ 135\ cm^{-1}\ ] = v_1(\ k)[\ 989\ cm^{-1}\ ] \tag{44b}$$

The temperature variation of γ for v_{10} can be then represented by

$$\gamma_{up}(v_{10}) = U_{(121)}[n_{(121)}] + U_{(135)}[n_{(135)}] \tag{45}$$

The curve 3u of Fig. 10b has been calculated [7] with $U_{(135)} = 0.45$ cm^{-1} and $U_{(121)} = 0.35$ cm^{-1}. As can be seen from eq. (42), up-processes are not active at $T = 0$. The v_{10} experimental data show, however, a width of 0.035 cm^{-1} at $T = 0$. Owing to the observed linear variation of γ with T, this small residual width cannot be due to fourth-order processes and is thus interpreted as due to scattering from impurities[11]. The full line curve of Fig. 10b is thus obtained by shifting the 3u curve by 0.035 cm^{-1}.

The variation with temperature of the width of the v_6 vibron is shown in Fig. 10d and is also linear with T. Again in this case down-processes cannot occur since the next vibron level at lower frequency (418 cm^{-1}) is not accessible through creation of a lattice phonon. As can be seen from Fig. 10d there are two possible up-processes

$$v_6[\ 606\ cm^{-1}\ ] + \omega_{lattice}(-k)[\ 90\ cm^{-1}\ ] = v_{11}(\ k)[\ 696\ cm^{-1}\ ] \tag{46a}$$

$$v_6[\ 606\ cm^{-1}\ ] + \omega_{lattice}(-k)[\ 105\ cm^{-1}\ ] = v_4(\ k)[\ 711\ cm^{-1}\ ] \tag{46b}$$

and the experimental data can be then reproduced by the equation

$$\gamma_{up}(v_6) = U_{(90)}[n_{(90)}] + U_{(105)}[n_{(105)}] \tag{47}$$

The curve 3u of Fig.10d corresponds to $U_{(90)} = 0.12$ cm^{-1} and $U_{(105)} = 0.16$ cm^{-1}. Also in this case there is a small residual width at $T = 0$ attributed to scattering from impurities.

The variation with temperature of the width of the v_9 vibron is shown in Fig. 10c and is not-linear with T. From Fig.9c are possible only the two down-processes

$$v_9[\ 1174\ cm^{-1}\ ] = v_{15}(\ k)[\ 1150\ cm^{-1}\ ] + \omega_{lattice}(-k)[\ 24\ cm^{-1}\ ] \tag{48a}$$

$$v_9[\ 1174\ cm^{-1}\ ] = v_{18}(\ k)[\ 1040\ cm^{-1}\ ] + \omega_{lattice}(-k)[\ 134\ cm^{-1}\ ] \tag{48b}$$

In addition to these we must consider also fourth-order processes to account for the non-linear T dependence of the bandwidth. The closest vibron level that can be reached by fusion of v_9 with two lattice phonons is v_{14}. The most probable process is thus

$$v_9[\ 1174\ \text{cm}^{-1}\ ] + 2\omega_{\text{lattice}}(\ -k)[\ 70\ \text{cm}^{-1}\ ]\ =\ v_{14}\ (\ k)[\ 1314\ \text{cm}^{-1}\ ] \qquad (49)$$

The temperature variation of γ for v_9 can be then represented by the equations

$$\begin{aligned}
\gamma_{\text{down}}(v_9) &= D_{(24)}[n_{(24)} + 1] + D_{(134)}[n_{(134)} + 1] + && (43b)\\
\gamma_{\text{up}}(v_9) &= N_{(70)}[n_{(70)}\ ]^2
\end{aligned}$$

where N is a fourth-order coupling coefficient. The curve 3d of Fig. 10c is calculated with $D_{(134)} = 0.15$ cm^{-1} and $D_{(24)} = 0$, and the curve 4u with $N_{(70)}$ = 0.165 cm^{-1}.

4b. KClO$_4$ and K$_2$SO$_4$

Both compounds crystallize in the orthorombic D_{2h}^{16} space group, with four formula units in the unit cell. Both the ClO$_4^-$ and the SO$_4^{--}$ ions have a tetrahedral structure (T$_d$ point group), and possess four internal modes v_1 (A$_1$), v_2 (E) , v_3 (F$_2$) and v_4 (F$_2$).

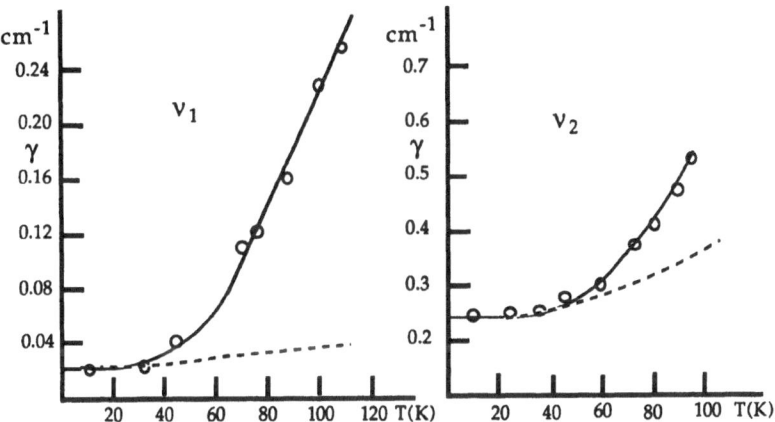

Figure 11. Variation with temperature of the bandwidth of the v_1 and v_2 vibrons of K$_2$SO$_4$

The lifetimes of the v_1 and v_2 vibrons have been determined in both cases as a function of temperature by picosecond spectroscopy[12,13]. The inverse lifetime (bandwidth) of these vibrons (A$_g$ component) is reported as a function of T in Fig. 11 for K$_2$SO$_4$ and in Fig. 12 for KClO$_4$.

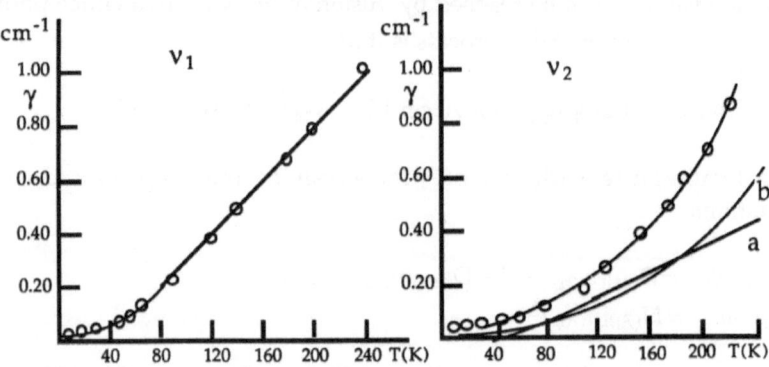

Figure 12. Variation with temperature of the bandwidth of
the V_1 and V_2 vibrons of $KClO_4$

From the pattern of crystal levels of K_2SO_4 shown in Fig. 13a it is clear
that only V_3 and V_4 can decay, through a third-order process involving a
high frequency lattice phonon, in the V_1 and in the V_2 levels, respectively,
according to

$$V_4[\ 617\ cm^{-1}\ ] = V_2(\ k)[\ 447\ cm^{-1}\ ] + \omega_{lattice}(\ -k)[\ 170\ cm^{-1}\ ] \qquad (44)$$
$$V_3[1145\ cm^{-1}] = V_1(\ k)[\ 983\ cm^{-1}\ ] + \omega_{lattice}(\ -k)[\ 162\ cm^{-1}\ ]$$

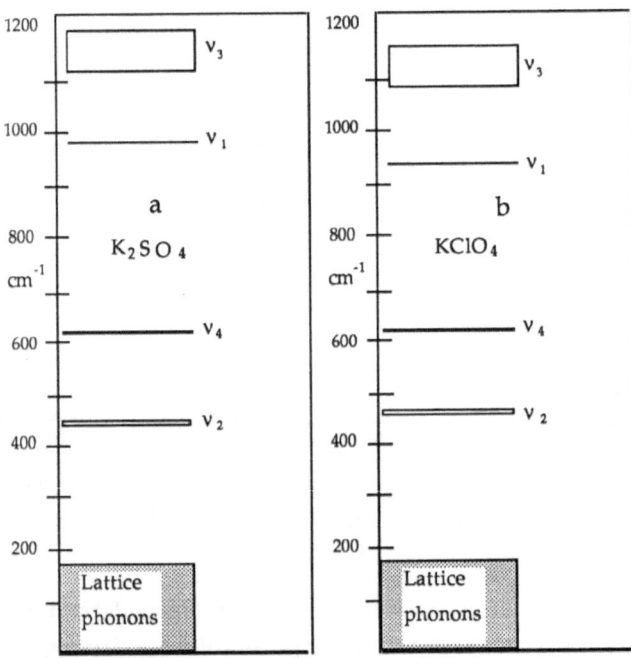

Figure 13. Scheme of the vibrational levels of K_2SO_4 and $KClO_4$ crystals.

For the v_1 and v_2 vibrons instead, only the three-phonon up-processes

$$v_1[\ 983\ cm^{-1}\ ] + \omega_{lattice}(\ -k)[\ 162\ cm^{-1}\ ] = v_3(\ k)[\ 1145\ cm^{-1}\ ] \qquad (45)$$
$$v_2[\ 447\ cm^{-1}\ ] + \omega_{lattice}(\ -k)[\ 170\ cm^{-1}\ ] = v_4(\ k)[\ 617\ cm^{-1}\ ]$$

are possible. As a consequence, the widths of v_3 and v_4 at T=0 are expected to be larger than that of v_1 and v_2 and this is perfectly verified by the experimental data shown below

vibron	freq. (cm^{-1})	$\gamma(T=0)$ (cm^{-1})
v_1	983	0.018
v_2	447	0.253
v_3	1145	> 0.6
v_4	617	> 0.6

Since up-processes are not effective at T=0, the residual width of v_1 and v_2, as well as the non-linear variation of $\gamma(T)$ for v_2, must be explained as due to quartic processes. The possible quartic processes are

$$v_1[\ 983\ cm^{-1}\ ] = v_4(\ k)[\ 619\ cm^{-1}\ ] + 2\omega_{lattice}(\ -k)[\ 182\ cm^{-1}\ ] \qquad (46)$$

and

$$v_2[\ 447\ cm^{-1}\ ] = 3\omega_{lattice}\ [\ 149\ cm^{-1}\ ]$$

The experimental data[12] of Fig. 11a are then reproduced by the equation

$$\gamma(T) = U_{(162)}[n_{(162)}\ ] + N_{(182)}[n_{(182)}\ + 1\ ]^2 \qquad (47)$$

and those of Fig. 11b by

$$\gamma(T) = U_{(170)}[n_{(170)}\ ] + 3N_{(149)}[n_{(149)}\ + 1/2\ ]^2 \qquad (48)$$

with $U_{(162)} = 1.06\ cm^{-1}$, $U_{(170)} = 1.25\ cm^{-1}$, $N_{(182)} = 0.018\ cm^{-1}$ and $N_{(149)} = 0.253\ cm^{-1}$.

Exactly the same situation occurs for the vibrons of $KClO_4$. The bandwidth of the A_g components of the v_1 and of the B_{2g} component of the v_2 vibron as a function of temperature are shown in Fig. 12. The corresponding three-phonon up processes are

$$v_1\ [\ 940\ cm^{-1}\ ] + \omega_{lattice}(\ -k)[\ 90\ cm^{-1}\ ] = v_3[\ 1030\ cm^{-1}\ ] \qquad (49a)$$
$$v_2\ [\ 461\ cm^{-1}\ ] + \omega_{lattice}(\ -k)[\ 160\ cm^{-1}\ ] = v_4[\ 621\ cm^{-1}\ ] \qquad (49b)$$

The non-linear variation of $\gamma(T)$ for the v_2 vibron is explained through the process

$$v_2\ [\ 461\ cm^{-1}\ ] + 2\omega_{lattice}(\ -k)[\ 80\ cm^{-1}\ ] = v_4[\ 621\ cm^{-1}\ ] \qquad (50)$$

The $\gamma(T)$ curve

$$\gamma_{up}(V_1) = U_{(90)}[n_{(90)}] \tag{51}$$

calculated from eq. 49a for the V_1 vibron is shown in Fig. 12a and fits perfectly the experimental data[13]. In the same way the processes (49b) and (50) lead to the equations

$$\gamma_{up}(V_2) = U_{(160)}[n_{(160)}] \tag{52}$$

$$\gamma_{up}'(V_2) = N_{(160)}[n_{(160)}]^2 \tag{53}$$

shown in Fig. 12b by the curves a and b, respectively.

4c. α- oxalic acid

α- oxalic acid crystallizes in the orthorombic system, space group P_{cab} (D_{2h}^{15}) with four molecules per unit cell on C_i sites. It possesses 9 internal modes. Each internal mode splits in the crystal into four components (A_g , B_{1g}, B_{2g}, B_{3g}). A full vibrational analysis of the vibron spectrum of α-oxalic acid is given in Ref. 14. Here we consider the temperature evolution of the bandwidth of only one vibron at 819 cm^{-1} assigned to the out of plane bending motion of the OH group {$\gamma(OH)$)}, which is the narrowest band in the spectrum. Fig. 14 shows the measured bandwidths[15] of this vibron as a function of temperature in the range 10 - 200 K. The variation is clearly non-linear with T and this is a proof that fourth-order processes are involved in the relaxation of this mode.

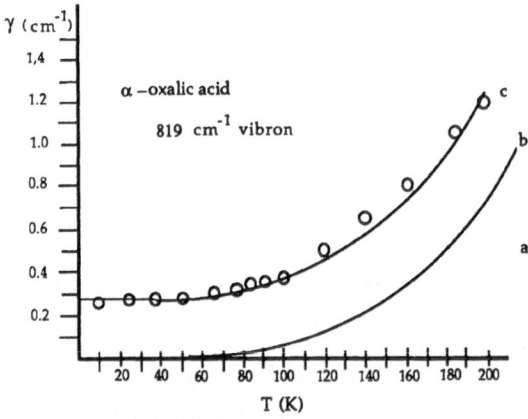

Figure 14. Temperature evolution of the bandwidth of the γ(OH) vibron of α- oxalic acid

From the scheme of vibron levels of α- oxalic acid shown in Fig. 15 it is easily seen that the only possible three-phonon down decay process can occur in the manifold of vibron states associated to the $\delta(O\text{-}C\text{=}O)$ infrared active mode extending from 660 to 690 cm^{-1}. according to the relation

$$\nu_{\gamma(OH)} \,[819cm^{-1}] = \nu_{\delta(O\text{-}C\text{=}O)}(k)[\,675\ cm^{-1}\,] + \omega_{lattice}(\,\text{-}k)[\,144\ cm^{-1}\,] \qquad (54)$$

which, from eq. 32, gives a contribution to the bandwidth of the type

$$\gamma_{down}(T) = D\,[n_{(675)} + n_{(144)} + 1] \qquad (55)$$

This equation is represented in Fig. 14 by curve a, with a coefficient D = 0.27 cm^{-1}.

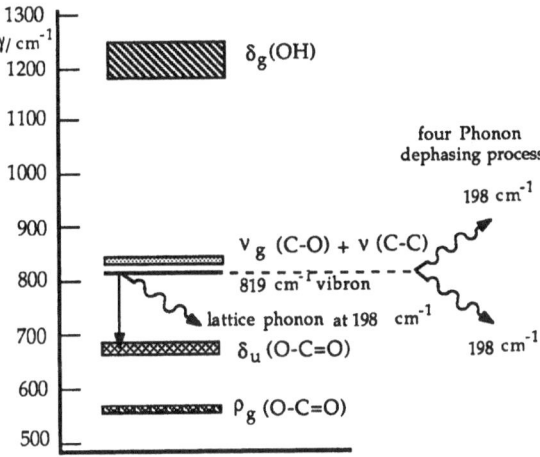

Figure 15. Scheme of the vibron levels of α- oxalic acid

In principle there is also possible a three-phonon up process of the type

$$\nu_{\gamma(OH)} \,[819cm^{-1}]\ + \omega_{lattice}(\,\text{-}k)[\,25\ cm^{-1}\,]\ = \nu_{g(C\text{=}O)}(k)[\,844\ cm^{-1}\,] \qquad (56)$$

Such a process, however, involves a very low-frequency phonon whose occupation number increases very fast with T, giving rise to a curve which is much steeper than the experimental one, even with a very small coupling coefficient. For this reason this kind of process has not been considered.

In order to account for the non-linear variation of the bandwidth with T, we consider now four-phonon processes. Several possibilities may occur:

i) four phonon decay in the 660-690 cm^{-1} manifold
ii) four phonon decay in the 550-570 cm^{-1} manifold
iii) four phonon pure dephasing processes.

The first two processes are represented by the equations

$$\nu_{\gamma(OH)}\ = \nu_{\delta(O\text{-}C\text{=}O)}(k_3)\ + \omega_{1(lattice)}(k_1) + \omega_{2(lattice)}((k_2) \qquad (57)$$

$$\nu_{\gamma(OH)} = \nu_{\rho(O-C=O)}(k_3) + \omega_{1(lattice)}(k_1) + \omega_{2(lattice)}((k_2) \qquad (58)$$

respectively, with $k_1 + k_2 + k_3 = 0$. Since both occupation numbers for the $\nu_{\delta(O-C=O)}$ and the $\nu_{\rho(O-C=O)}$ vibrons are practically zero, the contribution of these processes to the bandwidth is well represented by the equation

$$\gamma(T) = N[(n_1 + 1)(n_2 + 1)] \qquad (59)$$

which is obtained from eq. 34 by putting $n_3 = 0$. All attempts to fit the experimental data using this equation, with all possible choices of the two lattice phonons ω_1 and ω_2 were unsuccessful. Invariably the resulting curve has a slope different from that of the experimental data[15].

The third process is described by eq.38 which can be rewritten in the simpler form

$$\gamma(T) = N[n_1(n_2 + 1)] \qquad (60)$$

where n_1 and n_2 are the occupation number of the two bath phonons involved in the dephasing process (see diagram m). In contrast to the previous case, eq. 60 yields the correct slope for a good fit of the experimental data. Furthermore it involves two phonons at 198 cm^{-1}, i.e. in the region of the density of phonon states where this latter has its maximum. The corresponding curve, obtained with $\omega_1 = \omega_2 = 198$ cm^{-1} and $N = 1.80$ cm^{-1} is represented by curve b of Fig.14. The curve c which fits perfectly the experimental data is the sum of curves a and b[15].

REFERENCES

1. S. Califano and V. Schettino, Int. Rev. in Phys. Chem. **7**, 19, 1988
2. S. Califano, V. Schettino and N. Neto, *Lattice Dynamics of Molecular Crystals*, Lecture Notes in Chemistry, Vol. **26**, Springer (Berlin) 1981
3. S. Velko and R. M. Hochstrasser, J. Phys. Chem. **89**, 2240, 1985
   S. Velko and R. M. Hochstrasser, J. Chem. Phys. **82**, 2180, 1985
4. D. D. Dlott, Ann. Rev. Phys. Chem. **37**, 157, 1986
5. W. Demtröder, *Laser Spectroscopy*, Springer Series in Chemical Physics, Vol.5, Springer (Berlin) 1982
   J. R. Shen, *The Principles of Nonlinear Optics*, Wiley (New York) 1984
6. P. Ranson, R. Ouillon and S. Califano, Chem. Phys. **86**, 115,1984
7. R. F. and M. Balkansky, *Many-body Aspects of Solid State Spectroscopy*, North-Holland (Amsterdam), 1986
8. R. S. Tripathi and K. N. Patak, Nuovo Cimento, **21B**, 289, 1974
9. V. K. Jindal, R. Righini and S. Califano, Phys. Rev. B, **38**, 4259,1988
10. G. Taddei, H. Bonadeo, M. P. Marzocchi and S. Califano, J. Chem. Phys. **58**, 966, 1973
11. R. Torre, R. Righini, L. Angeloni and S. Califano, J. chem. Phys. 1990, in press
12. L. Angeloni, R. Righini, E. Castellucci, P. Foggi and S. Califano, J. Phys. Chem. **92**, 983, 1988
13. R. Righini, L. Angeloni,P. Foggi, E. Castellucci and S. Califano, Chem. Phys. **131**, 463, 1989
14. J. De Villepin, A. Novak and D. Bougeard, Chem. Phys. **73**, 291,1982
15. Nguyen Van Tien, E. Castellucci, M. Becucci, S. Califano and D. A. Dows, J. Mol. Structure 1990, in press.

NONLINEAR TIME RESOLVED SPECTROSCOPY IN CONDENSED MATTER

Christos Flytzanis

Laboratoire d'Optique Quantique du C.N.R.S.
Ecole Polytechnique
91128 Palaiseau cedex, France

I. INTRODUCTION

The study of relaxation processes in a material system that has been brought off equilibrium provides very crucial information about the way it interacts with its environment. The latter has at its disposal a very large, practically infinite, number of degrees of freedom and for most purposes can be visualized as an ideal bath that remains at a constant temperature for all relevant time scales and unaffected by the relaxation processes. These infinite many degrees of freedom introduce incessant perturbations, random in time and space, that act in the system throughout its evolution during the excitation process as well as afterward and insure its return to an equilibrium state after the external forces have been switched off. The return to equilibrium thus is not a single instantaneous act but occurs under time scales that depend on some specific features of the system and the fluctuating forces, the overall constraints and constants of motion as well as on the nature of the physical quantity that is being measured ; these span a most wide range of values and accordingly different techniques must be devised for their measurement.

For processes related to the relaxation of excitations in the molecular or atomic level the optical techniques that exploit the different features of the lasers, in particular their high selectivity and resolution, in frequency or time domain, have provided a unique wealth of information and constitute the most powerful tools that we have presently at hand to selectively investigate these processes. This information is essential for both fundamental and technological reasons since the relaxation processes constitute the central problem for understanding [1] the irreversible processes in general and also set in particular the ultimate limits for the implementation and exploitation of several optical processes in devices. Many of these techniques have reached a very high degree of technical sophistication and others are still under assessment and some will certainly be replaced by others and there is no point in enumerating and describing all of them here.

Below we will content ourselves to classify the most important ones that are currently in use and focus our attention on some of their most important aspects in particular those related to the invertigation of propagating excitations. Before doing this we pause to define and clarify certains elementary aspects of the relaxation processes in molecular or atomic excitations in condensed matter.

Applied Laser Spectroscopy, Edited by W. Demtröder and
M. Inguscio, Plenum Press, New York, 1990

II. ELEMENTARY RELAXATION PROCESSES

Our central problem is the relaxation processes and temporal evolution of an assembly of two level systems embedded in a condensed matrix. We formulate the problem and derive the equations for the standard case, the so called Bloch equations, [2,3] and proceed to cursively discuss the complications introduced by more realistic cases.

a. General case

We designate by a and b the low and high energy levels respectively and by $h\omega_0$ their energy spacing when the system is isolated ; but inside the matrix in principle $h\omega_o$ depends on the environment as well. These systems are uniformy distributed with number density N and are incessantly subject to random perturbations $V_r(t)$ through their coupling to the fluctuating degrees of freedom of the condensed environment. We will assume these perturbations to be stationnary markoffian processes [1,4,5,6] with

$$<V_r(t)> = 0 \tag{1}$$

$$<V_r(t)\ V_r(t')> = D^2\ e^{-|t-t'|/\tau_c} \tag{2}$$

where D is a measure of their strength and τ_c is a correlation time which for most purposes will be very short, much shorter than any relevant time of the problem so that the correlation function (2) practically can be replaced by a delta function of strength $\sigma = D^2\tau_c$, < > designates ensemble averages. The 2-level systems are brought off equilibrium by an external coherent perburbation $V_c(t)$ not correlated with $V_r(t)$; in addition these systems may interact mutually through V_t that provides excitation transfer from system to system. The perturbation $V_c(t)$ can be steady state, for instance sinusoidal, or pulsed and the relaxation processes accordingly can be studied in frequency or time domain respectively. In the latter case one has a direct measure of relaxation times while in the former one measures spectral widths [7] and line forms from which one can also extract relaxation times. The two approaches in principle are related through a Fourier transform ; in practice each one has its own advantages and range of usefulness. Here we limit ourselves in techniques related to real time domain which give direct information about the relaxation times.

The temporal evolution of the system is described by the density matrix operator equation

$$\frac{d\rho}{dt} = \frac{1}{ih}\left[h_o + V_r + V_t + V_c, \rho\right] \tag{3}$$

where h_o is the unperturbed hamiltonian which may depend on the environment. The relaxation processes are reflected in the temporal evolution of different physical quantities

$$A = Tr\rho\ A \tag{4}$$

where A is the quantum mechanical operator representing this quantity. The complexity of equation (3) can be greatly reduced [8,9] by appealing to the statistical features of the random potential V_r and its strenght relative to other perturbations.

b. Bloch equations

The standard case [2,3] corresponds to the systems being localized in identical environments and not mutually interacting : $V_t = 0$ and ω_0 identical for all systems (in space and time). In this case the infinite order matrix ρ is diagonalized in 2×2 identical matrices and equation (3) reduces to

$$\frac{d\rho}{dt} = \frac{1}{ih}\left[h_o + V_c,\rho\right] + \frac{d\rho}{dt}\bigg|_R \tag{5}$$

with

$$\frac{d\rho_{ab}}{dt}\bigg|_R = -\frac{\rho_{ab}}{T_2} \tag{6}$$

$$\frac{d\rho_{aa}}{dt}\bigg|_R = -\frac{\rho_{aa} - \rho_{aa}^{(o)}}{T_1} \tag{7}$$

with $\rho_{aa} + \rho_{bb} = \rho_{aa}^{(o)} + \rho_{bb}^{(o)} = 1$, $\rho_{ii}^{(o)}$ being the equilibrium values of the diagonal elements of ρ. For all purposes we assume that

$$V_c(t) = \mu F(t) \tag{8}$$

where μ is a material operator proportional to a "generalized coordinate" q of the system that couples to the externally applied "force" F(t) ; we also simplify by assuming that only the nondiagonal element of μ is nonvanishing.

In (6) and (7) T_2 and T_1 are the transverse and longitudinal relaxation times [3] respectively with $T_2 \ll T_1$. They are expressed [3,8,9] in terms of certain correlation functions and are related to two quite distinct processes, the coherence and energy relaxation processes respectively. The former, which is essentially classical, is the decay of the induced polarisation (or induced transition "dipole") $\mathscr{P} = \mu_{ab} (\rho_{ab} + \rho_{ba})$ and the latter which is essentially quantum mechanical corresponds to the decay of the population difference $\Delta\rho = \rho_{aa} - \rho_{bb}$ or equivalently the decay of the energy $\varepsilon = h\omega_0 \Delta\rho$ stored into the system through the excitation process. The inequality $T_2 \ll T_1$ implies that first the matrix ρ becomes diagonal in a time scale T_2 and then its diagonal elements revert to their equilibrium values in a time scale T_1. Intuitively one expects $T_2 < T_1$ because all random fluctuations can contribute to dephase the dipole but among them only those that can also accomodate the energy quantum $h\omega_0$, as required by energy conservation, are involved in the energy relaxation as well.

Equations (5) (6) and (7) can also be rewritten [10] in vector from in the space of 2x2 Pauli mattrices. One can also show that (5) (6) and (7) can be replaced by the following system of two coupled equations

$$\frac{d^2\mathscr{P}}{dt^2} + \frac{2}{T_2}\frac{d\mathscr{P}}{dt} + (\omega_0^2 + \frac{1}{T_2^2})\mathscr{P} = \mu_{ab}F\Delta\rho \tag{9}$$

$$\frac{d\Delta\rho}{dt} + \frac{\Delta\rho}{T_1} = \frac{1}{\hbar\omega_0} F \cdot \frac{d\mathcal{P}}{dt} \tag{10}$$

a damped harmonic oscillator and a diffusion equation respectively. Since $T_2 \ll T_1$ the damping of the harmonic oscillator is essentially determined by T_2 ; furthermore $T_2\omega_0 \gg 1$ and the T_2^{-2} shift in ω_0^2 can be neglected.

The validity conditions for equations (5) (6) and (7), or equivalently (9) and (10) are 3,8,9,11

* the bath is unaffected by the relaxation processes (infinite energy and phase sink reservoir)

* the correlation time τ_c is much shorter than any relevant time or

$$\Gamma\tau_c, \ \tau_c/T_i, \ \Delta\omega\tau_c, \ \Omega_R\tau_c/\hbar \ll 1$$

where Γ is the occurence rate of the individual random events ("collisions"), $\Delta\omega = \omega - \omega_0$ is the frequency mismatch ω being any characteristic frequency of the external force F and $\Omega_R(t) = \mu_{ab} F(t)/\hbar$ is a generalized Rabi frequency.

These conditions are essentially the same as those used in the "impact approximation" and in particular insure that the random perturbations are markoffian.

c. Extended cases

Despite the simplifying assumptions previously used to derive the Bloch equations in the standard case these seem to satisfactorily describe the relaxation in a very wide class of real quantum systems and over a wide spectral range : with slight modifications it can be extended to cover a far wider class of relaxation processes encountered in realistic complex systems. We shall attempt below to classify the most important cases.

- independent systems (homogeneous case) : this is essentially the case where the standard model strictly applies. We recall that the two-level systems are assumed localized in identical environments and do not mutually interact. By identical environment we understand that all systems are modified by the same amount irrespective of their position in the condensed matrix and furthermore $\tau_c D \ll 1$ (fast modulation [6]) ; this case is also termed homogeneous.

- independent systems (inhomogenous case) : if the two-level systems are not situated in identical environments the ω_0 will be shifted by amounts $\delta\omega_0$ that depend on their position inside the condensed matrix ; one has a distribution of h_0 in (5) or equivalently a distribution of ω_0 in (9) over a range $\Delta\omega^* = 2/T^*$ and this is the inhomogeneous case. A similar situation also occurs when $\tau_c D \gg 1$ (the slow modulation limit [6]). Equations (5) or (9) can still be used after one has performed an averaging process over the h_0 in (5), or the ω_0 in (9), and this essentially amounts in introducing an additional dephasing time T^* so that the total dephasing time is $1/T'_2 = 1/T_2 + 1/T^*$; if $T^* < T_2$ one has the inhomogenous case while for $T^* > T_2$ one recovers the homogenous one.

- interacting systems (random case) : if the two-level systems are mutually interacting V_t is nonvanishing and the infinite order matrix ρ in (3) cannot be diagonalized in 2 x 2 matrices each satisfying (5). The 2-level systems are coupled and in particular the excitation on one an be transfered to another along different pathways involving several intermediate real or virtual transfer steps. The problem can become tractable only under some very drastic simplifications by introducing a cut off on the extent of these pathways and suppressing any memory effects : for instance the excitation may not return in the initial point. One then obtains a population relaxation time T_3, in addition to T_1 and T_2, related to the so called cross relaxation [3,12] ; the apparent population relaxation time is $1/T'_1 = 1/T_1 + 1/T_3$

- interacting systems (periodic case) : if the mutually interacting two-level systems form a periodic array the situation changes drastically. With Floquet's theorem one shows that the eigenstates of such a periodic assembly are distributed [13] in bands or branches labelled by σ and within each such band or branch the states are labelled with a continuous index, the wave vector $\underline{k}$, that can take all values within the first Brillouin zone ; the latter has an extension of the order of $K = \pi/a$ where a is the smallest spatial period. One has propagating excitations with energies $h\omega_\sigma(k)$, which are analytic functions of $\underline{k}$ characterized by a density of states $J \approx \partial\omega/\partial\underline{k}$ and a group velocity

$$\underline{v}_g = \nabla_k \, \omega_\sigma \, (k)$$

which also measures the flatness of the dispersion relation, the relation between ω_σ and k. If the group velocity is very small over the whole of the B.Z., flat dispersion, one essentially recovers a localized state picture similar to the homogeneous case. If the group velocity is large, propagation effects may interfer with intrinsic relaxation processes and special techniques must be developped to desintagle them. This is in particular the case of the decay of the polariton [13], the polarization excitation mode close to a dipole allowed resonance of a phonon or exciton.

III. TIME RESOLVED SPECTROSCOPY

The central goal of time resolved spectroscopy is the identification of the different relaxation regimes and determination of the corresponding relaxation times.

a. Main scheme

Despite the apparent complexity and diversity of the techniques that are being used for directly studying the relaxation processes in real time domain there are some underlying principles and patterns common to all of them [14,17]. Indeed all these techniques employ the following scenario.

* excitation stage : the system is perturbed by an "instantaneous" external force $F_e(t)$ that resonantly couples to a "'coordinate" q of the system and drives it off equilibrium at time "t" = 0 ; as a consequence the coordinate acquires a phase and an amplitude.

* free precession : the system may be left to "precess" freely for a time interval t_d after the excitation stage till it settles in the relaxation regime we wish to study.

*interrogation stage : the system is probed at time t_d with an external weak force $F_p(t)$ that couples to a coordinate of the system q' that acquired phase and amplitude following the excitation but this coordinate may not necessarily be the same as the one directly driven off equilibrium in the excitation stage ; by measuring the signal as a function of the delay t_d one obtains the relaxation time appropriate to the chosen relaxation regime or channel.

The exciting and probing forces, $F_e(t)$ and $F_p(t)$ respectively, are powers of the electric field fixed by the degree of the nonlinear interaction involved in the excitation and probing stages respectively. Because the off equilibrium amplitude of the coordinate q' necessarily depends on the external force F_e applied at the excitation stage and in addition the signal also depends on F_p, time resolved spectroscopy is by essence nonlinear. Furthermore because the excitation and probing stages must be of short duration one must use short light pulse techniques. We anticipate that the time resolution is not fixed by the pulse duration but by the decay of the appropriate correlation [17] function of F_e and F_p ; this is actually determined by the decay time of F_e and rise time of F_p.

b. Principles

We wish to give a more quantitative [17] but still rough description of the different time resolved techniques. For this we remind that the optical properties of an assembly of identical microscopic systems (molecules) are described with the dielectric constant tensor

$$\varepsilon = 1 + 4\pi N \alpha \tag{11}$$

where α is the molecular polarizability tensor which can be expressed as a sum of harmonic oscillator contributions of type (9) if μ is the dipole operator, $\mu = ex$, and $F(t) = E(t)$ is the applied electric field ; we disregard local field corrections. As a general rule the polarizability α is an analytic function of certain coordinates q, for instance the vibrational or rotational coordinates, and of the electric field E present in the medium. If these acquire amplitude and phase through an external agency (excitation stage) we may formally set

$$\alpha(q,E) \equiv \alpha + \delta\alpha = \alpha_0 + \alpha_q^{(1)} q + \beta_E E + \frac{1}{2}\alpha_q^{(2)} q^2 + \ldots \tag{12}$$

α_0 is the electronic polarizability of the rigidly fixed molecule, $\alpha_q^{(1)} = \partial\alpha / \partial q$ and $\alpha_q^{(2)} = \partial^2 \alpha / \partial q^2$ are the Raman first and second order tensors, $\beta_E = \partial\alpha/\partial E$ is the second order electric polarizability.

The probing field $E_p(t)$, applied at time t_d after the excitation was switched off, sets up a polarisation

$$P = N\alpha_0 E_p + N\delta\alpha E_p = P_0 + \delta P \equiv P_0 + P_{NL} \tag{13}$$

which in addition to the linear term $P_0 = N\alpha E_p$ also contains oscillating terms in $P_{NL} = N\delta\alpha E_p \equiv \delta\varepsilon E_p$ which will radiate and generate fields E_s with different characteristics than E_p according to equation

$$\Delta E - \frac{1}{c^2}\frac{\partial^2}{\partial t^2}(\epsilon_0 E) = \frac{4\pi}{c^2}\frac{\partial^2 P_{NL}}{\partial t^2} \qquad (14)$$

One selects a particular component in P_{NL}

$$P_s(t) = \mathcal{P}_s(t) e^{ik_s r - i\omega_s t} \qquad (15)$$

and solves equation (14) as a function of t_d by setting

$$E_s = A_s(t) e^{ik_s r - i\omega_s t} \qquad (16)$$

The decay of the intensity $I_s \approx |E_s|^2$ as a function of t_d gives a direct measurement of the relevant relaxation times. The amplitude A_s is appreciable in the direction of phase matching

$$\underline{k}_e \approx \underline{k}_s \qquad (17)$$

and this is the essence of the coherent time resolved spectroscopy for the determination of T_2 or T^* ; note that the same information can be obtained by conventional high resolution spectroscopy in frequency domain. On the other hand the determination of T_1, the much longer energy relaxation time, is made by measuring the scattered A_s field at large angle with respect to (17), say 90° degrees and is the essence of the incoherent time resolved spectroscopy ; note that this information cannot be obtained by high resolution spectroscopy in frequency domain since there one measures the total width

$$\Delta\omega = \frac{1}{T_1} + \frac{2}{T_2} \approx \frac{2}{T_2} \qquad (18)$$

since $T_1 \gg T_2$.

For the coherent regime, which is essentially classical, unstead of using the quantum mechanical approach to derive equation (9) we may proceed along a classical approach by noticing that in the presence of an electric field one stores energy

$$W(Q) = -\frac{1}{2}\alpha(Q) E^2 \qquad (19)$$

per molecule which can be interpreted as a potential energy where Q is the classical quantity related to the coordinate q that affects the polarizability α ; its conjugated force being $F_Q = -\partial W/\partial Q$ the classical equation of motion of Q is

$$\ddot{Q} + \Gamma\dot{Q} + \Omega_0^2 Q = -\partial W/\partial Q \qquad (20)$$

where Ω_0 is the frequency for small amplitude motion of coordinate Q and Γ is its damping constant ; by proper identification of the different quantities and minor approximations equation (20) is identical to (9) and also gives a more intuitive description of the process. The above intuitive approach can actually be used for all coordinates Q that satisfy the criteria of the adiabatic approximation (Born-Oppenheimer approximation) and its generalizations.

c. Short light pulses

As we will see in the following section the nonlinear time resolved techniques have several advantages which are closely connected with the important improvements that recently occured in the art of producing short and very intense light pulses of any duration down to a few femtoseconds, possessing very clean and steep fronts, well defined forms and tunability over wide spectral ranges which can be additionally increased by nonlinear wave mixing techniques. These aspects are crucial in time resolved spectroscopy since, as we already stated, the time resolution is fixed by the appropriate correlation function of the exciting and probing pulses. There are several techniques [18] for producing short light pulses or shortening them and subsequently amplifying them and improving their form.

The breakthrough in the short light pulse techniques was made with the mode-locking that is achieved by inserting inside the laser cavity a nonlinear element that impresses on the cavity modes a fixed phase and amplitude relationship. There are two major mode-locking techniques, the active and the passive. The active one has been mainly used in Nd:YAG and Argon lasers ; the passive one has been used either in giant pulse lasers, like the Nd glass laser, or for continuous mode operation in dye lasers. These techniques produce very intense light pulses in the range of a few picoseconds ; one actually has a train of pulses out of which one selects and extracts and eventually amplifies a single pulse . The low rate of these intense few picosecond duration pulses is compensated by the very high intensity that these pulses carry.

Shorter pulses, down to the hundred femtosecond pulsewidth range were obtained in the colliding pulse mode locked dye lasers and their several improved versions. These pulses are weak and one exploits their high repetition rate to accumulate data ; one can also introduce an amplification stage which however considerably reduces the repetition rate. The tunability of these sources is very restricted.

In addition to these primary ultrashort pulse laser sources there are several others containing a pulse compression stage which allows to produce ultrashort light pulses down to few, less than ten, femtoseconds. In the compression stage one usually impresses a frequency sweep on the pulse which then is compressed by using a dispersive delay line ; to these sources we should also include the so-called soliton lasers.

Unfortunately not much progress has been achieved concerning the tunability range of these laser sources. For the interrogation stage one has used the different spectral components of white ultrashort pulses produced by focusing an intense ultrashort pulse in a liquid. In certain cases like in the impulsive stimulated Raman technique or the coherent infrared spectroscopy one can also exploit the spectral width $\Delta\omega \approx 2\pi/\tau_p$ related to the pulsewidth τ_p ; (for a pulse of say $\tau_p \approx 10$ fs $\Delta\omega_p \approx 300$ cm^{-1}).

IV. NONLINEAR TIME RESOLVED TECHNIQUES

Following the scenario outlined in the previous section by appropriate choice of the excitation and interrogation mechanisms a multitude [17] of time resolved nonlinear techniques can be devised that allow to address a most wide range of relaxation processes. Some of these techniques are more commonly used than others and have reached a high level of sophistification and flexibility and below we shall give a succint discussion of the latter. The time-resolved nonlinear optical techniques have several advantage which are continuously improved with the new ultrashort light pulse technology [18] ; some of them are

* extremely high selectivity achieved by exploiting appropriate nonlinear mechanisms, resonances and other means,

* high signal to noise discrimination,

* very clear separation of coherent and incoherent regimes by exploiting the phase matching conditions,

* high selectivity in time and space that in particular allows to study relaxation processes in any chosen space point inside the bulk or on the surface of the medium.

We classify the time resolved nonlinear optical techniques into local and non local.

a) Local time resolved techniques

These concern the study of relaxation processes in localized excitations.

The most developped [17] time resolved nonlinear technique for the study of vibrational relaxation proceses is the time resolved Coherent Antistokes Raman Scattering also designated CARS. Here two short intense light pulses of frequencies ω_1 and ω_2 interact simultaneously inside the medium and coherently drive a Raman active transition of frequency ω_R in the region of their spatial overlap ; the frequencies ω_1 and ω_2 are such that $\omega_1 - \omega_2 \approx \omega_R$. The evolution of the coherence and population difference between the two states involved in the transition is probed with a delayed weak short pulse of frequency ω_3 by measuring the antistokes intensity at frequency $\omega_p = \omega_3 + \omega_R$. In regard to the scheme outlined in section III and the notation used there we have

$$V_e \approx -\frac{1}{2} \alpha_q^{(1)} q \, E_1(t) \, E_2(t) \tag{21}$$

at the excitation stage or with the notations of section

$$III \, \mu = -\frac{1}{2} \alpha_q^{(1)} q \text{ and } F(t) = E_1(t) \, E_2(t) \text{ and}$$

$$V_p = -\frac{1}{2} \alpha_q^{(1)} q \, E_3(t) \, E_p(t) \tag{22}$$

at the interrogation stage. In the coherent regime the detection is made in the phase matching condition (17) and one measures T_2 while in the incoherent regime the detection is made at right angle to the phase matching condition and one measures T_1. At presently this constitutes the most powerful technique for measuring T_1. There are several parameters that can be varied to improve the selectivity and other performances of this technique. Note in particular that the excitation can be made at any point inside the bulk of the medium if the latter is transparent to the frequencies involved.

In addition, in molecules with several coupled modes, one may wish to study [17] intermode energy and coherence transfer. Such information can be obtained by detecting the antiStokes from another mode than the one that has been coherently driven in the excitation stage.

There are other extensions of the CARS technique. One of them is the time resolved Coherent AntiStokes Higher Order Raman Scattering [20,21], also designated by CAHORS, to study vibrational overtones in liquids and bound two-phonon (biphonon) states in molecular crystals or more complex vibrational states like the Fermi resonances. All these compound states are manifestations of strong anharmonic interactions and the study of the loss of their coherence and desintegration channels allow a very selective study of the different anharmonic terms of the lattice potential. Using the CAHORS techique it was established that there are two major paths for loss of coherence of the compound states in molecular crystals, the intrinsic [20] (temperature independent) and the extrinsic [21] (temperature dependent). Furthermore since these compound phonon states predominantly desintegrate into large wave vector phonons, one has for the first time the possibility to optically study the dynamics of the latter.

Clearly the CARS technique can only be used for Raman active modes. For infrared active (dipole allowed) modes one may replace [21] the coherent excitation stage by one photon (infrared) excitation resonant with the material transition followed by a delayed excitation to a fluorescent state. The measure of the fluroescence as a function of the delay between the two excitations gives information about the relaxation of the infrared active mode. Because of the lack of tunable infrared sources with short pulses this technique or similar ones [22] has not been used to the same extent as the CARS. A limitation of this technique is that the excitation and probe states are quite often restricted into the optical penetration depth of the sample near the surface which can be different from the bulk. Here we also mention the time resolved luminescence technique which has been successfully [23] used to study fast photocarrier relaxation processes in semiconductors but can also be used elsewhere. Here the fast luminescence signal subsequent to a first ultrashort pulse is sampled at a given frequency by up converting it with a second delayed ultrashort pulse inside a crystal without inversion center ; the intensity of the up converted signal as a function of the delay between the two pulses gives important information about the decay.

The CARS technique can also be used [17] for the study of relaxation proccesses in inhomogeneous broadened transitions. Here, however, techniques like photon echos or time resolved hole burning provide a more direct insight into this question. The photon echo technique [24] has been extensively developped to study coherent relaxation processes in the visible involving electronic transitions ; in the infrared where vibrational transitions occur the results have been scarce. There are several variants of the original photon echo technique like the stimulated echo [25] and accumulated echo [26] techniques. The time resolved hole burning technique is straight forward. One measures the transmission characteristics of a weak short pulse tuned in the spectral range of a inhomogenous broadened transition which has been previously excited by a very intense short pulse. The decay of the hole burned and its width give informations about population and coherence relaxation times.

A very powerful time resolved technique to study relaxation processes is the transient optical gratings technique [27-30] which can also be designated as real time holography. In its simplest configuration two pulsed beams of same frequency ω close to a transition of the medium interact in the medium via a refractive index or photoinduced absorption change ; a third pulsed beam at the same or a different frequency ω' delayed with respect to the first two diffracts off this pattern ; and the evolution of the diffracted intensity as a function of the time gives information about the amplitude and phase relaxation of the grating and by the same token about energy migration and loss of coherence. There are several variants and extensions of the original transient grating technique described above. One such a variant is the transient degenerate four wave mixing [31,32] or the transient optical phase conjugation. Two intense counterpropagating pulsed beams E_1 and E_2 of same frequency ω interact inside a medium with a third pulsed beam E_3 of same frequency ω and set up a third order poalrization $\underline{P}_c^{(3)}$ which generates a fourth beam of frequency ω counterpropagating and phase conjugated to E_3. One can show that this phase conjugated beam results from diffraction off two spatial transient gratings and a temporal grating. Indeed one can show [31] that

$$\underline{P}_c^{(3)} = a(\underline{E}_1 \cdot \underline{E}_3^*) \underline{E}_2 + b(\underline{E}_2 \cdot \underline{E}_3^*)\underline{E}_1 + c(\underline{E}_1 \cdot \underline{E}_2) \underline{E}_3^* \tag{23}$$

and the three terms in the right hand correspond to the three gratings : by appropriate delays between the pulsed beams and judicious choice of beam polarizations one can study different relaxation and diffusion processes of the material excitation at frequency ω. One may also use the so-called polychromatic optical phase conjugation [33], where E_3 has a freqency $\omega' \neq \omega$ in which case E_c has frequency $\omega_c = 2\omega - \omega'$; furthermore E_1 and E_2 are not counterpropagating but their propagation directions make a angle θ that is bisected by that of E_3 and E_c which are counterpropagating.

Another variant of transient optical grating technique is the impulsive stimulated Raman technique [34-37] with ultrashort pulses of duration τ_p in the few femtosecond range. Such pulses have a frequency width $\Delta\omega_p \approx 1/\tau_p$ which can extend up to few hundreds cm^{-1} and under certain conditions one can, with a single ultrashort pulse, coherently drive vibrational or librational or other low lying states of frequencies that fall within $\Delta\omega_p$; diffraction of a single ultrashort pulse off the interference pattern set by the first pulse gives information about the relaxation processes of these low lying states. There have been several investigations [34-39] with this technique or similar ones. One has observed in particular Raman quantum beats and studied relaxation processes of low lying vibrational modes.

All the previous techniques employ coherent pulses. There has been recently proposed and demonstrated [40] the use of incoherent light to study relaxation processes and in particular extract information concerning the relaxation time T_1 and T_2. Here one exploits the fact that a temporally incoherent light of wide spectral width possesses a vary short correlation time τ_c much smaller than the light pulse duration τ_p. In an autocorrelation measurement such a light appears like a single pulse of duration τ_c. Accordingly this technique consists in spatially splitting a pulsed light beam of frequency ω into two beams and temporally delaying the one relative to the other by τ before mixing them in a resonant medium. The energy of the output beams in the phase matching direction is measured as a function of τ to obtain a kind of correlation profile associated with both the incident light and the resonant material response. The time resolution of this technique is set up by the light correlation time τ_c and not by the duration of the light pulse itself. This technique has been applied in some simple cases but its exploitation in a larger scale and in more complicated cases is not that straightforward.

b) <u>Nonlocal time resolved techniques</u>

The previous techniques are essentially designed to study relaxation processes of localized excitations in a molecular entity embedded in a condensed matrix. Here the energy and coherence relaxation occur within the immediate environment of the molecule. For delocalized excitations in more or less periodic molecular systems the problem can be fondamentally different and also becomes more complex because the relaxation process within the immediate molecular environment can take place concurrently with the escape of the excitation from the excited molecule and its resonant transfer to other identical molecules. At presently there are two techniques that allow to address this problem and desintangle the relaxation from the propagation : the electro-optic Cerenkov effect [41-43] and the nonlocal (space resolved) time resolved CARS [41,45]. In addition some of the previous techniques, for instance the transient optical grating technique, can be used to study relaxation and diffusion processes of delocalized excitations.

The electro-optic Cerenkov technique with femtosecond pulses in the visible exploits [41] the optical rectification or inverse electro-optic effect in a non centrosymmetric crystal to generate an extremely fast far infrared electromagnetic transient, few femtoseconds in duration, and excite propagating infrared active excitations in the same crystal ; these are, for instance, the polaritons. This produces a Cerenkov cone of pulsed radiation since the relevant far infrared polarisation is created with a group velocity v'_g appropriate to the visible spectrum (dielectric constant ε_∞) while its far infrared radiation propagates with a group velocity v''_g approriate to the far infared (dielectric constant $\varepsilon_o >$ ε_∞) and one in general has $v''_g < v'_g$. The existence of this Cerenkov cone is subsequently exploited to study the propagation of the polariton in real space. This is achieved with a second femtosecond probe pulse parallel to the previous which measures the small birefringence that is due to the electrooptic effect induced by the electric field of the far infrared transient as it moves in synchronism with the Cerenkov wave front. This technique has been already demonstrated [43] for polaritons in ferroelectric crystals (Li Ta O_3) ; its selectivity is, however, limited and the separation of temporal and spatial features is not straight forward. Furthermore this method is restricted to low-frequency polaritons, with an upper frequency limit given by the infrared femtosecond pulse bandwidth.

The non local [44,45] time resolved CARS technique does not suffer from these limitations. This technique has been recently proposed [44] and demonstrated for the study of polariton pulses in crystal without inversion center. In this technique a picosecond-duration polariton wave packet of frequency ω_π is created at an initial instant and space point in the crystal by coherent Raman scattering using time-coincident picosecond pulses of frequencies ω_L and ω_s, such that $\omega_L - \omega_s = \omega_\pi$. This polariton wave packet subsequently propagates freely inside the crystal in a direction fixed by overall wave-vector conservation, and its temporal and spatial evolution is followed by coherent anti-Stokes Raman scattering of a picosecond probe pulse (ω_p) displaced in time (by t_d) and space (by X_d) with respect to the excitation. The measured dependence of the spatial displacement X_d, where the signal is maximum, on time delay t_d is directly related to the energy propagation characteristics of the polariton wave packet. This technique has been to used [44] to study dephasing of polariton packets both in perfect and partially disordered crystals, dressed [45] polaritons (Fermi resonance) and can also be extended to study surface polaritons.

Nonlocal time resolved techniques have only recently been demonstrated but it is expected to undergo rapid developments in the near future.

V. GENERAL REMARKS AND CONCLUSION

In this summary of the time resolved spectroscopic techniques we restricted ourselves to a very elementary presentation of their principles and outlined some of their features to study the dynamics of molecular excitations in real time in the picosecond and femtosecond timescale. The main effort was directed towards a simple unified presentation of these techniques. This however should not conceal either the intrinsic difficulties encountered in the quantitative description and application of these techniques or the numerous differences among themselves and differing scopes. Their relevance too for the study of irreversible processes at the molecular level should not be minimized. These are the most powerful techniques to study the evolution of molecular excitations in real time and selectively follow the relaxation processes. The real advantages of the spectroscopic techniques in time domain over the ones in frequency domain are now well appreciated although in principle the two approaches should be equivalent since they are related through Fourier transform operation. Furthermore the technology of ultrashort pulsed laser sources has advanced to an extent that makes these sources more accessible, reliable and flexible than the highly stable and tunable CW laser sources needed in high resolution spectroscopy in frequency domain for the study of line profiles. Consequently we expect important developments in the time resolved nonlinear techniques, both local and nonlocal, and over a wider spectral regions than presently, from the far infrared up to the X-ray region. In the later case they will allow us to follow the global intramolecular motion and not only the one according a single vibrational mode as the previous techniques that use visible sources do.

Acknowledgements

The author acknowledges many fruitful discussions with Geoffrey Gale and Fabrice Vallée on different aspects of the present material.

References

1. See for instance R. Kubo, M. Toda and N. Hashitsume, Statistical Physics II, Nonequilibrium Statistical Mechanics, Springer Verlag, Berlin 1978
2. F. Bloch, Phys.Rev. 70, 460 (1946)
3. See for instance, A. Abragam , Principles of Nuclear Magnetism, Oxford Un.Press London, 1961, or C.P. Slichter, Principles of Magnetic Resonance, Springer Verlag Berlin, 1980.
4. N. Bloembergen, E.M. Purcell and R.V. Pound, Phys.Rev. 73, 679 (1948)
5. P.W. Andersson and P.R. Weiss, Rev.Mod.Phys. 25, 269 (1953)
6. R. Kubo in Fluctuations, Relaxation and Resonance in Magnetic Systems, D. Ter Haar, ed. (Plenum, N.Y. 1962) p.23
7. See for instance. M.D. Levenson, Introduction to Nonlinear Laser Spectroscopy, Academic Press, New-York, 1982.
8. R.V. Wagness and F. Bloch, Phys.Rev. 89, 728 (1953)
9. A.G. Redfield, Phys.Rev. 98, 1787 (1955)
10. R.P. Feynman, F.C. Vernon and R.W. Hellwarth, J.Appl.Phys. 28, 49 (1957)
11. See for instance P.R. Berman, J.Opt.Soc. B3, 564 and 572 (1986)
12. G. Mourou, I.E.E.E. J.Quant.Electr. 11, 1 (1975)
13. See for instance C. Kittel, Introduction to Solid State Physics John Wiley, New-York 1966 or W. Ashcroft and N.D. Mermin, Solid State Physics, Holt Saunders, Tokyo, 1961
14. F. de Martini and J. Ducuing, Phys.Rev.Lett. 17, 117 (1966)
15. R. R. Alfano and S.L. Shapiro, Phys.Rev.Lett. 26, 1247 ; ibid 29, 1655 (1972)
16. A. Laubereau, D. van der Linde and W. Kaiser, Phys.Rev.Lett. 27, 802 (1971); ibid 28, 1162 (1972)
17. A. Laubereau and W. Kaiser, Rev.Mod.Phys. 50, 607 (1978). This paper contains very thorough discussion of several aspects and applications of nonlinear time resolved techniques for the study of vibrational relaxation in liquids and crystals.

18. For a very accessible and up to date review of optical pulse techniques see C.V. Shank, in Ultrashort Light Pulses and Applications,W. Kaiser Ed.,Springer Verlag, Berlin 1988

19. M.L. Geirnaert, G.M. Gale and C. Flytzanis, Phys.Rev.Lett. 52, 815 (1984)

20. G.M. Gale, P. Guyot-Sionnest, W.Q. Zheng and C. Flytzanis, Phys.Rev.Lett. 54, 823 (1985)

21. A. Laubereau, A. Seilmeier and W. Kaiser, Chem.Phys.Lett. 36, 232 (1975)

22. D. Ricard and J. Ducuing, J.Chem.Phys. 62, 3616 (1975)

23. See for instance J. Shah, T.C. Damen and B. Deveaud, Appl.Phys.Lett. 50, 1307 (1987)

24. A. Kurnit, I.D. Abella and S.R. Hartmann, Phys.Rev.Lett. 13, 567 (1964)

25. W. Mossberg, A. Flusberg, R. Kachru and S.R. Hartmann, Phys.Rev.Lett. 42, 1665 (1979)

26. H. Hesselink and D.A. Wiersma, Phys.Rev.Lett. 43, 91 (1979)

27. D.W. Phillion, D.J. Kuizenga and A.E. Siegman, Appl.Phys.Lett. 27, 85 (1975)

28. J.R. Salcedo, A.E. Siegman, D.D. Dlott and M.D. Fayer, Phys.Rev.Lett. 41, 131 (1978)

29. M.D. Fayer, Ann.Rev.Phys.Chem. 33, 63 (1982)

30. H.J. Eichler, Opt.Acta. 24, 631 (1977)

31. See for instance Optical Phase Conjugation, R. Fisher Ed. Acad.Press, New-York, 1985.

32. T. Yajima and Y. Taira, J.Phys.Soc.Japan, 47, 1620 (1979)

33. G. Mannenberg, J.Opt.Soc.Am. B3, 853 (1986)

34. See for instance S. Ruhman, A.G. Joly, B. Kohler, L.R. Williams and K.A. Nelson, Rev.Phys.Appl.(Paris) 22, 1717 (1987)

35. Yan Y.X. Gamble, E.B. and K.A. Nelson, J. Chem.Phys. 83, 5391 (1989)

36. J. Chesnoy and A. Mokhtari, Phys.Rev. A38, 3566 (1988)

37. A. Mokhtari and J. Chesnoy, Europh. Letters. 5, 523 (1988)

38. M.J. Rosker, F.W. Wise and C.L. Tang, Phys.Rev.Lett. 57, 321 (1986) and J.Chem.Phys. 86, 2827 (1987)

39. M. Mitsunaga and C.L. Tang, Phys.Rev. A35, 1720 (1987)

40. N. Morita and T. Yajima, Phys.Rev. A30, 2525 (1984)

41. D.H. Auston, Appl.Phys.Lett. 43, 713 (1983)

42. D.H. Auston, K.P. Cheung, J.A. Valdmanis and D.A. Kleinman, Phys.Rev.Lett. 53, 1555 (1984)

43. D. Auston and M.C. Nuss, IEEE, J.Quant.Electr. 24, 184 (1988)

44. F. Vallée, G. Gale and C. Flytzanis, Phys.Rev.Lett. 61, 2102 (1988)

45. G. Gale, F. Vallée, C. Flytzanis, Phys.Rev.Lett. 57, 1867 (1986)

BRILLOUIN GAIN SPECTROSCOPY IN GLASSES AND CRYSTALS

Gregory W. Faris

SRI International, Molecular Physics Department

Menlo Park, California 94025   USA

INTRODUCTION

Gain spectroscopy is a well established technique for the study of nonlinear scattering phenomena. Several capabilities of gain spectroscopy make the technique very powerful for frequency domain measurements of scattering processes. The use of gain spectroscopy is illustrated by the application of the technique to Brillouin measurements in solids.

GAIN SPECTROSCOPY

In the presence of large optical intensities such as those possible with focused lasers, significant stimulated gains can be realized for scattering processes including Raman, Brillouin, and Rayleigh scattering.[1,2] By overlapping a strong 'pump' laser beam and a weaker 'probe' laser beam in a sample, gain or loss may be induced on the probe beam. By scanning the frequency of either the pump or probe laser, spectroscopy can be performed on the resonant scattering modes of a sample. This technique, referred to here as gain spectroscopy, has been used most widely for studying Raman scattering, where it has also been referred to as stimulated Raman spectroscopy, stimulated Raman gain spectroscopy, and inverse Raman spectroscopy (for the case of loss on the probe beam).

The classical method for studying light scattering phenomena is to illuminate a sample with an intense, narrowband light source (typically a laser) and observe the scattered light with a spectrometer. The spectral resolution for these spontaneous scattering measurements is limited by the resolution of the spectrometer. Nonlinear scattering processes such as gain spectroscopy, however, have spectral resolution which is limited only by the linewidths of the lasers. Nonlinear scattering spectroscopy also yields optical signals in the form of intense coherent beams, which is an advantage for eliminating background signals. For Raman measurements, gain spectroscopy provides signals which are linearly proportional to the

Applied Laser Spectroscopy, Edited by W. Demtröder and
M. Inguscio, Plenum Press, New York, 1990

imaginary part of the third order nonlinear susceptibility which is in turn proportional to the Raman scattering cross section. This is in contrast to another popular nonlinear Raman technique called CARS[2,3] (coherent anti-Stokes Raman spectroscopy) for which the signal is proportional to the squared magnitude of the sum of the real and imaginary parts of the third order susceptibility. Thus gain spectroscopy is the more appropriate technique for measurements of Raman line shapes, linewidths and absolute gain coefficients.

Gain spectroscopy may be performed with either a cw[4,5,6] or pulsed[7,8,9,10] laser as a pump. A cw laser can give higher spectral resolution but because the small signal gain scales with the pump intensity, higher sensitivity may be obtained with pulsed lasers. Commercially available single-mode cw ring dye lasers have linewidths of 1 MHz, so the spectral resolution for a cw measurement can be quite high. To obtain higher pump laser powers for a cw measurement, a stabilized argon or krypton ion laser may be used as a fixed-frequency cw pump laser. Multi-pass cells have been used to improve the sensitivity for cw measurements.[4,6] The spectral resolution for a pulsed measurement is limited to the transform limit of the laser temporal pulse width.[11] Injection-seeded Nd:YAG lasers,[12] which provide near transform-limited pulses (50 MHz or less), are well suited for use as fixed-frequency pump lasers. For a tunable pump laser, the amplification of a single-mode cw dye laser with pulse dye amplifiers[13] (typically pumped by an injection-seeded Nd:YAG laser) can give linewidths of less than 100 MHz. Single-mode cw lasers are an optimal probe laser for either cw or pulsed measurements both because of their narrow linewidth and low noise. Small gains (on the order of 1%) can be measured with good signal to noise even for pulsed measurements because the measurements are performed in a bandwidth at high frequencies where there is less laser noise. For pulsed laser experiments, the probe laser is generally gated to illuminate the diode only during the pump pump to avoid saturation of the diode. An ion laser is typically a less noisy probe than a dye laser for pulsed measurements.

Gain spectroscopy has been used for a wide variety of measurements. For Raman scattering, gain spectroscopy has given information on the temperature and density dependence of linewidths[9,10,14], line shifts[9] and line shapes,[15] as well as the wavelength dependence of the Raman gain coefficient.[16] Raman gain spectroscopy has also been used to examine the nonresonant ac Stark effect.[17,18] Gain spectroscopy has been used to measure linewidths, lineshifts, and absolute gains for Brillouin scattering in gases[5,6] and solids.[19] Gain spectroscopy has allowed resolution of stimulated Rayleigh scattering.[6,19,2]

BRILLOUIN SCATTERING

Brillouin scattering refers to the scattering of light by acoustic waves.[1,21] Acoustic waves cause a variation in the density and hence the refractive index of a medium. In this manner, an acoustic wave acts as a moving refractive index grating and scatters light. The Bragg condition for scattering from the grating, together with the Doppler shift from the motion of the grating, dictate that the frequency shift of the Brillouin line depend on the acoustic velocity and the scattering angle. The width of the Brillouin line depends on the acoustic damping time. The Brillouin gain coefficient depends on the acoustic velocity, the acoustic damping time, and the electrostrictive coupling coefficient. Thus Brillouin spectroscopy provides information on a number of elastic properties.

Several aspects of Brillouin scattering make gain spectroscopy an appropriate technique for frequency domain Brillouin measurements. As the frequency shift for Brillouin scattering depends on the scattering angle, a large collection angle can cause broadening of the Brillouin line. Because the signal for gain spectroscopy is in a collimated beam, the collection angle is quite small, and broadening is negligible. Because Brillouin linewidths can be quite narrow, the fact that the resolution is only limited by the resolution of the lasers is important. For example, linewidths are well below 50 MHz for crystals and for gases at high pressure. Lastly, for stimulated scattering measurements, the Rayleigh gain peak is smaller than the Brillouin gain peak. This is in contrast to spontaneous scattering measurements, where the strong Rayleigh scattering peak can limit the sensitivity for Brillouin measurements.

BRILLOUIN GAIN SPECTROSCOPY IN GLASSES AND CRYSTALS

Recent measurements of Brillouin spectra in solids provide an example of gain spectroscopy measurements.[19] The apparatus, shown in Figure 1, may be used for Brillouin, Raman, or Rayleigh measurements in a variety of materials. The pump laser is a frequency-doubled home-built injection-seeded single-mode Nd:YAG laser, which provides pulses with a transform-limited bandwidth (less than 40 MHz). A half-wave plate and polarizer are used to vary the pump laser intensity. The probe laser is a commercial cw single-mode ring dye laser (Coherent 699-29) with a linewidth of 1 MHz. Both beams are passed through spatial filters to provide good spatial modes for accurate absolute measurements. A chopper wheel is used to chop out a 100 μs long pulse from the dye laser to avoid saturation of the photodiode. The pump and probe beams are overlapped at a small crossing angle in the sample. The geometry shown in Figure 1 is for a nearly collinear crossing geometry. A crossed beam geometry, where the pump laser is focused to a line with cylindrical lenses, is used for some measurements. After passing through the sample, the probe beam is passed through another spatial filter to discriminate against scattered light and the intensity is measured with a fast photodiode. To perform absolute gain coefficient measurements, the intensity of the pump beam must be known. For this purpose, an pyroelectric or thermopile energy meter is used to measure the pump pulse energy, a video camera is used to measure the beam spatial profile, and a photodiode is used to measure the temporal pulse width. Linewidths and lineshifts are measured by scanning the probe laser. An example of a scan performed in fused silica is shown in Figure 2. Both gain (positive) and loss (negative) peaks are present. The loss peaks correspond to transfer of energy from the probe beam to

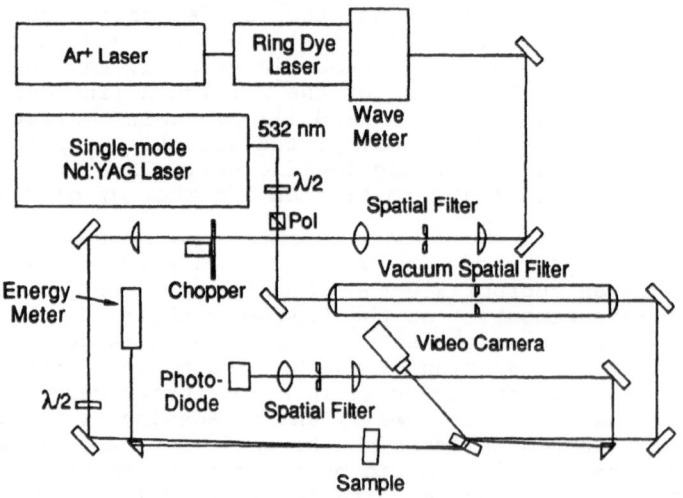

Figure 1. Apparatus used for Brillouin gain spectroscopy.

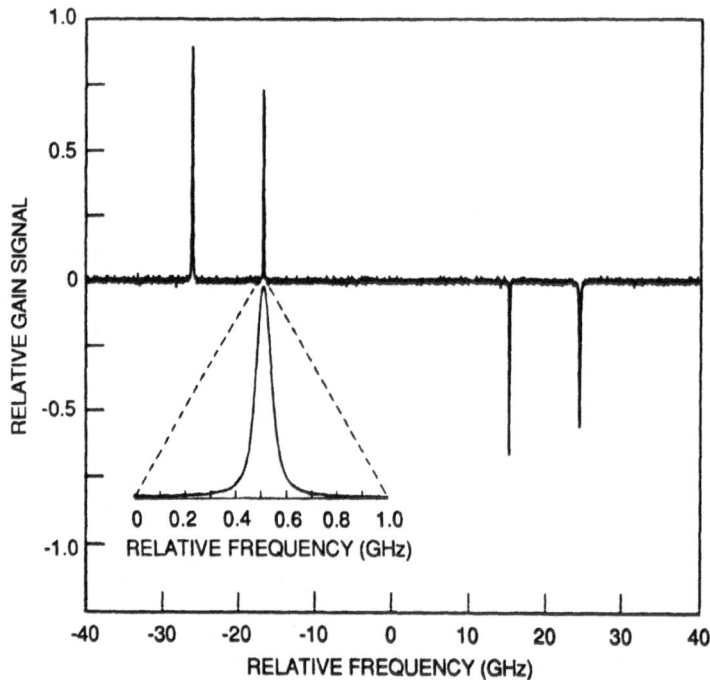

Figure 2. Brillouin gain spectrum for fused silica, showing
gain and loss peaks for longitudinal (outer peaks) and
transverse (inner peaks) acoustic waves.

the pump beam. This measurement was performed with a form of crossed beam geometry, and lines for both longitudinal (or compressional) and transverse (or shear) acoustic waves are present. Shown as an inset is a high resolution scan of the transverse gain peak. The data are shown as dots. A Voigt fit to the data is shown as a solid line. The linewidth of the pulsed laser is input as the Gaussian component of the Voigt. The Lorentzian component determined from the fit gives the Brillouin linewidth. The separation between the gain and loss peaks yields the acoustic velocity. For higher laser intensities, stimulated Rayleigh scattering is also seen. While the Brillouin lines give information on the elastic properties of the material, the Rayleigh line gives information on the thermal properties of the sample. Measurements similar to those shown in Figure 2 are being carried out in a variety of glasses and crystals.

ACKNOWLEDGEMENTS

The experimental results reported here were performed in collaboration with Mark J. Dyer, Leonard E. Jusinski, William K. Bischel, and A. Peet Hickman.

REFERENCES

1. W. Kaiser and M. Maier, Stimulated Rayleigh, Brillouin and Raman Spectroscopy, chapter E2 in. "Laser Handbook," vol. 2, F. T. Arecchi and E. O. Schulz-Dubois, eds., North-Holland Publishing Company, Amsterdam (1972).
2. Y. R. Shen, "The Principles of Nonlinear Optics," John Wiley & Sons, New York (1984).
3. G. L. Eesley, "Coherent Raman Spectroscopy," Pergamon Press, Oxford (1981).
4. A. Owyoung, CW Stimulated Raman Spectroscopy, pp. 281-320, in: "Chemical Applications of Nonlinear Raman Spectroscopy," A. B. Harvey, ed., Academic Press, New York, (1981).
5. S. Y. Tang, C. Y. She, and S. A. Lee, Continuous-Wave Rayleigh-Brillouin-Gain Spectroscopy in SF_6, Opt. Lett. 12:870 (1987).
6. G. C. Herring, M. J. Dyer, and W. K. Bischel, High Resolution Stimulated Rayleigh-Brillouin Spectroscopy of Xe and SF_6, to be published.
7. W. J. Jones and B. P. Stoicheff, Inverse Raman Spectra: Induced Absorption at Optical Frequencies, Phys. Rev. Lett. 13:657 (1964).
8. P. Esherick and A. Owyoung, High Resolution Stimulated Raman Spectroscopy , in: "Advances in Infrared and Raman Spectroscopy," R. J. H. Clark and R.E. Hester, eds., Heyden and Son, Ltd., London (1982).
9. G. C. Herring, M. J. Dyer, and W. K. Bischel, Temperature and Density Dependence of the Linewidths and Line Shifts of the Rotational Raman Lines in N_2 and H_2, Phys. Rev. A 34:1944 (1986).
10. L. A. Rahn and D. A. Greenhalgh, High-Resolution Inverse Raman Spectroscopy of the v_1 Band of Water Vapor, J. Mol. Spectrosc. 119:11 (1986).

11. A. E. Siegman, "Lasers," University Science Books, Mill Valley, California (1986), p. 335.

12. R. L. Schmitt and L. A. Rahn, Diode-Laser-Pumped Nd:YAG Laser Injection Seeding System, Appl. Opt. 25:629 (1986); M. J. Dyer, W. K. Bischel and D. G. Scerbak, High-Power 80-ns Transform-Limited Nd:YAG Laser, in. SPIE vol. 912 - "Pulsed Single-Frequency Lasers: Technology and Applications," the Society of Photo-Optical Intrumentation Engineers, Bellingham, Washington (1988).

13. R. Wallenstein and T. W. Hänsch, Powerful Dye Laser Oscillator-Amplifier System for High Resolution Spectroscopy, Opt. Commun. 14:353 (1975); P. Drell and S. Chu, A Megawatt Dye Laser Oscillator-Amplifier System for High Resolution Spectroscopy, Opt. Commun. 28:343 (1979).

14. G. C. Herring, M. J. Dyer, and W. K. Bischel, Temperature and Wavelength Dependence of the Rotational Raman Gain Coefficient in N_2, Opt. Lett. 11:348 (1986).

15. L. A. Rahn, R. E. Palmer, M. L. Koszykowski, and D. A. Greenhalgh, Comparison of Rotationally Inelastic Collision Models for Q-Branch Raman Spectra of N_2, Chem. Phys. Lett. 133:513 (1987).

16. W. K. Bischel and M. J. Dyer, Wavelength Dependence of the Absolute Raman Gain Coefficient of the Q(1) Transition in H_2, J. Opt. Soc. Am. B 3:677 (1986).

17. R. L. Farrow and L. A. Rahn, Optical Stark Splitting of Rotational Raman Transitions, Phys. Rev. Lett. 48:395 (1982).

18. W. K. Bischel, M. J. Dyer and L. E. Jusinski, Study of the ac Stark Effect for the Q(0) and Q(1) Vibrational Raman Transitions in H_2, to be published.

19. G. W. Faris, L. E. Jusinski, M. J. Dyer, W. K. Bischel and A. P. Hickman, High Resolution Brillouin Gain Spectroscopy in Glasses and Crystals, to be published.

20. C. Y. She, G. C. Herring, H. Moosmüller, and S. A. Lee, Stimulated Rayleigh-Brillouin Gain Spectroscopy in Pure Gases, Phys. Rev. Lett. 51:1648 (1983).

21. I. L. Fabelinskii, "Molecular Scattering of Light," Plenum Press, New York (1968).

Coherent Raman Spectroscopy: Techniques and Recent Applications

Joseph W. Nibler[*]

Department of Chemistry
Oregon State University
Corvallis, Oregon, 97330

INTRODUCTION

With the advent of tunable pulsed lasers of high power, a large variety of nonlinear optical mixing processes have become possible in recent years. Coherent Raman Spectroscopy refers to those interactions in which the frequency difference ω_1-ω_2 of two incident laser fields is tuned to match a vibrational- rotational energy difference in the ground electronic state of a molecule. The fields can thus drive the molecular oscillators in phase, causing a coherent scattering of the incident ω_1 beam at the $\omega_3 = 2\omega_1$-ω_2 anti-Stokes Raman frequency (CARS). The interaction will also pump molecules to the higher energy level and, due to energy conservation, thereby cause a loss (or gain) in the ω_1 (or ω_2) intensity, leading to Stimulated Raman Loss (SRL) or Gain (SRG) Spectroscopy (more generally, SRS). Subsequent energy relaxation to create an acoustic wave can alternatively be detected by sensitive microphones, giving rise to Photo-Acoustic Raman Spectroscopy (PARS). Other acronyms based on polarization changes also exist.[1]

The response of the medium to the fields is characterized by the third order electric susceptibility (chi-three) and a description of the physics of the interaction is given in the chapter by Taran in this book. Further details can be found in, for example, the references given in two recent reviews.[2,3]

Because these coherent Raman methods provide several orders of magnitude improvement in signal intensities and resolution compared to spontaneous Raman spectroscopy, they have proven useful in a number of "difficult" applications. Examples include the study of combustion systems, electric discharges, free jet expansions, and photofragmentation events. Some discussion of the first two areas can be found in the chapter here by Taran and in references 2, 3 and the literature cited therein. In this chapter, we concentrate on some experimental aspects for high resolution and low frequency shift studies and on applications involving the study of clusters formed in cold expansions and fragments formed by laser photolysis.

---

[*]1989-90 Visiting Fellow, JILA, Boulder, CO.

Applied Laser Spectroscopy, Edited by W. Demtröder and
M. Inguscio, Plenum Press, New York, 1990

EXPERIMENTAL

CARS Apparatus

The schematic of the CARS setup at Oregon State University shown in Fig. 1 is generally typical of that used by most workers. The apparatus is based on a pulsed Nd:YAG laser whose 532 or 355 nm harmonics pump a dye laser to provide a tunable ω_2 beam of 1-10 mJ. This is focussed with a comparable amount of the 532 nm light in either a collinear or, as shown in Fig. 1, a folded "BOXCARS"[4] fashion to produce the anti-Stokes signal at resonance. The latter geometry involves two 532 nm beams crossing at a few degrees in a plane perpendicular to the ω_2 and ω_3 beams. This crossed-beam geometry results in some signal loss due to a limited interaction length but achieves the required phase matching with the decided advantage of allowing spatial separation of the ω_3 signal from the much more intense input beams.

The BOXCARS geometry is especially useful for studies of low frequency Raman shifts, as shown in the comparison of Fig 2. Here, the rotational lines of N_2, cooled in a free jet expansion, are easily seen, and with the addition of a polarizer to block residual 532 nm light, scans through zero frequency shift can be readily achieved. In cases where parallel ω_1, ω_2, ω_3 polarizations are desirable (e.g. for low frequency <u>symmetric</u> vibrations), an alternative rejection scheme using an iodine absorption cell gives equivalent results. In this case an injection seeded single mode Nd:YAG laser is used which is temperature-tuned to match the second harmonic to a strong I_2 absorption line so that a simple absorption cell in front of the final monochromator can serve to filter out all scattered 532 nm light.

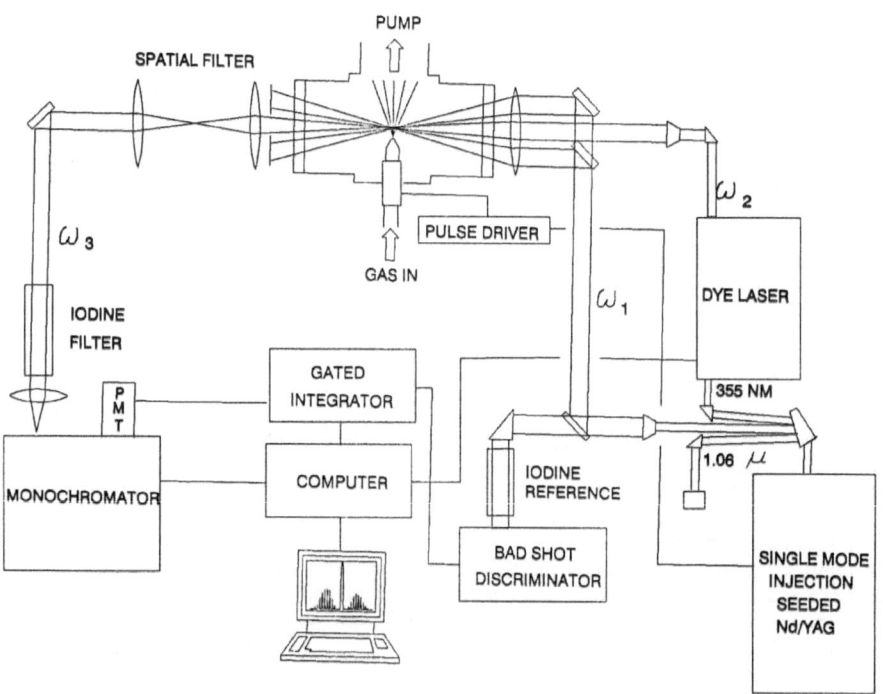

Fig. 1.   Schematic of CARS apparatus

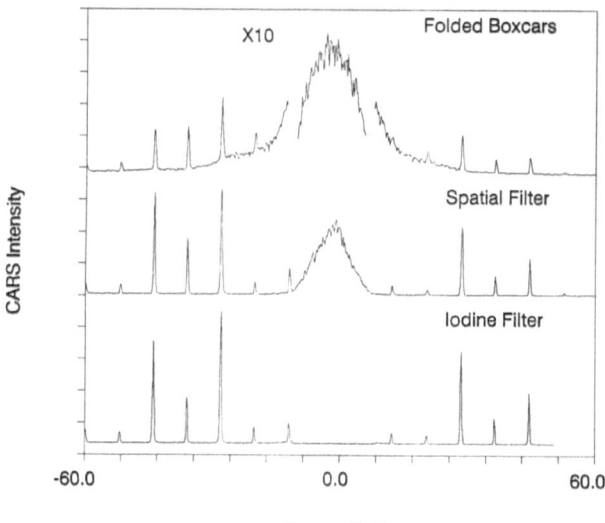

Neat Nitrogen Expansions

X10   Folded Boxcars

Spatial Filter

Iodine Filter

CARS Intensity

-60.0   0.0   60.0

Raman Shift

Fig. 2. Rotational CARS spectra of N_2 in a free jet expansion for
different discriminatory techniques. Spectra at negative shifts
are due to Coherent Stokes Raman Spectroscopy (CSRS) where the role
of ω_1 and ω_2 are interchanged.

SRS Apparatus

For high resolution SRS spectroscopy, the OSU system depicted in Fig.
3 is representative. Here a cw ring dye laser is pulse-amplified by a
single mode Nd:YAG laser to provide a tunable ω_2 pump beam of 5-10 mJ and
60-100 MHz linewidth. Single mode cw argon or krypton ion lasers provide a
probe beam of 100-500 mW whose intensity change at resonance is monitored
with an avalanche photodiode (EG100Q). To reduce the average power on this
detector and to minimize thermal heating in the dye amplifier chain, the cw
beams are chopped to give 100 microsecond pulses at 10 Hz. With care,
detection near the shot noise limit of the probe lasers, about 1 part in
10^5, is achieved. By use of various probe laser lines, the entire spectral
shift region from 0 to 5000 cm^{-1} can be covered with the convenient
rhodamine 6G dye tuning range. Spatial and polarization filtering are used
for scans at low frequency shifts. Additional details on the OSU CARS and
SRS spectrometers are given in references 2 and 3.

Because of their nonlinear power dependence, both CARS and SRS signals
benefit from high input power densities. For CARS, the principal
limitation is saturation broadening which is observable for P_1P_2 power
product densities in excess of about 30 GW^2/cm^4.[5,6] This effect is not a
problem for SRS due to the low probe power.

For high resolution SRS spectroscopy, a second limitation arises from
ac Stark shifts and broadening by the pump beam. For pure rotational

315

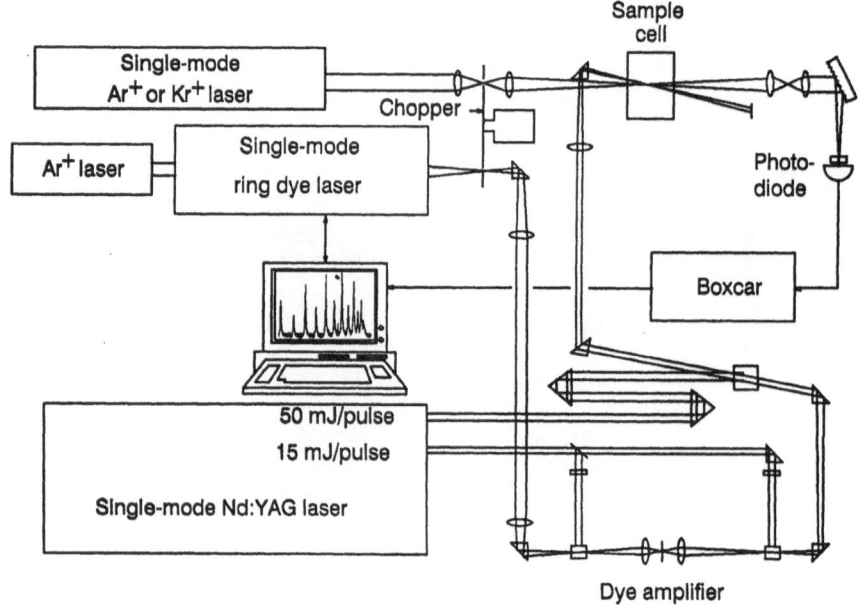

Fig. 3.   Schematic of SRS apparatus

transitions, the M degeneracy of the rotational levels is split by the dc component of the field perturbation term $H = -\gamma E^2/4$. Here γ is the polarizability anisotropy of the molecule and E is the pump laser field. Time dependent perturbation theory can be used to obtain the level shifts and intensities for various polarization orientations of the incident beams.[7-8] Fig. 4 shows experimental spectra for low J lines of N_2 at 2 and 8 mJ pump pulse energies (corresponding to intensities of about 2-8 GW/cm^2), along with calculated stick patterns and convolutions of these using Lorentzian lineshapes.[9] By using higher laser intensities and perhaps N_2 as a reference, such effects might be useful in determining polarizability anisotropies. However the results of Fig. 4 show that the broadening is small for pump energies of a few mJ and that it also decreases for higher J values. Thus the effect is not too restrictive for high resolution rotational spectroscopy.

For vibrational Q branch studies, the ac Stark shifts are about the same for $v = 0$ and 1 levels so that the transition splittings are negligible. There is however a second effect, a shift in the vibrational frequency $d\omega_v = - \omega_v (3E^2/16D_e\beta) (d\alpha/dR)$ caused by the field acting through the polarizability derivative $(d\alpha/dR)$.[9-10] Here $D_e\beta^2$ corresponds to the force constant for a Morse potential. Fig. 5 shows the effect on the Q branch of N_2 and again it is clear that the perturbation is small but sufficient to require correction if accurate transition frequencies are to be deduced from measurements at energies in excess of a few mJ. It may be noted that the direction and magnitude of the shift provide access to the sign as well as the magnitude of $d\alpha/dR$.

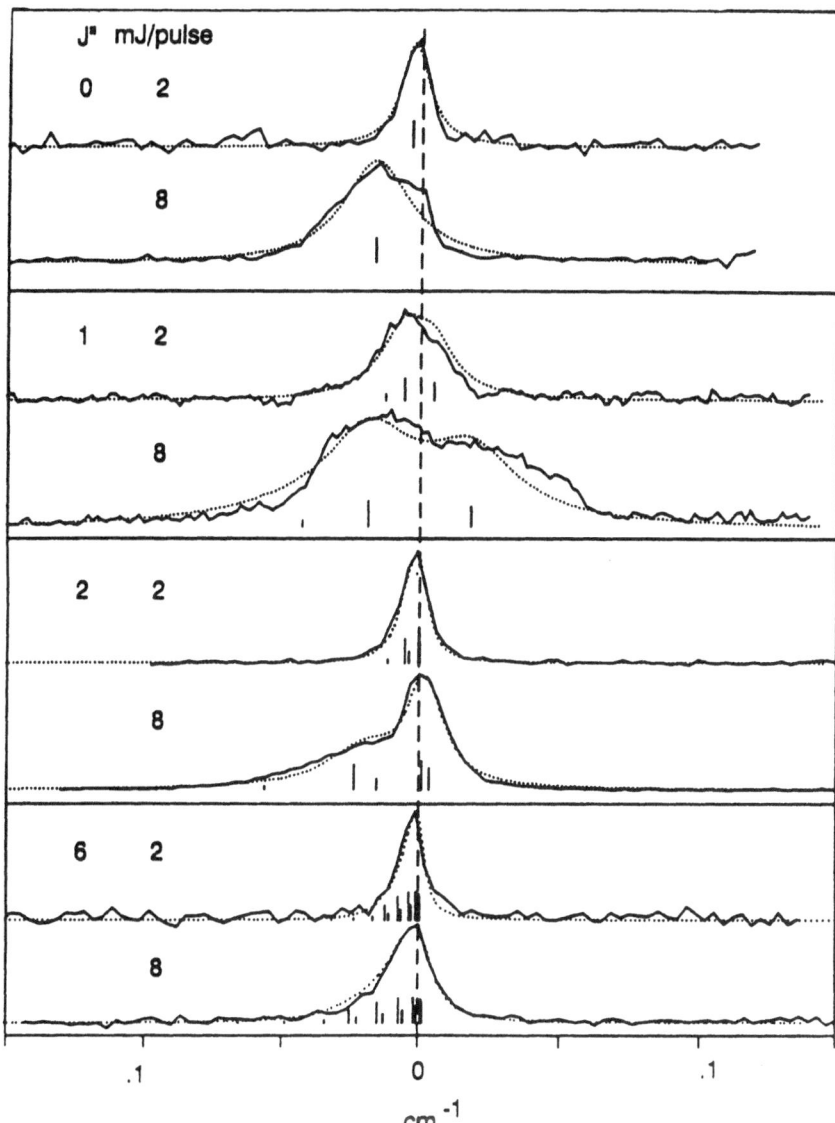

Fig. 4.   Optical Stark effect on pure rotational Raman transitions of
nitrogen.   The unshifted position (zero field) is indicated by
the vertical dashed line.   The vertical bars show the center
positions of the rotational sublevels (±M), split by the Stark
effect.   The dotted line shows a simulation of the Stark effect
assuming a Lorentzian lineshape for each component.

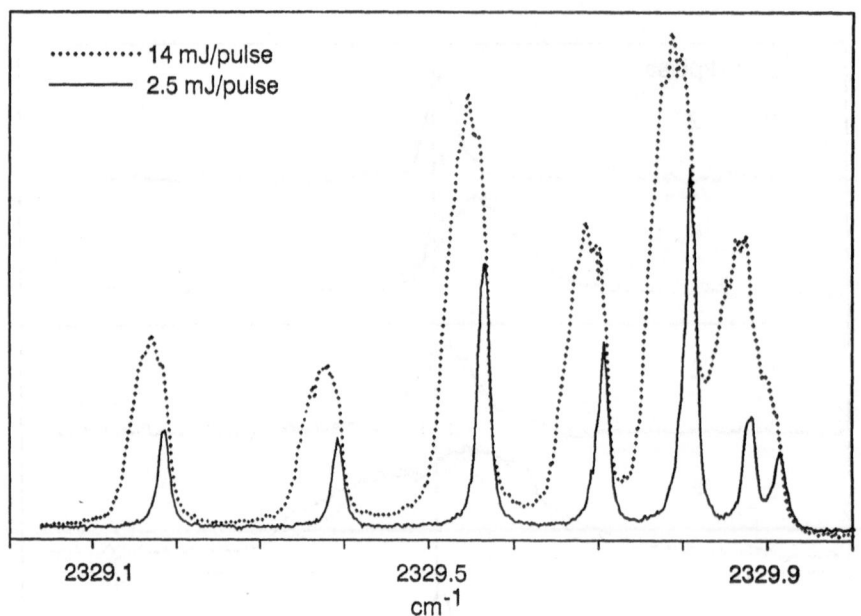

Fig. 5.   Optical Stark effect on the Q-branch spectrum of N_2 cooled in a free expansion jet.

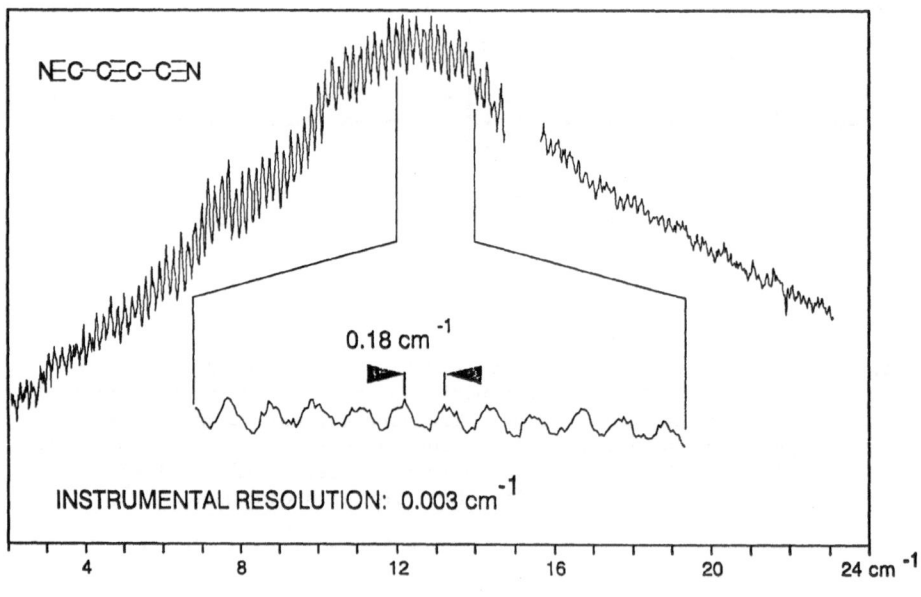

Fig. 6. Pure rotational Raman Loss spectra of C_4N_2.

High Resolution Studies

Schrotter et al.[11] have given an excellent survey of the use of CARS and SRS for high resolution molecular spectroscopy and a compilation of molecules studied up until 1987 is given in Ref. 2. Most of the work has centered on improved frequency or linewidth measurements for small stable molecules but extension to larger and more reactive molecules is likely. As one example, Brown et al.[12] have recently applied pure rotational SRS to the study of the dicyanoacetylene molecule C_4N_2. The rotational constants and bond lengths of this linear molecule were not previously available, perhaps due to its absence of a dipole moment and its tendency to photopolymerize. Part of the interest in the properties of the molecule stem from its postulated existence in the atmosphere of Titan and probably in interstellar space. Fig. 6 shows the SRS spectrum of this molecule, obtained at 50 Torr. Each spectral "line" is actually a composite of the many hot band transitions that result because the molecule has low frequency bending modes at 107 and 263 cm^{-1}. Analysis gave an average B value of 0.04487(2) cm^{-1} and an estimate of Bo of 0.0445(1) cm^{-1}, in good accord with a structure deduced in an electron diffraction measurement which was part of the same study.[12]

Cluster Studies

Small Complexes

The dramatic cooling provided in free jet expansions has led to an explosion of activity in the study of structures and spectroscopic properties of small van der Waals and hydrogen-bonded complexes. Much of the interest has been in understanding the structures, potential energy surfaces, and dissociation dynamics of these unusual molecules. Though much less sensitive than the more common microwave, IR, fluorescence, or ionization techniques, Raman methods too have been applied to the detection and vibrational characterization of such species. Examples include studies of Ar_2[13], $(HCN)_2$[14], $(CO_2)_2$[15], $(NH_3)_n$[16], $(H_2O)_2$[17], $(HCl)_2$[18], and $(C_6H_6)_2$[19]. In most cases, the symmetries of the complexes are such that Raman spectra provide vibrational information inaccessible by other techniques.

Vibrational Raman spectra are generally dominated by the Q branch and in no case yet has the rotational structure in this band been resolved for complexes. However, the simple measurement of the frequencies of the Raman active modes has proven useful in making structural conclusions based on symmetry and IR-Raman selection rules. For example, a C_{2h} planar symmetric structure was deduced for the CO_2 dimer from such data[15], a geometry confirmed by subsequent rotationally-resolved IR beam experiments.[20] In the case of the HCl dimer, tunneling motions yield a similar C_{2h} symmetry for the potential so that half the vibrational transitions are invisible to IR spectroscopy. The combined IR-Raman data yield tunneling splittings for both vibrational levels, showing interestingly that the tunneling potential barrier increases substantially in going from the v = 0 to the v = 1 state.

The low frequency van der Waals region is of special interest since transitions here relate most directly to the intermolecular potential of the binding pair. Experimental access to this region is difficult although pure rotational Raman spectra and the low frequency vibration of the Ar_2 dimer have been obtained in a *tour de force* spontaneous Raman experiment by Silvera.[13] Similar results have not yet been achieved using coherent Raman methods but, as a step in this direction, Fig. 7 shows the low frequency

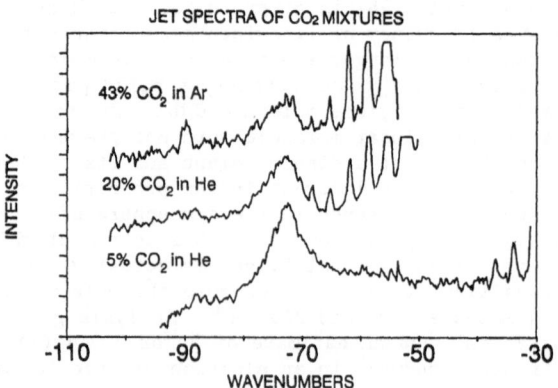

Fig. 7. CSRS spectra of clusters formed in CO_2 jet expansions.
($P_0 \approx 25$ atm, $T_0 \approx 258$ K, X/D ≈ 1.4)

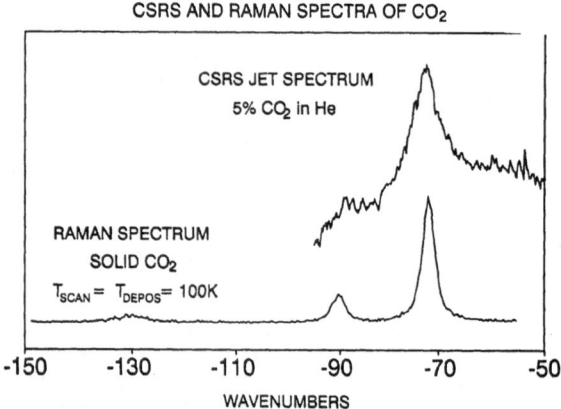

Fig. 8. Comparison of CSRS jet spectra of CO_2 clusters with spontaneous
Raman spectrum of solid CO_2.

CARS spectra of expansions of CO_2 obtained by Triggs.[21] The region near 0 cm^{-1} is dominated by the monomer rotational spectrum which extends to high J values since rotational cooling is incomplete so close to the nozzle. Of greatest interest is a weak cluster feature seen near 70 cm^{-1}. This is not attributed to the dimer since a channel nozzle was used in this experiment and the conditions were such as to favor large clusters. As seen in Fig. 8 the spectra resemble those of solid CO_2 in the lattice region where the expected three Raman active librational modes are seen. The similarity is evidence that, even though formed in less than a microsecond, the large clusters of CO_2 are not amorphous but rather have taken on a somewhat ordered form akin to that in the solid. Similar behaviour is seen in the SRS studies of nitrogen described below.

Micro-Clusters

If one cools a channel nozzle, it is possible to produce very large clusters and indeed, a spray of liquid droplets under extreme conditions. By manipulation of the driving conditions, Beck[9,22] has managed to produce jets of nitrogen micro-clusters which were crudely estimated from Mie scattering intensities to be in the range 5 to 500 nm. Fig. 9 shows the onset of such aggregation in the early stages of an expansion as monitored by SRS spectroscopy. The monomer Q branch dominates the spectrum and give a measure of the rotational temperature. The lines are clearly collision-ally broadened in the high density region near the nozzle and the growth of the weak cluster feature at 2326.5 cm^{-1} is apparent. Fig. 10 shows the

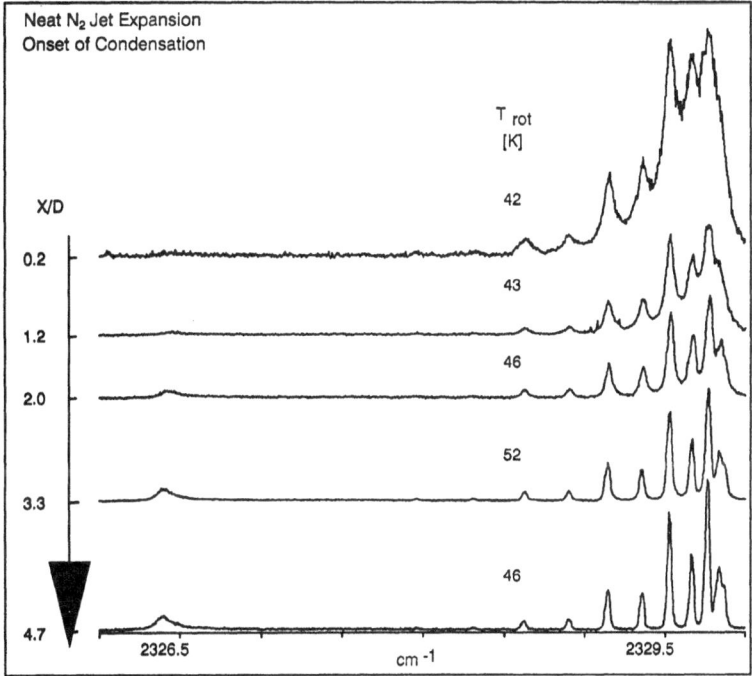

Fig. 9. SRL spectra of a N_2 expansion as a function of position in nozzle diameter units. Residual collisional broadening of the Q-branch rotational lines is seen at low X/D values.

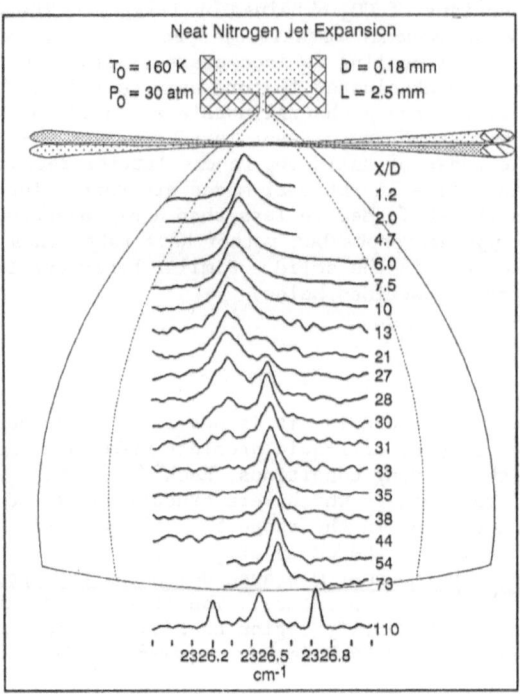

Fig. 10. SRL spectra of N_2 clusters as a function of position in a free jet
expansion. The signal is generated from an approximate cylinder of
0.05 mm diameter and 1 mm length.

evolution of this feature as one moves out in the expansion, ultimately
reaching the Mach disk boundary at about X/D — 100, D — 0.18 mm being the
nozzle diameter. The high resolution and point-probing characteristics of
the coherent Raman methods provide a unique capability for such studies.

The analysis of the jet spectra is aided by concurrent studies of
equilibrium samples of liquid and solid (both α and β phases) nitrogen
obtained with the same spectrometer and a simple cryogenic cell cooled by a
closed cycle helium refrigerator.[9,23] These data are displayed in Fig. 11
and serve as a basis for conversion of frequency to temperature for the
micro-clusters formed in the jets. This assumes of course that the size is
such that bulk properties obtain, a condition believed to be true for the
size range investigated.

The interpretation of the jet spectra is then that one sees the
formation of micro-drops of liquid nitrogen in the X/D region below about
5, these subsequently cool by evaporation until, at X/D — 20-30, freezing
to form β-phase solid occurs. These microcrystals cool further by
evaporation until they strike the shock boundary at X/D — 100 where they
warm in the hot background gas region, Here the J — 7 and 8 lines of the
warm (160K) background gas are also seen in the spectrum. Using the
equilibrium sample data, temperatures are deduced for the clusters and
displayed in Fig. 12, along with rotational temperatures measured for the
monomer. To the author's knowledge, these data represent the first clear
point-by point monitoring of the formation and freezing to a well-ordered
state for micro-clusters produced in such types of expansion.

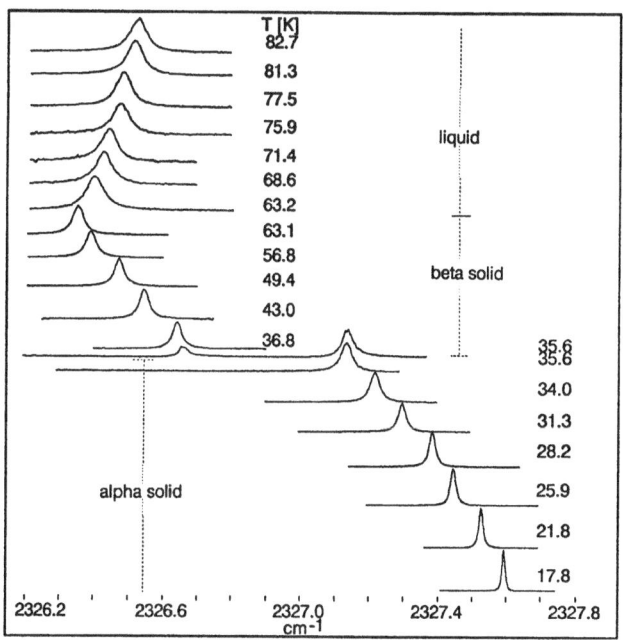

Fig. 11. SRL spectra of equilibrium samples of liquid, α- and β-solid phases of nitrogen at various temperatures.

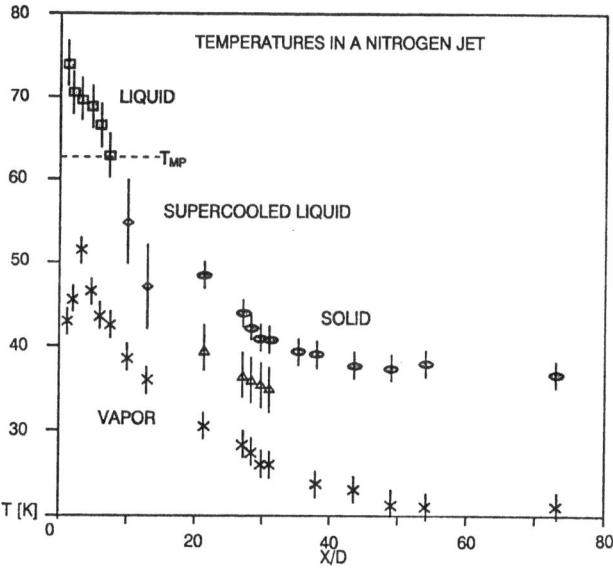

Fig. 12. Cooling curves deduced for the nitrogen expansion of Fig. 10.

Modeling of the evaporative cooling for the solid clusters in the low density region of the expansion can be used to obtain an improved estimate of the size of the clusters. Cooling data for the solid in Fig. 12 imply a terminal diameter range of 18 to 80 nm, with a value of 35 nm giving the best fit through the freezing region.[9] Similar modeling of the liquid cooling is complicated by the likely occurence of monomer-cluster collisions whose frequency is difficult to know precisely in such channel nozzle expansions. Assuming simple evaporative cooling of the liquid micro-drops from X/D = 6 outward indicates that a 40 nm drop would yield the terminal diameter of 35 nm for the solid and give a cooling curve that is in reasonable accord with the liquid observations.

These measurements suggest that a novel area of research in condensed phase properties exists in such cluster studies. It should be noted that the liquid nitrogen data indicate supercooling temperatures <u>far lower</u> than those achievable in cells. Estimates of these temperatures were obtained by assuming adiabatic freezing of the liquid to the solid, whose temperature is well-determined. Access to such unusual metastable conditions is unique to jet studies and is quite intriguing. Investigations of linewidths and frequency shifts would be a desirable complement to theoretical calculations of liquid ordering that may occur in supercooled liquids prior to freezing. Detection of solid-solid phase changes is also possible and, by cooling more rapidly via added He as a driving gas, the production of the low temperature α-phase of nitrogen has already been observed.[9] Further studies involving smaller cluster sizes may permit a distinction between surface and bulk molecules. Other novel aspects of interest might include the study of mixtures where phase separation and crystallization phenomena could be examined.

CARS Spectra of Molecular Photofragments

The short duration of the coherent Raman measurement with pulsed lasers has made the technique useful in the study of transient species such as those generated by laser or flash discharges. Molecules subjected to laser photolysis whose products have been examined by CARS include C_6H_5X[25,26], CH_3I[27,28], O_3[29,30], HI[31], CH_3NNCH_3[32], HNO_3[33], NH_3[34], H_2CO[35], CF_3NO[24], CS_2[35], and H_2O_2[36]. A summary of the activity in this field to 1987 can be found in Ref. 2, 3; here only a brief description of the photolysis of CF_3NO and CH_3I will be given.

CF₃ Photoproduct

The CF_3 species is a reactive radical that is difficult to monitor due to the absence of convenient visible - u.v. absorptions. Bozlee et al. first utilized CARS detection in a study of the visible photodissociation of CF_3NO with the objective of determining the vibrational-rotational energy distributions of the products.[24] Fig. 13 shows the Q-branch spectra obtained for nascent CF_3 and NO fragments produced by photolysis with the red Stokes CARS beam. In this wavelength region, the dissociation process is thought to involve direct dissociation on the S_1 potential energy surface. The NO fragment is rotationally hot ($T_{Rot} \approx 1100$ K) with a statistical rotational distribution that has been studied previously using laser induced fluorescence methods.[38] As for NO, the CARS spectrum of the CF_3 radical also indicates extensive rotational excitation with little or no excess vibrational energy. This is consistent with a simple picture of an impulsive dissociation along the CN bond which excites rotation of both fragments since the CNO linkage is nonlinear. The individual Q-branch lines of the CF_3 transition were not resolved but the overall contour was modeled to yield upper state rotational constants and a 1088.6 cm^{-1} value for the ω_1 CF symmetric stretching mode.

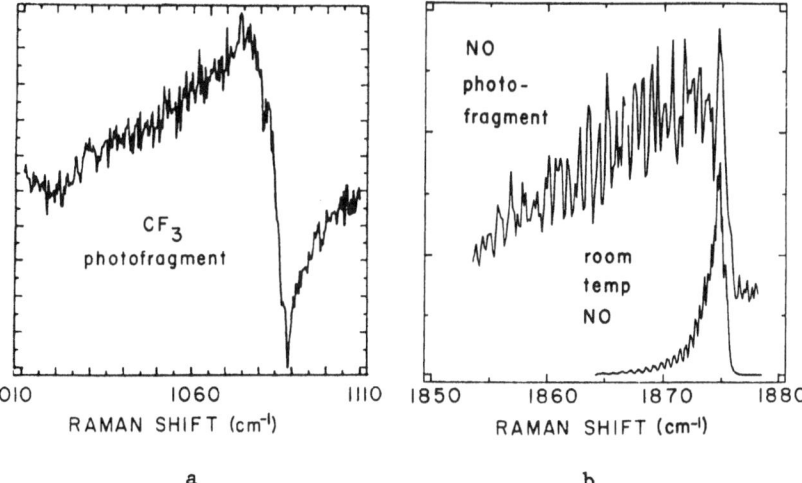

Fig. 13.   CARS spectra of the photofragments of CF_3NO.

   a. Q branch of rotationally excited CF_3 in the ω_1 symmetric C-F
      stretching region. Spectrum resulted from photolysis of CF_3NO
      by the CARS laser beams and is the average of five scans.

   b. Q branch of nitric oxide. Lower trace. Spectrum of 80 Torr
      pure NO at 295 K. Upper trace. Spectrum of rotationally
      excited NO formed in the photolysis of 30 Torr CF_3NO.

CH_3 Photoproduct

     CARS detection of the IR-forbidden ω_1 symmetric CH stretching mode of
the planar CH_3 radical has given a value of 3004.8 cm^{-1} for this mode.[32]
The photolysis of CH_3I has been much studied since it serves as an ideal
model for a prompt bond dissociation in which most of the internal fragment
energy is expected to be in the out-of-plane vibrational bending mode
rather than in rotation. Fig. 14 shows the loss of the CH_3I parent upon
266 nm laser photolysis and the appearance of a rich Q-branch spectrum of
CH_3 in the ω_1 region. The latter spectrum was recorded with a time delay
of 20 ns so that some rotational but no vibrational relaxation is expected.
Structure attributable to the $v_2 = 1$ excited bending level is apparent but
no indication of extensive population in higher levels is seen. This
result conflicts with earlier time-of-flight measurements interpreted in
terms of a $v_2=2/v_2=0$ ratio of 14[39] but is more in accord with recent
photoionization studies in which this ratio is found to be less than one.[40]
This vibrational distribution warrants more careful study since it can
serve as a critical test of theoretical results which have generally
predicted significant excitation in the higher levels.

     At zero time delay and lower pressure, the nascent rotational
distribution in the CARS spectrum is obtainable[28] and is found to be
simpler and more resolved than in the delayed case shown in Fig. 14.
Analysis of the spectral intensities is in accord with earlier
indications[40,41] that the CH_3 radical is preferentially produced in
rotational states in which most of the angular momentum is about the
symmetry axis. This occurs since initial CH_3I rotation about the

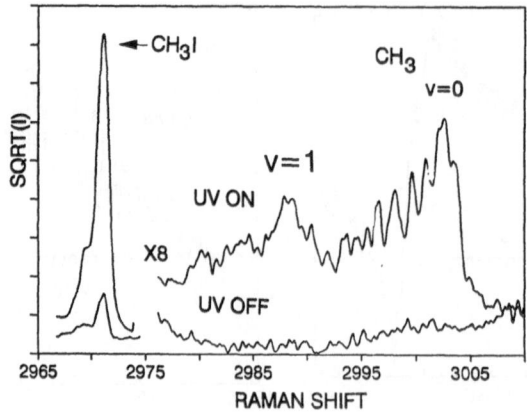

Fig. 14. Experimental CARS spectrum of the ω_1 Q branch of gas phase methyl
radicals formed by 266 nm photolysis of CH_3I.

perpendicular axes largely goes into orbital angular momentum of the
dissociating fragments. Subsequent collisions cause very rapid
redistribution of this population, a process which, because of the high
spectral resolution and ns probe times, is well suited for study by
coherent Raman methods.

SUMMARY

From the examples described in this chapter and in the section by
Taran, it is apparent that coherent Raman techniques represent a valuable
new spectroscopic tool for the study of molecular properties. These
methods have found widespread practical use for diagnostic purposes and
their value in the investigation of basic structural, dynamic, and
spectroscopic characteristics of molecules has also been demonstrated.
Especially promising are applications to the study of unusual and reactive
intermediates formed by photolysis, reaction, or aggregation due to
condensation in cold beams. The high resolution capability may prove
useful in the investigation of condensed phases at low temperatures that
can serve as a valuable complement to time domain studies of dynamic
processes in such media. Further improvements in sensitivity and
resolution in these coherent methods can be anticipated as laser sources
improve and as signal enhancements due to one photon resonances or combined
Raman-ionization techniques are implemented by the use of tunable sources
in the ultraviolet.

ACKNOWLEDGMENTS

Research support by the NSF and the AFOSR is appreciated and the author
also is grateful for a JILA Visiting Fellowship at the University of
Colorado where this chapter was written. He is especially thankful to his
co-workers at Oregon State University whose research results are described
herein: Rainer Beck, Kirk Brown, Brian Bozlee, Max Hineman, Kyung Hee Lee,
Nancy Triggs, and Ming Yang.

REFERENCES

1. M. D. Levenson, "Introduction to Nonlinear Laser Spectroscopy",
    Academic Press, New York, 1982.

2. J. Yang and J. W. Nibler, Annual Review of Physical Chemistry, Ed.
    by H. L. Strauss, Vol. 38, 349 (1987), Ann. Revs. Inc, Palo Alto.

3. J. W. Nibler and G. A. Pubanz, Advances in Non-Linear Spectroscopy,
    Ed. by R. J. H. Clark and R. E. Hester, Vol. 15, 1 (1988) Wiley, New
    York.

4. J. A. Shirley, R. J. Hall, and A. C. Eckbreth, Opt. Lett. 5, 380
    (1980).

5. M. Pealat, M. Lefebvre, J.-P. E. Taran, and P. L. Kelley, Phys. Rev.
    A38, 1948 (1988).

6. R. L. Farrow and R. P. Lucht, Opt. Lett. 11, 374 (1986).

7. R. L. Farrow and L. A. Rahn, Phys. Rev. Lett. 48, 395, (1982).

8. R. A. Hill, A. Owyoung, and P. Esherick, J. Mol. Spectrosc. 112, 233
    (1985).

9. R. Beck, Ph.D. thesis, Oregon State University, 1989.

10. L. A. Rahn, R. L. Farrow, M. L. Koszykowski, and P. L. Mattern,
    Phys, Rev. Lett. 45, 620 (1980).

11. H.W. Schrötter, H. Frunder, H. Berger, J.-P. Boquillon, B. Lavorel,
    and G. Millot, Advances in Non-Linear Spectroscopy, Ed. by R. J. H.
    Clark and R. E. Hester, Vol. 15, 97, (1988) Wiley, New York.

12. K. Brown, J. W. Nibler, L. Hedberg, and K. Hedberg, J. Phys. Chem.
    93, 5679 (1989).

13. H. P. Godfried and I. F. Silvera, Phys. Rev. A27, 3008 (1983).

14. M. Maroncelli, G. A. Hopkins, J. W. Nibler and T. R. Dyke, J. Chem.
    Phys. 83, 2129 (1985).

15. G. A. Pubanz, M. P. Maroncelli and J. W. Nibler, Chem. Phys. Lett.
    120, 313 (1985).

16. F. Huisken and T. Pertsch, Chem. Phys. Lett. 123, 99 (1986).

17. S. Wuelfert, D. Herren, S. Leutwyler, J. Chem Phys. 86, 3751 (1987).

18. A. Furlan, S. Wulfert and S. Leutwyler, Chem. Phys. Lett. 153, 291,
    (1988).

19. B. F. Henson, G. V. Hartland, V. A. Venturo, and P. M. Felker, J.
    Chem. Phys. 91, 2751 (1989).

20. K. W. Jucks, Z. S. Huang, D. Dayton, R. E. Miller, and W. J.
    Lafferty, J. Chem. Phys. 86, 4341 (1987).

21. N. E. Triggs, R. Rodriguez, M. Yang, K. H. Lee, and J. W. Nibler, Proc. of the XI Int. Conf. on Raman Spectroscopy, pp. 167-168, Ed. by R.J.H. Clark and D. A. Long, Wiley, New York (1988).

22. R. Beck and J. W. Nibler, Chem. Phys. Lett., 148, 271 (1988).

23. R. Beck and J. W. Nibler, Chem. Phys. Lett. 159, 79 (1989).

24. B. J. Bozlee, and J. W. Nibler, J. Chem. Phys., 84, 3798 (1986).

25. K. P. Gross, D. M. Guthals, and J. W. Nibler, J. Chem. Phys. 70, 4673 (1979).

26. K. Luther and W. Wieters, J. Chem. Phys. 73 (8), 4131 (1980).

27. J. W. Fleming, Optical Engin. 22, 317 (1983).

28. Nancy Triggs, Ph.D. thesis, Oregon State University, 1990.

29. J. J. Valentini, D. S. Moore, and D. S. Bomse, Chem. Phys. Lett. 83, 217 (1981).

30. J. J. Valentini, Chem. Phys. Lett. 96, 395 (1983).

31. D. P. Gerrity and J. J. Valentini, J. Chem. Phys. 79, 5202 (1983), 81, 1298 (1984).

32. P. L. Holt, K. E. McCurdy, R. B. Weisman, J. S. Adams, and P. S. Engel, J. Chem. Phys. 81, 3349 (1984), J. Amer. Chem. Soc. 107, 2180 (1985).

33. T. Dreier and J. Wolfrum, J. Chem. Phys. 80, 957 (1984).

34. T. Dreier and J. Wolfrum, Appl. Phys. B 33, 213 (1984).

35. D. Debarre, M. Lefebvre, M. Pealat, J.-P. E. Taran, D. J. Bamford and C. B. Moore, J. Chem. Phys. 83, 4476 (1985).

36 J. E. Stout, B. K. Andrews, T. J. Bevilacqua, and R. B. Weisman Chem. Phys. Lett. 151, 156 (1988).

37. G. J. Germann and J. J. Valentini, Chem. Phys. Lett. 157, 51 (1989).

38. R. D. Bower, R. W. Jones and P. L. Houston, J. Chem. Phys. 79, 2799 (1983).

39. R. K. Sparks, K. Shobatake, L. R. Carlson , and Y. T. Lee, J. Chem. Phys., 75, 3838 (1981).

40. R. Ogorzalek Loo, H,-P. Haerri, G. E. Hall, and P. L. Houston, J. Chem. Phys. 90, 4222, 1989.

41. J. F. Black and I. Powis, J. Chem. Phys. 89, 3986 (1988); Chem. Phys. 125, 375 (1988).

328

LASER DETECTION OF VERY RARE LONG-LIVED RADIOACTIVE ISOTOPES

Yu.A. Kudryavtsev, V.S. Letokhov, and V.V. Petrunin

Institute of Spectroscopy
USSR Academy of Sciences
142092, Troitsk, Moscow region, USSR

INTRODUCTION

Laser spectroscopy techniques have made it possible to solve the cardinal problems of optical spectroscopy: (1) the spectral resolution of the Doppler-free nonlinear spectroscopy techniques has already reached a value of $R = \nu/\Delta\nu \simeq 10^{11}$ at a spectral resonance width of $\Delta\nu \simeq 10^3\,Hz$, the development of methods for further reduction of $\Delta\nu$ being under way;[1] (2) where femtosecond mode-locked tunable lasers are used, the time resolution amounts to a few tens of femtoseconds, i.e. only a few tens of light oscillation periods;[2] (3) the sensitivity of some techniques, atomic and molecular photoionization spectroscopy in particular,[3] reaches ultimate values – single atoms and molecules. There is one more, perhaps the last, problem of optical spectroscopy that is still to be solved; (4) high-selectivity detection of trace atoms and molecules in a real environment, particularly the detection of trace rare isotope atoms in the presence of an abundant isotope or the detection of trace molecules of a certain species in a molecular mixture. Subject to intensive development are now being various approaches that can combine a maximum possible sensitivity with an exceptionally high detection selectivity. The present lecture treats in short of some possible ways to solve the first of these problem – to attain a high selectivity of optical detection of very rare isotopes. This problem was introduced in Ref.[4] and discussed in Refs.[5,6].

1. RARE ISOTOPES AND EXISTING METHODS FOR THEIR DETECTION

There are a fairly large number of rare isotopes of cosmic origin, particularly those formed in the upper atmosphere as a result of nuclear reactions under of effect of cosmic rays. They include such isotopes as ^{10}Be resulting from interaction of galactic cosmic rays with the N and O nuclei, ^{14}C formed in the reaction between secondary neutrons and N, and ^{26}Al produced as a result of splitting of the Ar nucleus. These isotopes form in the upper atmosphere, precipitate, and accumulate on the earth's surface and ocean bottom. The rate of their precipitation in the ocean can be considered to remain constant during a long period of time exceeding their half-life $T_{1/2}$. There also exist some technogenic radioactive isotopes being a product of human activity, such as ^{85}Kr and ^{90}Sr. Table I lists some rare isotopes along with their half-lives and concentrations relative to the content of their main stable isotopes.

Applied Laser Spectroscopy, Edited by W. Demtröder and
M. Inguscio, Plenum Press, New York, 1990

Table 1. Some rare cosmogenic isotopes

Isotope	$T_{1/2}$, y	Relative concentration
^{10}Be	1.5×10^6	10^{-10}
^{14}C	5.7×10^3	$10^{-12} - 10^{-16}$
^{26}Al	7.4×10^5	10^{-14}
^{36}Cl	3.1×10^5	10^{-17}
^{41}Ca	8.0×10^4	10^{-21}
^{81}Kr	2.1×10^5	5×10^{-13}
^{85}Kr	10.8	10^{-11}
^{90}Sr	28.5	10^{-10}

The best known radioisotope among these is radiocarbon, ^{14}C, which is used for estimating the age of objects of organic origin.[7] Radiocarbon, which is formed in the upper atmosphere in a concentration of $^{14}C/^{12}C = 10^{-12}$, is involved in the Earth's biochemical life cycle. After an organism has ceased to be living and participating in the carbon cycle, its content of ^{14}C decreases exponentially in accordance with the 5730 - years' half-life of this radioisotope. To date organic archaeological objects or events of 50 000 years, for example, it is necessary to detect ^{14}C in relative concentrations a low as 10^{-15}. By utilizing other, longer-lived isotopes, radioisotope dating can be extended to cover millions of years in the past.

At present, there exist two universal methods for detecting cosmogenic radioactive isotopes in low concentrations. the first, most popular one consists in measuring the specific radioactivity of the sample under analysis and comparing it with that of a specimen of zero age. This standard quantity for ^{14}C, for example, is well known and amounts to 15.3 beta-decay events per minute per gram of the natural mixture of carbon isotopes. To realize this method requires fairly large samples (around 5 g) and a long observation time (approximately 1 day). Therefore, a major part of a very valuable sample has frequently to be sacrificed in order that its age can be determined. Serious measures should also be taken to ensure proper protection against background activity. The nuclear method of detecting rare isotopes is disadvantageous since it depends for its operation on radioactive transformations of the isotopes, which occur extremely seldom. Therefore, to have a reasonable observation time (a few days), the sample must, in principle, contain a large number of the rare isotope atoms of interest.

The second method for detecting rare isotopes consists in utilizing a linear[8] or cyclic[9] accelerator as a high-resolution mass spectrometer.[10] The principal difficulty in implementing this method is the ned to suppress background noise due to abundant isotopes and isobaric atoms such as ^{14}N in the case of ^{14}C. This problem can be solved for atoms having negative ions (C^-) and lacking negative ions of isobaric atoms (N^-). The method handles much smaller samples (down to 5 mg). The cost of the equipment required by the method is rather high.

It is clear that the shortcomings of the both generally accepted methods open up a wide field of application for laser techniques, for these are, in principle, capable of tackling the very difficult task of detecting a few rare isotope atoms against the background of 10^{10} to 10^{20} atoms of the most abundant isotope of the same atomic species.

2. POSSIBLE LASER TECHNIQUES. LIMITATIONS AND WAYS TO OVERCOME THEM

All the existing laser spectroscopy methods whose ultimate sensitivity lies at a level of single atoms[5,11] can, in principle, be employed to effect a highly selective detection of rare isotope atoms. Figure 1 presents simplified schemes of the three main laser techniques for detecting single atoms. These techniques take advantage of the effects due to resonant interaction of the atom with photons: (1) spontaneous reradiation of many photons absorbed from the laser beam; (2) photoionization of the atom as a result of absorption of few photons; (3) changes of the atomic coordinate and velocity consequent upon reradiation of a large number of photons.

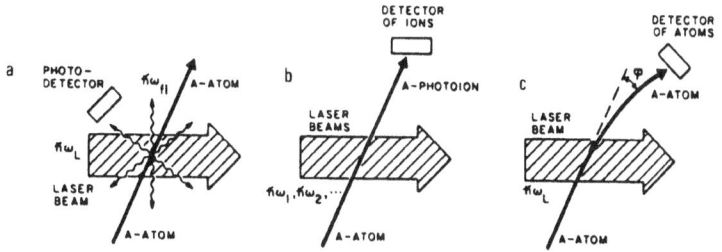

Figure 1.   Various mechanisms of laser-atom interaction which can be used for atomic detection: (a) fluorescence; (b) photoionization; (c) deflection.

The main question relating to the applicability of these methods to detecting rare isotopes is their ultimate selectivity S, i.e. the ability to detect a small number (N_B) of the rare isotope atoms B in the presence of much greater number (N_A) of the main isotope atoms A:

$$S = N_B / N_A \qquad (1)$$

The selectivity of these techniques stems from the presence of a small isotope shift $\Delta = \omega_A - \omega_B$ of the spectral line of one or several consecutive resonant transitions of the atom from its ground state to an excited

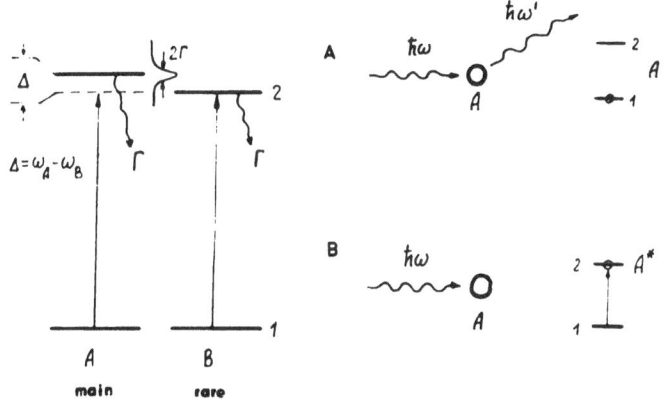

Figure 2.   Limitation of detection selectivity if rare B atoms in presence of abundant A atoms with close spectral line: (a) method of fluorescent detection; (b) method of excitation detection.

state (Figure 2). The fact that the width of any spectral line is finite naturally limits the selectivity because of the overlapping of the wings of the close spectral lines of the atoms A and B, but the character of the limitation largely depends on the particular technique used.

With the fluorescence technique, the fluorescence excitation line usually has a Lorentzian shape, i.e. the probability that a photon will be scattered by the atoms A when the laser frequency coincides with the center of the spectral line of the rare atoms B is defined (in excitation conditions far from saturation) by the expression:

$$W_{scat}^A = \sigma_o \, I\mathcal{L}(\Delta\Gamma) \simeq (1/2)(\mu_{12}\mathcal{E}/\hbar\Delta)^2 \Gamma \qquad (2)$$

where I is the laser radiation intensity (in photons/cm^{2}s), σ_o the cross section of the radiative transition $1 \rightarrow 2$ at a maximum, Γ the natural half-width or the rate of spontaneous decay of the atom into its initial ground state, μ_{12} the dipole moment of the transition $1 \rightarrow 2$, $\mathcal{E}$ the electric field strength of the light wave, and $\mathcal{L}(x) = 1/(1 + x^2)$ the Lorentzian function. By virtue of Eq.(2), the selectivity of the fluorescence detection of rare atoms is limited to the level of:

$$S_{fl} = W_{scat}^B / W_{scat}^A = (\Delta/\Gamma)^2 \qquad (\Delta \gg \Gamma) . \qquad (3)$$

For typical isotope shift and radiative linewidth values, $S_{fl} \simeq 10^4 - 10^6$, i.e. it is much lower than the required values indicated in Table 1.

The photoionization technique detects excited atoms by their subsequent transition into an ionized state. Its selectivity is therefore governed by the probability of the atomic excitation and not by that of photon reradiation. The probability of the atom A being excited on the wing of its spectral line is determined by the probability of absorption of two photons with the frequency $\omega = \omega_B$ and concurrent spontaneous reradiation of a photon with the shifted frequency $\omega_{f1} = 2\omega - \omega_A = \omega + \Delta$:[12]

$$W_{exc}^A \simeq (\mu_{12}\mathcal{E}/\hbar\Delta)^4 \Gamma \propto \Delta^{-4} . \qquad (4)$$

This expression differs from Eq.(2), which is the one commonly used for estimation purposes, by a stronger dependence of the excitation rate on the frequency shift Δ, the difference being substantial, but the selectivity of the method depends materially on the type of subsequent ionization of the excited atoms A. In the case of nonresonant simultaneous ionization where the difference between the energy of the first absorbed photon and the atomic excitation energy, $\hbar\Delta$, can be compensated for by the second absorbed photon, the selectivity is reduced to the former level defined by Eq.(3). Where the atomic ionization is a resonance process or delays from excitation pulse, a higher selectivity can be attained.[12]

It follows from the above simple estimates that none of the laser techniques in its simplest version can provide for the very high detection selectivity required. However, each of the techniques can be modified so as to ensure a substantial increase in selectivity. The methods that are possible here may be divided into two groups:

(1) Methods based o n a repeated resonance interaction of the atom with the laser light, wherein the atom reradiates a large number of photons. In this case, there occurs what may be called a "selectivity accumulation" on account of an effect which is possible only where a single atom reradiates a large number of photons following its repeated resonance excitation. This effect is made use of in the method of "fluorescent bursts"[13,14] and that of laser deceleration and cooling of atoms.[15]

(2) Methods based on a multistep resonance excitation of the atom in a multiple-frequency laser field, wherein use is made of the isotope shift on several consecutive resonant transitions. As a result of such a multistep resonance excitation, the selectivities S_i attained at each excitation and ionization stage are multiplied.[16]

The method of "fluorescent bursts" can be used to design a photon burst atom counter at the inlets of mass-spectrometers or AMS machines in order to increase the selectivity and overcome isobar interferences.[17,18]

The creation of high-power single-frequency CW dye lasers has allowed resonance multistep excitation to be applied for isotope-selective ionization of atoms at the outlets of mass spectrometers.[19,20] To increase the detection selectivity the method of laser resonance depletion of the intermediate state was suggested[21] and realized.[22]

The idea of multiplication of isotopic selectivity at each excitation stage is difficult to realize for the most interesting rare isotopes (Table 1) because it is hard to find for them a series of consecutive upward transitions with noticeable isotope shifts, such shifts being characteristic of the ground state only.

3. METHOD OF MULTISTEP COLLINEAR PHOTOIONIZATION OF ACCELERATED ATOMS

A universal way to overcome this difficulty and make the method of multistep resonance photoionization really applicable to the detection of rare isotopes was suggested in [23]. The idea of the method is based on collinear stepwise photoionization of a beam of accelerated atoms. It is common knowledge that acceleration of atoms in the form of ions under a given potential difference U and subsequent neutralization of the ions into atoms lead to the bunching of the longitudinal ionic velocities, hence to the narrowing of the Doppler width ο f all the spectral lines of the given atomic species (if viewed in a collinear fashion) as compared with the initial Doppler width $\delta \nu_D(0)$ at an ion source temperature of T:[24]

$$\delta \nu_D(V) / \delta \nu_D(0) = (1/2)(kT/eU)^{1/2} . \tag{5}$$

At $U = 10^4$ V the narrowing factor reaches 10^3. What is important is that the atoms in this case group in smaller volume of the phase space and their Doppler-free spectroscopy is effected without any loss of sensitivity.

Along with the narrowing of the Doppler width, there also occurs the Doppler shift of all the spectral lines of the accelerated atoms, which depends on the mass of the ion. As a result, there occurs an artificial "mass" shift on any spectral transition of the atom:

$$\delta \nu_{sh} / \nu_o = (1/c)(\sqrt{2eU})(1/\sqrt{M_1} - 1/\sqrt{M_2}) . \tag{6}$$

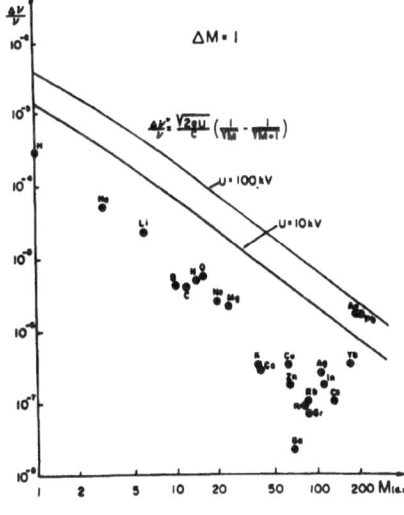

Figure 3. Values of relative isotope shift of resonance lines of various elements (dots). The kinematic isotope shift of accelerated atoms with energies 10 and 100 keV as a function of atomic mass (solid lines) for $\Delta M = 1$.

The circles in Fig.3 indicate the values of the natural isotope shift at resonant transitions of various elements and the artificial shift with the accelerating voltage. U = 10 kV and 100 kV. It can be seen that the kinematical isotopic shift may increase the isotope shift by 10 to 500 times. It is particularly important to apply the kinematic shift to elements in the middle of the periodic system.

What is most important is that such a kinematic isotope shift occurs on any spectral transition of the atom, this making it possible to realize in a natural way the idea of selectivity multiplication in multistep isotope-selective excitation.

Another important characteristic of this method is that it is possible to realize multistep isotope-selective excitation and ionization of elements with a high (~ 20 eV) ionization potential with the use of the available dye lasers. In this case the atoms should have high-lying metastable states.

General scheme of realization of this general idea is shown on Fig.4. A beam of fast atoms is formed by charge exchange of the ion beam with the atoms of the gas cell (alkali metal vapor, for instance). In the case of resonant charge exchange, when the ionization potential of the target atom is equal to the electron binding energy in the state, fast atoms are formed in the ground electronic state. In order to produce fast atoms with a high I.P. in the metastable state, one should choose such a gas cell that its ionization potential be close to the electron binding energy in this metastable state.

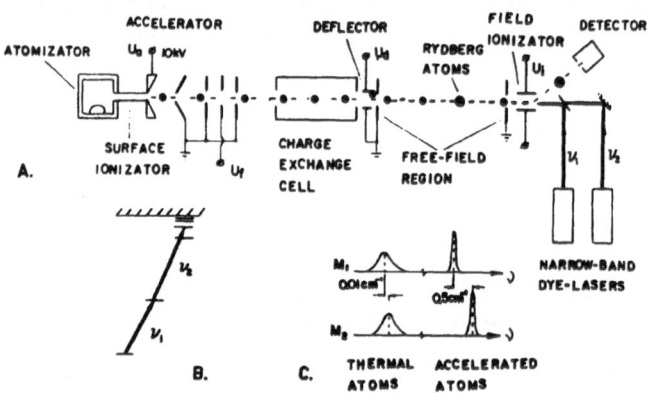

Figure 4. Laser detection of rare isotopes based on multistep collinear ionization.

4. EXPERIMENTAL DETECTION OF ^{40}K AND ^{3}He

The experiments on two-step collinear excitation of accelerated K isotopes to Rydberg states with their subsequent ionization in an electric field have proved the possibility of attaining the detection selectivity of the isotope ^{40}K equal to 10^{6}.[25] The isotope-selective excitation in this case was performed only at one step.

The K atom has two stable isotopes, ^{39}K and ^{41}K, with their relative concentrations of 93.2 and 6.8 %, respectively, and the radioactive isotope ^{40}K ($T_{1/2}$ = 1.3x10^{9} years) with its relative concentration of 0.012 %. A beam of neutral K atoms was formed in the ground electronic state $4S_{1/2}$ by resonant charge exchange of accelerated K ions on K atoms. A beam of K ions was formed by ionizing the neutral atoms on the surface of tantalum foil heated to 1100 K. When the atoms are accelerated to an energy of 4 keV, the absorption line Doppler width of the transition $4P_{3/2}$ $-21D_{5/2}$ is reduced to $\delta\nu_{D}$ = 5.9 MHz.

The atoms were ionized by an electric field from the Rydberg state $21D_{5/2}$ upon two-step excitation at the transitions $4S_{3/1}-4P_{3/2}$, $4P_{3/2}-21D_{5/2}$ by laser radiation propagating to meet the atomic beam. The values of natural isotope shift at these transitions are 127 and 220 MHz, respectively. At the first transition the isotope shift is smaller than the hyperfine level splitting. After the atoms are accelerated to 4 keV, the absorption spectra of the isotopes are fully separated and the isotope shift increases to 2.5 GHz at the first transition and to 4.1 HHz at the second one.

In the experiments all the isotopes were excited at the transition $4S_{3/2}-4P_{3/2}$ by laser radiation with a spectral width of 0.8 cm^{-1}. the isotope-selective excitation of the atoms to Rydberg states was performed only at the second transition $4P_{3/2}-21D_{5/2}$ using a narrow-band laser with a spectral width of 0.015 cm^{-1}.

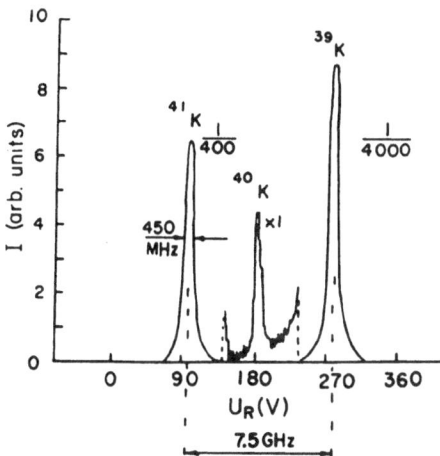

Figure 5. The ion signal of K isotopes as a function of detuning laser frequency and atomic resonance controlled by velocity variation of atoms.

The effect of multiplication of excitation selectivities at every step $S = S_1 \times S_2$ would be next natural experiment for a radical increase of the detection selectivity. Yet, it has been found, there is quite a number of elementary collisional processes to be taken into account in estimating the ultimate selectivity value.[26,27] The isotope ^{3}He, its natural concentration relative to the basic isotope ^{4}He being 10^{-6} was chosen to study these processes.

The ionization potential of the He atom E_1 = 24.6 eV, so the existing lasers cannot ensure multistep ionization from the ground state. The He atom has two long-lived metastable states": singlet 2^1S and triplet 2^3S. The charge exchange of accelerated ions on alkali metal vapor permits formation fast neutral atoms in metastable states. K atom is good candidate for such charge exchange. The electron binding energy of the states 2^1S and 2^3S of the He atom is close to the I.P. of K from the ground state $4S_{1/2}$. As a result of such charge exchange about 75 % of atoms are formed in the triplet metastable state 2^3S.[28]

The isotope-selective laser excitation of accelerated He atoms to the Rydberg n^3S and n^3D states with their subsequent ionization by an electric field was realized through the intermediate 3^3P level. To suppress this background noise, the method can be used in conjunction with various mass spectrometry techniques. in the present work, this was achieved through an additional time-of-flight separation of ^{4}He and ^{3}He, for which purpose use was made of an additional pulse intensity modulation of the continuous ion beam.[29] This made it possible to reduce the background noise due to ^{4}He$^+$

ions by a factor of 10^4 and to detect ^{3}He with a relative abundance as low as 10^{-8}. This limit was set by the number of the ^{3}He atoms excited by the low-repetition-rate tunable dye lasers used in the experiment and not by the background noise.

Figures 6a and b illustrate the isotopic selectivity of the process of ionization of He atoms. Figure 6a shows the ion signal obtained while varying the frequency ν_2 of the second-step excitation laser, the first-step excitation laser frequency $\nu_1 = 25671.4$ cm^{-1} being in resonance with ^{4}He ($t_d = 5.2$ μs). The background ions in the spectrum of ^{4}He are due to the two collisional processes.[29] The ion signal from ^{3}He with a relative content of 10^{-5} is shown in Fig.6b.

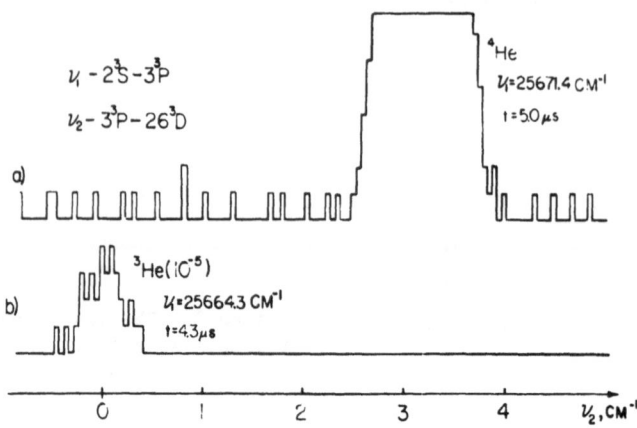

Figure 6.   Ion signal as a function of the second-step excitation laser frequency ν_2. (a) The first-step laser frequency is in resonance with ^{4}He. (b) The first-step laser frequency is in resonance with ^{3}He. Averaged over 64 pulses.

The described technique for detecting the rare isotope ^{3}He, based on combining the collinear isotope-selective photoionization of fast atoms with the time-of-flight separation of isotopes, makes it was possible to measure the relative content of ^{3}He as low as 10^{-10}. The determination of the isotopic composition of helium is essential to the solution of many problems in geochemistry and geophysics, cosmochemistry, oceanology, and so on.[30] The isotopic ration [^{3}He]/[^{4}He] in various objects ranges between 10^{-4} and 10^{-10}. The method of collinear laser ionization of fast atoms can be used to solve these problems and to detect other rare isotopes as well.

5. POSSIBILITY OF DETECTION OF ^{81,85}Kr

Let us consider here the possibility of detecting ^{81,85}Kr isotopes by means o f laser collinear ionization in a fast atomic beam. As seen from Table 1, the detection of these isotopes must be realized with the selectivity $S = 10^{10}$-10^{12}. Such a selectivity can be provided in the case of two-step excitation by narrow-band laser radiation due to a kinematic isotope shift. The experiments on He isotope detection, however, show that the collisional processes of excitation and ionization of fast atomic beams on the molecules of residual gas represent a considerable limitation. The time-of-flight isotope separation due to atomic beam modulation cannot be used for such heavy elements as Kr . Yet it is rather convenient to apply collinear ionization in combination with preseparation in a mass separator. For this purpose, at the mass separator outlet the ions of the isotope to be detected separated from the basic isotope are directed into a charge exchange cell where they are transformed to atoms in the ground or the metastable states. If the attenuation of the line wing of the basic isotope on the mass of the isotope being detected due to mass separator $S_M = 10^4$, the spectral selectivity should be 10^6 to 10^8 to detect these isotopes.

The Kr atom has six stable isotopes: ^{86}Kr (17.3 %), ^{84}Kr (57 %), ^{83}Kr (11.5 %), ^{82}Kr (11.6 %), ^{80}Kr (2.25 %), and ^{78}Kr (0.35 %). The ionization potential of Kr I = 14 eV. Like in the case of He, the high-lying metastable states $1S_3$ or $1S_5$ can be used, too, as starting ones for laser multistep ionization. Pairs of alkali metals, such as K, Rb, and Cz, whose ionization potential is similar to the binding energy of electron in the state $1S_5$. The most effective transitions in the Kr atom where the radiation of dye lasers pumped by Cu lasers can be used. When the first-step radiation has a wavelength of 760.2 nm (the first-stage wavelengths are given for thermal atoms), the green line of Cu laser (λ_2 = 510.6 nm) can be used at the second step. The resonance tuning to the transition to a Rydberg state with n - 24,25 is realized by varying the atomic velocity. The nuclear spin of all the stable Kr isotopes, except ^{83}Kr, is equal to zero. The spectra o;f the rare isotopes ^{81}Kr (I = 7/2), ^{85}Kr (I = 9/2), and the stable isotope ^{83}Kr (I = 9/2) are characterized by a hyperfine structure. The magnitude of the natural isotope shift is about 50 MHz per one atomic mass[31] which is much smaller than that of hyperfine splitting.[32] The spectra of absorbed isotopes at the first stage (λ_1 = 788.5 nm) accelerated to an energy of 50 keV. The hyperfine splitting constants for this line are known only for the isotopes ^{83}Kr and ^{85}Kr. In this case the absorption spectra of the isotopes are fully separated. The magnitude of isotope shift of the most efficient component of ^{85}Kr is 3.7 GHz relative to ^{84}Kr and 1.4 GHz relative to ^{86}Kr. The radiative absorption linewidth is 6.12 MHz (τ_{rad} = 26 ns) and the laser excitation selectivity S_1 will be 2.1×10^5. The selectivity attained at the second excitation stage may be not less in magnitude.

REFERENCES

1. V. S. Letokhov and V. P. Chebotayev, "Nonlinear Laser Spectroscopy", Vol.4, Springer Series in Optical Sciences, Springer-Verlag, Berlin, Heidelberg, New York (1977).
2. D. H. Auston and K. B. Eisenthal, eds., "Ultrafast Phenomena IV", Vol.38, Springer Series in Chemical Physics, Springer-Verlag, Berlin, Heidelberg, New York, Tokyo (1984.
3. V. S. Letokhov, "Laser Photoionization Spectroscopy", Academic Press, Orlando (1987).
4. V. S. Letokhov, in "Tunable Lasers and Applications", A. Mooradian, T. Jaeger, and P. Stokseth eds., Springer Series in Optical Sciences, Vol.3, p.122, Springer-Verlag, Berlin, Heidelberg, New York (1976),
5. V. S. Letokhov, in "Chemical and Biochemical Applications of Lasers", Vol.5, p.1, C. B. Moore, ed., Academic Press, New York, London etc. (1980).
6. V. S. Letokhov, Comm. Atom. Molec. Phys., 10:257 (1981).
7. W. F. Libby, "Collected Papers", Vol.1, Tritium and Radiocarbon, R. Berger and L. M. Libby eds., Geo Science Analytical, Santa Monica (1980).
8. D. E. Nelson, R. G. Korteling, and W. R. Scott, 198:507 1977).
9. R. A. Muller, Science, 196:489 (1977).
10. Proc. 4th Int'l Symp. on Accelerator Mass spectromety, Niagara-on-the-Lake, Ontario, Canda (1978), Nucl. Instr. Meth. in Phys. Res., B29:1-445 (1987).

11. V. I. Balykin, G. I. Bekov, V. S. Letokhov, and V. I. Mishin,
    Uspekhi Fiz. Nauk (Russ), 132:293 (1980),
    [Sov. Phys. Usp., 23:651 (1980)].
12. A. A. Makarov, Zh. Eksp. Teor. Fiz. (Russ), 85:1192 (1983).
13. G. W. Greenless, D. L. Clark, S. L. Kaufmann, D. A. Lewis, J. F. Tonn,
    and S. L. Broadhurst, Opt. Comm., 23:236 (1977).
14. V. I. Balykin, V. S. Letokhov, V. I. Mishin, and V. A. Semchishen,
    Pis'ma JETF (Russ), 26:492 (1977);
    Zh. Eksp. Teor. Fiz. (Russ), 77:2221 (1979).
15. V. I. Balykin, V. S. Letokhov, and V. G. Minogin,
    Appl. Phys., B33:247 (1984).
16. V. S. Letokhov and V. I. Mishin, Opt. Comm., 29:168 (1979).
17. R. A. Keller, D. S. Bomse, and D. A. Cremers, Laser Focus,
    October:75 (1981).
18. W. M. Fairbank Jr., Nuclear Instr. and Meth., B29:407 (1987).
19. B. A. Bushaw and G. K Gerke, in "Resonance Ionization Spectroscopy",
    T. B. Lucatorto and J. E, Parks eds., Institute of Physics Conference
    Series Number 94:277 (1988).
20. B. L. Fearey, D. C. Parent, R. A. Keller, and C. M. Miller,
    JOSA, B6:   (1989).
21. A. A. Makarov, Appl. Phys., B29:287 (1982);
    Sov. J. Quantum Electron. (Russ), 10:1127 (1983).
22. G. R. Janik, B. A. Bushaw, and B. D. Cannon,
    Optics Lett., 14:266 (1989).
23. Yu. A. Kudriavtzev and V. S. Letokhov, Appl. Phys., B29:219 (1982).
24. K. R. Anton, S. L. Kaufmann, W. Klempt, G. Moruuzi, R. Heugart,
    E. W. Otten, and B. Schinzler, Phys. Rev. Lett., 40:642 (1978).
25. Yu. A. Kudriavtzev, V. S. Letokhov, and V. V. Petrunin,
    Pis'ma JETF (Russ), 42:23 (1985), [JETP Lett., 42,26 (1985)].
26. Yu. A. Kudriavtzev and V. V. Petrunin, Sov. Phys. JETP, 67:691 (1988).
27. Yu. A. Kudriavtzev, V. S. Letokhov, and V. V. Petrunin,
    Opt. Commun., 68:25 (1988).
28. C. Reynaud, Y. Pommier, V. N. Tuan, and M. Barat,
    Phys. Rev. Lett., 43:579 (1979).
29. Yu. A. Kudriavtzev, V. V. Petrunin, V. M. Sitkin, and V. S. Letokhov,
    Appl. Phys., B48:93 (1989).
30. B. A. Mamyrin and I. N. Tokstikhin, "Helium Isotopes in Nature",
    Elsevier, Amsterdam (1984).
31. D. A. Jackson, JOSA, 69:503 (1979).
32. H. Gerhard, F. Jeschonnek, W. Makat, E. Matthias, H. Rinneberg,
    F. Schneider, A. Timmerman, R. Wenz, and P. J. West,
    Hyperfine Interac.,9:175 (1981).

NUCLEAR GROUND-STATE PROPERTIES FROM LASER AND MASS

SPECTROSCOPY

H.–Jürgen Kluge

Institut für Physik, Universität Mainz, D-6500 Mainz
Fed. Rep. of Germany

INTRODUCTION

Atomic physics played an important role in establishing our present-day knowledge on the atomic nucleus. Especially mass spectrometry and optical spectroscopy were the main sources of information on nuclear properties in the early days of nuclear physics. Still now, precise information on nuclear masses (or binding energies) are obtained by mass spectrometry whereas mass differences between two isotopes are usually determined by nuclear-spectroscopy techniques via a determination of the Q-value of nuclear reactions or decay. Almost all our information on the nuclear spins I, the nuclear magnetic dipole moment μ_I, the spectroscopic quadrupole moment Q_s, and the changes in the mean-square charge radii $\delta < r^2 >$, which characterize a nuclear ground-state, stem from atomic spectroscopy [1]. These quantities are accessible via a determination of the hyperfine structure (HFS) in atomic levels or of the isotope shift (IS) in optical transitions. The development of on-line mass separators where long chains of isotopes are available for investigation, the invention of tunable lasers, and recently the development of ion traps coupled to isotope separators led to a renaissance of atomic spectroscopy applied to nuclear physics problems.

LASER SPECTROSCOPY OF SHORT-LIVED ISOTOPES

Laser spectroscopy of short-lived isotopes started some 15 years ago. Today, charge radii [2], spins, and moments [3] have been measured for about 30 elements in long isotopic chains for about 500 nuclear ground or isomeric states (Fig.1). Several techniques have been applied:

- laser-excited fluorescence in resonance cells [4,5]

- laser-excited fluorescence from collimated atomic beams [6,7]

- laser-induced optical pumping of an collimated atomic beam with a Stern-Gerlach analyzer [7,8]

- resonance ionization (mass) spectroscopy with and without pulsed-laser induced desorption [9,10]

- collinear laser spectroscopy with fluorescence detection, particle detection or detection of the nuclear polarization induced by optical pumping [11-13].

Applied Laser Spectroscopy, Edited by W. Demtröder and
M. Inguscio, Plenum Press, New York, 1990

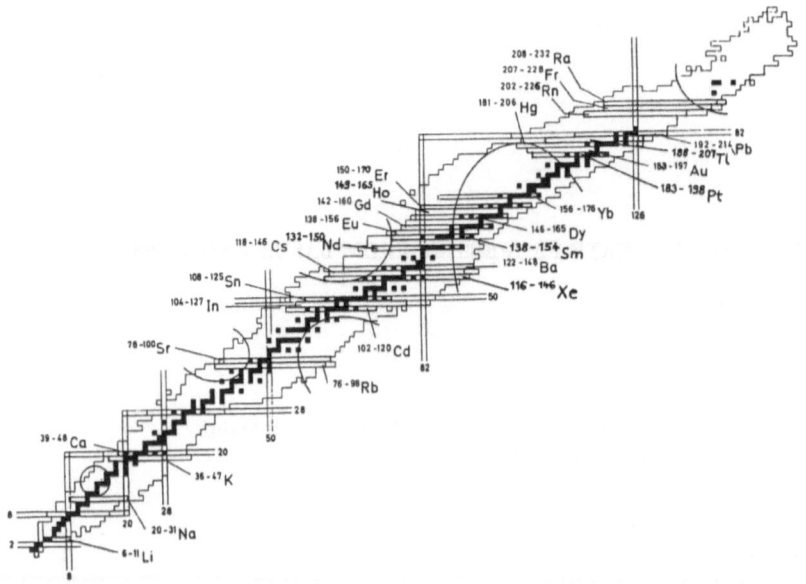

Fig.1. Chart of nuclei. Black squares represent stable nuclei. The boundary line encloses those nuclei which are known to exist. The neutron and proton shell closures are indicated as well as the regions of strong nuclear deformations (curved line). Those isotopes are marked which have been investigated by optical spectroscopy in long isotopic chains leading far off stability.

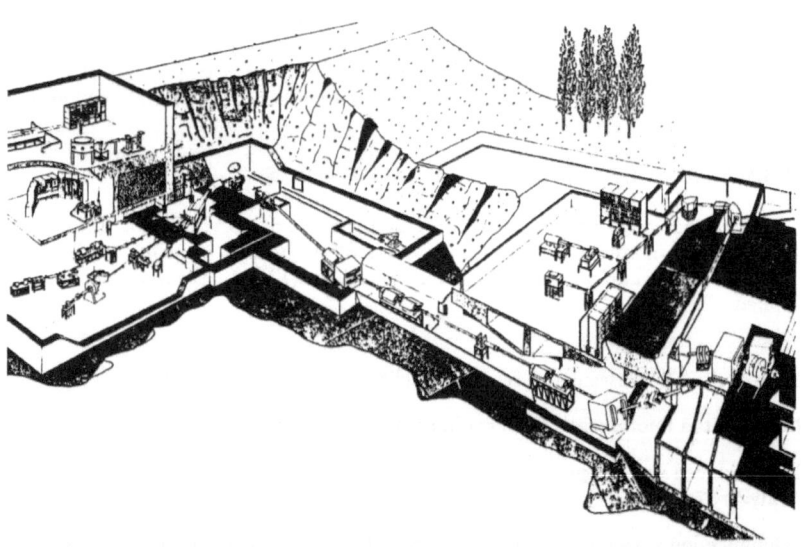

Fig.2. On-line isotope separator facility ISOLDE at CERN/Geneva. At right, the synchrocyclotron is shown, which delivers 600 MeV protons to ISOLDE–2 (left) where the mass and collinear laser experiments are placed, and the ISOLDE–3 (middle,right) where resonance ionization mass spectroscopy on Au and Pt was performed.

Since laser spectroscopy of short-lived isotopes is the subject of several recent reviews [1,4-13] and was addressed in a conference in 1982 dedicated to lasers in nuclear physics [14], only some few examples will be given in the following.

On-line isotope separation

Figure 2 shows the ISOLDE facility at CERN/Geneva. It is an on-line isotope separator [15] which delivers mass-separated ion beams of almost every element below uranium with intensities of up to 10^{11} particles per second and per mass number [16]. In some cases, over 40 isotopes of the same element are available for investigation. The synchro-cyclotron (SC) at CERN delivers protons with an energy of 600 MeV and an intensity of some μA to the ISOLDE target, where radioactive isotopes are produced in a spallation or fission reaction. These products diffuse out of the target matrix into an ion source. After ionization, acceleration to 60 keV and mass separation, the ions are delivered through beam lines to the different experiments.

Collinear laser spectroscopy

Collinear laser spectroscopy is the technique which produced most of the data on short-lived isotopes. It was developed by Kaufman [17] and independently by Wing et al. [18]. It's advantage is that it works directly on the ion beam delivered by an on-line isotope separator (Fig. 3) and combines high sensitivity with high resolution. This is due to the shrinking of the longitudinal Doppler width $\Delta\nu_D$ in the fast ion or (after charge exchange) fast atomic beam as compared to the Doppler width $\Delta\nu_D(0)$ before acceleration. The reduction is given by

$$\Delta\nu_D = \frac{1}{2}\sqrt{\frac{kT}{eU}}\,\Delta\nu_D(0) \tag{1}$$

and reaches about a factor of 1000 for T = 2000 K and U = 60 kV. Typical line widths as obtained in experiments range from 10 to 50 MHz. Minimum ion beam intensity of about 10^4 particles per second is required in case of fluorescence detection as shown in Fig.3. The

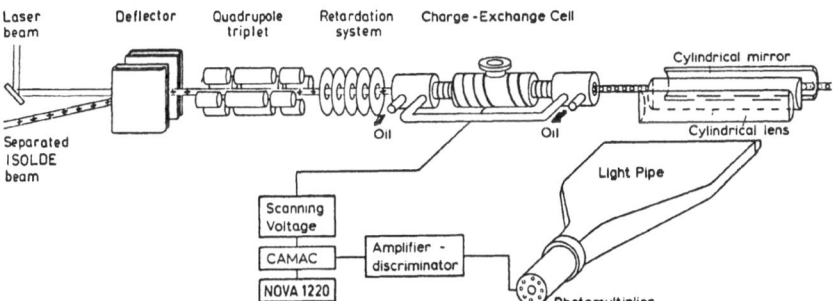

Fig.3. Schematic diagram of a collinear-laser spectroscopy experiment with fluorescence detection. The ion beam is made collinear with the cw dye laser beam. The ions are neutralized in a charge exchange cell containing alkaline vapour. After neutralization the fast atomic beam is excited by laser light and the fluorescence is observed via a light guide by a photomultiplier. Instead of scanning the laser wavelength, the Doppler shift is tuned by applying a voltage to the charge exchange cell.

sensitivity can be increased drastically by detection of particles instead of the fluorescence photons or by looking for the asymmetry in the decay of polarized nuclei obtained by optical pumping in collinear geometry [1,13].

Resonance ionization mass spectroscopy

Resonance ionization spectroscopy (RIS) was first applied to the study of short-lived isotopes at Leningrad [19]. The method is described in more detail in my second talk at this Summer School. Figure 4 shows the experimental set-up installed at ISOLDE for the investigation of gold and platinum isotopes [20]. A similar experiment has been performed by a Montreal–Orsay collaboration [21]. Since these isotopes were not available at ISOLDE as on-line beam but only in the decay of a primary Hg ion beam, collinear spectroscopy could not be applied. Instead, a mass-separated Hg ion beam is implanted into a graphite target. After decay to the Au or Pt daughter, the radioactivity is released by pulsed-laser induced desorption.

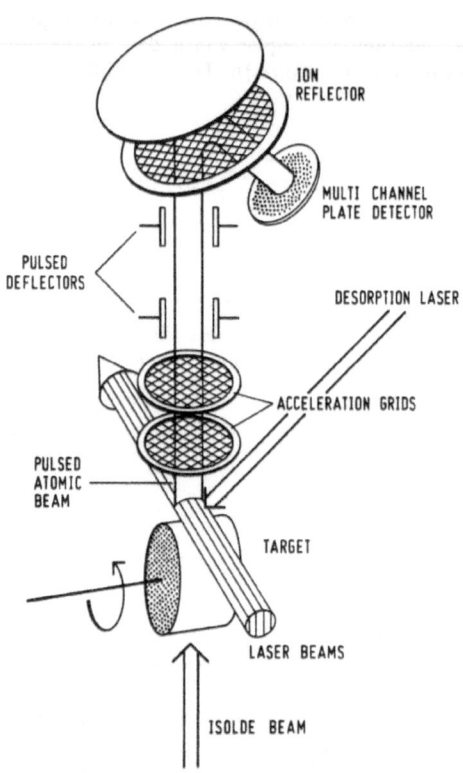

Fig.4. Schematic diagram of resonance ionization mass spectroscopy for the investigation of Au and Pt isotopes. The mass-separated Hg ion beam is implanted into a wheel made of graphite. After decay to the daughter nucleus Au or Pt the wheel is turned. The atoms are evaporated by pulsed-laser induced desorption, resonantly step-wise excited, and ionized, and finally detected with the help of a time-of-flight spectrometer.

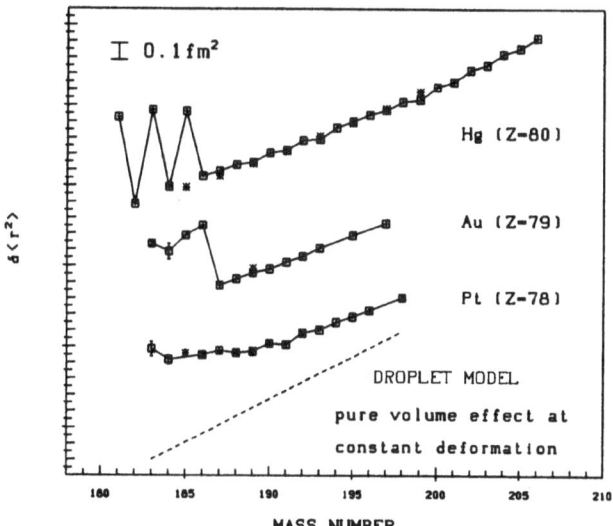

Fig.5. Changes of mean-square charge radii in the isotopic sequences of Hg, Au, and Pt as a function of mass number. Isomeric states are indicated by asterics. The dashed line indicates the variation of a hypothetically spherical nucleus according to the droplet model.

The pulsed atomic beam is then step-wise resonantly excited and finally ionized by the light of three pulsed tunable dye lasers. The photoions created in resonance are detected mass-selectively by a time-of-flight (TOF) spectrometer. This completes the method to resonance ionization mass spectroscopy (RIMS).

Figure 5 shows the charge radii of the Au and Pt isotopes measured by RIMS together with the corresponding data on mercury. These results show the usual shrinking of the nucleus as a function of decreasing neutron number. The dashed line indicates what is expected for a spherical nucleus in the droplet model (which is a refinement of the liquid drop model). An unexpected break in the rather smooth shrinking is observed in the very light Hg and Au isotopes: Suddenly, the charge radius increases drastically and flips forth and back in the case of the very neutron-deficient Hg isotopes. This behaviour is due to a nuclear shape transition from nearly spherical nuclear shape to strong deformation resulting in an increased nuclear radius as observed clearly via the IS. This shape transition is closely related to nuclear-shape coexistence at almost degenerate energies. This can clearly be seen in case of ^{185}Hg where two different nuclear shapes are found within the same nucleus.

MASS SPECTROMETRY

The mass or the binding energy of a quantum mechanical system is one of, or even the most fundamental property of this system because it reflects all forces acting. Masses of nuclei far from stability are measured generally as mass differences (Q-values) between isotopes in nuclear decays or reactions. They have the disadvantage that the nuclear level schemes have to be known and the mass of a nucleus is determined by a sum of many mass differences. Direct mass determinations have only be performed by the Orsay group with the help of a Mattauch-Herzog double-focusing mass spectrometer [22,23]. Another, new approach is to measure the mass M of an radioactive isotope via a determination of the cyclotron frequency of the ion in a homogenous magnetic field by

$$\omega_c = \frac{e}{M}B \ .$$

(2)

The resolving power $R = \omega_c/\Delta\omega_c$(FWHM) should be for highest precision as large as possible, i.e., the cyclotron frequency should be as high and the linewidth $\Delta\omega_c$ as narrow as possible. The ultimate linewidth is given by the Fourier limit

$$\Delta\omega_c \cdot \Delta T = 2\pi$$

(3)

for an interaction time ΔT of the ions with the radio frequency (RF) field. Hence, a high magnetic field and long interaction time are required for high resolving power. This can be achieved by storing the ions in a Penning trap placed in the field of a superconducting magnet. With B = 6 T, ΔT = 1 s one calculates for a medium-mass ion a resolving power of R = 10^6 exceeding that of all double-focusing conventional mass spectrometers [24].

Such an experiment has been set-up at ISOLDE. Figure 6 shows the principle. As in a big accelerator complex, many devices with different tasks are coupled together. However, in the case of mass measurements in a Penning trap, we start with high-energy particles and try to have at the very end ions with very low energy. The purpose of the Paul or RF quadrupole traps is retardation of the 60 keV ions, capture, cooling, and bunched ejection [25]. This part of the set-up is just under test at CERN. For the time being, the radioactive ions are stopped on a foil and then surface-ionized. The ions are captured in a Penning trap. Here cooling of heavy ions is performed by a novel cooling mechanism [26]: It is a combination of buffer gas cooling and centering of the ion cloud by an azimuthal quadrupole field at the frequency ω_c. This cooling technique has in addition the advantage of being mass-selective.

The ions are ejected out in a bunch, transfered and captured with high efficiency in a second Penning trap [27]. Here, the cyclotron motion of the stored ions is excited. The cyclotron

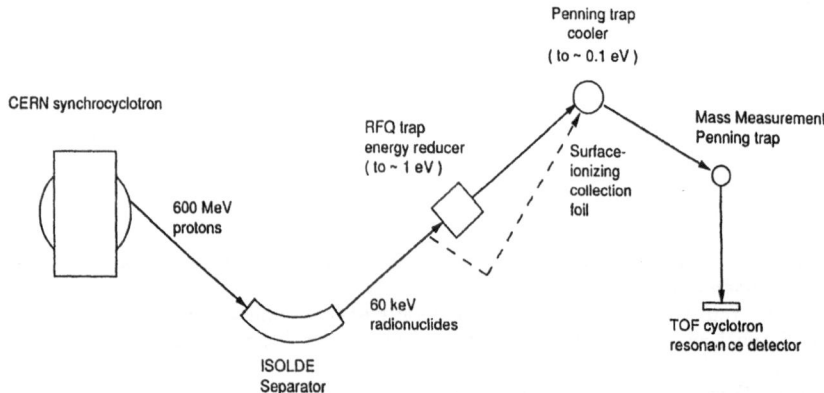

Fig.6. Schematic diagram of the Penning trap mass spectrometer set up at ISOLDE/CERN. For details see text.

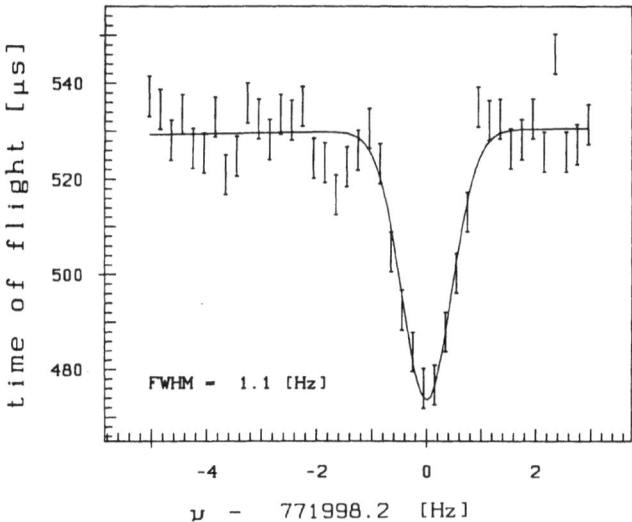

Fig.7. Cyclotron resonance of ^{118}Cs ($T_{1/2} = 14$ s) at $\nu_c = 771999.2$ Hz. The solid line represents a fit with a Gaussian. The linewidth of $\Delta\nu = 1$ Hz is the Fourier limit due to the interaction time of the RF with the stored ions of $\Delta T = 0.9$ s. A resolving power of R $= 700\,000$ is obtained.

resonance is detected by a TOF technique [28]. A mass resolving power of maximum R $= 2.3 \cdot 10^6$ was obtained for ^{133}Cs. Radioactive Cs isotopes could be measured in the mass range $137 > A > 118$ with a resolving power between 700 000 and 1 400 000 depending on the interaction time of the ions with the RF. As an example, Fig. 7 shows a cyclotron resonance of ^{118}Cs ($T_{1/2} = 14$ s). An accuracy of $2 \cdot 10^{-7}$ could be obtained corresponding to about 20 keV which is much better than achieved by other techniques far off stability.

CONCLUSION

On-line laser spectroscopy and mass spectrometry in combination with powerful iso-tope production and separation facilities allow to study the nuclear properties of ground and isomeric states of short-lived, exotic isotopes which are only available in small quan-tities. Since the model-independent data are collected in long isotopic chains reaching far away from the valley of nuclear stability, these data are valuable for testing nuclear mo-dels. Furthermore, new phenomena might be discovered which do not show up in stable or long-lived isotopes or which can only be detected as deviation from a systematic trend of neighbouring isotopes. Examples are the shape transition detected by optical spectroscopy in the Hg and Au isotopes, the effect of the coupling of the valence neutron to the nuclear core as observed in the spectroscopic quadrupole moments of the I = 13/2 Hg isomers, or the octupole deformation observed first by mass spectrometry [1].

The first isotopes which have been investigated by laser spectroscopy and mass spec-trometry were those elements which are easily ionized and have high vapour pressure. Now the first isotopes of refractory elements were studied by RIMS. Still a lot has to be done by laser spectroscopy on low-Z isotopes. Here, only few elements are studied until now because small HFS splitting and only little contribution of the volume effect to the IS require both high resolution. On the other hand, all isotopes delivered by on-line separators will be accessible to mass spectrometry, if the Paul trap works for retarding and accumulating ISOL beams.

Many surprises might be expected.

References

[1] E.W. Otten. *Treatise on Heavy Ion Science 8: Nuclei far from Stability (Ed.: D.A. Bromley)*, page 515. Plenum Press, New York, 1989.

[2] P. Aufmuth, K. Heilig, and A. Steudel. *Atomic Data and Nuclear Data Tables*, 37:455, 1987.

[3] P. Raghavan. Table of nuclear moments. *Atomic Data and Nuclear Tables*, 42:189, 1989.

[4] H.-J. Kluge. *Optical Spectroscopy of Short-Lived Isotopes*, chapter 17, pages 727–765. Progress in Atomic Spectroscopy B. Plenum Press, New York, 1979.

[5] H.-J. Kluge. Optical Measurements of Ground State Properties of Short–Lived Nuclei in Resonance Cells. In *Proc. of Int. Workshop on 'Hyperfine Interactions', Kanpur, Indien, 1984*. Hyperfine Interactions 24:69, 1985.

[6] G. Meisel, K. Bekk, H. Rebel, and G. Schatz. *Hyperf. Interact.*, 38:723, 1987.

[7] P. Jaquinot and R. Klapisch. *Rep. Progr. Phys.*, 42:773, 1979.

[8] C. Thibault. Laser Spectroscopy and Mass Measurements on Alkali Isotopes. *Hyperf. Interact.*, 24:95, 1985.

[9] G.D. Alkhazov, A.E. Barzahk, V.P. Denisov, V.S. Ivanov, I. Ya. Chubukov, V.N. Buyanov, V.S. Letokhov, V.I. Mishin, S.K. Sekatsky, and V.N. Fedoseev. Investigation of nuclear charge radii and electromagnetic moments of rare earth elements by laser spectroscopy. In I.S. Towner, editor, *Proc. of 5th Int. Conf. on 'Nuclei far from Stability', Rousseau Lake, Ontario, Canada (1987)*. AIP conference proceedings 164:415, 1988.

[10] H.-J. Kluge. Resonance Ionization Mass Spectroscopy for Nuclear Research and Trace Analysis. In *Proc of Workshop on 'Modern Optics, Lasers and Laser Spectroscopy', Kanpur, India, Januar 1987*. Hyperfine Interactions 37:347, 1987.

[11] E. W. Otten. *Nucl. Phys.*, A 354:471c, 1981.

[12] R. Neugart. *Hyperf. Interact.*, 24:159, 1985.

[13] R. Neugart and the ISOLDE Collaboration. *Collinear fast-beam laser spectroscopy, in: Progress in Atomic Spectroscopy D (Ed.: H.-J. Beyer and H. Kleinpoppen)*. Plenum Press, New York, 1987.

[14] C. E. Bemis and H. K. Carter, editors. *Lasers in Nuclear Physics*. Harwood Academic Publishers, New York, 1982.

[15] H. L. Ravn and B. W. Allardyce. *On–Line Mass Separators (Ed.: D.A. Bromley)*, page 363. Plenum Press, New York, 1989.

[16] H.-J. Kluge. *ISOLDE User's Guide*. CERN, Geneva, CERN Yellow Report 86–05 edition, 1986.

[17] S.L. Kaufmann. *Optics Communications*, 17:309, 1976.

[18] W. H. Wing, G. A. Ruff, W. E. Lamb, and J.J. Spezewski. *Phys. Rev. Lett.*, 36:1488, 1976.

[19] G.D. Alkhazov et al. *JETP Lett.*, 37:274, 1983.

[20] U. Krönert, St. Becker, G. Bollen, M. Gerber, Th. Hilberath, H.-J. Kluge, G. Passler, and the ISOLDE Collaboration. Observation of Strongly Deformed Ground-State Configurations in [184]Au and [183]Au by Laser Spectroscopy. *Z. Phys.*, A331:521-522, 1988.

[21] H. T. Duong, J. Pinard, S. Liberman, G. Savard, J. K. P. Lee, J. E. Crawford, G. Thekkadath, F. Le Blanc, P. Kilcher, J. Obert, J. Oms, J. C. Putaux, B. Roussiere, J. Sauvage, and the Isocelle Collaboration. Shape transitions in neutron deficient Pt isotopes. *Phys. Lett.*, B 217:401, 1985.

[22] M. Epherre, G. Audi, C. Thibault, R. Klapisch, G. Huber, F. Touchard, and H. Wollnik. *Phys. Rev.*, C19:1504, 1979.

[23] G. Audi, A. Coc, M. Epherre-Rey-Campagnolle, G. Le Scornet, C. Thibault, F. Touchard, and the ISOLDE Collaboration. *Nucl. Phys.*, A449:491, 1986.

[24] G. Bollen, P. Dabkiewicz, P. Egelhof, T. Hilberath, H. Kalinowsky, F. Kern, H. Schnatz, L. Schweikhard, H. Stolzenberg, R. B. Moore, H.-J. Kluge, G.M. Temmer, G. Ulm, and the ISOLDE Collaboration. First Absolute Mass Measurements of Short-Lived Isotopes. In *Proc. of 'Workshop on Modern Optics, Lasers and Laser Spectroscopy', Kanpur, India, 1987 in: Hyperfine Interactions 38:793*, 1987.

[25] R. B. Moore and S. Gulick. 'The transfer of continuous beams and storage ring beams into electromagnetic traps' in: Workshop and Symposium on the Physics of Stored and Trapped Particles, AFI Stockholm june 1987. *Phys. Scr.*, T22:28-35, 1988.

[26] G. Savard, St. Becker, G. Bollen, H.-J. Kluge, R. B. Moore, L. Schweikhard, H. Stolzenberg, and U. Wiess. A new cooling mechanism for heavy ions in a Penning trap. *submitted to Phys. Rev. Lett.*, 1989.

[27] H. Schnatz, G. Bollen, P. Dabkiewicz, P. Egelhof, F. Kern, H. Kalinaowsky, L. Schweikhard, H. Stolzenberg, H.-J. Kluge, and the ISOLDE Collaboration. In-Flight Capture of Ions into a Penning Trap. *Nucl. Instrum. Meth.*, A251:17, 1986.

[28] G. Gräff, H. Kalinowsky, and J. Traut. A Direct Determination of the Proton Electron Mass Ratio. *Z. Phys.*, A297:35-39, 1980.

RESONANCE IONIZATION MASS SPECTROSCOPY FOR TRACE ANALYSIS

H.–Jürgen Kluge

Institut für Physik, Universität Mainz, D-6500 Mainz
Fed. Rep. of Germany

INTRODUCTION

My first lecture at this Summer School on Applied Laser Spectroscopy dealt with the determination of nuclear ground-state properties, i.e. atomic mass M, the nuclear spin I, the magnetic dipole moment μ_I, the spectroscopic quadrupole moment Q_s, and the changes in the mean-square charge radius $\delta\langle r^2\rangle^{A,A'}$ between isotopes with mass number A and A′. These quantities can be determined for stable, long-, or short-lived isotopes by mass spectrometry and optical spectroscopy. In the latter case, the hyperfine structure (HFS) and the volume effect of the isotope shift (IS) are determined in atomic levels or optical transitions. The state of the art mainly concerning short-lived nuclei is described in a recent review[1].

If these quantities are known, we can use them as fingerprints for an atom searched for. The atomic mass, the optical transition wavelengths, HFS, and IS are unique characteristics for each isotope and might be therefore used for identification. However, sensitive and selective methods are required. Since trace analysis is a well established subject in chemistry, physics, and radiochemistry, it is not easy to beat already existing detection techniques. Only new revolutionary, technical developments enable to do so. Such an invention is that of the laser which combines high intensity with small bandwidth, spatial coherence, and tunability. These outstanding properties and the large cross section σ for optical transitions with wavelength λ given by

$$\sigma = \lambda^2/2\pi \tag{1}$$

will push the detection limit for many isotopes and hence elements from the present nanogram or picogram region ($\simeq 10^{13} - 10^{10}$ atoms) to or even below the femtogram limit ($\leq 10^7$ atoms).

Three requirements have to be met for trace analysis in the femtogram regime:

1. Extreme sensitivity, i.e. detection efficiency.

2. High selectivity against interfering contaminations in the sample under investigation.

3. Very low background.

A technique which fulfills these conditions was proposed as early as 1972[2].

Applied Laser Spectroscopy, Edited by W. Demtröder and
M. Inguscio, Plenum Press, New York, 1990

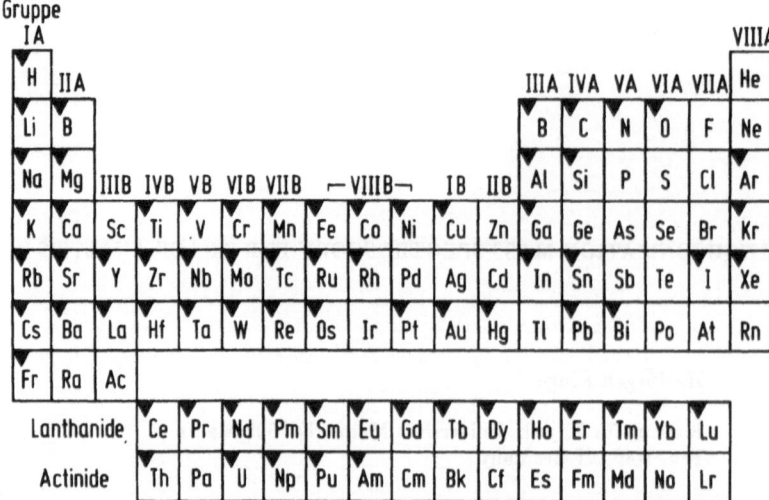

Figure 1. Periodic table of elements. The black triangles indicate those elements where resonance ionization spectroscopy has already been performed.

It is resonance ionization spectroscopy (RIS) or resonance ionization mass spectroscopy (RIMS): An atom is step-wise, resonantly excited and finally ionized (RIS). Quite often the ions created in resonance are mass-separated before detection (RIMS). In the meantime this technique has been applied to the investigation of a large number of elements as shown in Fig. 1 and is well documented in books[3,4] and the proceedings of the conference series "Resonance Ionization Spectroscopy"[5]. The next symposium of this kind will be held in September 1991 in Varese, Italy.

PRINCIPLE OF RESONANCE IONIZATION SPECTROSCOPY

Figure 2 shows the scheme for resonance ionization: A three-colour, two- or three-step resonant excitation is taken as an example. The three schemes shown differ in the last step, the ionization step. This represents usually the bottleneck of the excitation path due to the low cross section for non-resonant photoionization (left) which is of the order of $\sigma = 10^{-17} - 10^{-19}$ cm^2. Two to four orders in cross section can be gained by exciting to a Rydberg level (middle), which is ionized by an electric field, or to an autoionizing state (right). Since the electron has to be promoted to the next excited atomic level before decay by resonance fluorescence takes place ($\tau \simeq 10^{-8}$ s), pulsed tunable dye lasers have to be applied in general. But in selected cases, where optical pumping can be avoided, and small laser bandwidth is required, also cw tunable lasers might be applied.

The selectivity of a spectroscopic technique is defined as the supression of an unwanted contamination in respect to the spezies under study. If the laser is tuned to resonance of the trace element at frequency ν_1 resulting in a signal of hight $H_1(\nu_1)$, an interference (Fig. 3) might be produced by the tail of a neighbouring resonance at ν_2 of another element or isotope with an signal height $H_2(\nu_1)$. The selectivity is defined by $S = H_1(\nu_1)/H_2(\nu_1)$ under the assumption of an equal number of atoms of both spezies. The general shape of the resonance will be a Voigtian, a convolution of a Gaussian profile (Doppler broadening) and a Lorentzian (natural line width).

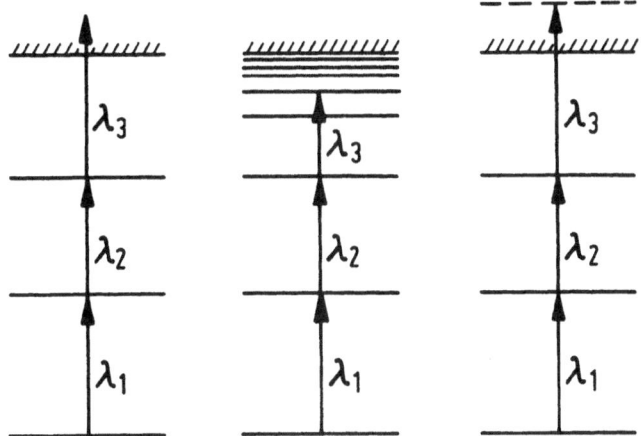

Figure 2. Scheme of resonance ionization spectroscopy for the case of a three-colour exitation to the continuum (left), to a Rydberg level (middle), or to an autoionizing state (right).

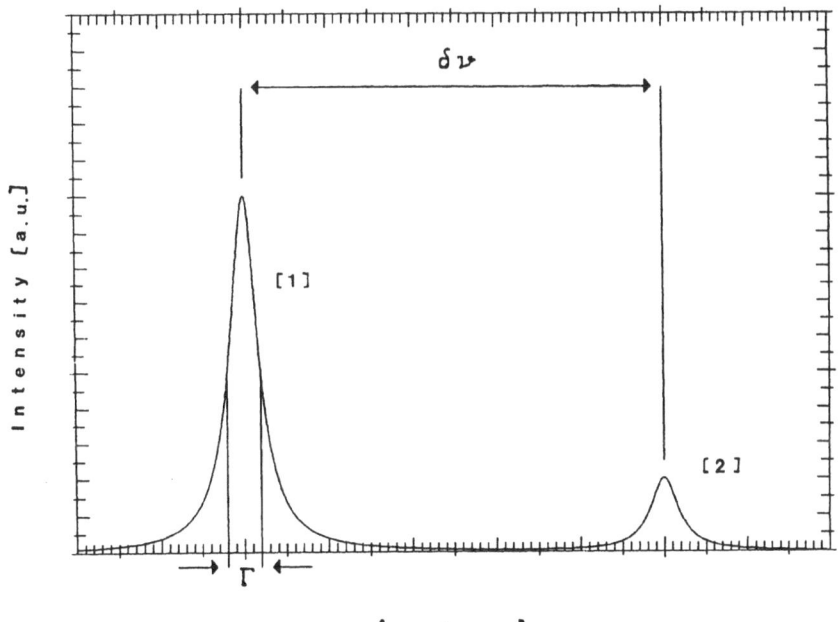

Figure 3. Selectivity in spectroscopic experiments limited by the long tail of the Lorentzian resonance profile.

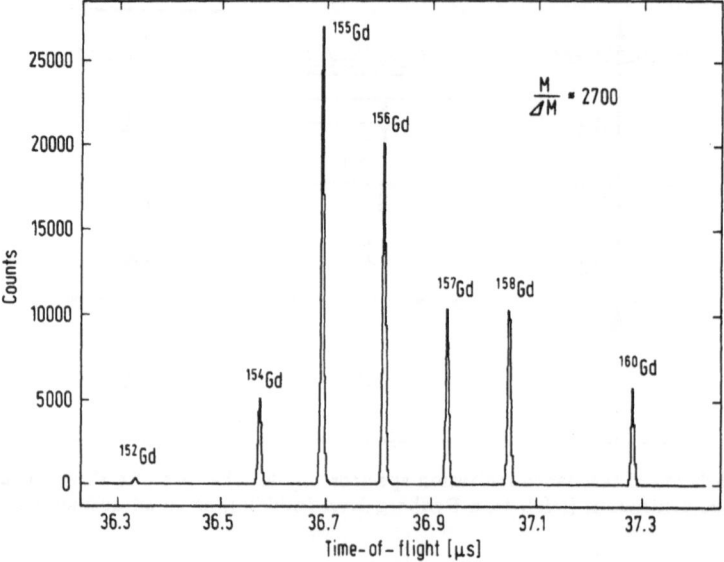

Figure 4. Time-of-flight mass spectrum of a stable gadolinium sample. Resonance ionization mass spectroscopy was applied by use of a dye laser system pumped by a copper vapor laser.

If selectivity against isotopes of the same element is required, the HFS and IS are usually to small in order to achieve efficient discrimination. Here, a pulsed laser system offers the cheap but efficient possibility to add a time-of-flight (TOF) mass spectrometer.

Figure 4 shows an example for the case of a sample of stable gadolinium isotopes. A mass resolving power of $M/\Delta M(\text{FWHM}) = 2700$ is obtained with a drift length of the TOF spectrometer of $l = 2$ m.

If much higher selectivity is required one may induce an artificial mass shift as in the Sr experiment described below. In this case and in the case of interfering resonances of other elements, the separation $\Delta\nu$ between those resonances is much larger than the linewidth. Since the Gaussian profile drops of very quickly, only the Lorentzian width $\Gamma = (2\pi\tau)^{-1}$ is of concern. Then, for $\Delta\nu > 10\Gamma$, the selectivity is given by

$$S = H_1(\nu_1)/H_2(\nu_1) = (2\Delta\nu/\Gamma)^2. \tag{2}$$

The elemental selectivity of RIS is extremely high because of the low level density in atoms which can be reached by allowed electric dipole transitions (about one level per 1 eV at low excitation energies and one level per 10^{-2} eV near the ionization limit at principal quantum number $n \simeq 20$). This has to be compared with the natural linewidth of $\Gamma = 7 \cdot 10^{-8}$ eV for a typical atomic lifetime of $\tau = 10^{-8}$ s.

If more than one resonant, step-wise excitation is induced as in Fig. 2 (i=3), the selectivities in the different resonant transitions j can be multiplied to obtain the total selectivity S_{tot} in the RIS process

$$S_{tot}^{RIS} = \prod_{j=1}^{i} S_j . \tag{3}$$

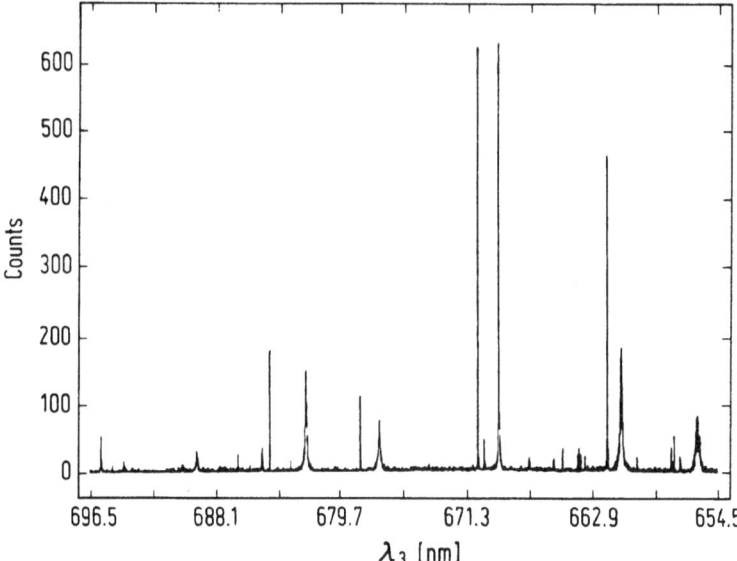

Figure 5. Autoionizing states of technetium. A three-colour RIS excitation was applied with third dye laser scanned (λ_3). A copper vapor laser was used for pumping the dye lasers.

In case of RIMS, where a mass spectrometer is added with an enhancement factor of typically $S_{MS} \simeq 1000$, the total selectivity is given by

$$S_{tot}^{RIMs} = S_{MS} \cdot \prod_{j=1}^{i} S_j \; . \qquad (4)$$

The excitation path for RIS has to be carefully chosen in respect to the available laser system, the possibility to saturate the optical transition (sensitivity), and possible interferences (background). This task generally involves the determination of cross sections and the search for autoionizing states. The determination of cross section is extensively discussed in Ref. 6. The result of a search for autoionizing states is shown in Fig. 5 for the case of technetium. Almost an enhancement in ionization efficiency by three orders of magnitude is obtained by use of the transition to an autoionizing state located at $\lambda_3 = 580.5$ nm.

LASER SYSTEMS FOR RESONANCE IONIZATION SPECTROSCOPY

Almost all available laser systems have been applied for RIS or RIMS[5], but generally pulsed lasers are used because of the ease of saturating the transitions and in order to avoid optical pumping effects. In this case, however, the duty cycle and hence the repetition rate of the laser system play an important role. Most RIS experiments are performed by continous evaporation of the sample atoms. The duty cycle or the temporal overlap ϵ_{temp} is determined by the diameter of the laser beams d_L, the mean velocity $\bar{v}$ of the atoms under investigation and the pulse repetition rate of the lasers ν_{rep} with

$$\epsilon_{temp} = \left(d_L / \bar{v} \right) \cdot \nu_{rep} \; . \qquad (5)$$

For $d = 5$ mm and $\bar{v} = 500 m/s$, $\epsilon_{temp} = 10^{-5} \nu_{rep}$. Therefore, a laser system with ν_{rep} as high as possible is required.

Dye lasers pumped by copper vapor lasers[7] offer today the highest repetition rate ($\nu_{rep} \simeq 10$ kHz) and sufficient power (tunable light of several Watts). The green and yellow output of the Cu lasers, however, restricts the usable wavelength range.

Another possibility to increase the temporal overlap is the use of a pulsed atomic beam instead of continuous one[8,9]. This can be achieved by evaporation of the sample atoms by pulsed-laser induced desorption (PLID). With a power density of the order of 10 MW/cm² a temporarily well defined pulsed atomic beam is produced which contains mainly neutral atoms and can be ionized efficiently by synchronously fired laser light used for RIS. Hence, even laser systems with low repetition rate (e.g. Nd:YAG lasers) might be applied for sensitive resonance ionization. In this case, laser light in the blue or ultraviolet spectral region is easily obtained.

EXPERIMENTS AND RESULTS

Four experiments will be described in the following. Since experimental details and most of the results are published, this chapter tries to give an overview. The reader is refered to the original publications.

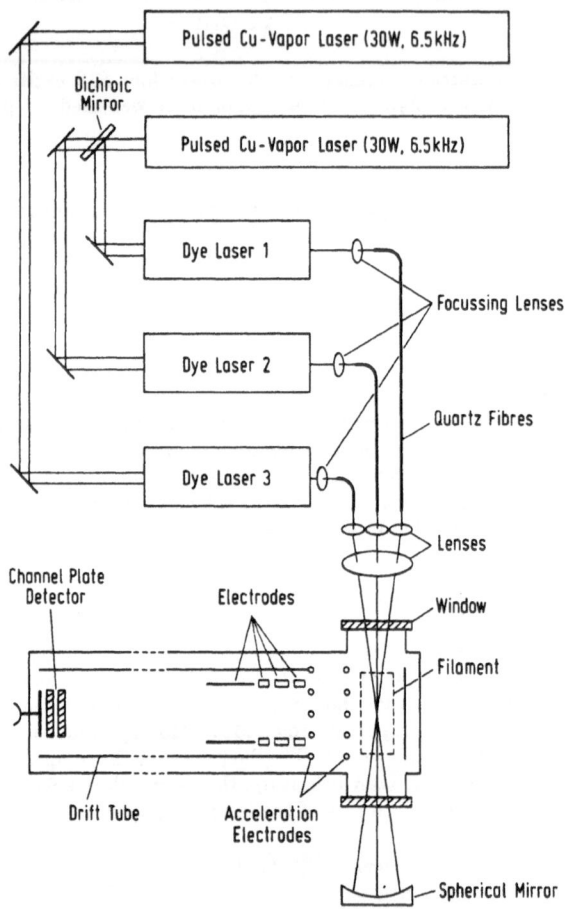

Figure 6. Set-up for trace analysis of plutonium and technetium evaporated from a filament. The atoms are resonantly step-wise exited and ionized by the light of three dye lasers pumped by two copper vapor lasers. The ions are analyzed by a time-of-flight mass spectrometer.

Trace Analysis of Plutonium and Technetium on a Thermal Atomic Beam Evaporated from a Filament[6,7,10-12]

Plutonium and technetium are toxic, radioactive elements released in accidents of nuclear power plants or in atomic-bomb tests (fall-out Pu). Until today, trace detection of Pu (Tc) is mainly done by α- (β-) spectroscopy with a detection efficiency of $4 \cdot 10^8$ (10^{11}) atoms and a very long measuring time. RIMS has opened up the possibility to lower the detection limit by several orders of magnitude. Furthermore it enables to determine the isotopic composition of a sample with an accuracy of about 10%, and therefore to trace the source of pullution and to study the migration.

The experimental set-up is shown in Fig. 6. It consists of three dye lasers pumped by two copper vapor lasers and a vacuum chamber for evaporation of the sample from a filament, for resonant ionization and for TOF detection.

An overall detection efficiency of the instrument of $2 \cdot 10^{-6}$ was found for Pu as well as for Tc corresponding to a detection limit of some 10^6 atoms per isotope. The experimental efficiency is mainly caused by the poor temporal overlap ($\epsilon_{temp} = 5 \cdot 10^{-2}$ for $\nu_{rep} = 6.5$ kHz) and spatial overlap ($\epsilon_{spat} \simeq 10^{-2}$)[6] of the laser beams with the atomic beam evaporated from the filament where the sample is electrolytically deposited. Further losses are due to the transmission of the TOF spectrometer (0.6), the detection efficiency of the channel plate detector (0.3), population of the atomic ground state (0.5 for Pu, 0.6 for Tc), incomplete saturation of the transitions and possibly molecular evaporation of the sample atoms.

Laser Ion Source: Trace Analysis of Technetium [11,13-15]

As shown above the most limiting factors for the detection efficiency are the temporal and spatial overlap of the atomic beam with the laser beams, even at a very high repetition rate of the lasers. To improve the sensitivity, the sample is placed in a hot cavity with a small hole in it for injecting the laser beams for RIS and extracting the photo ions by an electric field. In this case, the atoms have the chance to pass the interaction zone for the laser light several times, before they escape out of the cavity [13,16]. A similar device was used by Andreev et al. for the investigation of Fr isotopes [17]. The efficiency of such a set-up is determined by the rate of photo-ionization R_I, the rate of diffusion R_D out of the hole as neutral atoms, and the rate of escape R_S as surface-ionized ions. The photo-ionization efficiency is given by

$$\epsilon = R_I/(R_I + R_D + R_S) . \tag{6}$$

For a cylindrical cavity with a length l and a diameter $D = 2l$ and neglecting surface ionization as justified in case of Tc ($\epsilon_{surf} \simeq 10^{-6}$ at $T = 2400K$, Wo cavity) one obtains

$$\epsilon = (\nu_{rep} \cdot \epsilon_{photo})/(\nu_{rep}\epsilon_{photo} + \bar{v}/4l) \tag{7}$$

with the photo-ionization probability ϵ_{photo} in resonance. The mean free path λ of the atoms in the cavity between two wall collisions was taken as $\lambda \simeq l$.

As can be seen from (7) the efficiency does not depend on the size of the hole in the cavity as long as it is completely filled by the laser beams. Therefore, its diameter can be adjusted to the available laser power, pumping speed and penetration of the electric field. Using (7) one calculates a photo-ionization efficiency of $\epsilon_{theo} = 0.2$ for ^{99}Tc at 2400K, with a repetition rate of the laser pulses of $\nu_{rep} = 6.5$ kHz, $\epsilon_{photo} = 0.6$ (corresponding to the ground-state population at complete saturation), $\bar{v} = 650$ m/s, and a cavity length $l = 1$ cm. Hence, an improvement by several orders of magnitude is expected by use of a hot cavity instead of evaporation of the sample from a filament.

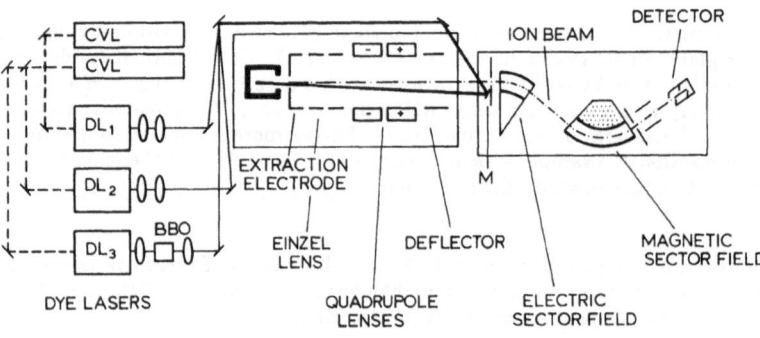

Figure 7. Laser ion source for trace detection of technetium. The sample is put into a hot cavity. The atoms are ionized by resonance ionization, extracted by an electric field, and analyzed and detected by a Mattauch-Herzog-mass spectrometer.

Figure 7 shows the set-up of the laser ion source equiped again with a Cu vapor laser pumped dye laser system and a Mattauch-Herzog mass spectrometer [18] for diagnosis of the ion beam. The laser light is injected into the hot cavity through a hole with a diameter of 2.5 mm. The ions are extracted through this hole by a penetrating field caused by the extraction electrode (10 kV relative to cavity). The ions are imaged by an einzel lens and a quadrupole doublet to the entrance slit of the double-focusing mass spectrometer.

The efficiency of the laser ion source was measured to be $\epsilon_{exp} = 2 \cdot 10^{-3}$ (not including the transmission of the mass spectrometer and the efficiency of the detector). This value is two orders of magnitude lower than expected. It can be improved by more laser power, since the photo-ionization of Tc could not be saturated with the laser power available. A further possible improvement might be a more homogeneous line profile of the laser beams covering the full hole. Finally, it can not be excluded that part of the ^{99}Tc atoms remained in the hot cavity even after extensive heating. This might be due to traps on the surface of the cavity having higher heat of desorption or due to diffusion into the walls of the cavity. This has to be checked in further experiments by β- or γ-spectroscopy, as well as means for reduction of the background of surface-ionized molybdenum present as impurity in the construction material of the cavity, i.e. tungsten. Figure 8 shows this effect with lasers off (top), lasers on and in resonance for RIS of ^{99}Tc (middle), and with gated detection of the ions (bottom) by use of the pulse structure of the photo ions. As can be seen from Fig. 8, the ratio of the ion signals of ^{99}Tc to ^{98}Mo is improved by a factor of 2400 by use of RIS instead of surface ionization.

If the Mo background can be reduced by purer construction material for the cavity or by material with lower work function, the laser ion source can be applied for a integral, geochronical measurement of the solar neutrino flux. As pointed out by Cowan and Haxton [19], ^{97,98}Tc is produced by a (ν_e, e^-) reaction from molybdenum ore. Tc samples have been taken after extensive chemical treatment from the Mo ore of the Henderson Rock Mine/Colorado, USA [20] and are waiting for a technique able to detect 10^8 atoms of each ^{97}Tc and ^{98}Tc in the presence of 10^{15} Mo atoms.

Laser light in the ultraviolet spectral region is required for RIMS of inert elements which exhibit a large energy gap between the ground state and the first excited level. Here, for frequency doubling, pump lasers are required with more power than delivered by copper vapor lasers and/or ultraviolet output. The trade-off is the usually low repetition rate of these lasers. As mentioned above, PLID might be used for improved sensitivity [8,9].

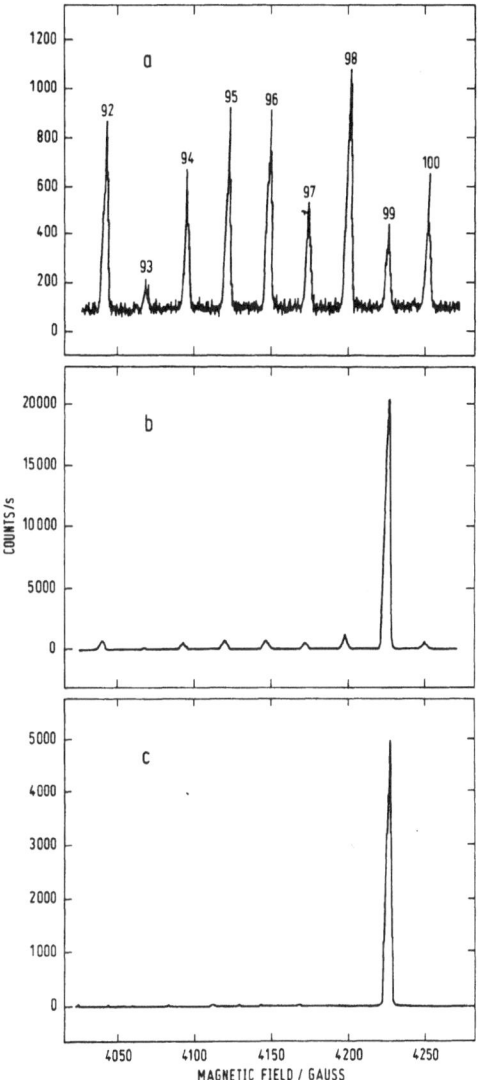

Figure 8. Mass spectrum of ^{99}Tc and the stable molybdenum isotopes in the range $92 \leq A \leq 100$. Top: With lasers shut off. Middle: With lasers on and resonant for photoionization of ^{99}Tc. Bottom: With lasers resonant for photoionization of ^{99}Tc and gated detection (gate length: $1.5\,\mu$s).

Figure 9 shows the experimental set-up originally developed for the determination of nuclear ground-state properties of short-lived nuclei at the on-line isotope separator ISOLDE/CERN, Geneva (see the first lecture, this issue).

It can also be used for trace analysis of gold and platinum. During test experiments searching for the best matrix material for implantation and desorption, we found in all materials investigated trace amounts of Au and Pt. Even spectroscopy-grade, pyrolytic graphite delivered a strong signal of stable Pt isotopes as shown in Fig. 10. Since the overall concentration of Au and Pt is of the order of 10^{-8} to 10^{-9}, it is not surprising to detect everywhere these elements by use of a technique with high sensitivity and selectivity.

In case of a RIMS/PLID-combination care has to be taken to suppress the ions and electrons created during the desorption process even at low desorption intensities. For this purpose several electrodes, deflectors, and diaphrams are installed near the interaction region and in the drift tube of the TOF spectrometer. These measures resulted in a background within a time window of 100 ns for a selected mass, which is as low as 1 event per 1000 laser shots with graphite as implantation target.

The overall efficiency turns out to be $\epsilon = 10^{-8}$, if an oven is used for evaporation of a thermal atomic beam of Au, and increases by PLID to $\epsilon = 10^{-5}$ for Au and $\epsilon = 10^{-6}$ for Pt. The high performance of the set-up might enable in the future to investigate the amount, the accumulation and the migration of platinum in the environment. About 1 μg corresponding to $3 \cdot 10^{15}$ Pt atoms are exhausted per 100 km out of the catalyst of modern cars. Until now, very little is known about distribution of Pt and about health risks due to microscopically distributed platinum with its catalytic power.

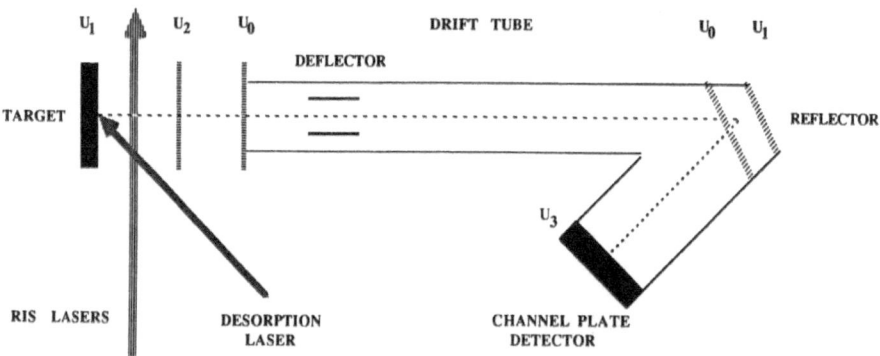

Figure 9. Set-up for trace detection of gold and platinum for resonance ionization mass spectroscopy combined with pulsed-laser induced desorption. The sample atoms are implanted into graphite. The atoms are desorbed by the frequency-doubled output of a Nd:Yag laser, resonantly ionized by the light of three dye lasers pumped by a second Nd:Yag laser, and detected by a time-of-flight mass spectrometer.

Resonance Ionization Spectroscopy on a Fast Atomic Beam in Collinear Geometry: Application to Trace Analysis of ^{90}Sr and ^{89}Sr [25]

A challenge for trace analysis is the fast and sensitive detection of the radioactive isotopes ^{89}Sr and ^{90}Sr. Both are not naturally occuring but produced man-made as fission products in nuclear-weapon tests or in case of reactor accidents. Especially the long-lived isotope ^{90}Sr ($T_{1/2} = 28.5$ a) is dangerous as it is accumulated in the human bones as chemical equivalent of Ca. The radiochemical detection of ^{90}Sr is slow, of the order of two weeks, because it is detected via the β-decay of the daughter nucleus ^{90}Yt. In order to have immediate information on the pollution by ^{90}Sr in an accident such as Chernobyl, an alternative method is searched for. Since the pollution is firstly transported by air, air samples should be investigated. A typical sample as extracted from 1000 m^3 air contains only about 10^8 atoms of ^{89}Sr and ^{90}Sr but also up to 1 mg stable strontium corresponding to about 10^{19} particles.

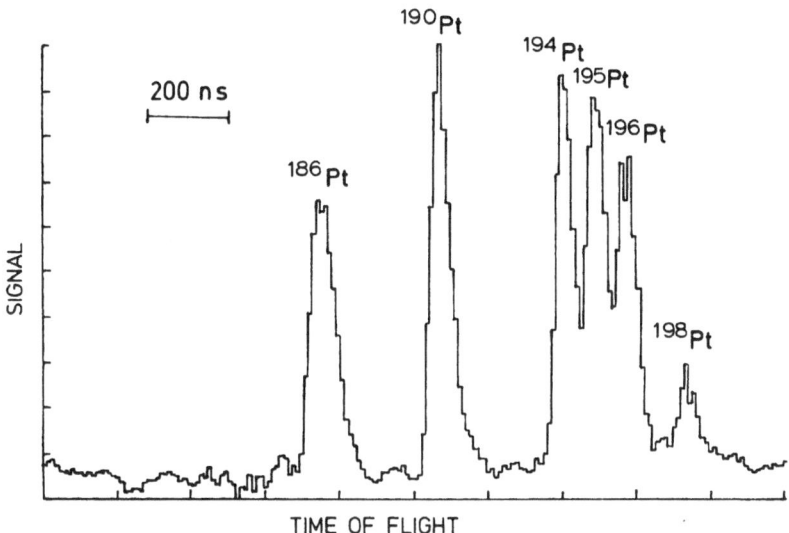

Figure 10. Mass spectrum of stable platinum isotopes obtained from spectroscopy-grade, pyrolytic graphite by resonance ionization mass spectroscopy combined with pulsed-laser induced desorption. ^{186}Pt $(T_{1/2} = 2.0$ h) and ^{190}Pt $(10^{-4}$ abundance in natural Pt) was implanted by use of an on-line isotope separator.

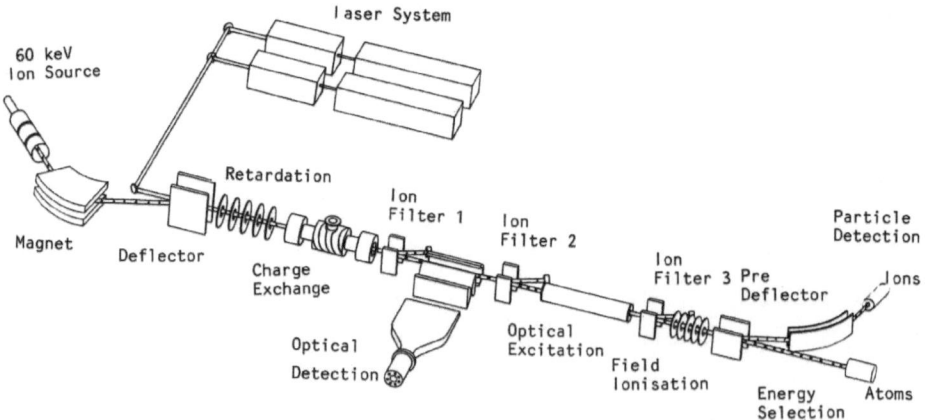

Figure 11. Experimental set-up for trace detection of 10^8 atoms of ^{90}Sr in the presence of 10^{19} atoms of stable Sr.

The requirements which have to be met by a radioactivity-independent and competitive detection scheme are therefore:

1. The selectivity in suppression of stable Sr isotopes must exceed a value of $S = 10^{11}$.

2. The detection efficiency must surpass a value of about 10^{-5} for reasonable statistics of the trace analysis of only about 10^8 particles.

3. The total measuring time including sample preparation must not exceed one day.

These are stringent demands which can be met hopefully by RIS on a fast, charge-exchanged atomic beam in collinear geometry and by excitation using cw dye laser light [26]. Figure 11 shows the set-up which is now under construction at Mainz. The Sr sample is chemically extracted out of an air sample and put into a surface ionization source (eventually after pre-enrichment in a medium- current isotope separator). The ions are accelerated to 60 keV, mass-separated and charge-exchanged.

Since the metastable $^3P_2$-state is populated with cesium as medium for charge exchange, the metastable atoms are excited with the help of light from one or two cw dye lasers to an 5sns- or 5snp-Rydberg state with principal quantum number around $n \simeq 20$. Then they are ionized again by an electric field, deflected from the neutral beam mainly containing stable Sr, and detected.

A selectivity $S > 10^{11}$ is obtained by inducing an artifical mass shift via the different Doppler shifts of ^{90}Sr or ^{89}Sr and the stable Sr isotopes. This shift amounts to 5.5 GHz for neighbouring Sr isotopes in a typical optical transition of $\lambda \simeq 700$ nm and an acceleration voltage of U = 60 kV. Inserting these values into (2) leads to a single-step selectivity of the order of $S = 10^9$. Taking the enhancement factor of the mass separator into account, one obtains theoretically a selectivity well above the required milestone of $S = 10^{11}$.
The background in this experiment will be produced by the much more abundant stable Sr isotopes, mainly ^{88}Sr. Most cumbersome are collision-induced transitions to Rydberg states. Hence, three ion filters will be installed in the path of the fast atomic beam. In addition, the field ionization region is constructed in such a way, that ultra-high vacuum conditions are guaranteed, and the area where field ionization of ^{89,90}Sr takes place is very well defined. An estimate of the signal-to-background ratio yields $S/B = 300$ corresponding to 1000 detected ^{90}Sr ions and 3 background events of ^{88}Sr from a typical sample of 10^8 atoms of ^{90}Sr and 10^{19} atoms of ^{88}Sr.

360

CONCLUSION

Resonance ionization mass spectroscopy has been proved to push the detection limit to or even below the femtogram regime. Four examples were discussed in this talk which demonstrate the potential of modern laser spectroscopic techniques: Trace analysis by RIMS of Pu evaporated from a filament is 100 times, that of Tc 10^5 times more sensitive than the conventionally applied method of radiochemistry. Trace analysis of Tc in a hot cavity pushes this limit further down by another factor of about 1000. In these experiments, the high pulse repetition rate of Cu vapour pump lasers is essential. The laser ion source has some unique features: High efficiency, isobaric purity, and pulsed ion beam. Therefore, in addition to the improvements in sensivity for trace analysis, other applications are possible like substitution of conventional ion sources, high-purity beams for implantation, injection into storage rings or ion traps, or use at mass separators on-line with accelerators for radio-isotope production.

The third example of Au or Pt detection demonstrates that RIMS reaches the limit of the overall concentration of these rare elements in the earth crust. Here, even high sensitivity was obtained for a pump laser of low repetition rate by pulsed-laser induced desorption. All three experiments summarized above are examples for the high elemental selectivity offered by RIMS.

The final experiment is an example for ultra-high isotopic selectivity since it concerns trace detection of a rare radioactive isotope in the presence of an overwhelming number of stable atoms of the same element. In this case, all the experience gained at Mainz and at CERN in the work concerning the determination of nuclear ground-state properties of short-lived isotopes at on-line isotope separators has to be combined to meet the requirements.

ACKNOWLEDGEMENT

The trace analysis experiments were done in collaboration with F. Ames, G. Herrmann, U. Krönert, C. Mühleck, E.W. Otten, P. Peuser, D. Rehklau, J. Riegel, H. Rimke, W. Ruster, P. Sattelberger, F. Scheerer, and N. Trautmann. The project is funded by the German Ministerium für Umwelt, Naturschutz und Reaktorsicherheit (BMU). The laser ion source was developed together with F. Ames, A. Becker, Th. Brumm, K. Jäger, R. Kirchner, H. Rimke, W. Ruster, B.M. Suri, N. Trautmann, and A. Venugopalan. This project is funded by the Deutsche Forschungsgemeinschaft (DFG). Resonance ionization spectroscopy in combination with pulsed-laser induced ionization was performed at ISOLDE/CERN together with St. Becker, G. Bollen, M. Gerber, Th. Hilberath, U. Krönert, and G. Passler and funded by the Bundesministerium für Forschung und Technologie (BMFT). The ^{90,89}Sr trace detection experiment is set up by by K. Christian, G. Haub, G. Herrmann, F. Janß, S. Köhler, L. Monz, E.W. Otten, G. Passler, P. Senne, J. Stenner, N. Trautmann, K. Walter, K. Wendt, and K. Zimmer with funds from the BMU and supported by the programme Umwelt und Neue Technologien of the state Rheinland-Pfalz.

References

[1] E.W. Otten. *Treatise on Heavy Ion Science 8: Nuclei far from Stability (Ed.: D.A. Bromley)*, page 515. Plenum Press, New York, 1989.

[2] R.V. Ambartsumyan and V.S. Letokhov. *Appl. Opt.*, 11:354, 1972.

[3] V.S. Letokhov. *Laser Photoionization Spektroscopy.* Academic Press, London, 1987.

[4] G.S. Hurst and M.G. Payne. *Principles and Applications of Resonance Ionisation Spectroscopy.* Adam Hilger, Bristol and Philadelphia, 1988.

[5] See for example:

G.S. Hurst and C.G.Morgan, editors. *Resonance Ionization Spectroscopy 1986, Proc. Third International Symposium on 'Resonance Ionization Spectroscopy and its Applications', Swansea (1986)*, Bristol, 1987. Institute of Physics, Institute of Physics Conference Series Nr. 84.

and

T.B. Lucatorto and J.E. Parks, editors. *Resonance Ionization Spectroscopy 1988, Proc. Fouth International Symposium on 'Resonance Ionization Spectroscopy and its Applications', Gaithersburg (1988)*, Bristol, 1988. Institute of Physics, Institute of Physics Conference Series Nr. 94.

[6] W. Ruster, H.-J. Kluge, E.W. Otten, F. Ames, D. Rehklau, G. Herrmann, M. Mang, Ch. Mühleck, J. Riegel, H. Rimke, P. Sattelberger, and N. Trautmann. A Resonance Ionization Mass Spectrometer as an Analytical Instrument for Trace Analysis. *Nucl. Instrum. Meth.*, A281:547, 1989.

[7] P. Peuser, G. Herrmann, H.Rimke, P. Sattelberger, N. Trautmann, W. Ruster, F. Ames, J. Bonn, H.-J. Kluge, U. Krönert, and E.W. Otten. Trace Detection of Plutonium by Three-Step Photoionization with a Laser System Pumped by a Copper Vapor Laser. *Appl. Phys. B*, 38:249–253, 1985.

[8] H.-J. Kluge. Optical Measurements of Ground State Properties of Short–Lived Nuclei in Resonance Cells. In *Proc. of Int. Workshop on 'Hyperfine Interactions', Kanpur, Indien, 1984*. Hyperfine Interactions 24:69, 1985.

[9] U. Krönert, St. Becker, Th. Hilberath, H.-J. Kluge, C. Schulz, and the ISOLDE Collaboration. Resonance Ionization Mass Spectroscopy with a Pulsed Thermal Atomic Beam. *Applied Physics*, A44:339–345, 1987.

[10] U. Krönert, J. Bonn, H.-J. Kluge, W. Ruster, K. Wallmeroth, P. Peuser, and N. Trautmann. Laser Resonant Ionization of Plutonium. *Appl. Phys. B*, 38:65–70, 1985.

[11] H.-J. Kluge. Resonance Ionization Mass Spectroscopy for Nuclear Research and Trace Analysis. In *Proc of Workshop on 'Modern Optics, Lasers and Laser Spectroscopy', Kanpur, India, Januar 1987*. Hyperfine Interactions 37:347, 1987.

[12] W. Ruster, F. Ames, M. Mang, Ch. Mühleck, D. Rehklau, H. Rimke, P. Sattelberger, G. Herrmann, H.-J. Kluge, E.W. Otten, and N. Trautmann. Determination of trace elements by resonant ionization mass spectrometry (RIMS). In *Regensburger Symposium 'Massenspektrometrische Verfahren der Elementspurenanalyse', Regensburg, 1987 in: Fresenius' Z. Anal. Chem. 331:182–185*, 1988.

[13] H.-J. Kluge, F. Ames, W. Ruster, and K. Wallmeroth. Laser Ion Sources. In L. Buchmann and J.M. D'Auria, editors, *Proc. of the accelerated radioactive beams workshop, Parksville, Canada, 1985*, page 119. TRIUMF, Canada TRI-85-1, 1985.

[14] F. Ames, A. Becker, H.-J. Kluge, H. Rimke, W. Ruster, and N. Trautmann. A Laser Ion Source for Trace Analysis. In *Regensburger Symposium 'Massenspektrometrische Verfahren der Elementspurenanalyse', Regensburg, 1987 in: Fresenius' Z. Anal. Chem. 331:133*, 1988.

[15] F. Ames, T. Brumm, K. Jäger, H-J. Kluge, and B.M. Suri. A High-Temperature Laser Ion Source for Trace Analysis and other Applications. *submitted to Appl. Phys. B*, 1989.

[16] S.V. Andrew, V.I. Mishin, and V.S. Letokhov. High-efficiency laser resonance photoionization of Sr atoms in a hot cavity. *Optics Communications*, 57:317–320, 1986.

[17] S.V. Andreev, V.S. Letokhov, and V.I. Mishin. Laser Resonance Photoionization Spectroscopy of Rydberg Levels in Fr. *Phys. Rev. Lett.*, 59:1274–1276, 1987.

[18] M. Epherre, G. Audi, C. Thibault, R. Klapisch, G. Huber, F. Touchard, and H. Wollnik. *Nucl. Phys.*, A340:1, 1980.

[19] G.A. Cowan and W.C. Haxton. Solar Neutrino Production of Technetium–97 and Technetium–98. *Science*, 216:51–54, 1982.

[20] D.J. Rokop, N.C. Schroeder, and K. Wolfsberg. High Sensitivity Technetium Analysis using Negative Thermal Ionization Mass Spectrometry. *Proc. 11th Int. Mass. Spectr. Conf., ed. P. Longevialle Bordeaux 1988 in: Advance in Mass Spectrometry*, Heyden & Son Ltd., London, page 1788, 1988.

[21] K. Wallmeroth, G. Bollen, A. Dohn, P. Egelhof, J. Grüner, F. Lindenlauf, U. Krönert, J. Campos, A. Rodriguez Yunta, M.J.G. Borge, A. Venugopalan, J.L. Wood, R. B. Moore, H.-J. Kluge, and the ISOLDE Collaboration. Sudden change in the nuclear charge distribution of very light gold isotopes. *Phys. Rev. Lett.*, 58:1516–1519, 1987.

[22] U. Krönert, St. Becker, G. Bollen, M. Gerber, Th. Hilberath, H.-J. Kluge, G. Passler, and the ISOLDE Collaboration. Observation of Strongly Deformed Ground-State Configurations in ^{184}Au and ^{183}Au by Laser Spectroscopy. *Z. Phys.*, A331:521–522, 1988.

[23] Th. Hilberath, St. Becker, G. Bollen, M. Gerber, H.-J. Kluge, U. Krönert, G. Passler, and the ISOLDE Collaboration. The Charge Radii of $^{198}Pt-^{183}Pt$. *Z. Phys.*, A332:107–108, 1989.

[24] K. Wallmeroth, G. Bollen, A. Dohn, P. Egelhof, U. Krönert, M.J.G. Borge, J. Campos, A. Rodriguez Yunta, K. Heyde, C. de Coster, J.L. Wood, H.-J. Kluge, and the ISOLDE Collaboration. Nuclear shape transitions in light gold isotopes. *Nucl. Phys.*, A493:224, 1989.

[25] K. Wendt et al. Quantitive Detection of Strontium-90 and Strontium-89 in Environmental Samples by Laser Mass Spectrometry. Proc. Intern. School of Quantum Electronics, Erice, September 1989, *to be published by Plenum Press, London (1990)*.

[26] Yu. A. Kudriavtsev and V.S. Letokhov. Laser Method of Highly Selective Detection of Rare Radioactive Isotopes through Multistep Photoionization of Accelerated Atoms. *Appl. Phys. B*, 29:219–221, 1982.

CARS SPECTROSCOPY AND APPLICATIONS

Jean-Pierre Taran

Office National d'Etudes et de Recherches Aerospatiales
BP 72, 92322 Chatillon Cedex, France

INTRODUCTION

The possibility of carrying out temperature and concentration measurements in gases by Raman spectroscopy was suggested and demonstrated about a decade ago. Following some early publications on this subject,[1,2] a massive effort was undertaken in order to evaluate the potential of spontaneous Raman scattering in the important areas of atmospheric sounding and combustion diagnostics. In these experiments, the concentrations of the molecular species are deduced from the intensities of their respective Raman bands and the temperature is obtained from the contour of any one of these bands. Detailed accounts of early experimental work can be found, amont other publications, in several Project SQUID and AIAA workshop proceedings.[3-5] Important instrumental developments were accomplished, with improved collection efficiencies and signal to noise ratio enhancement. The fields of applications were rapidly delineated and it appeared that spontaneous Raman scattering could prove valuable for the investigation of cold or warm aerodynamic jets, which are quite easy to analyse, but was of limited potential in low pressure gases, fluorescent samples or in luminous reactive media.

A non-linear optical technique capable of performing Raman spectroscopy with much improved signal strength was then proposed as a competitor to spontaneous Raman scattering in the specific area of combustion diagnostics.[3-6] This technique is based on a four-wave-mixing process called Coherent Anti-Stokes Raman Scattering, or CARS. CARS, which is one of many well-known third-order processes, was actually observed as early as 1963,[7,8] and has since then been applied to crystal spectroscopy[9-11] and to the measurement of third-order susceptibilities in gases.[12-14] Raman spectroscopy by CARS received a considerable impetus in the early seventies when reliable tunable sources of good optical quality were developed. Progress was made in three important areas, which we shall review in turn.

The first area is that of the theoretical understanding of CARS. Particular attention has been paid to the creation of the non-linear source polarization, to the birth and growth of the signal electromagnetic wave and to energy exchange processes within the material.

In the second chapter, we discuss the application of CARS to practical temperature and concentration measurements in reactive media. We also present a prospective study of electronic resonance enhancement in CARS, which shows great promise for a novel form of molecular spectroscopy and for sensitive detection of trace species.

CARS THEORY

General presentation

In gases, CARS is observed when two collinear light beams with frequencies ω_1 and ω_2 (hereafter called laser and Stokes respectively, with $\omega_1 > \omega_2$), traverse a sample with a Raman active vibrational mode of frequency $\omega_v = \omega_1 - \omega_2$. A new wave is then generated at the anti-Stokes frequency $\omega_3 = \omega_1 + \omega_v = 2\omega_1 - \omega_2$ in the forward direction, and collinear with the pump beams (fig. 1).

This new wave results from the inelastic scattering of the wave at ω_1 by the molecular vibrations, which are coherently driven by the waves at ω_1 and ω_2 (hence the name of the effect). A similar wave at $2\omega_2 - \omega_1$ is also created by the same mechanism (CSRS for Coherent Stokes Raman Scattering). This wave has been observed and is sometimes used for spectroscopic purposes, in spite of the difficulties connected with background light rejection and poorer detector efficiencies. In the following, we shall give a presentation of the CARS theory and a description of the physical mechanisms.

We recall that, because of nonlinearities, it is customary to write the polarisation created in a medium by intense optical beams in terms of a power series in the field amplitudes. We have for the polarisation vector $\mathcal{P}$ at point r:

$$\mathcal{P}(r,t) = \mathcal{P}^{(1)}(r,t) + \mathcal{P}^{(2)}(r,t) + \cdots + \mathcal{P}^{(n)}(r,t) + \cdots, \tag{1}$$

The main linear contribution $\mathcal{P}^{(1)}(r,t)$ and those of higher order (which are smaller in general) all can given an explicit expansion. To this end, we take the applied radiation in the form of monochromatic plane waves.

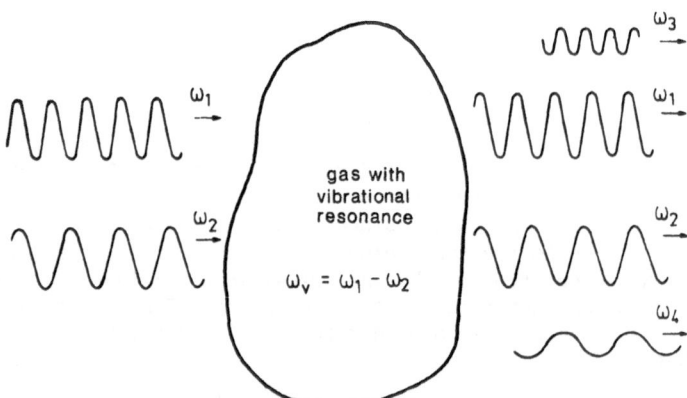

Fig. 1. CARS and CSRS.

Under steady-state conditions, we can expand the total field vector $\mathcal{E}$ as a function of its m distinct frequency components ω_i of wave vector $\mathbf{k}_i$:

$$\mathcal{E} = \Sigma_i \; \mathcal{E}_i(\mathbf{r},t) \tag{2}$$

$$\mathcal{E}_i(\mathbf{r},t) = \frac{1}{2} E(\mathbf{r},\omega_i) \; e^{-i(\omega_i t - \mathbf{k}_i \mathbf{r})} + c.c. \tag{3}$$

The higher-order components of the polarisation expansion (1) now can be written. If ω_s is the frequency of one such component, we write the latter as:

$$\mathcal{P}^{(n)}(\mathbf{r},t,\omega_s) = \frac{1}{2} P^{(n)}(\mathbf{r},\omega_s) \; e^{-i\left(\omega_s t - \mathbf{k}_s' \cdot \mathbf{r}\right)} + c.c., \tag{4}$$

with the phenomenological expansion:

$$P^{(n)}(\mathbf{r},\omega_s) = \left(\frac{1}{2}\right)^{(n-1)} \underline{\underline{\chi}}^{(n)}\left(-\omega_s, \ell_1\omega_{j_1}, \ell_2\omega_{j_2}, \cdots, \ell_n\omega_{j_n}\right)$$
$$\times E_{\ell_1}(\mathbf{r},j_1) \; E_{\ell_2}(\mathbf{r},j_2) \; \cdots \; E_{\ell_n}(\mathbf{r},j_n) \tag{5}$$

and with $\omega_s = \displaystyle\sum_{i=1}^{n} \ell_i\omega_{j_i}$, $\mathbf{k}_s' = \displaystyle\sum_{i=1}^{n} \ell_i\mathbf{k}_{j_i}$; here, $\underline{\underline{\chi}}^{(n)}$ is the susceptibility tensor of order n (the rank of this tensor is n+1); we also specify $\ell_i = \pm 1$ and $1 < j_i < m$, and:

$$E_{\ell_i}(\mathbf{r},j_i) = E(\mathbf{r},\omega_i) \qquad \text{if } \ell_i = +1$$
$$E^*(\mathbf{r},\omega_i) \qquad \text{if } \ell_i = -1,$$

where the symbol * denotes the complex conjugage. $\mathcal{P}^{(1)}$ is associated with linear effects (dispersion and absorption); $\mathcal{P}^{(2)}$, which is responsible for such effects as frequency doubling or parametric conversion, vanishes in media processing inversion symetry, e.g. centrosymmetric crystals, gases and liquids; all other even-order terms also vanish in these media; $\mathcal{P}^{(3)}$ stands for a large class of effects such as third harmonic generation, and three-wave mixing via two-photon and Raman nonlinearities.

The source polarisation component of frequency ω_s gives birth to an electromagnetic wave at ω_s. This wave is a solution of the wave equation, which can be written:

$$\left(\nabla^2 - \frac{n^2}{c^2}\frac{\partial^2}{\partial t^2}\right) \mathcal{E}_s(\mathbf{r},t) = \frac{4\pi}{c^2}\frac{\partial^2}{\partial t^2} \mathcal{P}^{(n)}(\mathbf{r},t,\omega_s) \tag{6}$$

for a non-magnetic homogeneous medium. We assume here that $\mathcal{P}^{(n)}(\mathbf{r},\omega_s)$ as given in (4), (5) is the only source term at frequency ω_s. Its spectral properties or, in other words, the spectral dependence of the nonlinear optical susceptibility tensor $\underline{\underline{\chi}}^{(n)}\left(-\omega_s, \ell_1\omega_{j_1}, \ell_2\omega_{j_2}, \cdots, \ell_n\omega_{j_n}\right)$ as a function of the applied field frequencies ω_1, ω_2, etc, are directly reflected in the rate of growth of the signal wave.

There are numerous instances where the source polarisation component of Eq. (6) receives contributions from quite distinct physical processes, with the same order in the nonlinearity, which are associated with a different susceptibility tensor. Terms with a higher order in the nonlinearity are also possible. All such terms giving contributions at ω_s have to be added to $\mathcal{P}^{(n)}(r,\omega_s)$ in the right hand side of (6).

For instance, if two fields at ω_1 and ω_2 are applied with $\omega_1 > \omega_2$, one has two third-order polarisation terms at the frequency $\omega_3 = 2\omega_1 - \omega_2$.

$$\mathcal{P}^{(3)}(r,t,\omega_3) = \mathcal{P}^{(3)CARS}(r,t,\omega_3) + \mathcal{P}^{(3)SRS}(r,t,\omega_3) \tag{7}$$

The first one is the CARS component:

$$\mathcal{P}^{(3)CARS}(r,t,\omega_3) = \frac{1}{2} P^{(3)CARS}(r,\omega_3)\, e^{i\left(k_3' \cdot r - \omega_3 t\right)} + c.c. \tag{8}$$

with

$$P^{(3)CARS}(r,\omega_3) = \frac{1}{4}\underset{=}{\chi}^{(3)CARS}(-\omega_3,\omega_1,\omega_1-\omega_2)\, E^2(r,\omega_1)\, E^*(r,\omega_2)$$

and $k_3' = 2k_1 - k_2$, while the second one reflects the stimulated Raman scattering interaction between the waves at ω_3 and ω_1:

$$\mathcal{P}^{(3)SRS}(r,\omega_3) = \frac{1}{4}\underset{=}{\chi}^{(3)SRS}(-\omega_3,\omega_1,-\omega_1,\omega_3)\, |E(r,\omega_1)|^2$$

$$\times E(r,\omega_3) \times e^{i(k_3 \cdot r - \omega_3 t)} + c.c.. \tag{9}$$

The latter is an inverse Raman process and represents a loss at ω_3. It is negligible in a CARS experiment, since the susceptibility components are of comparable magnitude and since

$$|E(r,\omega_3)| \ll |E(r,\omega_1)|, \ |E(r,\omega_2)| \ ; \tag{10}$$

k_3 is here the wave vector of the anti-Stokes wave. Similarly, the third-order polarisation terms at ω_1 and ω_2 can also be broken down into equations similar to (6). For these, however, the stimulated Raman scattering term is the stronger.

In conclusion, we have two separate problems to solve in nonlinear optical spectroscopy:

• derivation of all the relevant nonlinear susceptibility terms,
• calculation of the electric field solution of the wave equation.

The derivation of the nonlinear susceptibility terms, which is essential in predicting the spectral properties of a medium, can be done through several distinct approaches. In the case of third-order Raman-type nonlinearities, the classical Placzek model of molecular polarisabilities leads to a rapid calculation of essential results. It gives good insight into the physical mechanisms, but is inadequate for the case where one or more of the light waves is in resonance with one-photon absorption frequencies of the Raman-resonant species. Quantum mechanical derivations are more accurate. Those based on a wavefunction representation are often sufficient and have been employed extensively using a perturbative treatment of the electric field interactions. We prefer the density operator formalism which, in association with a Feynman-like diagrammatic representation, leads to a rapid derivation of all relevant susceptibility terms and to an easy interpretation of the physical mechanisms involved. The tensor properties of the susceptibility components also follow easily.

The search for the wave equation solution is the second major problem. This solution reveals the important properties of the signal generation: phase-matching, energy exchange between the light waves and the matter, pulse shape characteristics and spatial resolution of CARS measurements using focused beams.

Derivation of the susceptibility

Our purpose in this section is not to give a complete derivation of the susceptibility, but only an outline of the principles. The quantum state of the scattering molecules is represented, at point r, as is conventional, by the density operator ρ with the well-known equation of motion:

$$\frac{\partial}{\partial t} \rho(r,t) = -\frac{i}{\hbar} [H_0 + V(r,t), \rho(r,t)] + \frac{\partial \rho}{\partial t}\bigg|_{damp} . \tag{11}$$

H_0 is the free molecule Hamiltonian with a discrete spectrum of eigenstates $|n>$ corresponding to eigenenergies $\hbar \omega_n$; the Hamiltonian describing the interaction of the molecules with the radiation field is $V(r,t) = -p \cdot E(r,t)$ in the dipolar approximation; p is the dipole moment operator; $(\partial \rho/\partial t)|_{damp}$ is the damping term, which is determined by stochastic processes such as spontaneous emission of light and collisions between molecules.

We assume the perturbation $V(r,t)$ to be weak enough to allow the solution of (11) to be expanded in successive powers of $V(r,t)$. The density operator is then obtained to any order ℓ by the familiar series expansion:

$$\rho(r,t) = \rho^{(0)}(r,t) + \rho^{(1)}(r,t) + \cdots + \rho^{(\ell)}(r,t) \tag{12}$$

The ℓth order term $\rho^{(\ell)}(r,t)$ is proportional to $V^{\ell}(r,t)$ and is obtained by ℓ iterative applications of (11). The term responsible for the CARS polarisation is of order 3, and the polarisation is given by:

$$\mathscr{P}^{(3)CARS}(r,t,\omega_3) = N \, \text{Tr} \left[\rho^{(3)CARS}(\omega_3,r,t) \, p \right] \tag{13}$$

where $\rho^{(3)CARS}(\omega_3,r,t)$ labels the CARS Fourier component of $\rho^{(3)}(r,t)$ at frequency ω_3. Identification between (8) and (13) eventually yields the expression for the CARS susceptibility tensor.

The entire derivation is straightforward but time consuming. Recently, diagrammatic representations have been introduced for the treatment of nonlinear optical processes. These representations give useful insight into the microscopic physical mechanisms.[15-24] They are applied with a set of simple rules which allow one to rapidly calculate all the relevant susceptibility terms.[17-24]

Similar representations are used in nuclear physics. In our representation, we use the fact that the density operator at any specified order can be shown to result from a number of contributions; each of these is associated with a specific time sequence of perturbations to the density operator, or to the ket vector $|\psi\rangle$ and its complex conjugate $\langle\psi|$ (in the pure state case); the time-ordering of the perturbations to $|\psi\rangle$ with respect to those to $\langle\psi|$ is of crucial importance in the case of collisional relaxation. Each of these elementary time-ordered contributions can be visualized by means of a double-sided Feynman-like diagram. Ordinary Feynman diagrams[25] have been used in nonlinear optics.[26-29] However, their application is limited to the case where simplifying assumptions on collisional rate are made[17,18,30] and they do not depict the physical processes as clearly. In a double-sided diagram, the time evolution of the density matrix is depicted along two parallel vertical bars (one for each subscript of the density matrix) with time increasing upwards. Each interaction with the electromagnetic field is represented by a segment pointing downwards from a vertex if it corresponds to a term oscillating as $e^{-i\omega_j t}$ in the interaction Hamiltonian $V(t)$ and pointing upwards if the term oscillates as $e^{+i\omega_j t}$. The vertex is on the left- or right-hand-side vertical bar depending on whether the left- or right-hand-side subscript of the density matrix element is changed through the interaction. The eigenstates between which the interaction Hamiltonian is operating are indicated below and above each vertex. Examples of these diagrams are shown in Fig. 2.

In CARS, one must combine two vertices at ω_1 and one at ω_2 in order to get the polarisation component $P^{(3)}$ (as given in (8)); its $e^{-i\omega_3 t}$ dependence implies two segments pointing down from the vertex for the interactions with ω_1 and one pointing up for ω_2. If we use a set of four molecular levels a, b, n, n' as shown in the energy level diagram of Fig. 3 (n and n' being of parity opposite to that of a and b), and if the sequence of interactions is applied to be unperturbed density operator $\rho_{aa}^{(0)}$ then there are 24 time-ordered possibilities for this sequence of

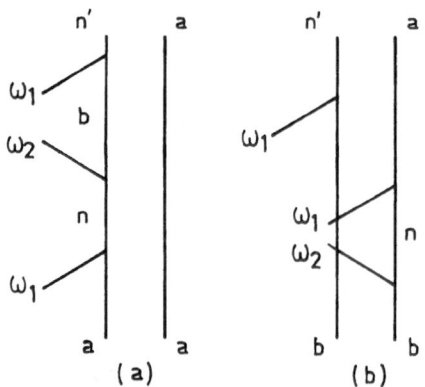

Fig. 2. Two of the time-ordered diagrams representing the density operator evolutions. (a) term proportional to $\rho_{aa}^{(0)}$; (b) term proportional to $\rho_{bb}^{(0)}$. Indices a, b, n, n' refer to molecular eigenstates.

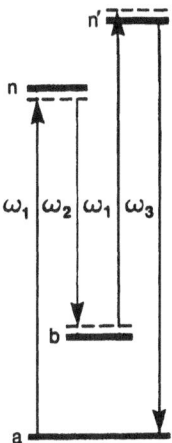

Fig. 3. CARS energy level diagram showing the
transitions in closest resonance with
the fields.

interactions. Each possible time sequence of interactions gives a distinct contribution to the susceptibility. Another set of 24 terms proportional to $\rho_{bb}^{(0)}$ is also found if state b is populated. It is beyond the scope of this overview to present a detailed account of the rules one uses to derive the susceptibility term from its associated diagram. These rules are found in several publications.[17-24] together with their justification.

The CARS susceptibility can be written:

$$\underset{=}{\chi}^{(3)CARS}(-\omega_3, \omega_1, \omega_1, -\omega_2) = \underset{=NR}{\chi} + \underset{=R}{\chi}^{a,b,n,n'} \tag{14}$$

where $\underset{=R}{\chi}^{a,b,n,n'}$ is the Raman-resonant part associated with the Raman-active transition of frequency ω_{ba} between a and b, for the set of two levels n and n'. The tensor element pertaining to a particular set of field polarisations described by unit vectors e_1, e_2, e_3 and pertaining to the waves at ω_1, ω_2, ω_3, respectively, is:

$$e_3 \underset{=R}{\chi}^{a,b,n,n'} e_1 \, e_1 \, e_2 = \frac{N}{\hbar^3} \frac{1}{(\omega_{ba} - \omega_1 + \omega_2 - i\Gamma_{ba})} \times (A + B)$$
$$\times \left[\rho_{aa}^{(0)}(\alpha + \beta) - \rho_{bb}^{(0)}(\gamma + \delta) \right] \tag{15}$$

with: $A = \mu_{an'} \, \mu_{n'b} \, (\omega_{n'a} - \omega_3 - i\Gamma_{n'a})^{-1}$,
$ B = \mu_{an'} \, \mu_{n'b} \, (\omega_{n'b} + \omega_3 + i\Gamma_{n'b})^{-1}$,
$ \alpha = \mu_{bn} \, \mu_{na} \, (\omega_{na} + \omega_2 - i\Gamma_{na})^{-1}$,

$$\beta = \mu_{bn}\,\mu_{na}\,(\omega_{na} - \omega_2 - i\Gamma_{na})^{-1},$$
$$\gamma = \mu_{bn}\,\mu_{na}\,(\omega_{nb} - \omega_2 + i\Gamma_{nb})^{-1},$$
$$\delta = \mu_{bn}\,\mu_{na}\,(\omega_{nb} + \omega_1 + i\Gamma_{nb})^{-1}.$$

Here, N is the number density of active molecules. The absorption frequencies from states $|a\rangle$ and $|b\rangle$ to state $|n\rangle$ are ω_{na} and ω_{nb} respectively, and the Γ's are the corresponding damping factors; μ_{an} is the matrix component of the dipole moment operator $\mu_{an} = \langle a|\; p \cdot e_1 |n\rangle$; μ_{bn}, $\mu_{n'b}$, $\mu_{an'}$ involve interactions with the ω_2, ω_1, ω_3 fields respectively. If more than four levels are present, a summation must be taken and the vibrationally-resonant part becomes:

$$\underset{=R}{\chi} = \sum_{a,b,n,n'} \underset{=R}{\chi^{a,b,n,n'}}$$

Molecular spectroscopy by CARS consists in carrying out an analysis of the spectral properties of $\underset{=}{\chi}^{(3)CARS}(-\omega_3,\omega_1,\omega_1,-\omega_2)$. As shown in (14), the latter contains two parts. In mixtures, part $\underset{=NR}{\chi}$, which is called nonresonant, is contributed both by probed molecules and by the non-Raman-resonant molecular species (diluent molecules). It is composed of terms analogous to those of $\underset{=R}{\chi^{a,b,n,n'}}$ but with nonresonant two-photon sum or difference denominators in place of the Raman resonance denominator. In the usual case where the number density N of the probed molecules is small compared to that of the diluent molecules, $\underset{=NR}{\chi}$ is mainly contributed by the latter and is therefore a frequency-independent real tensor (provided that there are no one- or two-photon electronic resonances in the diluent molecules). The presence of this nonresonant part is one of the most severe problems in the application of CARS spectroscopy.

We are particularly interested in the spectral properties of the Raman-resonant part (Eq. (15)). Off electronic resonance, i.e. $\omega_1,\omega_2,\omega_3 \ll \omega_{nb},\omega_{n'a},\omega_{n'b}$, all the coefficients A, B, α, β, γ, δ are of similar magnitude and depend only weakly on the electric field frequencies. If we assume for simplicity, that all fields have the same polarisation e_1, the relevant tensor element then reduces to:

$$e_1\,\underset{=R}{\chi^{(3)}}\,e_1 e_1 e_1 = \sum_{ab} \frac{Nc^4}{\hbar^3 \omega_1 \omega_2^3}\left(\rho_{aa}^{(0)} - \rho_{bb}^{(0)}\right)\frac{d\sigma}{d\Omega}\,\frac{1}{\omega_{ba} - \omega_1 + \omega_2 - i\Gamma_{ba}} \quad (16)$$

where $d\sigma/d\Omega$ is the spontaneous Raman scattering cross section. The spectral analysis thus reveals the Raman resonances ω_{ba} contained in the denominator of (16). Identification of these resonances and monitoring of their amplitudes are active research areas in analytical chemistry. Furthermore, the other tensor elements of $\underset{=R}{\chi}$ can be measured by an adequate choice of field polarisations.

When the electronic resonances are approached, only two terms in (15) become large. Thus we have:

$$\underset{=R}{\chi} \simeq \frac{N}{\hbar^3}\,\frac{1}{\omega_{ba} - \omega_1 + \omega_2 - i\Gamma_{na}}\left(\rho_{aa}^{(0)}\,A\,\beta - \rho_{bb}^{(0)}\,A\,\gamma\right). \quad (17)$$

We give in Fig. 2 the time-ordered diagrams from which these two terms have been obtained. Since A, β and γ all undergo large enhancements and large variations as the field frequencies are varied, the spectral analysis is complicated somewhat. This particular problem of resonance enhanced CARS has been treated in detail.[17,20]

In addition to these main vibrationally resonant terms given in (15), one has to consider terms which contain vibrational resonances in the excited electronic state and which are generally left out with $\chi_{=NR}$. These terms have been identified and treated as corrections to the main terms in (15) by Druet et al. and Yee et al.[17,18,24] Their spectroscopic and physical importance has been recognized by Bloembergen and coworkers,[30-33] who drew attention to the fact that they represent vibrational contributions from states that have no initial population, and who experimentally demonstrated their existence in Na vapor.

An example of these corrective terms has been treated by Druet and Taran (Ref. 24, Appendix I). This example is different from that of Bloembergen et al.[33] but also lends itself to experimental checks in molecular spectroscopy. It is based on the two terms shown in Fig. 4 (with their corresponding resonance denominators) which can be combined with the one in $\rho_{bb}^{(0)}$ of Fig. 2, yielding a susceptibility contribution of the form:

$$N\rho_{bb}^{(0)} \frac{\mu_{an'}\mu_{n'b}\mu_{bn}\mu_{na}}{\hbar^3} \frac{1}{(\omega_{bn} + \omega_2 - i\Gamma_{bn})(\omega_{ba} - \omega_1 + \omega_2 - i\Gamma_{ba})}$$

$$\times \frac{1}{\omega_{n'b} - \omega_1 - i\Gamma_{n'b}} \tag{18}$$

$$\times \left[1 + \frac{i(\Gamma_{n'a} - \Gamma_{n'b} - \Gamma_{ba}) + i(\Gamma_{nn'} - \Gamma_{n'b} - \Gamma_{bn}) \frac{\omega_{ba} - \omega_1 + \omega_2 - i\Gamma_{ba}}{\omega_{n'n} - \omega_1 + \omega_2 - i\Gamma_{n'n}}}{\omega_{n'a} - \omega_3 - i\Gamma_{n'a}} \right]$$

where the terms have been grouped as in Ref. 30.

We can see that the 3 terms in (18) sum to give one term when there is no damping (Γ = 0) or if:

$$\Gamma_{n'a} - \Gamma_{n'b} - \Gamma_{ba} = \Gamma_{nn'} - \Gamma_{n'b} - \Gamma_{bn} = 0 \tag{19}$$

This particular equation is what we call the damping approximation. It is satisfied when there is no collisional elastic broadening and if the lifetimes of states |a> and |b> are much longer than those of |n> and |n'>. Combining (18) and (19), we then obtain for the susceptibility:

$$N\rho_{bb}^{(0)} \frac{\mu_{an'}\mu_{n'b}\mu_{bn}\mu_{na}}{\hbar^3}$$

$$\times \frac{1}{(\omega_{bn} + \omega_2 - i\Gamma_{bn})(\omega_{ba} - \omega_1 + \omega_2 - i\Gamma_{ba})(\omega_{n'b} - \omega_1 - i\Gamma_{n'b})} \tag{20}$$

373

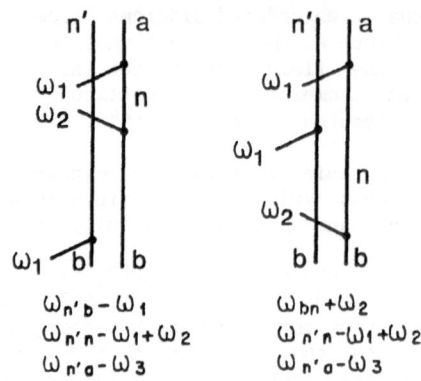

$$\omega_{n'b} - \omega_1$$
$$\omega_{n'n} - \omega_1 + \omega_2$$
$$\omega_{n'a} - \omega_3$$

$$\omega_{bn} + \omega_2$$
$$\omega_{n'n} - \omega_1 + \omega_2$$
$$\omega_{n'a} - \omega_3$$

Fig. 4. Diagrammatic representation of $\rho_{bb}^{(0)}$ contributions to $\underset{=}{\chi}^{(3)CARS}$ having the resonant denominator $\omega_{n'n} - \omega_1 + \omega_2$.

We notice that the resonances at $\omega_{nn'} - \omega_1 + \omega_2$ vanish and that the terms have combined to merely result in a shift of the electronic resonance in the $A\gamma\rho_{bb}^{(0)}$ term of (15) from $\omega_{n'a} - \omega_3$ to $\omega_{n'b} - \omega_1$. The latter expression (20) is usually obtained from perturbation theory in the absence of damping.[29,34]

These terms allow interesting Raman spectroscopic information to be collected about unpopulated states, i.e. without having to populate those states beforehand; they differ in the nature of their broadening by the Doppler effect;[23] they finally pose delicate problems about their true nature and their relation to collisional broadening, as pointed out by Grynberg.[35]

Wave propagation

We here treat the problem of CARS in gases. The amplitude of the signal wave in CARS is obtained from the wave equation (6). We take the pump fields in the form of collinear, travelling plane waves as given by (3). Looking for an anti-Stokes wave also travelling in the forward direction along the z axis, we can reduce the degree of (6), to obtain a steady-state equation:

$$\frac{\partial}{\partial z} E_3 = \frac{i\pi\omega_3}{2c} \left| \underset{=}{\chi}^{(3)CARS}(-\omega_3, \omega_1, \omega_1, -\omega_2) E_1^2 E_2^* e^{i\delta k z} \right.$$

$$\left. + \underset{=}{\chi}^{(3)SRS}(-\omega_3, \omega_1, -\omega_1, \omega_3) E_1 E_1^* E_3 \right| \tag{21}$$

with $\delta k = k_3' - k_3 = 2k_1 - k_2 - k_3$ and $k_i = |k_i|$ (i = 1, 2, 3) ; we have taken the refractive index $n \simeq 1$ and we assume the gas to be homogeneous (i.e., $\underset{=}{\chi^{(3)}}$ is independent of r). We have written for simplicity $E_1 = E(r,\omega_1)$, $E_2 = E(r,\omega_2)$, $E_3 = E(r,\omega_3)$ and assumed that E_3 is a slowly varying function of z:

$$\left|\frac{\partial}{\partial z} E_3\right| \ll k_3 |E_3|$$

Similar equations also hold at ω_1 and ω_2:

$$\frac{\partial}{\partial z} E_1 = \frac{i\pi\omega_1}{2c} \left| \underset{=}{\chi^{(3)CARS}}(-\omega_1,\omega_2,\omega_3,-\omega_3)E_1^* \ E_2 \ E_3 e^{-i\delta kz} \right.$$
$$+ \underset{=}{\chi^{(3)SRS}}(-\omega_1,\omega_3,-\omega_3,\omega_1)E_1 \ E_3 \ E_3^* \tag{22}$$
$$+ \left. \underset{=}{\chi^{(3)SRS}}(-\omega_1,\omega_2,-\omega_2,\omega_1)E_1 \ E_2 \ E_2^* \right|$$

$$\frac{\partial}{\partial z} E_2 = \frac{i\pi\omega_2}{2c} \left| \underset{=}{\chi^{(3)CARS}}(\omega_2,\omega_1,\omega_1,-\omega_3)E_1^2 \ E_3^* \ e^{i\delta kz} \right.$$
$$+ \left. \underset{=}{\chi^{(3)SRS}}(\omega_2,\omega_1,-\omega_1,\omega_2) \left|E_1^2\right| \ E_2 \right| \tag{23}$$

Equations (21-23) can only be solved numerically. However, if we assume the coupling to be weak, E_1 and E_2 can be taken as constants and the SRS term in (21) can be neglected. Then the latter equation is integrated readily. With boundary condition $E_3|_{z=0} = 0$, we have:

$$E_3 \simeq \frac{i\pi\omega_3}{2c} \underset{=}{\chi^{(3)CARS}}(\omega_3,\omega_1,\omega_1,-\omega_2) \ E_1^2 \ E_2^* \ e^{i\delta kz} \ \frac{\sin(\delta kz/2)}{\delta k/2} \tag{24}$$

$$I_3 = \frac{16\pi^4\omega_3^2}{c^4} \left| \underset{=}{\chi^{(3)CARS}}(-\omega_3,\omega_1,\omega_1,-\omega_2) \ e_1 \ e_1 \ e_2 \right|^2 \ I_1^2 \ I_2$$
$$\times \left(\frac{\sin |\delta kz/2|}{\delta k/2}\right)^2 . \tag{25}$$

Equations (24) and (25) constitute the basis for the interpretation of the anti-Stokes wave properties. The most important ones are the following:

1 - The anti-Stokes field polarisation vector is oriented along the vector:

$$f_3 = \underset{=}{\chi^{(3)CARS}}(-\omega_3,\omega_1,\omega_1,-\omega_2) \ e_1 \ e_1 \ e_2 ,$$

which depends on the applied field polarisations as well as on the tensor properties of the susceptibility. This vector has two independent components associated with the non-resonant and the

Raman-resonant parts of the susceptibility (as we have mentioned at the end of the Section on susceptibilities). This property can be used for non-resonant background cancellation in the spectra.

2 - The CARS signal intensity, which is the parameter directly measured using photodetectors, is proportional to $|f_3|^2$ (see Eq.(25)), and, therefore, to the squared number density of the medium. It also has a sinusoidal dependence vs z; in gases, however, we have $\delta k \simeq 0$ because the dispersion is weak, so that the behavior is parabolic over long distances. Pump depletion would eventually limit this parabolic growth.

Energy exchange between the light waves can be analysed by recasting (21-23) into equations for the rates of change of photon number per unit volume and also considering the rate equation for the molecular population change $N \frac{\partial}{\partial t}\left(\rho_{aa}^{(4)} - \rho_{bb}^{(4)}\right)$. This discussion has been conducted in detail elsewhere[20-24] and we only summarize the conclusions here for the off-electronic resonance case.

The stronger process is the SRS coupling between the ω_1 and ω_2 waves: one photon at ω_1 is converted into a Stokes photon at ω_2 and a quantum of molecular vibration. The rate of this process is proportional to the imaginary part of $\underline{\underline{\chi}}^{(3)SRS}$. Although it is not specifically a CARS interaction, this process is important because vibrational population changes can result; this can in turn bring higher-order corrections to $\underline{\underline{\chi}}^{(3)CARS}$ and bias the results. If the corrections should become of the same order as the main terms, then the approximation that allows one to write the density operator as a series expansion (Eq. (12)) is no longer valid. The use of Bloch equations becomes necessary to arrive at a solution which is exact to all orders. This has been done recently to treat saturation in CARS.[36,37] Saturation causes measurement errors:[36] (i) the lines are broadened and shifted; (ii) their amplitudes are reduced. Therefore, unless precautions are taken to correct the experimental results, concentrations and temperatures measured by CARS will be biased; note, however, that rotational temperature measurements are little influenced by saturation.

The CARS generation mechanism per se is made up of two distinct processes:

(i) a "parametric" process whereby two laser photons at ω_1 are converted into a Stokes photon at ω_2 and an anti-Stokes photon at ω_3; the molecules are, on the average, returned to their ground state after the interaction; this process can be reversed if the phases of the waves are changed and its rate is proportional to the real part of the susceptibility;

(ii) a "Raman like" process whereby a Stokes photon is converted into an anti-Stokes photon and two vibrational quanta, on the average, are taken away from the molecules; this process has a rate proportional to the imaginary part of the susceptibility and can be reversed by changing the phases of the waves.

376

It is noteworthy that the second process is the only one responsible for the anti-Stokes generation exactly on vibrational resonance, since the real part of the susceptibility then vanishes. Yet, the so-called "parametric" process has often been erroneously cited as being the only CARS mechanism. This belief has originated from the fact that the energy level diagram of Fig. 3 gives the misleading impression that the CARS interaction returns the molecules to their initial state. It should be emphasized that such energy level diagrams should be used in nonlinear optics to depict only the establishment of polarisations. It is only in the case of processes like Raman scattering or multiphoton absorption that they can also be used to depict energy exchange without ambiguity. Finally the above-mentioned considerations on net energy exchange and molecular population changes cannot be dissociated from the quantum processes themselves. In effect, on the microscopic scale, molecules can undergo sequences of interactions which either return them to their initial state after the final interaction with the anti-Stokes field (e.g. Fig. 2a) or place them in a different vibrational state (e.g. Fig. 2b).

PRACTICAL APPLICATION OF CARS

General considerations

The laws governing the signal growth and the spectral properties of CARS have been established in the preceeding chapter. We here show how CARS can be used for practical measurements and what level of performance can be obtained.

Spatial resolution. Unfocussed parallel beams with large diameters are seldom used because no spatial resolution is possible in this geometry. Since the growth of the power density I_3 is proportional to $I_1^2 I_2$, it seems advantageous to focus the beams to a small diameter and to use high peak power sources. If the condition $\delta k = 0$ is assumed, then it can be shown that:

1 - the anti-Stokes flux is contained within the same cone angle as the pump beams energing from the focal region;

2 - this flux is generated for the most part within the focal region (where $I_1^2 I_2$ is large);

3 - the total power in the anti-Stokes beam some distance beyond the focus is independent of beam diameter and focal length and is approximately given by:

$$P_3 = \left(\frac{2}{\lambda}\right)^2 \left(\frac{4\pi^2 \; \omega_3}{c^2}\right)^2 \; |f_3|^2 \; P_1^2 \; P_2 , \qquad (26)$$

where refractive indices were taken as unity, where $\lambda = 2\pi \, c/\omega$ with $\omega \simeq \omega_1 \simeq \omega_2 \simeq \omega_3$ and where P_1 and P_2 are the powers at ω_1 and ω_2 respectively. This expression was obtained by assuming that all the signal is generated from a small cylindrical volume about the focus having a length equal to the confocal parameter h of the beams. If Gaussian beams are used, the beam waist at the focus is $\Phi = 4\lambda f/\pi d$ where f is the focal length of the lens and d the beam diameter in the plane of the lens: we also have $h = \pi \Phi^2/2\lambda$. In reality, 75 % of P_3 are generated from a volume of length 6h as shown by numerical calculations.

In practical experiments, the spatial resolution is on the order of 1 to 20 mm with laser beams of good optical quality. This may still be insufficient in some experimental situations where higher spatial resolutions are needed. A particular beam arrangement called BOXCARS has been proposed for better resolution.[38] In this arrangement, the beams are crossed at a small angle so that the polarisation wave vector k_3' remains equal to the anti-Stokes signal wave vector k_3 (Fig. 5). The beam configuration is shown in Fig. 6. The spatial resolution then can be adjusted in the range 1-10 mm by varying the beam crossing angle.

Spectral information. CARS spectroscopy can be accomplished in various manners depending on the application envisioned (e.g. high-resolution spectroscopy or chemical analysis). The spectra are usually retrieved by holding ω_1 fixed, varying ω_2 so that $\omega_1 - \omega_2$ is swept across the resonances of interest while monitoring the anti-Stokes flux. In gas mixtures, the following information is obtained from the spectra using (16).

• composition since each molecular species has a particular set of vibrational resonances which can seldom be confused with that of other species;

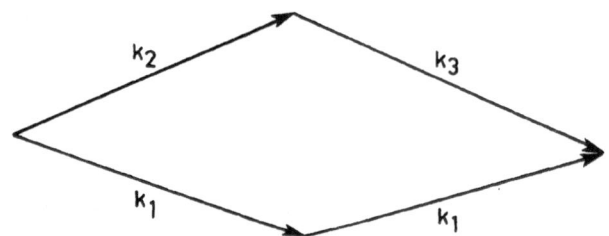

Fig. 5. Wave vector diagram for crossed-beam phase-matched CARS or BOXCARS.

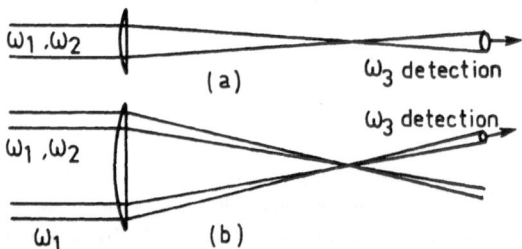

Fig. 6. Experimental beam arrangement: a) conventional CARS; b) BOXCARS. In BOXCARS, a conventional CARS beam is also emitted in the direction of ω_2; this beam is actually 10 to 50 times stronger than the BOXCARS beam.

- temperature since the frequency of any particular vibrational mode depends slightly on the rotational quantum number (notably in Q-branch transitions, which are generally used in CARS) and on the vibrational quantum number; the resulting splitting can be used to monitor populations in distinct rovibrational states and to deduce the rotational and vibrational temperatures using the corresponding Boltzmann coefficients.

We note, however, that the existence of the nonresonant susceptibility poses a problem with the detection of trace species in mixtures, since the non-resonant contribution from the diluent gases may swamp the Raman-resonant part of the trace species of interest. As a matter of a fact, detection sensitivities are in the range of 10^2 to 10^4 ppm for most cases of interest. These figures can be improved by a factor of about 30 if advantage is taken of the different tensor properties of $\underset{=NR}{\chi}$ and $\underset{=R}{\chi}$ (polarisation CARS).[39]

Advantages of CARS. CARS offers many advantages over other optical methods for non-intrusive spatially-resolved diagnostics of gases and reactive media:

- Spatial resolution is excellent,

- the signal is emitted in a well-collimated beam, which makes discrimination against stray light easier,

- the spectral position of the signal, to the anti-Stokes side of the pump, makes it easier to reject fluorescence interference (which usually lies to the Stokes side of the pump),

- the signal strength is considerable; using conventional pulsed solid-state and dye lasers, the number of photons collected in a typical experiment is about 10^5 to 10^{15}, i.e. 5 to 10 orders of magnitude larger than that collected in a spontaneous Raman scattering experiment.

All these advantages justify the introduction of CARS as a diagnostics tool in reactive media. This application is to date the most important one. Other applications, such as high resolution molecular spectroscopy or chemical analysis of biological samples, also are attractive but shall not be discussed here.

Instrumental details

We here describe the CARS spectrometer in use at ONERA. The optical components for the laser sources and the beam combining optics are bolted onto a light-weight, portable, rigid 50 cm × 150 cm cast aluminium table (Fig. 7). The passively Q-switched Nd:YAG oscillator with two amplifiers and one frequency doubler delivers 200 mJ at 532 nm in 10 ns pulses at 1 to 10 Hz (ω_1 beam). The output is single frequency over 95 % of the shots and presents a spectral jitter under $\pm$ 0.01 cm^{-1}. Comparable or slightly improved characteristics can also be obtained with injection seeding. About 40 mJ of green are used to pump the dye chain. This one is composed of a dye laser and one amplifier stage and produces the "Stokes" beam at ω_2. The dye laser can be tuned with a fixed, high-incidence grating and a rotating mirror. The linewidth is 0.7 cm^{-1}; it can be reduced to 0.07 cm^{-1}

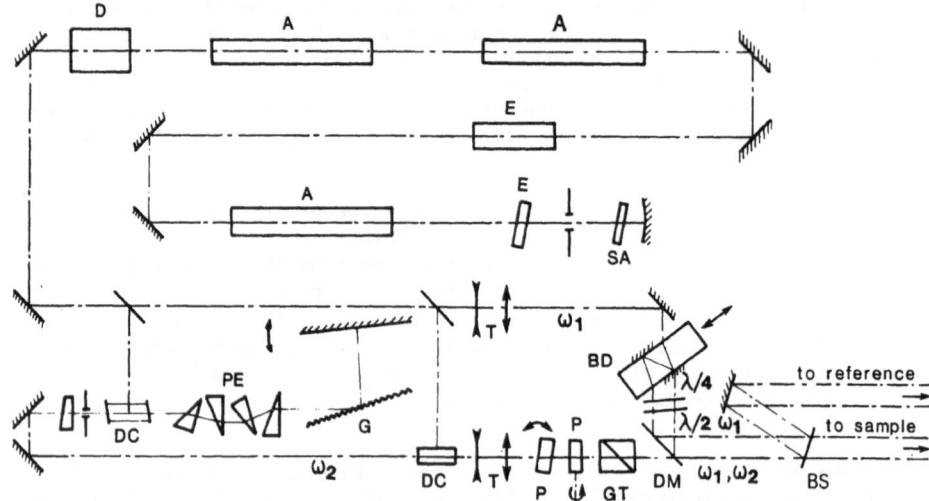

Fig. 7. Laser source assembly. A: Nd:YAG amplifier; BD: parallel plate for
production of parallel beams for BOXCARS (BOXCARS arrangement
shown, translation of the plate allows passage to collinear
arrangement without loss of alignment); BS: beam splitter for
reference channel; D: DKP doubler; DC: dye cell; DM: dichroic
mirror; E:Fabry-Perot etalon; G: grating; GT: Glan-Thompson prism;
P: AR coated parallel plate for beam translation; PE: prism
expander; SA: saturable absorber; T: telescope; $\lambda/4$ and $\lambda/2$ are
quarter-wave and half-wave plates respectively; ω_1: "laser" beam;
ω_2: "Stokes beam".

through insertion of a prism beam expander. The dye chain delivers from 1
to 5 mJ of tunable radiation in a diffraction-limited beam over the useful
CARS range of 560-700 nm. A broadband mode of operation is also provided
for the dye laser, giving about 100 cm^{-1} linewidth; this mode is used for
multiplex CARS experiments in conjunction with a spectrograph and an
optical multichannel analyser.[40] In this case, an interference filter is
used for the tuning. Various components are used for beam-matching and
superposition, and for simultaneous non-resonant background cancellation
and improved spatial resolution using BOXCARS.[41]

The detection assembly, including the reference leg, is installed on
a separate table of area 50 cm × 150 cm (Fig. 8). All focusing lenses are
AR coated air-spaced achromats. The anti-Stokes signals are filtered by
means of compact double monochromators preceeded by dichroic filters to
prevent breakdown on the entrance and intermediate diaphragms; detection
is done with PM tubes which are mounted in the same bay as the signal
processing electronics (see below) to avoid RF interference. Light is
piped to them by means of 1 mm-diameter optical fibers. The anti-Stokes
signal levels in the sample and reference channels are adjusted at
approximately 10^4 photoelectrons per shot, which corresponds to a Poisson
uncertainty of about 1 %. Higher fluxes may cause saturation, lower fluxes
result in unacceptable uncertainty levels. The sample and reference
channels are matched carefully, especially in BOXCARS experiments.

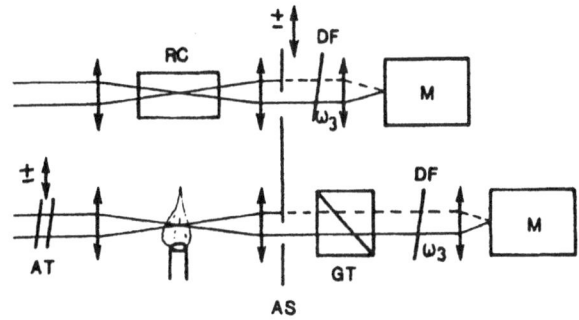

Fig. 8. Schematic of sample and reference channels;
AS: movable aperture stop for operation with
parallel beams or crossed beams (cross-beam
position shown here); AT: movable
attenuators; DP: dichroic filters; M:
monochromator and detector; RC: reference
cell.

In both channels, the signal level is maintained at its prescribed
level of 10^4 photoelectrons by adjusting the pump powers with attenuators.
The photocurrent pulses are treated by an electronic device that gates
them, calculates their ratios, square roots and average for a fixed number
n of shots (n = 1 to 10 in practice). The electronics unit also rejects
shots which do not fall with ± 35 % of the mean in the reference leg, and
tunes the dye laser after the n shots have been collected.

For multiplex CARS, a spectrograph and an optical multichannel
analyser (EGG OMA3) are used for both the signal and the reference. The
dispersive element in the spectrograph is a 2100 lines/mm,
aberration-corrected, concave holographic grating with f = 750 mm. The net
spectral resolution is 0.7 cm^{-1}. Both signal and reference spectra are
recorded simultaneously on the vidicon and ratioed channel by channel;
square roots, and averages if necessary, are subsequently calculated.
Recording the reference spectrum is a vital requirement since the dye
laser spectrum is not reproducible and exhibits appreciable modulation.

The technique of background suppression using the tensor properties
of the non-linear susceptibility has been studied in detail for collinear
beams, and several possible polarisation arrangements have been
described.[39,40] In BOXCARS, some flexibility is afforded by the
availability of two spatially distinct "laser" beams at ω_1, which can have
different polarisations. A discussion of that problem is found in Ref. 41.

Results

The feasibility of concentration measurements in flames by CARS was
shown in 1973.[6] Since then, many experiments have been carried out and
many species have been detected.

The bulk of the effort in combustion has concerned internal
combustion engines[43-46] and simulated jet engine combustors.[47-49]
Meanwhile, studies of rarefied gases in discharges were also initiated and
shown to yield very useful informations, especially on the homonuclear

diatomics. Finally, investigations were also undertaken quite early on the subject of electronic resonance enhancement, which greatly improves the detection sensitivity of CARS. All three subjects will now be covered in turn.

Studies of premixed turbulent combustion. Premixed turbulent flames offer interesting possibilities to study turbulent combustion and permit confrontation between experiments and calculations. In addition, there is direct application to the design of afterburners for turbojet engines, if the flow velocity is fast enough. An attempt to measure some aspects of the temperature profile was made using multiplex CARS at ONERA.[50] It has permitted measurement of the temperature and of its fluctuations. These results are presented here.

The experimental device is a two-dimensional duct, shown in Fig. 9. Most of the tests have been performed with a flame stabilized by a 2D hot jet, parallel to the main stream. The fuel is methane, at different equivalence ratios. The most characteristic feature is the high level of turbulence in the flow, even without combustion, because the velocities on both sides of the separating plate differ by 55 m/s, and also because the turbulence of the incoming premixed stream is itself very large: 4 to 8 m/s of velocity fluctuation (R.M.S.) compared with the laminar flame speed of methane and air, which is of the order of 0.5 m/s. This turbulent flame had been studied at first by wall pressure measurements.[51] Results show that turbulent combustion is probably controlled by both turbulence and chemistry. The structure of the flame as revealed by high speed shadowgraphy appears to be a thick grainy zone with slow undulations. When the turbulence intensity becomes very high, small scale fluctuations begin to appear within the flame front itself. This phenomenon could lead to the thickening of the turbulent flame by small scale turbulence while this flame remains wrinkled by the largest fluctuations.[52]

Prior to presenting the data, we recall that the temperature accuracy deduced from CARS measurements in a furnace ranges from 20 K at 300 K to 50 K at 1600 K.[53] In the present experiment, there are two locations in the burner where the temperature should be constant: at the exit of the pilot flame duct and in the fresh mixture. We found respectively 2000 K ± 170 K and 560 K ± 35 K. The slight increase with respect to the furnace values is explained by the turbulence in the flow.

Figure 10 shows the pdf obtained for the pure mixing of hot gases from the pilot with the main stream without methane. As the observation point is moved through a vertical section, the pdf shows one peak and progressively another one. In the middle, the pdf displays a very wide shape, in which it is possible, sometimes, to distinguish two broad bumps.

Figure 11 shows the pdf found at the same positions, when methane is added to the main stream and the combustion occurs. The comparison with figures 10a and 10b is interesting. In the first section, 42 mm after the beginning of the mixing, only a very small amount of fuel has been burnt: the pdf's at y = 25 mm and 10 mm look very similar with and without combustion; the same tendency, but less pronounced, is visible for those at 15 mm. These reductions in peak intensity are due to combustion; consequently, an increase of probability density is noticed at a higher temperature, but this increase does not appear at the adiabatic temperature which is here 2350 K; for y = 15 mm, a peak appears at 1800 K, and for y = 20 mm, no well defined peak is seen and the probability density is increased over all the range between 1000 and 1800 K.

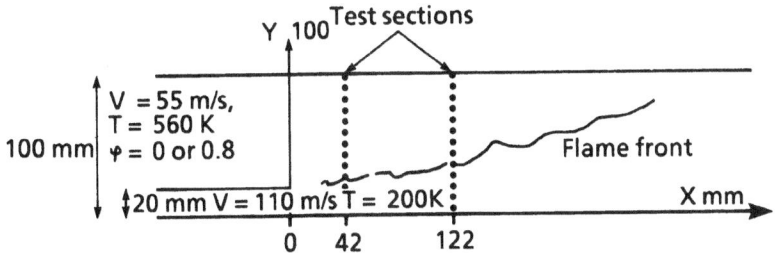

Fig. 9. Diagram of turbulent combustor.

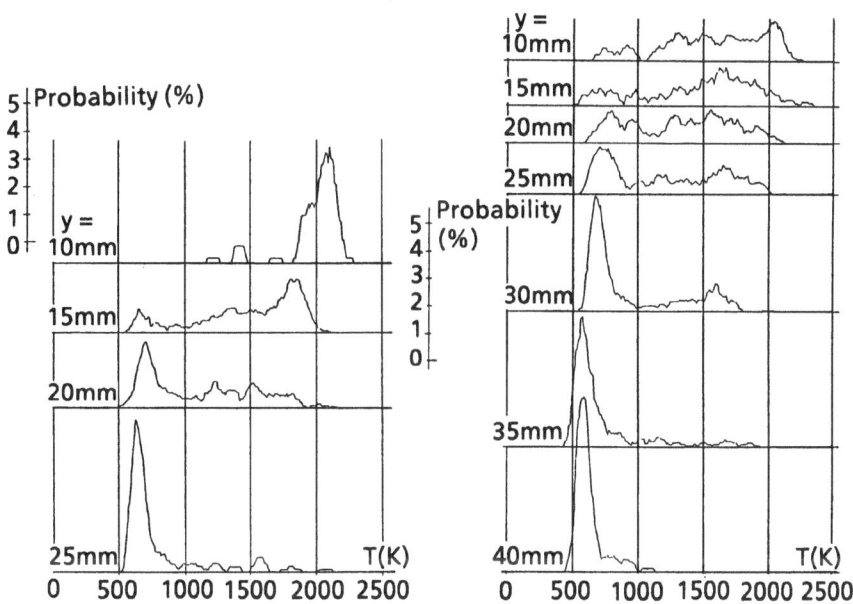

Fig. 10. Temperature pdf obtained for the pure mixing of hot gases from the pilot with the main stream.

In the second section, the influence of the combustion is quite significant. The low temperature peaks are shifted towards high temperatures in a manner quite comparable to the non combusting case, but their areas are clearly smaller. High temperature peaks are appearing also, but they are again very wide and their location is not the adiabatic flame temperature; they are still centered at a lower temperature. Heat losses can be invoked close to the walls, for instance for the y = 10 mm pdf. But this effect is not expected to be pronounced in the center of the channel at y = 20 mm and higher. Moreover radiation losses are not likely to be strong. In fact, results indicate the reaction rate to be too small: the hot peaks in the section x = 42 mm are found at about 1800 K, and at x = 122 mm they are at 2000 K. They would probably reach closer to 2350 K farther downstream.

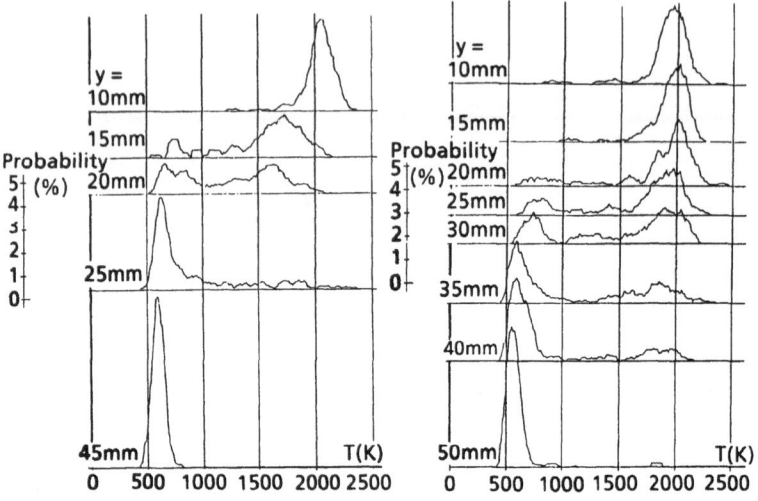

Fig. 11. Temperature pdf obtained in the turbulent flame stabilized by the pilot flame ; φ = 0.8 in the main stream.

In both sections, the characteristic time of the chemistry is then not short enough to make the flame purely wrinkled; this tends to be less true in the second section, where the turbulence characteristic time is probably larger (the length scale being larger and the turbulence kinetic energy not larger).

Studies of low pressure gases. The application of CARS to the study of rarefied gases has many features in common with the application to combustion studies. However the use of scanning CARS is required for optimal instrumental sensitivity. Recently CARS was used for discharge diagnostics.[54,55] Two homonuclear diatomics have been studied: i) N_2 in a glow discharge,[56] ii) H_2 in a magnetic multicusp.[57] The latest results are presented in this Section.

Research grade N_2 was flowed through a 2-cm-diameter pyrex tube. The discharge length was 40 cm, the current 80 mA. The spectra of figure 12 are typical. They were recorded in the discharge region at 5 mbar pressure using collinear CARS for all vibrational levels but ϑ = 0-5; for the latter BOXCARS was employed. The rotational temperature was 550 K for all of these levels, with an uncertainty ranging from ± 20 K for the first 10 vibrational levels and degrading to ± 100 K and beyond for the highest ones (Fig. 13). The vibrational distribution was clearly non-Boltzmann

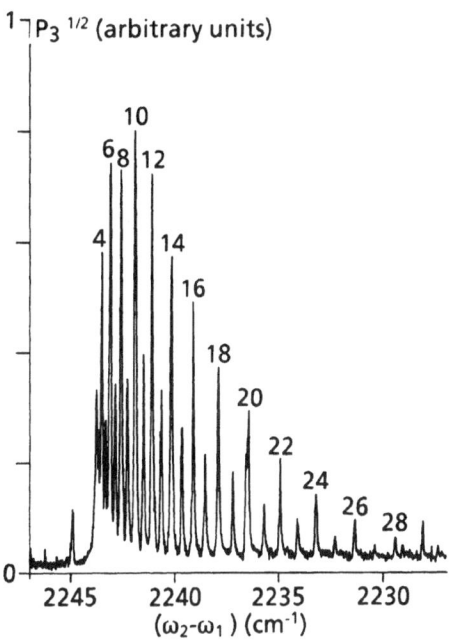

Fig. 12. Collinear beam CARS spectrum of
N_2 (ϑ = 3) in glow discharge at
5 mbar pressure. The discharge
current is 80 mA. Q-lines are
labelled by their J-numbers.

(Fig. 14). Note that CARS only gives population differences, which leaves some uncertainty on the absolute number densities of the high ϑ's. The most reasonable assumption (conservation of the slope between ϑ = 13 and 14) was used to project the value of ϑ = 15. Population calculations were also done assuming slope change of ± 10 % (adjacent lines) to show the most extreme distributions compatible with the instrumental error. The experimental data can be compared with the simple model of Treanor as modified by Gordiets and coworkers.[58] The discrepancy appears large at high ϑ; it can be explained by the effect of wall and electronic collisions which are not accounted for in the model.

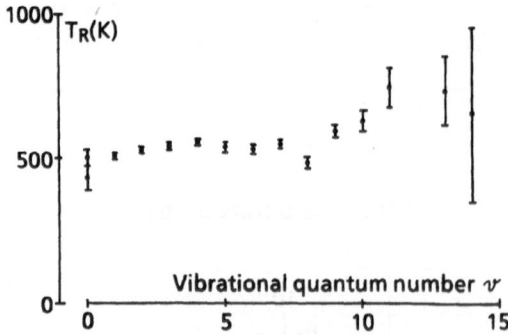

Fig. 13. Rotational temperature for each
vibrational state obtained under
conditions of Fig. 12. Vertical bars
indicate measurement precision.

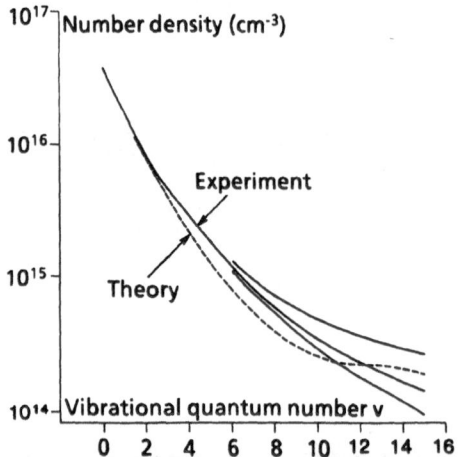

Fig.14. Vibrational distribution obtained
with the same conditions as in
fig. 12.

The analysis of a low-pressure H_2 discharge with magnetic multipolar confinement of primary electrons emitted by a tungsten filament has begun several years ago in our laboratory.[59,60] Measurements were done in a 16-cm-dia., 20 cm-high reactor, at 55 μbar pressure, and using a 90 V, 10 A discharge. The main conclusions of these previous studies are the following:

1) the rotational distribution strongly deviates from the Boltzmann law for $J \geqslant 5$; the rotational temperature deduced from $J < 5$ is about 530 K;

2) the vibrational distribution nearly obeys the Boltzmann law up to $\vartheta = 3$, highest vibrational level detected; the vibrational temperature is about 2400 K;

3) the H_2 partial pressure is about 70 % of the pressure measured by the gauge, indicating some level of dissociation;

4) the atomic fraction of H atoms is small ($\simeq$ 5 %), but their apparent translational temperature is high ($\simeq$ 4500 K), yielding a high partial pressure.

The complex set of rovibrational number densities measured is presented in Fig. 15.

New measurements have been attempted to assess the inhomogeneities within the reactor and the kinetics of molecular excitation by electrons and relaxation by walls. These are illustrated in Figs. 16 and 17. Figure 16 shows the rise and fall of the rotational temperature in $\vartheta = 0$ and of the vibrational temperature (derived from the densities of $\vartheta = 0$ and $\vartheta = 1$) under excitation by a rectangular discharge pulse of 10 A current and 1 ms duration. Figure 17 presents the radial steady-state distributions of the vibrational and rotational temperatures and their rise under excitation by the pulse leading edge. The difference between them is striking, as the rotational temperature presents a strong gradient between the center and the wall, while the vibrational distribution is flat. This is due to the difference in deactivation rates at the wall. A discussion of these phenomena can be found elsewhere.[61]

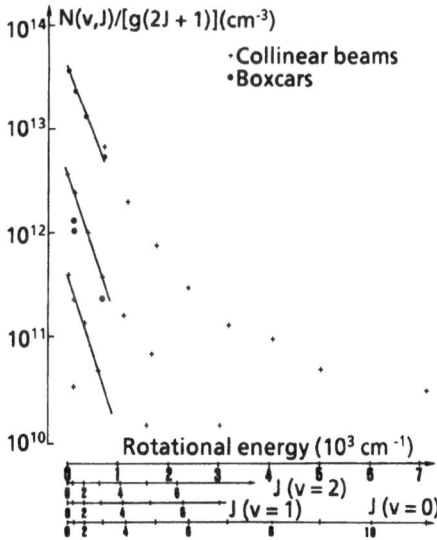

Fig. 15. H_2 rovibrational population distribution at the center of the generator for 55 μbar and with 90 V, 10 A discharge.

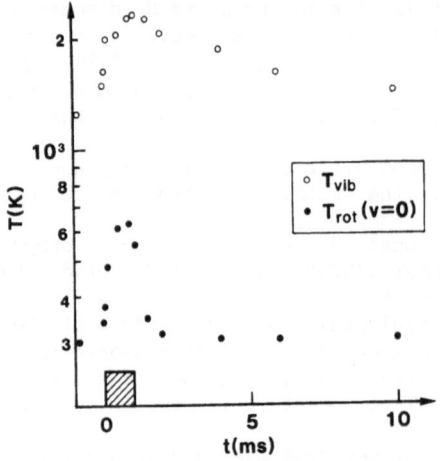

Fig. 16. Response of temperature values to 1 ms, 10 A pulse; 55 μbar pressure; notice overshoot of rotational temperature above steady-state value of 550 K.

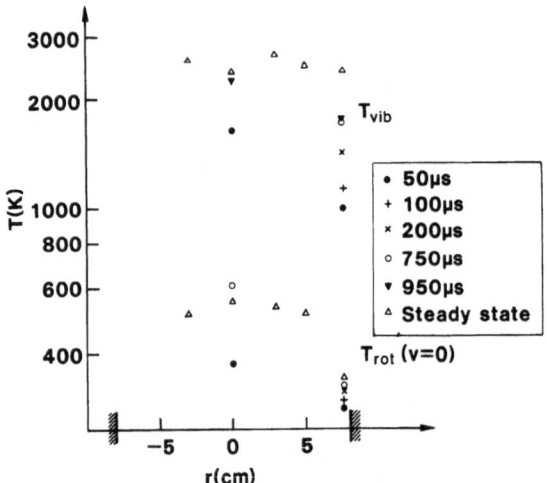

Fig. 17. Radial profiles and temporal response of rotational and vibrational temperatures; same conditions as for Fig. 16; the origin of times is the front edge of the discharge pulse.

Studies of OH by resonance-enhanced CARS. The presence of the non-resonant background susceptibility is a major difficulty experienced with CARS in trying to detect radicals in reacting gas mixtures. Electronic resonance-enhancement of the CARS susceptibility improves the detectivity as was theoretically[62] and experimentally[63,64] demonstrated. The OH radical is of particular interest for combustion. The study was conducted in a high-pressure burner.

The resonance CARS spectrometer used is a three-color system consisting of two frequency-doubled homemade fixed-frequency dye lasers and a commercial frequency-doubled tunable dye laser all pumped by a Nd:YAG laser chain. A Mac-Kenna flat flame burner is mounted at the bottom of a high-pressure vessel. It is 20 mm in diameter and is surrounded by a porous ring of 50 mm external diameter used to generate a shrouding flow of N_2 which prevents the flame from heating the vessel walls. The laser frequencies were tuned close to the $P_1(0-0)7.5$, $R_1(1-0)5.5$ and $R_1(1-1)5.5$ lines of the $A^2\Sigma^+ - X^2\pi$ electronic transition. The variation of OH concentration versus the distance from the burner surface and the pressure has been obtained using scanning CARS. Spectra were obtained at triple resonance.

The OH profiles (Fig. 18) show that OH is formed in the reaction zone. Its concentration rapidly decreases and then stabilizes at an equilibrium value. Also, the OH concentration does not increase proportionally to the pressure. It seems to depend linearly on the methane flow. Resonant CARS has here a sensitivity of 10^{13} cm^{-3} ($\simeq$ 1 ppm) at atmospheric pressure on the OH radical. The sensitivity remains excellent up to 10 atmospheres, and probably acceptable to 30-100 atmospheres.

CONCLUSION

CARS has become an essential diagnostics tool for combustion chemists and engineers, and has also allowed significant breakthroughs to be made

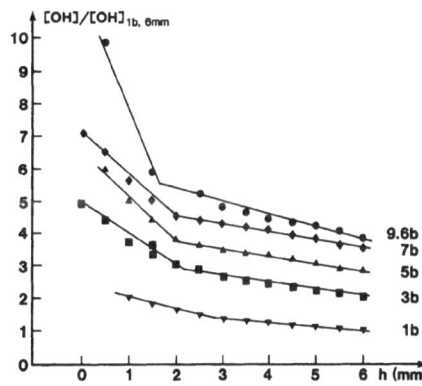

Fig. 18. OH profiles versus the distance from the burner for different vessel pressures: a) 1 bar; b) 3 bar; c) 5 bar; d) 7 bar; e) 9.6 bar.

in the Raman spectroscopy of gases. Its success stems from its high luminosity, which enables one to extract quantitative information on the composition and on the thermodynamic equilibrium in reactive media even if these produce a considerable amount of stray light; furthermore, this information is obtained with excellent spatial resolution and without causing aerodynamic or chemical perturbations.

In a typical turbulent combustion experiment, the CARS instrument can now measure temperature in 10 ns, with a spatial resolution of 2 mm and a temperature uncertainty of 40 K at 2000 K using N_2 spectra. Under the same conditions, it will also detect about 0.1 % CO. These measurements can be made at the rate of 2 Hz; this rate is currently limited by the detector. The rates actually needed for real time measurements in turbulent flows are very high (1 to 10 kHz) and are not feasible with current technology.

With the analysis of stable media, scanning CARS presents substantial gains in sensitivity (10 to 100 typically) and in temperature measurement accuracy. A detection sensitivity of 10^{11} cm^{-3} has been obtained with H_2 near 500 K in a plasma study.

CARS indeed has become an essential factor of technical progress in the industry and in the research laboratory. The number of its applications will undoubtedly grow. Also, resonance-enhanced CARS which offers gains in detection sensitivity of several orders of magnitude, thanks to susceptibility enhancement at one-photon electronic resonance, should rapidly expand the number of applications.

REFERENCES

1. G.F. Widhopf and S. Lederman, AIAA J., 9: 309 (1971).
2. M. Lapp, L.M. Goldman and C.M. Penney, Science, 175: 1112 (1972).
3. "Laser Raman Gas diagnostics", Proceedings of the Project SQUID Laser Raman Workshop on the Measurement of Gas Properties, May 10-11, 1973, Schenectady, M. Lapp and C.M. Penney, eds., Plenum Press, New York, London (1974).
4. "Proceedings of Project SQUID Workshop on Combustion Measurements in Jet Propulsion Systems", R. Goulard, ed., Purdue University, Lafayette, Indiana (1975).
5. "Experimental diagnostics in gas phase combustion system", Progress in Astronautics and Aeronautics, 53, B.T. Zinn, ed., Martin Summerfield Series Editor (1977).
6. P.R. Regnier and J.P.E. Taran, Appl. Phys. Lett., 23: 240 (1973).
7. R.W. Terhune, Bull. Amer. Phys. Soc., 8: 359 (1963).
8. P.D. Maker and R.W. Terhune, Phys. Rev., 137: A801 (1965).
9. E. Yablonovitch, N. Bloembergen, and J.J. Wynne, Phys. Rev., B3: 2060 (1971).
10. S.A. Akhmanov, V.G. Dmitriev, A.I. Kovrigin, N.I. Koroteev, and A.I. Kholodnykh, JETP Lett., 15: 425 (1972).
11. M.D. Levenson, C. Flytzanis, and N. Bloembergen, Phys. Rev., 6: B3962 (1972).
12. W.G. Rado, Appl. Phys. Lett., 11: 123 (1967).
13. G. Hauchecorne, F. Kerherve, and G. Mayer, J. Phys., 32: 47 (1971).
14. F. De Martini, G.P. Giuliani, and E. Santamato, Opt. Comm., 5: 126 (1972).
15. J. Fiutak and J. Van Kranendonk, Can. J. Phys., 40: 1085 (1962).
16. A. Omont, E.W. Smith, and J. Cooper, The Astroph. J., 175: 185 (1972).
17. S. Druet, B. Attal, T.K. Gustafson, and J.P.E. Taran, Phys. Rev., A18: 1529 (1978).
18. S.Y. Yee, T.K. Gustafson, S.A.J. Druet, and J.P.E. Taran, Opt. Comm., 23: 1 (1977).

19. S.Y. Yee and T.K. Gustafson, Phys. Rev., A18: 1597 (1978).
20. S.A.J. Druet and J.P.E. Taran, "Coherent anti-Stokes Raman Spectroscopy", in "Chemical and Biochemical Applications of Lasers", vol. 4, C.B. Moore, ed., Academic Press, New York (1979).
21. C.J. Borde, J.L. Hall, C.V. Kunasz, and D.G. Hummer, Phys. Rev., A14: 236 (1976).
22. J. Borde and C.J. Borde, J. Mol. Spectrosc., 78: 3530 (1979).
23. S.A.J. Druet, J.P.E. Taran, and C.J. BordE, J. Phys., 40: 819 (1979); addendum, ibidem, 41: 183 (1980).
24. S.A.J. Druet and J.P.E. Taran, Progress in Quant. Elec., 7: 1 (1981).
25. R.P. Feynman, Quantum Electrodynamics, Benjamin, New York (1962).
26. J.F. Ward, Rev. Mod. Phys., 37: 1 (1965).
27. A. Yariv, IEEE J. Quant. Elect., QE13: 943 (1977).
28. D.C. Hanna, D. Cotter, and M. Yuratich, "Non-linear optics of free atoms and molecules", D.L. McAdam, ed., Springer Series in Optical Sciences, 17, Springer Verlag, Berlin, Heidelberg, New York (1979).
29. G.L. Eesley, J.Q.S.R.T., 22: 507 (1979).
30. N. Bloembergen, H. Lotem, and R.T. Lynch, Indian J. Pure Appl. Phys., 16: 151 (1978).
31. Y. Prior, A.R. Bogdan, M. Dagenais, and N. Bloembergen, Phys. Rev. Lett., 46: 111 (1981).
32. A.R. Bogdan, Y. Prior, and N. Bloembergen, Opt. Lett., 6: 82 (1981).
33. N. Bloembergen, A.R. Bogdan, and M.W. Downer, in "Laser Spectroscopy V" McKellar, Oka and Stoicheff, ed., Springer Verlag, Berlin, Heidelberg, New York (1981).
34. L.A. Carreira, L.P. Gross, and T.B. Malloy, J. Chem. Phys., 69: 855 (1978).
35. G. Grynberg, J. Phys. B. Atom. Mol. Phys., 14: 2089 (1981).
36. M. Pealat, M. Lefebvre, J.P. Taran, and P.L. Kelley, Phys. Rev., 38: 1948 (1988).
37. F. Ouellette and M.M. Denariez-Roberge, Can. J. Phys., 60: 877 (1982);
38. A.C. Eckbreth, Appl. Phys. Lett., 32, 421 (1978).
39. L.A. Rahn, L.J. Zych, and P.L. Mattern, Opt. Comm., 30: 249 (1979).
40. W.B. Roh, P.W. Schreiberand, and J.P.E. Taran, Appl. Phys. Lett., 29: 174 (1976).
41. B. Attal, M. Pealat, and J.P.E. Taran, J. Energy, 4: 135 (1980).
42. J.J. Song, G.L. Eesley, and M.D. Levenson, Appl. Phys. Lett., 29: 567 (1976).
43. I.A. Stenhouse, D.R. Williams, J.B. Cole, and M.D. Swords, Appl. Opt., 18: 3819 (1979).
44. D. Klick, K.A. Marko, and L. Rimai, Appl. Opt., 20: 1178 (1981).
45. L.A. Rahn, S.C. Johnston, R.L. Farrow, and P.L. Mattern, in "Temperature, its measurement and control in science and industry", 5: 609 (1982), J.F. Schooley ed., American Institute of Physics, New York.
46. G.C. Alessandretti and P. Violino, J. Phys. D: Appl. Phys., 16: 1583 (1983).
47. B. Attal, M. Pealat, and J.P. Taran, J. Energy, 4:, 135 (1980).
48. G.L. Switzer, W.M. Roquemore, R.P. Bradley, P.W. Schreiber, and W.B. Roh, Appl. Optics, 18: 2343 (1979).
49. A.C. Eckbreth and R.J. Hall, Combustion and Flame, 36: 87 (1979).
50. P. Magre, P. Moreau, G. Collin, and M. Pealat, Combustion and Flame, 71: 147 (1988).
51. V.P. Singh, R. Borghi, and P. Moreau, Flammes contre rEactions dans les Ecoulements, 2eme Symposium International sur la Dynamique des REactions Chimiques, Padoue, December 1975.
52. R. Borghi, in "Recent advances in the aeronautical sciences" (C. Bruno, C. Casci eds.), Plenum Press (1984).
53. M. Pealat, P. Bouchardy, M. Lefebvre, and J.P. Taran, Appl. Opt., 24: 1012 (1985).
54. V.V. Smirnov and V.I. Fabelinskii, J.E.T.P. Lett., 28: 427 (1978).

55. S.I. Valyanskii, A. Vereshchaghin, V. Vernke, A. Yu. Volkov, P.P. Pashinin, V.V. Smirnov, V.I. Fabelinskii, and P.L. Chapovskii, Soviet J. Quantum Electronics, 14: 1226 (1984).

56. B. Massabieaux, G. Gousset, M. Lefebvre, and M. PEalat, J. Phys., 48: 1939 (1987).

57. M. Pealat, J.P. Taran, M. Bacal, and F. Hillion, J. Chem. Phys., 82: 4943 (1985).

58. C.F. Treanor, J.W. Rich, and R.G. Rehm, J. Chem. Phys., 48: 1798 (1968); B.F. Gordiets, Sh. Mamedov and L.A. Shelepin, Soviet Physics J.E.T.P., 48: 648 (1974).

59. M. Pealat, J.P. Taran, J. Taillet, M. Bacal, and A.M. Bruneteau, J. Appl. Phys., 52: 2687 (1981).

60. M. Pealat, J.P. Taran, M. Bacal, and F. Hillion, J. Chem. Phys., 82: 4943 (1985).

61. M. Lefebvre, M. Pealat, J.P. Taran, M. Bacal, and R.J. Hutcheon, "Coherent anti-Stokes Raman Scattering Study of the Dynamics of a Multipolar Plasma Generator", to be published.

62. S.A.J. Druet, B. Attal, T.K. Gustafson, and J.P.E. Taran, Phys. Rev., A18: 1529 (1978).

63. B. Attal, D. Debarre, K. Muller-Dethlefs, and J.P. Taran, Rev. Phys. Appl., 18: 39 (1983).

64. B. Attal-Tretout, S.C. Schmidt, E. Crete, P. Dumas, and J.P. Taran, J.Q.S.R.T., in press.

LASER SPECTROSCOPY APPLIED TO COMBUSTION

Antonio D'Alessio[*], Antonio Cavaliere[**]

[*] Dipartimento di Ingegneria Chimica,Università di Napoli
Federico II - P.le V. Tecchio - 80125 NAPOL1
[**] Dipartimento di Ingegneria Chimica, Università di Pisa
Via Diotisalvi 2 - 56100 PISA

INTRODUCTION

Researches on combustion and spectroscopy have been closely related since the very beginning at the time of Kirchhoff, Bunsen and Faraday. Since then every generation of spectroscopists gave its contribution to the understanding of combustion and used these sources for a deepening of their field. The purpose of this short paper is to highlight the actual relationship between these two fields of research.

Today combustion may be defined as the study of chemically reactive turbulent multiphase flows. Therefore when such media are investigated with laser beams there are some peculiar requirements that should be clearly defined in order to contribute significantly to the progress of the researches in this area.

The main route of the oxidation of the fuel at high temperature goes through the progressive pyrolysis of the complex hydrocarbons to small radicals, as C_2H_5 or CH_3, which react subsequently with small radicals or atoms, as OH, O, H, N, thus forming final products of combustion as H_2O, CO_2, CO, NO_x. However, particularly when the fuel and oxygen are not premixed, additional pyrolytic routes of the fuels should be considered which produces carbon polymorphs of high molecular mass and very complex structures. Polycyclic aromatic hydrocarbons, tar and soot particles are all different stages of this process.

Therefore both the spectroscopy of small radicals or atoms and that of large carbon containing structures are necessary tools for the chemical characterization of the combustion process. The main exothermic reactions which are responsible for the heat release, are very fast compared to diffusional rates of heat and species, responsible for the propagation of the flame. This means that in laminar conditions these reactions evolve in very small regions of the whole reactor.

In premixed flames the thickness and the velocity of the flame front is univocally determined by the initial conditions in the unburnt gas. Typical values for stoichiometric mixtures of hydrocarbon fuels with air are of the order of 100 um for the flame thickness and of 100 usec for the transit time of the flame front. In turbulent conditions fluid-dynamic fluctuations interact with the flame front according to their time and length scales. Different regimes can be identified which range from simple convolutions of the flame front (flamelet regime) to thickening of the flame (distributed reaction) which can degenerate in local quenching of the flame. Therefore the spatial distribution of the reacting species is of great interest in order to identify the different mechanisms of flame propagation. Few selected species are sufficient in order to determine the whole behaviour; for instance some radicals like OH, CH, C_2 can be a good signature of this type of investigation.

In diffusion flames there are not temporal or spatial scales linked to the initial state of the system; local chemical composition and temperature as well as spatial gradients of these quantities determine respectively the chemical and diffusion time scales, so that they cannot be known independently from the fluid-dynamic pattern. Therefore it is questionable whether flamelet structures are still present in turbulent regimes [1,2] because of uncertainty of these scales.

Therefore the study of turbulent diffusion flames has to be performed with great flexibility in terms of spatial and temporal resolution and higher the number of species detected better is the knowledge of the flame structure. The turbulent aspects of combustion require that the emphasis of laser diagnostics is on two dimensional, and hopefully three dimensional, quantitative imaging methods. In addition the time resolution of the measurements becomes of the greatest importance.

In practical combustion systems the flow is laden with liquid droplets or solid particles whose physical and chemical characterization is also a target for the applied spectrocopist.

These statements will be qualified in the following sections of the paper through reviewing the recent work done in the field of combustion by applied spectroscopists and relating examples among the researches carried out by the authors at the University/CNR group of Naples.

The first section will be devoted to the effects and methods applied to homogeneous combustion of simple gaseous fuels. The following section will be focused on the optical diagnostics of the carbon polymorphs produced in regions and/or conditions where the ratio between the fuel and the oxidant is below the stoichiometric one. Finally the possible optical and spectroscopic diagnostics of liquid droplets and other solid particles will be shortly addressed.

LASER SPECTROSCOPY OF SMALL MOLECULES PRODUCED DURING COMBUSTION

The work on the laser spectroscopy of combustion has been reviewed in a recent book by Alan Eckbreth [3] which covers the literature up to the end of 1987. The author discusses the most popular methods for determining temperature and species concentration namely: Spontaneous Raman and Rayleigh Scattering, Coherent Anti-Stokes Raman Spectroscopy (CARS) and Laser Induced Fluorescence (LIF). Only the final chapter of his book is devoted to field techniques where he summarizes the literature on 1-D and 2-D Imaging and on optical tomography.

Rayleigh scattering from gases is a very strong effect with typical angular cross of the order 10^{-27} cm^2/sr but it is not a specific signature for different components in the gaseous mixture, because it is an elastic effect. Furthermore it suffers of the interference of the elastic Mie scattering due to liquid or solid particles eventually present in the scattering volume although this inconvenience can be overcome using photon-correlation techniques [4]. Nevertheless Rayleigh Scattering has been a precious tool for the study of the structure of turbulent jets and flames. Dibble and coworkers carried out point measurements at high count rate and were able to follow fluctuations of temperature and density in turbulent premixed and diffusion flames up to 15 KHz [5].

Their measurements were a direct experimental verification of diffusion against the gradient as predicted by the "flamelet" theory of turbulent combustion by Bray, Libby and Moss [6]. More recently Rayleigh scattering has been used by Long and coworkers [7] to obtain the time-resolved three dimensional concentration field in a turbulent jet. Their data have constituted the basis of a critical reevaluation by Bilger [2] of the flamelet theories in different non premixed structures of flames.

Rayleigh scattering has also been employed in an original way in the characterization of the early phases of hydrocarbons ignition with ultrashort discharges by Borghese and coworkers [8]. They were able to distinguish, in the first hundrenth nanoseconds, the Thompson scattering due to free electrons from the Rayleigh one due to atoms and molecules and, at a later stage, to measure the dissociation fraction of nitrogen through the measurement of the depolarized components of the scattered light.

Spontaneous Raman cross sections are three orders of magnitude weaker than the Rayleigh ones but Raman spectrum is a specific signature for many species present in combustion systems. The Raman spectra have been also employed for determining the temperature inside flames because, in a good approximation, the vibronic levels are populated according to the Boltzmann statistics.

Again the most interesting application of this effect is the 2-D mapping of turbulent flames or time resolved measurements inside engines. The group of Dibble and coworkers at Sandia has carried out a very systematic analysis of the composition and temperature of turbulent H_2-air and CH_4-air diffusion flames [9,10].

An other interesting application of Raman scattering has been the determination of the fuel-air ratio before ignition in internal combustion engines [11] and in high pressure diesel-like bombs [12].

The main inconvenience of Spontaneous Raman is that it can not be applied to combustion regimes where carbon polymorphs of pyrolytic origin are present. In this case there is the interference with the broadband fluorescence emitted by the large aromatic molecules, which will be discussed later, and, in the case of pulsed laser, the incandescence due to soot particles [13].

These inconveniences presented by Spontaneous Raman are minimized employing non linear Raman effects, among which the most popular, in the combustion community, is the Coherent Anti-Stokes Raman Spectroscopy (CARS). This method has been widely employed in laboratories as well as in furnaces and engines for local measurements of temperature and concentration of major constituents of the gas. An almost complete review of all the theoretical and experimental aspects of this technique, pioneered by Taran [14] at ONERA in the midseventies, is contained in the book by Eckbreth quoted before.

Laser Induced Fluorescence (LIF) is certainly a more promising method for time and space resolved determination of the temperature, composition and, hopefully, also of the velocity in combustion systems; consequently a large number of papers have been devoted to this approach. Recent reviews by Eckbreth [3,15] and some papers by Hanson [16,17], who covers the 2-D applications are introductive guides to this technique. Since LIF is based on dipole allowed transitions, the cross sections of this effect are in the order 10^{-20} cm^2/sr and therefore also the detection of pollutants with small concentration is feasible. The principal inconvenience of LIF has been that only few atomic molecules of interest for combustion chemistry have electronic transitions in the energy range in which high power laser were commercially available. Therefore when single photon absorption effects are considered this method has been applied to the detection of radicals as OH, CH, CN, C_2, or to molecules as HO, NO_2.

This problem can be overcome with laser powerful enough to produce two or multiphotons absorption processes. This approach has been followed

mainly by Goldsmith at Sandia, Livermore and by Alden and coworkers at
Lund which were able to detect atomic hydrogen, oxygen and nitrogen.
However caution has to be taken in the interpretation of the results of
two photons fluorescence because photochemical effects may take place
simultaneously to it, as it has been pointed out by Goldsmith [18].

Nevertheless the most serious limitation of LIF is the intrinsic
slowness of the spontaneous emission stage compared to the collisional
deactivation channels, therefore it is not possible to correlate easily
the emission intensity to the ground state population via a Boltzmann
statistics. The aspects of the collisional quenching rate of
electronically excited OH, NH and CH radicals in flames has been recently
reviewed by Garland and Crosley [19] and other authors [20,21].

Saturation of the upper state has been used in order to avoid this
effect in low pressure flames [22,23] whereas, at higher pressure,
picosecond lasers have been employed for obtaining the temporal
resolution necessary to measure directly the quenching rate [24,26].
However quite recently the introduction of tunable excimer lasers allowed
what seems to be a real breakthrough in the instantaneous 2-D measurement
of the composition and temperature fields in atmospheric or higher
pressure flames. In fact Andresen and coworkers [27] have proposed a method
in which the effect of quenching is avoided by exciting the molecules in
a predissociating state with a lifetime shorter than the typical
collisional ones and called the method LIPS (Laser Induced
Predissociation Fluorescence). In their paper they show applications of
this concept by measuring separately the excitation spectra of OH, O_2 and
H_2O and discussing different schemes for the evaluation of rotational
temperature for O_2 and OH with one laser on a single shot basis. Later
on, LIPF has been successfully applied to bidimensional characterization
of the combustion inside a gasoline engine where fuel, OH, O_2 and NO
concentrations were followed in real time [28]. Great part of the works on
LIF and LIPF has been performed to detect one transition of a single
species and this can be a limit to the technique, so that some efforts
have been devoted to multispecies detection with one or two laser
excitation [29-31].

The presence of condensed phases in the reactive system can be a
further limit in the rejection of the elastically scattered light also
far from the excitation wavelength. This problem is partially overcome
using very high magnification ratio in the detection optics and spatial
filtering. With this type of approach LIF has been used for detection of
small radicals like OH and CH in presence of fuel droplets [32].

Another spectroscopic technique which presents a great sensitivity
is Resonance Enhanced MultiPhoton Ionization (REMPI) [33] however it has
limitations in turbulent flame studies for the interference due to the
probe which extract the ions.

Finally it is worthwhile to note that LIF is promising also for
determining instantaneously the velocity field in flames by using doping
compounds with long fluorescence or photophorescence lifetime.
Preliminary effort in this direction has been recently discussed in
Lambda Highlights [34].

SPECTROSCOPY OF CARBON POLYMORPHS

When the atomic C/O ratio in hydrocarbons flame exceeds the
stoichiometric value the oxydation is not complete and part of the fuel
follows a pyrolytic route. This chemistry is very complex indeed; in a
first stage an aromatic ring is formed starting from acetylenic
fragments and then Polycyclic Aromatic Hydrocarbons (PAH) are easily
formed. Later on the aromatic rings agglomerate and grow further for
addition of small molecules, mainly acethylene. In the last stage of the

process very large aggregates with thousand carbon atoms are formed which behave as a solid particle (soot). There is really an almost continuous mass distribution of aromatic structures and therefore the optical and spectroscopical diagnostics of rich flames is more complex and less specific than that of the small molecules and radicals, previously reviewed. An early monograph which considered both fluorescence by aromatics and light scattering by soot particles, was written in 1982 by one of the authors [35].

An intense broadband fluorescence is easily excited in rich premixed flames and on the fuel side of diffusion flames employing visible c.w. or pulsed laser. This effect is a nuissance for Raman scattering and LIF but it contains information on the chemistry of rich flames. More recent analysis in diffusion flames carried out from the near u.v. to the visible shows that fluorescence spectra have a structure which allows their attribution to classes of aromatic compounds [36,37]. Particularly the fluorescence excited in the visible has to be attributed not to 3-4 rings PAH but to heavier molecules with polar aromatic nature [38].

At a less phenomenological level fluorescence by aromatic molecules has been proposed as a tool for determining the flame temperature; Laurendeau et al. in fact has employed the dual fluorescence of pyrene, injected in a H_2-O_2 flame, for this purpose [39].

Although the spectroscopy of rich flames has not progressed systematically, particularly respect to the employment of powerful tunable lasers or picosecond lasers, a great contribution on the field came recently quite surprisingly from astrophysics. In fact this community has shown a great interest in PAH and, more generally, in structures with many carbon atoms as possible constituents of interstellar media. The literature and the debate in the field, as summarized up to 1987, can be found in the book edited by Leger et al. [40]. The other carbon polymorphs of great interest in combustion are solid carbonaceous submicronic particles. The knowledge of their size concentration and shapes is important for radiative heat transfer and for pollution problems. Furthermore the reactivity of soot particles depends upon their electronic structure and there is considerable interest in inferring it from optical and spectroscopical measurements inside the flame, using methods similar to those employed in solid state physics.

Laser light scattering in the u.v.-visible and diffusion broadening spectroscopy have been employed for this purpose and it was shown that soot particles formed in low pressure flames have a desordered graphitic like structure [41-43].

Also the problem of the structure of carbon particle has attracted interest in Astrophysics and in the production of diamond-like carbon in plasmas and in flames in condition of disequilibrium [44,45].

Other aerosols of inorganic nature can be found in flames when more complex fuels as carbon or heavy residual oil are used. These "flying" ashes are originated by nucleation or condensation of the volatile part of the inorganic matter contained as impurities into the original fuels. So far no systematic spectroscopical study has been carried out in the field of combustion for this class of aerosols despite of their toxic proporties particularly in the inceneration processes. But it is easy to predict that the large body of knowledge obtained in related fields as aerosol and atmospheric physics will be applied to this problem in a next future [46].

OPTICAL CHARACTERIZATION OF DROPLETS AND SPRAYS

A great part of the fuels used in practical systems are liquid mixtures of complex hydrocarbons and therefore combustion scientists are engaged in studying the behaviour of sprays of droplets, during their atomization, vaporization and dispersion in the air. All these processes take place in turbulent conditions as in the case of homogeneous combustion and therefore there is the need for a space and time resolution of the measurements.

Light scattering by droplets has been the object of extensive research summarized in different books and an overview of the field with literature updated to 1987 has been given by Kerker[47]. Most of the optical theoretical and experimental studies are devoted to the elastic light scattering effects more or less in the framework of the classical electromagnetic theory of Lorenz and Mie.

However there are studies also on spontaneous Raman and fluorescence by molecules embedded in small particles and droplets[48,49]. A systematic study of non-elastic nonlinear optical effects inside the droplets has been carried out by Chang and coworkers at Yale[50,51] who observed a series of four waves mixing processes, as CARS and Coherent Raman mixing, as well as Stimulated Raman Scattering from individual droplets. Their results demonstrate that both chemical specification and size determination of individual droplets are possible.

This kind of sophisticated spectroscopy has not find yet application in studies of sprays in burning environment which were mainly based upon elastic light effects.

The typical size of droplets are between 1 um and 300 um and it can be determined by light scattering measurements at different angles. The size of an individual droplet can be easily obtained by the absolute intensity of the scattered light or by the relative angular distribution in the diffraction lobe[52] or in the near forward[53]. Size distribution of droplets and/or their momenta are obtained by measuring the phase shift of the scattered light or from their polarization in the side scattering region around $\theta = 90°$[54-56].

At the moment commercial instruments for the simultaneous measurement of the size and velocity distribution functions in spray are also available.

2-D light scattering methods had also been employed for the study of stationary and intermittent spray[57,58]. In this case the same methodology of single-point measurement is extended to two-dimensional patterns using two detection systems focused on the same field[59]. It is also of interest in combustion systems burning liquid fuels to follow contemporarily the fuel concentration of liquid and gas phases. Some peculiar[60] two-dimensional techniques combining scattering-extinction effects[60] and fluorescence by "exciplex" species[61] have been used for this kind of characterization. Light scattering methods found also an application in the study of chemical reactions inside droplets at high temperature. The imaginary part of the refractive index of the liquid, i.e. its absorption coefficient, have been obtained inside a single droplet by the measurement of the polarization ratio in the forward scattering region; this method has been also used for the determination of the u.v.-visible absorption spectra of spray of diesel oil at high temperature. Laser light scattering has been used also for the evaluation of the temperature of the droplets determining the real part of the refractive index of the medium. This measurement can easily be obtained from the angular shift of the rainbow position[62] or the fall off of the polarization ratio near to the limit angle[63].

REFERENCES

1. N. Peters, Twenty-first Symp. (Int.) on Comb., p.1231, The Combustion Institute, Pittsburgh (1987).
2. R.W. Bilger, Twenty-second Symp. (Int.) on Comb., p.475, The Combustion Institute, Pittsburgh (1989).

3. A.C. Eckbreth, "Laser diagnostics for combustion temperature and species", A.K. Gupta and D.G. Lilley Eds., Energy and Engineering Science Series, Vol.7, Abacus Press, Tunbridge Wells (Kent) and Cambridge (MA), (1988).
4. J. Haumann, A. Leipertz, Opts. Letts. 9:487 (1984).
5. R.W. Dibble, R.E. Hollenbach, Eigtheenth Symp. (Int.) on Comb., p.1489, The Combustion Institute, Pittsburgh (1981).
6. K.N.C. Bray, P.A. Libby, J.B. Moss, Comb. Flame 56:199 (1984).
7. B. Yip, M.B. Long, Opts. Letts. 11:64 (1986).
8. A. Borghese, A. D'Alessio, M. Diana, C. Venitozzi, Twenty-second Symp. (Int.) on Comb., p.1251, The Combustion Institute, Pittsburgh (1989).
9. A.R. Masri, R.W. Dibble, Twenty-second Symp. (Int.) on Comb., The Combustion Institute, Pittsburgh (1989).
10. A.R. Masri, R.W. Bilger, R.W. Dibble, "Local structure of turbulent nonpremixed flames near extinction", in Comb. Flame to be published.
11. S.C. Johnstone, "An experimental investigation into the application of spontaneous Raman scattering to spray measurements in an engine", Proc. of ASME Fluids Engineering Conf., Morrept Ed., p.107, ASME, N.Y. (1981).
12. F. Pischinger, U. Reuter, E. Scheid, "Self-ignition of diesel sprays and its dependence on fuel properties and injection parameters", ASME paper, 88-ICE-14 (1988).
13. A.C. Eckbreth, J. Appl. Phys. 48:4473 (1977).
14. J.P. Taran, "Coherent Anti-Stokes Raman Spectroscopy", Proc. V Int. Conf. on Raman Spectroscopy, p.595, Schulz, Freiburg, FRG (1976).
15. A.C. Eckbreth, Eigtheenth Symp. (Int.) on Comb., p.1471, The Combustion Institute, Pittsburgh (1981).
16. R.K. Hanson, Twenty-first Symp. (Int.) on Comb., p.1677, The Combustion Institute, Pittsburgh (1987).
17. G. Kychakoff, R.K. Hanson, R.D. Howe, Appl. Opt. 23:704 (1984).
18. Goldsmith, Appl. Opt. 26:3566 (1987).
19. N.L. Garland, D.R. Crosley, Twenty-first Symp. (Int.) on Comb., p.1693, The Combustion Institute, Pittsburgh (1987).
20. P.W. Fairchild, G.P. Smith, D.R. Crosley, J. Chem. Phys. 79:1795 (1983).
21. G. Zizak, J.A. La Nauze, J.D. Winefordner, Comb. Flame 65:203 (1986).
22. K. Kohse-Höinghause, R. Heindenreich, Th. Just, Twentieth Symp. (Int.) on Comb., p.1177, The Combustion Institute, Pittsburgh (1985).
23. N.M. Laurendeau, Opts. Letts. 14:280 (1989).
24. R. Schwarzwald, P. Monkhouse, J. Wolfrum, Twenty-second Symp. (Int.) on Comb., p.1413, The Combustion Institute, Pittsburgh (1988).
25. N.S. Bergamo, P.A. Jaanimagi, M.M. Salour, J.H. Bechtel, Opts. Letts. 8:443 (1983).
26. R. Schwarzwald, P. Monkhouse, J. Wolfrum, Chem. Phys. Lett. 142:15 (1987).
27. P. Andresen, B. Bath, H.W. Lülf, G. Meijer, J.J. Ter Meulen, Appl. Opt. 27:365 (1988).
28. P. Andresen, G. Meijer, H. Schlüter, "Fluorescence imaging inside an internal combustion engine using tunable excimer lasers", to be published in Appl. Opt..
29. J.B. Jeffries, R.A. Copeland, G.P. Smith, D.R. Crosley, Twenty-first Symp. (Int.) on Comb., p.1709, The Combustion Institute, Pittsburgh (1987).
30. M. Aldén, H. Edner, S. Svanberg, Appl. Phys. 29:93 (1982).
31. M. Aldén, H. Edner, S. Wallin, Opts. Letts. 10:529 (1985).
32. M.G. Allen, R.K. Hanson, Twenty-first Symp. (Int.) on Comb., p.1755, The Combustion Institute, Pittsburgh (1987).

33. M.N.R. Ashfold, Mol. Phys. 58:1 (1986).
34. Lamba Highlights n°.15/16: a publication by Lambda Physik, December 1989.
35. A. D'Alessio, "Laser light scattering and fluorescence diagnostics of rich flames produced by gaseous and liquid fuels", in "Particulate Carbon: Formation during Combustion", D.C. Siegla and G.W. Smith Eds., p.207, Plenum Press, N.Y. (1981).
36. F. Beretta, V. Cincotti, A. D'Alessio, P. Menna, Comb. Flame 61:221 (1985).
37. L. Petarca, F. Marconi, "Fluorescence spectra and polycyclic aromatic species in a n-Heptane diffusion flame", to be published in Comb. Flame.
38. R. Barbella, F. Beretta, A. Ciajolo, A. D'Alessio, M.V. Prati, A. Tregrossi, "Optical and chemical characterization of carbon polymorphs formed during spray combustion of hydrocarbons", to be published on Comb. Sci. Tech..
39. D.L. Peterson, F.E. Lytle, N.M. Laurendeau, Appl. Opt. 27:2768 (1988).
40. A. Leger, L. d'Henecourt and N. Boccara Eds., "Polycyclic Aromatic Hydrocarbons and Astrophysics", NATO ASI Series, D. Reidel Publishing Company, Dordrecht (NL) (1986).
41. B.M. Vaglieco, A. D'Alessio, F. Beretta, "Determination of the optical properties in the u.v.-visible of carbonaceous matter produced in rich flames by scattering and extinction measurements", in "Experiments on Cosmic Dust Analogues", E. Bussoletti et al. Eds., Kluwer Academic Publishers, p.181 (1988).
42. B.M. Vaglieco, A. D'Alessio, F. Beretta, "In-situ evaluation of the soot refractive index in the u.v.-visible from the measurement of the scattering and the extinction coefficient in rich flames", to be published in Comb. Flame.
43. T.T. Charalumpopoulos, J.D. Felska, Comb. Flame 68:283 (1987).
44. D.R. Huffman, "Methods and difficulties in laboratory studies of cosmic dust analogues" in "Experiments on Cosmic Dust Analogues", Bussoletti et al. Eds., p.25, Kluwer Academic Publishers (1988).
45. L.M. Hanssen, W.A. Carrington, J.E. Butler, K.A. Snail, Materials Letts. 7:289 (1988).
46. Proc. 2nd Int. Aerosol Conf. Berlin on "Aerosol: Formation and Reactivity", Pergamon Press (1986).
47. M. Kerker, "Light scattering theory: a progress report", in "Proceedings of Int. Symp. on Optical Particle Sizing: Theory and Practice", University of Rouen, may 1987.
48. H. Chew, Phys. Rev. A 19:2137 (1979).
49. M. Kerker, D.S. Wang, H. Chew, Appl. Opt. 19:4159 (1980).
50. J. Zang, D.H. Leach, R.K. Chang, Opts. Letts. 13:270 (1988).
51. S. Quian, J.B. Snow, R.K. Chang, Opts. Letts. 10:499 (1985).
52. J.M. Béer, D.S. Taylor, D. Abbott, G.C. McCreath, "A laser diagnostics for the measurement of droplet and particle size distribution", AIAA 14th Aerospace Sciences Meeting, Washington D.C., Paper N°.76-79 (1977).
53. G. König, K. Anders, A. Frohn, J. Aerosol Sci. 17:157 (1986).
54. F. Beretta, A. Cavaliere, A. D'Alessio, Comb. Flame 49:183 (1983).
55. F. Beretta, A. Cavaliere, A. D'Alessio, Comb. Sci. Techn. 36:19 (1984).
56. F. Beretta, A. Cavaliere, A. D'Alessio, Twentieth Symp. (Int.) on Comb., p.1249, The Combustion Institute, Pittsburgh (1985).
57. A. Cavaliere, R. Ragucci, A. D'Alessio, C. Noviello, Proc. of the "Conference on Heat and Mass Transfer in Gasoline and Diesel Engine", Spolding Ed., Hemisphere Publishing Co., N.Y. (1988).

58. A. Cavaliere, R. Ragucci, A. D'Alessio, C. Noviello, Twenty-second Symp. (Int.) on Comb., p.1973, The Combustion Institute, Pittsburgh (1989).

59. A. Cavaliere, R. Ragucci, A. D'Alessio, P. Massoli, Proc. 4th Int. Conf. "Computational Methods and Experimental Measurements", CMEM'89, S. Carlomagno Ed., p.189, CUEN, Naples (1989).

60. J.M. Tishkoff, D.C. Hammond, A.R. Chraplyny, J. Fluids Eng. 104:313 (1982).

61. L.A. Melton, J.F. Verdieck, Twentieth Symp. (Int.) on Comb., p.1283, The Combustion Institute, Pittsburgh (1985).

62. N. Roth, K. Anders, A. Frohn, "Temporal evolution of size and temperature measurements of burning ethanol droplets for different initial temperatures", Joint Meeting of the German and Italian Sections of the Combustion Institute, p.2.3, CUEN, Naples (1989).

63. P. Massoli, private communication.

TRACE GAS DETECTION WITH INFRARED GAS LASERS

Jes Henningsen, Ari Olafsson, and Mads Hammerich

Physics Laboratory, University of Copenhagen
H.C.Ørsted Institute, Universitetsparken 5
DK-2100 Copenhagen, Denmark

INTRODUCTION

The growing concern about our environment has led to a demand for methods and equipment which can perform a wide variety of monitoring tasks. The detection of specific molecules in the atmosphere may be motivated by the need for monitoring the emission of toxic chemicals used by industry, or by the need to perform a general control of the air we breathe, in particular in areas which are subject to large scale emission caused by human activities, such as traffic, livestock breeding, and energy production. In addition, it may be motivated by the need to get a better understanding of global trends in the concentration of molecules which are of importance in connection with the greenhouse effect and ozone destruction in the stratosphere, or it may simply be necessary in order to get a deeper insight into the huge number of physical and chemical processes which occur in the atmosphere, and which act together to produce what we commonly denote as weather and climate.

In the following sections, we shall consider some monitoring techniques which share the common feature that they rely on transitions between molecular vibration-rotation states, induced by infrared gas lasers. We shall focus on extractive monitoring, as opposed to remote sensing, with main emphasis on photoacoustic detection.

WHY THE INFRARED?

Monitoring molecules through their interaction with electromagnetic radiation may be performed over a wide spectral range. In the microwave and mm wave region, source and detection techniques are highly developed. The relevant molecular energy levels are the low lying rotational states, and since many molecules have a large permanent dipole moment, they offer large transition rates. However, the energy level separation is small compared with thermal energies at room temperature, and the absorption coefficient thus suffers from the almost equal lower and upper state populations.

Higher lying rotational energy levels are addressed in the far-infrared or sub-mm region, but despite decades of development, this spectral region still remains a challenge to applied spectroscopy, owing to the absence of tunable lasers, and the absence of fixed frequency lasers exceeding the mW power level. On top of this comes the detection difficulties associated with the fact that on one hand, the photon energy is low compared with thermal energies, so that quantum detection is not possible, and on the other hand, the frequency is so high that extrapolation of mm wave techniques are technically demanding.

Moving up into the mid-infrared region, the situation changes, and several useful laser sources are available for communicating with the vibration- rotation transitions, which dominate this spectral region. These sources include lead salt diode lasers, which cover the 3-30 μm region quasi continuously at the 0.1-1 mW power level, as well as line tunable gas lasers like CO_2, N_2O, and CO, which produce cw radiation at the 0.1-100 watt level in the 9-11 μm region and around 5-6 μm. In addition, pulsed versions of these gas lasers may reach peak powers in the MW range, which makes them useful for remote sensing.

Near-infrared room temperature diode lasers have undergone a forced technical development owing to commercial applications such as fiber communication and compact disc players. They are available in certain wavelengths ranges around 0.75-0.9 μm and 1.3-1.5 μm, and can be tuned by varying the temperature, or the injection current. They have an attractive potential in connection with monitoring, but since molecular transitions in this range are weak overtone transitions, one cannot hope for the same sensitivity as in the mid infrared region. At visible, and shorter wavelengths, electronic transitions dominate the molecular spectra, and a wide variety of lasers are available. However, dye lasers are too expensive to be of much practical importance in connection with monitoring, except in special cases, and other lasers suffer from a lack of tunability. Thus, at visible and UV frequencies, practical monitoring at present mostly relies on white light sources.

Returning to the mid infrared region one notes that the vibration-rotation spectrum constitutes a highly specific spectral signature, which may be used as a molecular fingerprint. Thus, in particular when monitoring under conditions of strong interference, the infrared region offers optimal conditions. For obvious reasons a tunable laser source is advantageous, and the use of lead salt lasers is at present the subject of a strong effort [1]. However, these lasers must be cooled to 15-100 K, with a stability of 1 mK, they are multimode devices which require mode filters for precision spectroscopy, their low power level of 0.1-1 mW precludes the efficient use of very sensitive detection techniques such as photoacoustics, and the very fact that they are tunable, means that the problem of selecting and verifying the operating frequency is by no means trivial. In comparison, a gas laser like the CO_2 laser, can be tuned to any one of about 90 lines, and by using waveguide techniques, it is possible to achieve a fine tunability relative to the line centers ranging from about $\pm$ 500 MHz in the strongest lines to about $\pm$ 200 MHz in the weakest. Knowing the operating frequency to an accuracy of 10 MHz is essentially trivial, and with power levels exceeding 1 W, full advantage can be taken of sensitive techniques, such as photoacoustic detection. Using such a laser amounts to observing the molecular spectra through a large number of narrow spectral windows, and since a large number of molecules have dense spectra in the range of the CO_2 laser, enough can be seen through these windows to make this laser very useful for monitoring.

THE ABSORPTION COEFFICIENT

For an isolated spectral line of a trace gas, present at a low concentration c_x, the absorption coefficient is given by

$$c_x \alpha = \frac{(N_i - N_f)\pi\omega <\mu^2>}{\varepsilon c\hbar} \cdot \frac{1}{\Delta\omega_D} \cdot \sqrt{\frac{\ln 2}{\pi}} \cdot Re\{w(x + iy)\}$$

$$N_{i,f} = \frac{f_{i,f} \cdot c_x \cdot p}{kT}$$

$$w(z) = exp(-z^2)erfc(-iz)$$

$$x = \frac{\omega - \omega_0}{\Delta\omega_D} \cdot \sqrt{\ln 2}$$

$$y = \frac{\Delta\omega_L}{\Delta\omega_D} \cdot \sqrt{\ln 2}$$

where α is the hypothetical absorption coefficient for a trace gas concentration of 100%, but with collision parameters corresponding to those of the diluted limit. In these expressions, $<\mu^2>$ is the orientationally averaged squared transition dipole moment, $\Delta\omega_D$ and $\Delta\omega_L$ are the HWHM Doppler and Lorentz widths respectively, c_x denotes the trace gas concentration, p the total pressure, and $f_{i,f}$ is the relative population of the initial and final state. Apart from entering explicitly in $N_{i,f}$, the temperature enters in $f_{i,f}$ through the Boltzmann factor and through the temperature dependence of the partition function. At room temperature, $h\nu \simeq 5kT$ for CO_2 laser radiation, so that N_f may be neglected, but it must be included when monitoring at elevated temperature. In the literature, α is frequently given in units of $cm^{-1}atm^{-1}$, indicating that the actual absorption coefficient is proportional to the trace gas partial pressure. Some representative values are given in Table I. The quoted values refer to the limit $c_x \ll 1$ and a total pressure of 1 atmosphere, where in general several absorption lines contribute.

Table I.   Representative trace gas absorption coefficients. The values quoted are measured for dilute mixtures in air or N_2 at 1 atmosphere, and subsequently scaled to a concentration of $c_x = 1$.

molecule	laser line	abs.coeff. cm^{-1}	source
acetonitrile	9P16	.15	[2]
acetylene	10P14	33	[2]
acroleine	10R14	2.8	[3]
ammonia	9R30	56	[5]
benzene	9P30	2.0	[6]
butane	10R14	.5	[2]
carbon dioxide	9R18	0.003	[7]
chloroprene	10R18	9.2	[6]
dimethylamine	9P42	0.86	[4]
dimethylhydrazine (asym)	10P42	4.2	[4]
ethylacetate	9P6	12.0	[2]
ethylacrylate	9R12	7.3	[3]
ethylene	10P14	32	[6]
ethylene glycol dinitrate	9P14	2.6	[2]
ethyl mercaptane	10R28	0.57	[6]
freon 11	9R28	35	[6]
freon 12	10P42	92	[8]
freon 113	9P26	20	[2]
freon 114	9P14	28	[2]
furane	10R30	4.0	[2]
hydrazine	10P40	7.5	[4]
isopropanol	10P10	3.8	[2]
methanol	9P34	21.8	[4]
methylamine	9P24	0.88	[2]
methylchloroforme	9R24	9	[2]
methyl ethyl ketone	10P22⁻	1.2	[2]
monochloroethane	10R16	3.3	[6]
monomethylhydrazine	10R8	3.5	[4]
nitroglycerine	9P14	140	[2]
o-xylene	9P18	0.87	[9]
ozone	9P14	12	[10]
perchloroethylene	10P42	31	[2]
styrene	10P44	2.8	[3]
sulphur dioxide	9R26	0.11	[6]
sulphur hexafluoride	10P16	564	[2]
t-butanol	10P34	3.8	[2]
toluene	9P28	0.67	[9]
trichloroethylene	10P20	14.5	[3]
trimethylamine	9P24	4.2	[4]
vinyl bromide	10P22	4.1	[3]
vinyl chloride	10P22	8.8	[6]
vinylidenechloride	9R32	13.3	[3]
water	10R20	0.00083	[11]
1,1-difluoroethylene	10P22	19	[2]
1,2 dichloroethane	10P20	0.52	[6]

The homogeneous contribution to the linewidth $\Delta\omega_L$ may be expressed as

$$\Delta\omega_L = \sum \kappa_i \cdot c_i \cdot p$$

where p is the total pressure, c_i denote the concentrations of the different molecules present in the gas mixture, and κ_i are the respective collision broadening coefficients. At low trace gas concentration in the atmosphere, $\Delta\omega_L$ will be dominated by collisions with N_2. However, for high concentrations it should be kept in mind that the rate of self broadening may be significantly different from that for foreign gas broadening, and that $\Delta\omega_L$ then depends not only on the total gas pressure, but also on the gas composition. This is particularly pronounced for polar molecules where dipole-dipole interaction contributes strongly to the scattering cross section, and this is the reason why α will in general not be the absorption coefficient measured in an actual experiment at 100% concentration. A typical value for N_2 broadening is 5 MHz/mBar, leading to a HWHM width at 1 atmosphere of about 5 GHz.

The Doppler width is given by

$$\Delta\omega_D = \frac{\omega}{c}\sqrt{\frac{2kT\ln 2}{M}}$$

and thus a modest sized molecule with a mass of 80 amu has a HWHM Doppler width of about 20 MHz, so that the transition from the pressure broadened limit to the Doppler limit occurs at about 4 mBar. In the pressure broadened limit, the expression for the absorption coefficient reduces to

$$\alpha \simeq \frac{N_i\omega < \mu_{ij}^2 >}{\epsilon c\hbar}\frac{\Delta\omega_L}{(\Delta\omega_L)^2 + (\omega - \omega_0)^2}$$

Note that for $\omega = \omega_0$, i.e. on resonance, α is independent of pressure, since both N_i and $\Delta\omega_L$ are proportional to p. Thus, as long as monitoring can be performed at the line center, the pressure can be reduced by almost a factor of 100 without loss of sensitivity. The importance of this becomes clear if we consider the absorption at $\omega = \omega_0$ from a line of an interfering molecule, located far from the line being monitored. For this interfering line, α is proportional to p^2, so that a pressure reduction of a factor 100 may lead to a reduction of interference by a factor of 10^4.

The conclusions reached above, hold for isolated absorption lines only. If the line considered is part of a multiplet where the spacing between components is large compared with the Doppler width, several components may contribute at high pressure, and there will be a loss of sensitivity with reduced pressure until overlap has been eliminated. Furthermore, for heavy molecules the line density may be so large that loss of absorption will continue all the way to the Doppler limit. In these cases, the increased immunity to interference is thus payed for by a loss of sensitivity. The situation is illustrated in Fig.1, where the absorption coefficient is calculated as a function of pressure for small concentrations of CO_2, NH_3, and O_3 in a background of N_2, illustrating the isolated line, the multiplet, and the dense spectrum respectively. In addition, a calculation for water is included in order to demonstrate the complete elimination of interference from the Lorenz wings of distant absorption lines.

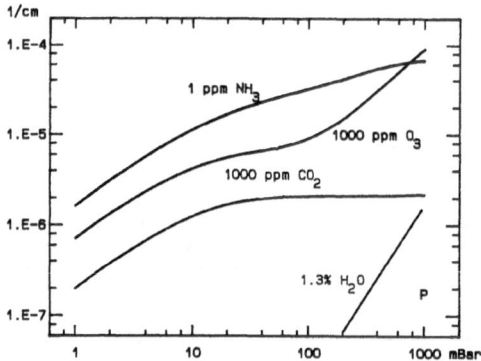

Fig.1   Absorption coefficient as a function of pressure for selected lines in CO_2, NH_3, and O_3

Extensive tabulations of line center frequencies, line strength, and broadening rates for molecules of atmospheric relevance are found in the databases AFGL [12], AFGL-HITRAN [13], and GEISA [14].

The accuracy in most cases is on the order of a few hundred MHz, which is more than adequate for monitoring at ambient pressure, but barely so at the Doppler limit. Note that Table I refers to a total pressure of 1 atmosphere, and cannot be used for predicting whether a given molecule can be detected at low pressure. Thus, an investigation of the absorption signatures in the CO_2 laser windows is required in each individual case. Some have been performed for methyl cyanide [15] and methanol [16], but no systematic database exist at present.

DETECTION PRINCIPLES

The absorption from trace gas molecules in a gas mixture may be monitored by detecting the attenuation of the laser beam over a fixed length L. According to the Lambert-Beer law, the transmitted intensity, in the absence of saturation, is given by

$$I(L) = I(0)exp(-c_x\alpha L)$$

so that

$$c_x = -\frac{1}{\alpha L}\ln\frac{I(L)}{I(0)}$$

For given L, the detection limit is given by the smallest relative change $\Delta I_{min}/I$ that can be measured in the transmitted signal. The most sensitive method employs frequency modulation and harmonic detection. The sensitivity depends on the linewidth, and for atmospherically broadened lines Reid et.al. [17] have reached $\Delta I_{min}/I \simeq 10^{-4}$ in a diode laser spectrometer. With a path length of 100 m, this correponds to a sensitivity of 10^{-8} cm^{-1}, and for a strong absorber like ethylene with $\alpha = 30$ cm^{-1}, this leads to a minimum detectable concentration of

$$c_{x,min} \simeq 330ppt$$

At low pressure the reduced linewidth leads to a larger harmonic response, and earlier work, using a White cell at 10 Torr, yielded $\Delta I_{min}/I \simeq 8 \cdot 10^{-6}$, corresponding to a sensitivity of $8 \cdot 10^{-10}$ cm^{-1} for 100 m optical path length [18].

In linear detection, the sensitivity is limited by fluctuations in the laser power, and a considerable improvement can be obtained with dark background methods, where one measures a quantity which is directly proportional to the absorption, rather than that part of the laser power which is not absorbed. In the visible, this can be done by monitoring the fluorescence from the upper level of the transition. In the infrared, however, the spontaneous emission rate is too low, and most of the excess vibrational energy is converted to heat through inelastic collisions. This heat may be detected by using a laser beam which is modulated at an audio frequency in the kHz range. The absorption then gives rise to temperature oscillations, and the associated pressure oscillations may be detected by a microphone (Fig.2). This socalled photoacoustic effect was discovered in 1881 by Bell [19], Tyndall [20], and Röntgen [21] but at that time the radiance of available light sources was too low to allow practical applications, and the effect fell into oblivion. The situation changed dramatically after the advent of the lasers, and the photoacoustic effect now plays an important part in a wide variety of spectroscopic applications [23] [24].

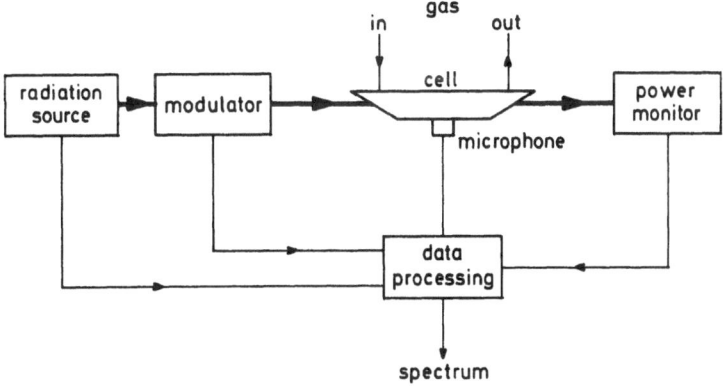

Fig.2   Principle of photoacoustic spectroscopy.

407

An improvement in sensitivity can be obtained by using an absorption cell which is acoustically resonant [22], and some designs which have been used, are shown in Fig.3. The most widely used, shown in Fig.3a, is a longitudinal organpipe resonator of length $\lambda/2$ with a typical Q value of 20 [25]. The microphone is mounted at the central pressure maximum, and since the resonator is open at both ends, it allows for easy introduction of the laser beam.

In photoacoustic monitoring, the detection limit is usually determined by the coherent acoustic signal generated by absorption in the windows of the cell. For the organpipe resonator, however, the windows are not an integral part of the resonator, and the presence of large buffer volumes between the resonator ends and the windows automatically yields a good suppression of this noise source

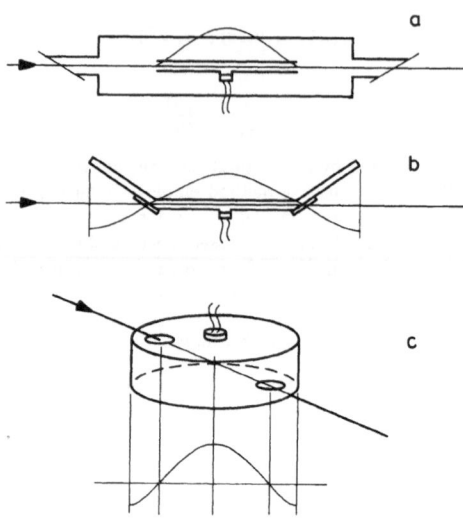

Fig.3   Different photoacoustic resonator designs. (a) longitudinal organpipe resonator, excited in first longitudinal mode [25]. (b) closed longitudinal resonator excited in second longitudinal mode. [27] (c) cylindrical resonator excited in first radial mode. [28]

Gandurin et.al. [26] use a cell of similar geometry, but operated at a frequency far below the first longitudinal resonance. They place two identical cells coaxially, one containing the trace gas, the other containing a reference gas, and window absorption is dealt with by electronic subtraction of the signals from the two channels.

Bernegger and Sigrist [27] use a longitudinal resonator of length λ, closed at both ends (Fig.3b). In order to allow introduction of the laser beam, the resonator is bent at a distance of $\lambda/4$ from the ends, and the windows are located at the bends, where the pressure has nodes.

A different resonator type consisting of a large diameter cylindrical resonator, excited in the first radial mode, was first used by Dewey et.al. [22], and later by Gerlach and Amer [28] (Fig.3c). The windows are located at pressure nodes on the end faces, and the microphone is mounted at the center of the end face, where the pressure amplitude is maximum. Owing to its significantly larger Q value of about 500, this resonator gives a higher acoustic amplification than the longitudinal resonators. However, since the resonance frequency is a sensitive function of parameters such as pressure, gas temperature, and gas composition, this also calls for a more precise control of the modulation frequency and phase of the laser beam.

The minimum detectable concentration for a photoacoustic resonator is inversely proportional to the laser power as long as saturation can be neglected, and as long as the coherent noise generated by absorption in the windows of the cell does not dominate the trace gas signal. The best reported sensitivity is around $10^{-9} cm^{-1} W$, and for ethylene in air at 1 atmosphere, this corresponds to a minimum detectable concentration of

$$c_{x,min} = 6 ppt$$

at 5 Watt of laser power. This is significantly better than for the linear absorption scheme, and in addition to having the advantage of a much simpler optical configuration, the smaller volume of the photoacoustic cell results in improved temporal resolution for constant flow rate.

At low pressure the sensitivity in linear detection improves, owing to improved harmonic response in frequency modulation, while the sensitivity of photoacoustic detection is reduced, partly because a smaller fraction of the absorbed energy goes into the translational degree of freedom, partly because of increasing acoustic losses in the boundary layer at the walls. Nevertheless, photoacoustics may still be preferable, owing to its simplicity and to its potential for faster response. At very low pressure, the absorbed energy may be carried directly by free flight of the molecules from the interaction region to a bolometer, in the socalled photothermal detection scheme. An analysis of the relative sensitivity of fluorescence, photoacoutic, and photothermal detection under various conditions has been given by Bordè [29].

MONITORING AT AMBIENT PRESSURE

As noted above, typical HWHM linewidths of molecular vibration-rotation lines are on the order of 5 GHz at a pressure of 1 atmosphere. Although the separation between adjacent CO_2 laser lines ranges from $\sim$ 35 GHz in the 9R band to $\sim$ 55 GHz in the 10P band, a linewidth of 5 GHz is large enough to ensure that essentially all molecules which have an absorption band overlapping the 9-11 μm region will show strong absorption in at least a few of the available laser lines. By the same token, however, when monitoring on a mixture of gases, the absorption in a given CO_2 line will usually contain contributions from several constituents.

Since the tunability of CO_2 lasers inside a single line is much smaller than the trace gas linewidth at ambient pressure, there is little variation in the absorption over the tuning range. Measurements may therefore just as well be performed at the laser line center, and identification of the trace gas molecules must be made by comparing the response in at least as many laser lines as there are absorbing constituents in the gas mixture. If the trace gas is the only relevant absorber, measurements can be performed by fast switching between two laser lines, corresponding to strong and weak trace gas absorption respectively. This method was used by Gandurin et.al.[26] for measuring NH_3 and C_2H_4 down to 1-10 ppb with a CO_2 laser, and 0.1 -1 ppb NO_2 and 1-10 ppb NO with a CO laser.

Multicomponent analysis has been performed by Bernegger et.al. [30]. They analyzed the exhaust gas from a jeep for several hydrocarbons by measuring the absorption profile for the individual constituents in a large number of CO lines between 5.2 and 6.4 μm, and fitting the measured profile of the exhaust gas to a linear combination of the individual profiles. In the CO laser range, absorption from water vapor presents a particular problem, and this was corrected for by a separate measurement in a reference cell. Following essentially the same philosophy, Loper et.al. [31] used a Gerlach-Amer cell (Fig.3c) to study low level monitoring of mixtures of different hydrazines and other toxic gases produced by rocket fuel.

A particular interference problem is posed by CO_2, which absorbs in all CO_2 laser lines, and which is present at a concentration of 340 ppm in the atmosphere. Perlmutter et al [32] have analyzed the effect of imperfect knowledge of the identity or of the absorption coefficient of interfering molecules on the minimum detectable concentration of C_2H_4 in air. They find a limit of 0.5 ppb if C_2H_4 is known to be the only absorbing molecule. Allowing for the natural abundance of H_2O and CO_2 in the atmosphere, increases the detection limit to 2 ppb, and for 1% CO_2, as is commonly found in greenhouses, it increases to 15 ppb. Allowing finally for a realistic uncertainty of 5-10% in the photoacoustic responsivity to CO_2, the final detection limit for C_2H_4 climbs to 50 ppb.

By removing the atmospheric CO_2 with a KOH scrubber, and traces of hydrocarbons with a catalyst consisting of a copper tube with platinized aluminium oxide heated to 350°C, Harren et.al. [33] have obtained a detection limit for C_2H_4 of 20 ppt. They use a 7 Watt CO_2 laser with a 30 cm long $\lambda/2$ organpipe resonator, and have succeeded in detecting the C_2H_4 emitted from a single flower during the withering process. By further placing the photoacoustic resonator inside the laser cavity, in order to profit from the higher circulating intracavity intensity, the detection limit for C_2H_4 has recently been pushed to 6 ppt [34].

Owing to the pressure dependence of the linewidth, interference problems can often be essentially eliminated by reducing the pressure by a factor of 100, so that the spectral lines approach the Doppler limit. A necessary condition is, however, that the spectral windows of the CO_2 laser are wide enough that monitoring can be performed at or close to the trace gas absorption line center. If, furthermore, interference from CO_2 is to be avoided, this center should be separated by at least a few Doppler widths from the CO_2 line center, leading to a tuning requirement of at least ±200 MHz. This is well outside the reach of conventional open structure CO_2 lasers, which typically have a tuning capability of about ±50 MHz, but the requirements can be met by using waveguide techniques.

Apart from reducing interference, an additional advantage associated with monitoring with a widely tunable laser at reduced pressure is that a characteristic spectral profile can now be recorded. This effectively eliminates any uncertainty associated with identifying the absorbing molecule, even if measurements are restricted to only a single CO_2 line. Furthermore, it reduces the problems associated with window absorption, since this remains constant as the laser is tuned, and hence is easily distinguished from the trace gas absorption.

The advantages fade away as the molecules get heavier, so that the average separation between the absorption lines becomes comparable to the Doppler width. The situation is illustrated in Fig.4, which shows the spectrum over ±250 MHz in selected CO_2 laser lines for pure methanol, pure ethanol, and for a mixture of 0.3 ppm methanol and 50 ppm ethanol in air at a total pressure of 5 mBar [35]. Ethanol absorption is relatively strong in 9P44 and 9P14, whereas methanol has strong absorption in 9P16 and 9P34. Signatures for both molecules are clearly seen in the mixture by choosing the appropriate CO_2 lines. Note, however, that while the contrast in the methanol signatures is very good, the slightly heavier ethanol has a significant unresolved background.

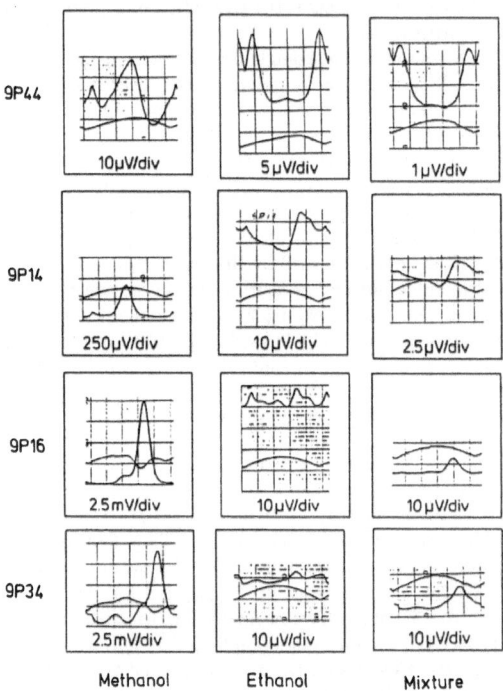

Fig.4   Photoacoustic spectra in selected CO_2 laser lines of pure methanol, pure ethanol, and a mixture of 0.3 ppm methanol and 50 ppm ethanol in air at 5 mBar. The lower trace of each graph shows the CO_2 power transmitted through the photoacoustic cell [35].

The technique of monitoring at reduced pressure has been used by Olafsson et al [36] to monitor NH_3 in power plant emission with a CO_2 waveguide laser. The NH_3 is injected into the burner in order to convert NO_x, which is one of the sources of acid precipitation, into molecular nitrogen and water. This process may proceed catalytically or non-catalytically, and in particular for the SNCR (Selective Non Catalytic Reduction) scheme, monitoring of NH_3 in the smokestack is necessary for process control, since the reduction process runs by itself in a fairly narrow temperature window only. The specific requirements thus are that NH_3 should be monitored at 100°C with a detection limit of 1 ppm, on a background of 12% CO_2, 6% H_2O, 50-500 ppm SO_2, dust, etc.

The NH_3 was monitored on the sR(5,0) transition, which is located 190 MHz below the 9R30 line center, and the requirement of $> \pm 200$ MHz single line and single mode tunability were met by constructing the laser according to the guidelines given by Tang and Henningsen [37]. Owing to the high CO_2 concentration, in combination with the fact that increased temperature leads to increased CO_2 absorption (hot band transition) and reduced NH_3 absorption (low initial state energy), the detection limit for NH_3 is determined by fluctuations in the CO_2 signal, despite the operating pressure of 12 mBar. Cell sensitivity was therefore not crucial, and a small, non-resonant cell was used. Laser, photoacoustic cell, and A/D interface were placed 30 m above ground level in the smokestack, and smoke was extracted through a heated sampling line and flowed through the cell. A data line transmitted data to the computer, which was located 100 m from the measuring unit.

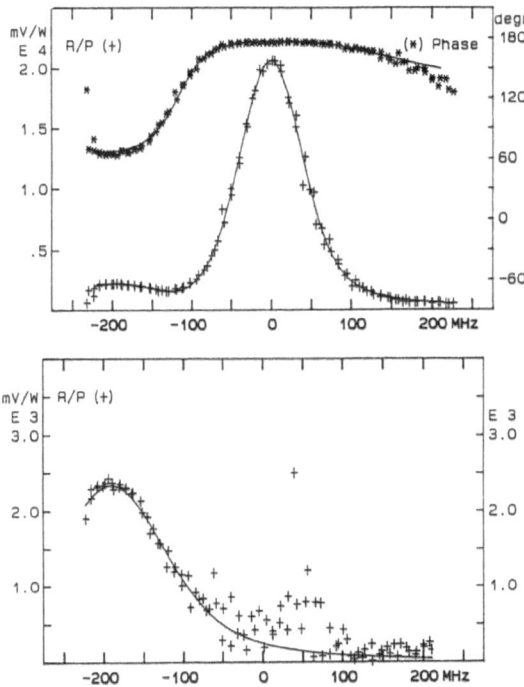

Fig.5   Screen print for concentration measurements of 9.2 ppm of NH_3 and 11.7% of CO_2 [36].

Fig.5 shows a screen print corresponding to a single concentration measurement at 9.2 ppm of NH_3 and 11.7% of CO_2, performed during a field test at a Danish power plant. The laser is scanned over two free spectral ranges in one minute, and the photoacoustic amplitude and phase, as well as the CO_2 laser power is recorded. An automatic algoritm transforms the displacement axis into a linear frequency axis, and the resulting signal, normalized with respect to the laser power, is seen in the upper frame. This signal is then fitted to a sum of two Voigt profiles with complex amplitudes, representing the NH_3 and CO_2 absorptions respectively, and a complex constant representing the coherent signal from window absorption and from scattered radiation. In the lower frame, the CO_2 signal has been digitally subtracted, and it is seen that the signal to noise ratio at the NH_3 line center corresponds to a detection limit of about 0.5 ppm. It is instructive to notice that the large fluctuations in the CO_2 signal around zero offset would effectively increase the detection limit by about a factor of 10, if monitoring had to be performed at or close to the CO_2 line center.

A crucial feature of photoacoustics on gas mixtures is the molecular dynamics involved in the conversion of internal molecular energy to heat. This is particularly important when dealing with mixtures involving CO_2 and N_2. The near degeneracy between the fundamental asymmetric stretch of CO_2 and the N_2 vibration, leads to a large cross section for resonant energy transfer (Fig.6).

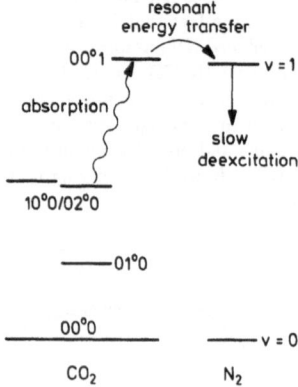

Fig.6   Energy levels in CO_2 and N_2, giving rise to kinetic cooling.

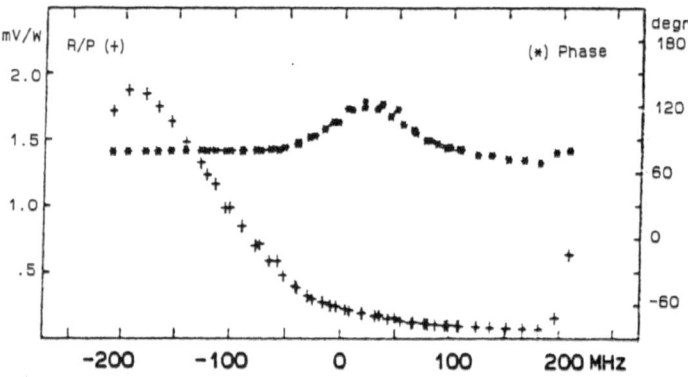

Fig.7   Trace amounts of CO_2 is visible in the photoacoustic phase only [40]

In the CO_2 laser this mechanism is used to advantage by adding N_2 to the gas mixture in order to increase the pump rate by energy transfer from vibrationally excited N_2 to CO_2 in the ground state. In our case, the situation is reversed. Thus, following absorption of CO_2 laser radiation, an excited CO_2 molecule will transfer its excitation energy to N_2, where it will reside for a long time owing to the metastable character of the excited N_2 levels. Since the CO_2 molecule was initially taken out of an excited state, the transition being a hot band transition, there is now a non-equilibrium situation among the CO_2 vibrational levels, and the equilibrium will be restored at the expense of translational energy. Thus, following absorption of radiation, a transient cooling takes place of the CO_2 gas, and the effect is therefore referred to as kinetic cooling [38,39]. In trace gas detection, this means that the photoacoustic phase of the CO_2 signal will be significantly different from the phase of the trace gas signal, where no kinetic cooling is involved. This phase contrast, which is clearly seen in Fig.5, is a very important aid in the analysis of mixtures where one of the components is strongly dominant. Fig.7 shows the photoacoustic amplitude and phase for 1.5 ppm NH_3 in N_2. The presence of trace amounts of CO_2 can not be inferred from the amplitude, but is clearly visible in the phase around zero offset [40].

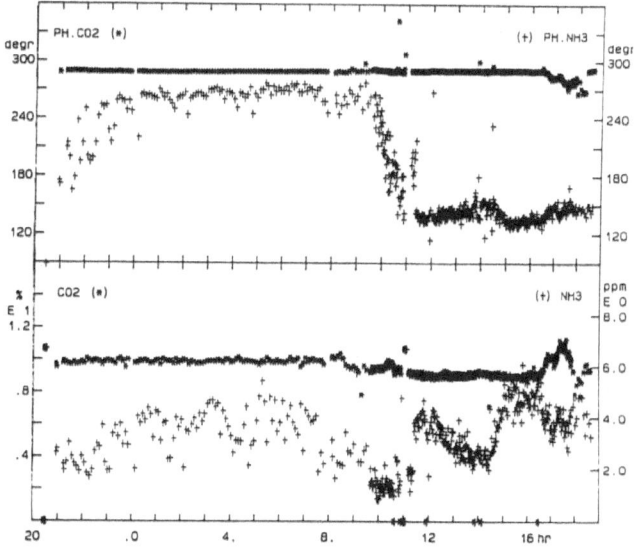

Fig.8   Data from 24 hour run on smokestack emission at low NH_3 level. Phase information is used
to discriminate against spurious concentration measurements [36].

In the NH_3 monitor the phase contrast is used for discriminating against spurious readings. Fig.8 shows the results of an automatic 24 hour run on stack emission at low NH_3 concentrations . During night hours, when no NH_3 is present in the gas mixture, small irregularities in the amplitude of the CO_2 signal will be interpreted as an NH_3 signal. That this is spurious is clear, however, if one considers the phase, which is close to the CO_2 phase and very different from the NH_3 phase. At 9am the monitor begins to smell real NH_3, and over the next hour, the phase gradually changes to that appropriate to NH_3, while the concentration readings become less erratic and eventually settle at 1 ppm. In this case the phase responds to significantly lower concentrations, but below 1 ppm the response cannot be quantified.

Phase effects may also aid in discriminating between the CO_2 absorption at the line center and close lying trace absorption lines, as seen in Fig.9. Here, the phase contrast between the central absorption in CO_2 and an C_2H_4 absorption located 100 MHz above the CO_2 line center, leads to destructive interference, and produces a deep minimum in the photoacoustic amplitude between the two lines. For comparison, the stipled line shows the signal which would be observed in linear absorption [41].

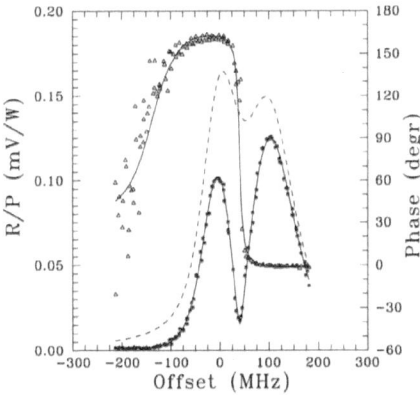

Fig.9   Photoacoustic amplitude (Δ) and phase ($*$) for a mixture of CO_2 and C_2H_4. Destructive
interference leads to improved discrimination between the two lines [41].

413

The situation is further complicated if water is present in the gas mixture, since water molecules are effective in deexciting the metastable N_2 levels, and hence act to reduce the phase contrast. This may be turned to advantage, since a quantitative analysis of the phase contrast may give information about the H_2O concentration [42].

SAMPLING AND CALIBRATION

Whenever monitoring is performed by flowing the gas mixture through a cell, a crucial question is whether the measured signal, which represents the trace gas concentration in the interaction region, also reflects the concentration at the source. On their way to the cell, the different components of the gas mixture may react with each other, they may form clusters or aerosols, and they may be react with or be adsorbed on particles present, or on the walls of the sampling line and the cell.

Adsorption problems are particularly severe for polar molecules with large dipole moments, such as water and NH_3, but they can be reduced py proper choice of material. Loper et.al. [31] have studied the properties of 304 stainless steel, gold, paraffin, and teflon when exposed to C_2H_4 and NH_3. They measured the time to reach 90% of the steady state photoacoustic signal, and found 2.5 min. in all cases for C_2H_4, but 99, 34, 2.5, and 2.8 min. respectively for NH_3. Sauren et.al. [43] used a leaf chamber for studying the adsorption of NH_3 on surfaces of aluminium, silicon coated aluminium, brass, and teflon. They found that teflon had the lowest affinity for adsorption, but they also noticed a seemingly constant loss of NH_3 in the chamber, which they ascribed to reaction with residual water in the air. Despite its inferior properties in relation to adsorption, Olafsson et.al. [36] used a stainless steel cell for detecting NH_3 and they found that the $100°C$ operating temperature in combination with the water content of the smoke led to a very significant reduction of NH_3 adsorption. Apparently, the water molecules stick to the walls even more efficiently than NH_3, and when exposed to smoke, the cell walls are effectively coated with water.

In the absence of sampling problems, calibration at high or intermediate concentration levels proceeds in a straightforward way through the use of certified gas mixtures. At low concentrations, permeation tubes can be used, where a pressurized gas leaks at a constant rate through a teflon membrane. The permeation tube is introduced in a stream of carrier gas with known flow rate, and if mixing is perfect, the trace gas concentration can be calculated. Determination of the leak rate is done by weighing the tube at regular intervals. The tube can be stored by cooling, but since thermal cycling may change the leak rate, it must then be recalibrated.

For molecules which show a strong tendency to wall adsorption, calibration is less straightforward. Olafsson et.al. [36] found time constants of several hours when attempting to calibrate their cell with a 25 ppm mixture of NH_3 in dry N_2. The problem was overcome by instead using air bubbled through a solution of a known amount of NH_3 in water. The concentration of NH_3 in the gas phase was determined from tables given in the literature [44], and an absolute calibration was performed by bubbling all of the NH_3 out of the water at constant air flow, integrating the measured response over time, and relating this to the known initial amount of NH_3. These two calibrations were mutually consistent to within 30%.

CONCLUSION

Atmospheric trace gas detection with infrared lasers is a rapidly expanding field owing to the high sensitivity and selectivity offered by laser spectroscopy. The use of tunable diode lasers in conjunction with long path absorption cells allows for monitoring of any molecule absorbing in the infrared. The disadvantage with this scheme is that equipment using lead salt diode lasers is quite complex to operate, and despite two decades in science, lead salt lasers still have not made it in large scale commercial applications. Tunable diode lasers based on GaAs technology are cheap and convenient radiation sources, but at present they are only available in the near infrared region where molecular spectra show weak vibrational overtone transitions. Infrared gas lasers are simple to work with, but their restricted tunability means that only certain molecules can be monitored. On the other hand, if their higher output power is utilized in conjunction with photoacoustic detection, monitoring with infrared gas lasers offers a very attractive combination of sensitivity, selectivity, and simplicity.

ACKNOWLEDGEMENT

The authors acknowledge the support of the Danish Science Researh Council under grants no 5.17.4.6.19 and 5.17.4.1.23, and the support and assistance from the Danish Power Utility Companies (ELSAM).

References

[1] "Monitoring of gaseous pollutants by tunable diode lasers", Eds. R.Grisar, G.Schmidtke, M.Tacke, and G.Restelli, Kluwer Academic Publishers, ISBN 0-7923-0334-2 (1989).

[2] G.A.West, J.J.Barrett, D.R.Siebert, and K.V.Reddy, Rev.Sci.Instr. 54, 797 (1983).

[3] G.L.Loper, G.R.Sasaki, and M.A.Stamps, Appl.Optics 21, 1648 (1982).

[4] G.L.Loper, A.R.Calloway, M.A.Stamps, and J.A.Gelbwachs, Appl.Optics 19, 2726 (1980).

[5] R.J.Brewer and C.W.Bruce, Appl.Optics 17, 3746 (1978).

[6] A.Mayer, J.Comera, H.Charpentier, and C. Jaussaud, Appl.Optics 17, 391 (1978), and ibid. 19, 1572 (1980).

[7] A.D.Devis and U.P.Oppenheim, Appl.Optics 8, 2121 (1969).

[8] W.Schnell and G.Fischer, Appl.Optics 14, 2058 (1975).

[9] P.Anderson and U.Persson, Appl.Optics 23, 192 (1984).

[10] R.R.Patty, G.M.Russwurm, W.A.McClenny, and D.R.Morgan, Appl.Optics 13, 2850 (1974).

[11] P.L.Meyer, M.W.Sigrist, F.K.Kneubühl, and J.Hinderling, Infrared Phys. 27, 345 (1987).

[12] L.S.Rothman et al., Appl. Optics 22, 2247-2256 (1983).

[13] L.S.Rothman et al., Appl. Optics 26, 4058 (1986).

[14] N.Husson et al., Ann.Geophys. 4, 185 (1986).

[15] S.T.Kornilov, I.V.Ostrejkovskij, E.D.Potsenko, V.M.Mikhailov, and S.N.Murzin, Int.J.Infrared and MM Waves 10 (1989).

[16] M.Inguscio, N.Ioli, A.Moretti, F.Strumia, and F.d'Amato, Int.J.Infrared and MM Waves 5, 1615 (1984), F.Tang, A.Olafsson, and J.Henningsen, Appl.Phys. B 47, 47 (1988).

[17] D.T.Cassidy and J.Reid, Appl.Optics 21, 1185 (1982).

[18] J.Reid, J.Shewchun, B.K.Garside, and E.A.Balik, Appl.Optics 17, 300 (1978).

[19] A.G.Bell, Phil Mag. 11, 510 (1881).

[20] J.Tyndall, Proc.Roy.Soc. (London) 31, 307 (1881).

[21] W.C.Röntgen, Phil.Mag. 11, 308 (1881).

[22] C.F.Dewey, R.D.Kamm, and C.E.Hackett, Appl.Phys.Lett. 23, 633 (1973).

[23] V.P.Zharov and V.S.Lethokov "Laser Optoacoustic Spectroscopy", Springer Verlag, Heidelberg (1986).

[24] "Photoacoustic, Photothermal, and Photochemical Processes in Gases", Topics in Applied Physics 46, Ed. P.Hess, Springer Verlag (1989).

[25] E.Kritchman, S.Shtrikman, and M.Slatkine, J.Opt.Soc.Am. 68, 1257 (1978).

[26] A.L.Gandurin et al., Sov.J.Appl.Spectrosc. 45, 769 (1986), engl.transl. 45, 886 (1987).

[27] S.Bernegger and M.W.Sigrist, Appl.Phys. B **44**, 125 (1987).

[28] R.Gerlach and N.M.Amer, Appl.Phys **23**, 319 (1980).

[29] Ch.J.Bordè, Journ. de Phys., suppl. no. 10, **44**, C6-593 (1983).

[30] St.Bernegger, P.L.Meyer, C.Widmer, and M.W.Sigrist, in "Photoacoustic and Photothermal Phenomena", Eds. P.Hess and J.Pelzl, Springer Verlag, Heidelberg (1988).

[31] G.L.Loper, J.A.Gelbwachs, and S.M.Beck, Can.J.Phys. **64**, 1124 (1986).

[32] P.Perlmutter, S.Shtrikman, and M.Slatkine, Appl.Optics **18**, 2267 (1979).

[33] F.J.M.Harren, J.Reuss, D.D.Bicanic, and E.J.Woltering, in "Photoacoustic and Photothermal Phenomena", Eds. P.Hess and J.Pelzl, Springer Verlag, Heidelberg (1988).

[34] F.J.M.Harren, F.G.C.Bijnen, J.Reuss, L.A.C.J.Voesenek, and C.W.P.M.Blom, in "Monitoring of gaseous pollutants by tunable diode lasers", Eds. R.Grisar, G.Schmidtke, M.Tacke, and G.Restelli, Kluwer Academic Publishers, ISBN 0-7923-0334-2 (1989).

[35] A.Olafsson and J.Henningsen, 11th Int. Conf. on Infrared and Mm Waves, Tirrrenia, Pisa (1987).

[36] A.Olafsson, M.Hammerich, J.Bülow, and J.Henningsen, Appl.Phys. B **49**, 91 (1989).

[37] F.Tang and J.O.Henningsen, Appl.Phys. B **44**, 93-98 (1987).

[38] A.D.Wood, M.Camac, and E.T.Gerry, Applied Optics **10**, 1877-1884 (1971).

[39] F.G.Gebhardt and D.C.Smith, Applied Phys. Lett. **20**, 129 (1972).

[40] M.Hammerich, A.Olafsson, and J.Henningsen, European Quantum Electronics Conference, Hannover, september 1988, paper MoCD2.

[41] A.Olafsson, unpublished.

[42] R.A.Roth, A.J.L.Verhage, and L.W.Wouters, 6th Int. Topical Meeting on Photoacoustic and Photothermal Phenomena, Baltimore, july/aug (1989).

[43] H.Sauren, B. van Hove, W.Tonk, H.Jalink, and D.D.Bicanic, in "Monitoring of gaseous pollutants by tunable diode lasers", Eds. R.Grisar, G.Schmidtke, M.Tacke, and G.Restelli, Kluwer Academic Publishers, ISBN 0-7923-0334-2 (1989).

[44] "Chemical Engineers Handbook", Ed. John H.Perry, p.14-4, McGraw-Hill.

ENVIRONMENTAL MONITORING USING OPTICAL TECHNIQUES

Sune Svanberg

Department of Physics
Lund Institute of Technology
P.O. Box 118
S-221 00 Lund, Sweden

INTRODUCTION

Advanced techniques are needed to monitor our threatened environment, to evaluate pollution levels and developmental trends. Tropospheric pollution has obvious manifestations in terms of health problems, water and soil acidification, and forest damage. Human-induced stratospheric changes in the ozone layer, as evidenced by the occurrence of "ozone holes" at the polar caps, may have much more far-reaching consequences[1-6]. Laser spectroscopy provides powerful means for remote sensing of molecules in the atmosphere, yielding information on pollution levels as well as meteorological conditions. Laser-induced fluorescence provides interesting possibilities for remote monitoring of marine pollution and land vegetation. In the present paper we will consider remote sensing of the atmosphere as well as the marine environment using optical techniques. We will start with atmospheric monitoring.

ATMOSPHERIC MONITORING USING OPICAL TECHNIQUES

There are two major kinds of laser methods applicable in remote sensing[7-15]:

* Long-path absorption monitoring

* Lidar (Light detection and ranging), with subdivisions:

   Fluorescence lidar
   Raman scattering lidar
   Mie scattering lidar
   Differential absorption lidar (dial)

Long-path absorption techniques are based on the same principles as spectrophotometry. However, by using laser beams of low divergence it is possible to use a pathlength of several km instead of the 1 cm cuvette used in the chemical laboratory. A single-ended arrangement can be achieved by utilizing a corner cube retroreflector at the end of the light path and collecting the back-reflected light with a telescope. Tunable diode lasers and CW line-tunable CO_2 lasers are useful in this

Applied Laser Spectroscopy, Edited by W. Demtröder and
M. Inguscio, Plenum Press, New York, 1990

417

approach. Since all detected photons have travelled the same path no range resolution is obtained and only average concentrations can be determined. The technique has a powerful non-laser counterpart in doas (differential optical absorption spectroscopy)[16-18], where a distant high-pressure xenon lamp is used in combination with fast spectral scanning detection to overcome limitations posed by atmospheric scintillation. A doas set-up is illustrated in Fig. 1 and an atmospheric recording of ambient SO_2 over a path length of 700 m is shown in Fig. 2.

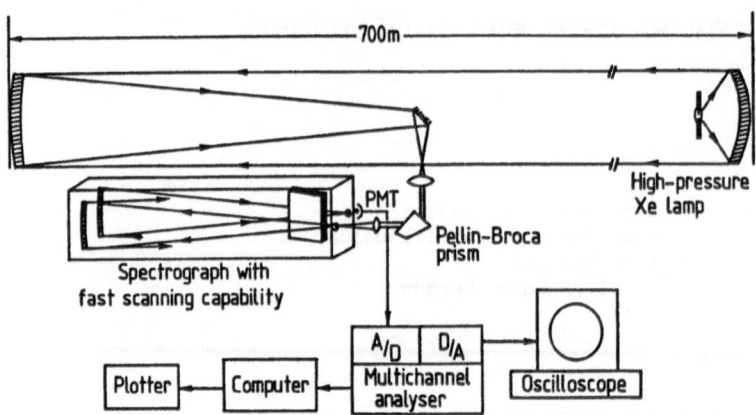

Fig. 1.   Set- up for differential optical absorption spectroscopy
         (doas). (From Ref. 18.)

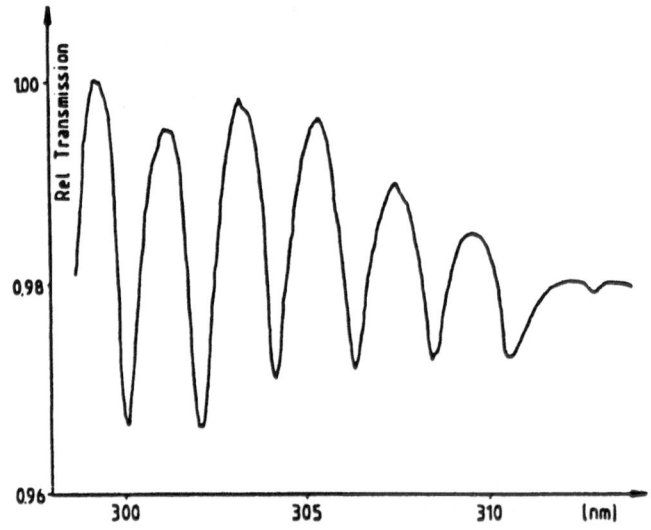

Fig. 2.   Doas recording of atmospheric SO2 using a 700 m path-
         length. The mean concentration is 22 ppb. (From Ref.
         18.)

In the lidar approach, a laser pulse is transmitted into the atmosphere and backscattered radiation is detected as a function of time by an optical receiver, in a radar-like fashion. In the case of fluorescence lidar a laser tuned to the absorption line of an atmospheric species is used and fluorescence light is detected in the subsequent decay. Fluorescence lidar is a powerful technique for measurements at mesospheric heights where the pressure is low and the fluorescence is not quenched by collisions. The technique has been used extensively to monitor layers of various alkali and alkaline earth atoms (Li, Na, K, Ca and Ca+) at a height of about 100 km[19-21]. Raman lidar has many attractive features but suffers from severe restrictions due to the weakness of the Raman scattering process. Only high gas concentrations can be detected at short ranges using high-power lasers and optical detection in a passband corresponding to the particular Raman shift. Water vapor profiles can be

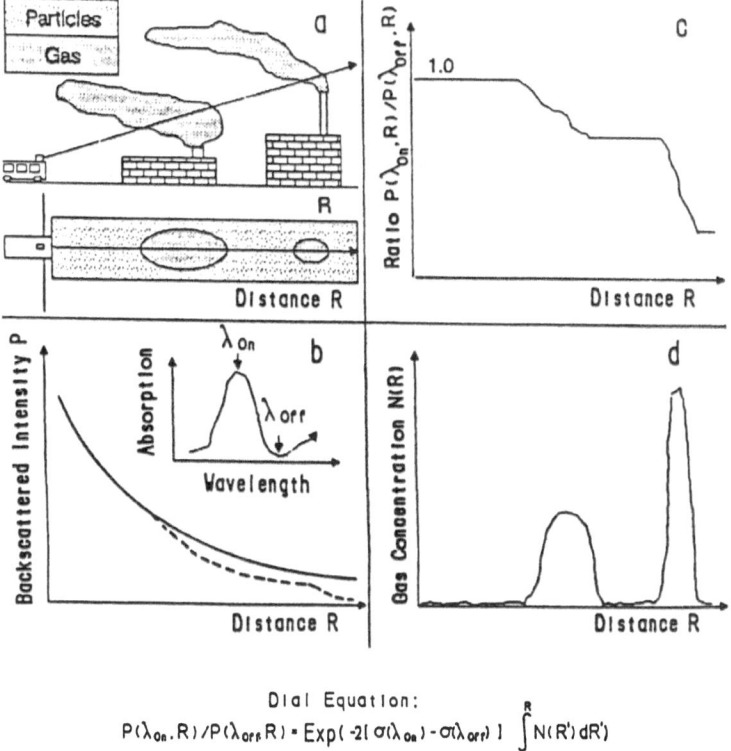

Dial Equation:

$$P(\lambda_{On}, R)/P(\lambda_{Off}, R) \cdot \mathrm{Exp}(-2[\sigma(\lambda_{On}) - \sigma(\lambda_{Off})] \int_0^R N(R') dR')$$

Fig. 3.   Illustration of the principle of differential absorption lidar (dial): a) pollution measurement situation, b) back-scattered laser intensity for the on- and off-resonance wavelengths, c) ratio (dial) curve, d) evaluated gas concentration. (From Ref. 23.)

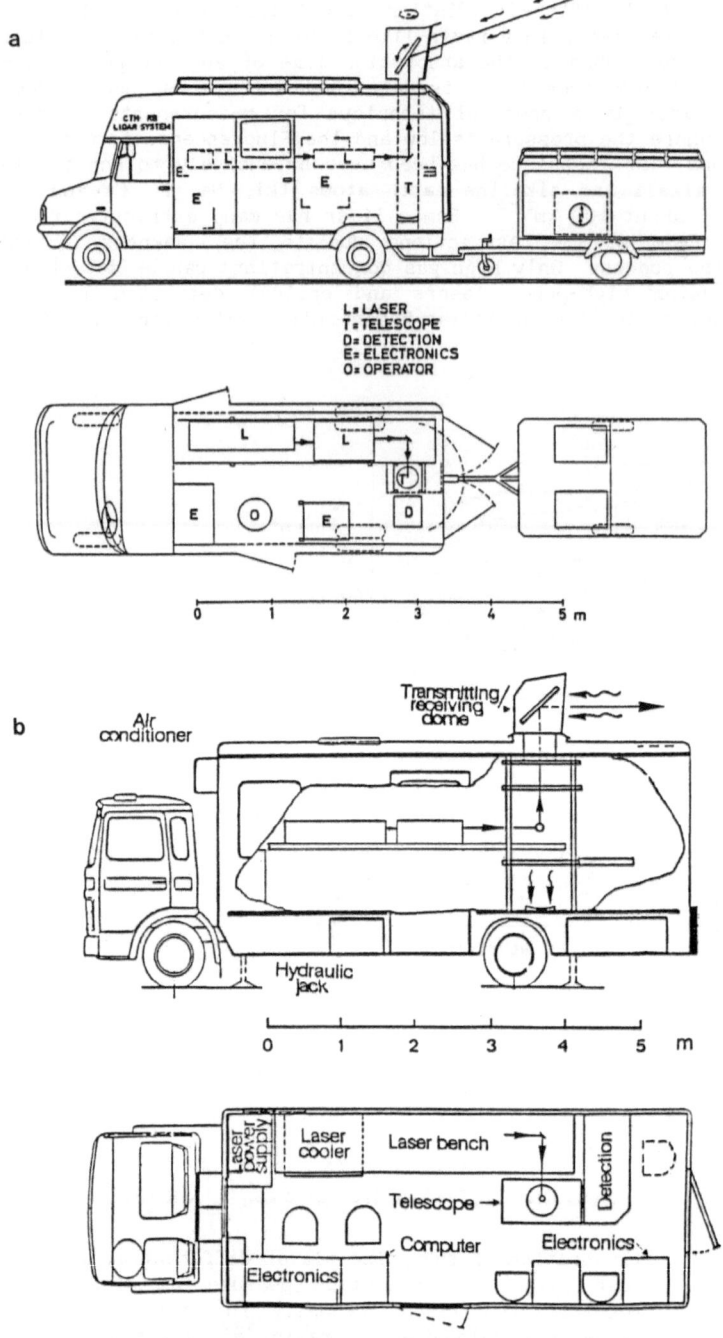

Fig. 4. Schematic views of two Swedish mobile lidar systems. a) System described in 1981. (From Ref. 24.), b) System described in 1987 (From Ref. 23.)

obtained in vertical soundings up to several km in height, and pressure profiles up to tens of km are measurable using Raman signals from atmospheric N_2.

Mie scattering from particles provides strong signals allowing mapping of the relative distribution of particles over large areas. Stratospheric dust from volcanic eruptions can also be studied[22]. However, since Mie scattering theory involves many normally inaccessible particle parameters quantitative results are difficult to obtain. Mie scattering is, however, extremely useful in providing the "distributed mirror" needed in differential absorption lidar (dial).

DIFFERENTIAL ABSORPTION LIDAR

The principles for dial are schematically represented in Fig. 3 (Ref. 23). Laser light is alternately transmitted at a wavelength where the species under investigation absorbs, and at a neighboring, off-resonant wavelength. In the presence of an absorbing gas cloud the on-resonance signal is attenuated through the cloud and the off-resonant one is not. By dividing the two lidar signals by each other, most troublesome and unknown parameters are eliminated and the gas concentration as a function of the range along the beam can be evaluated. Mobile laser radar systems for research and operational measurements have been constructed. In Fig. 4 two Swedish systems are shown[23,24], one constructed at the Chalmers Institute of Technology and the other one at the Lund Institute of Technology. A schematic of the optical arrangements in the older system is shown in Fig. 5[24].

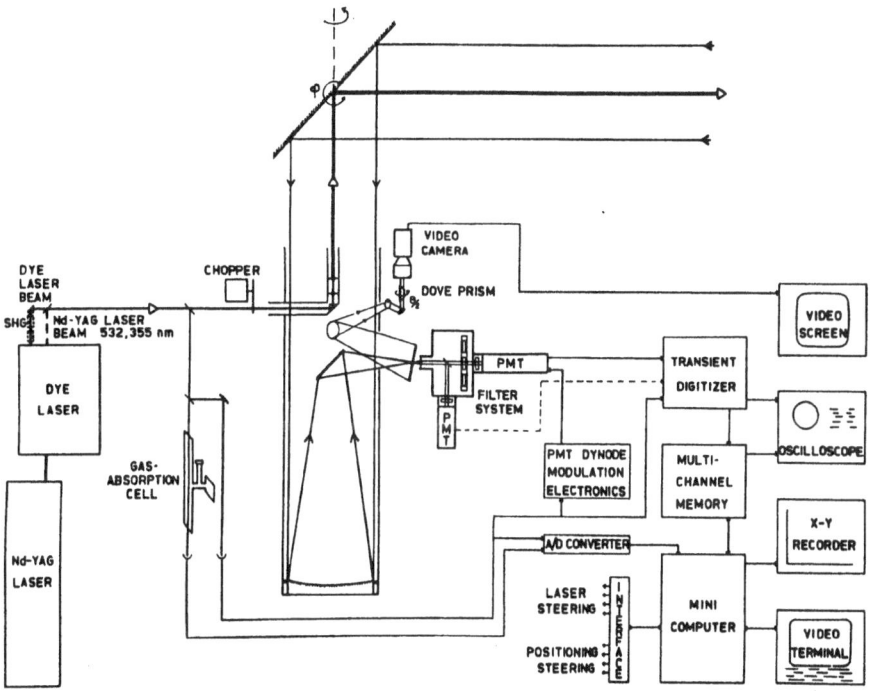

Fig. 5.   Optical arrangement for a mobile lidar system. (From Ref. 24.)

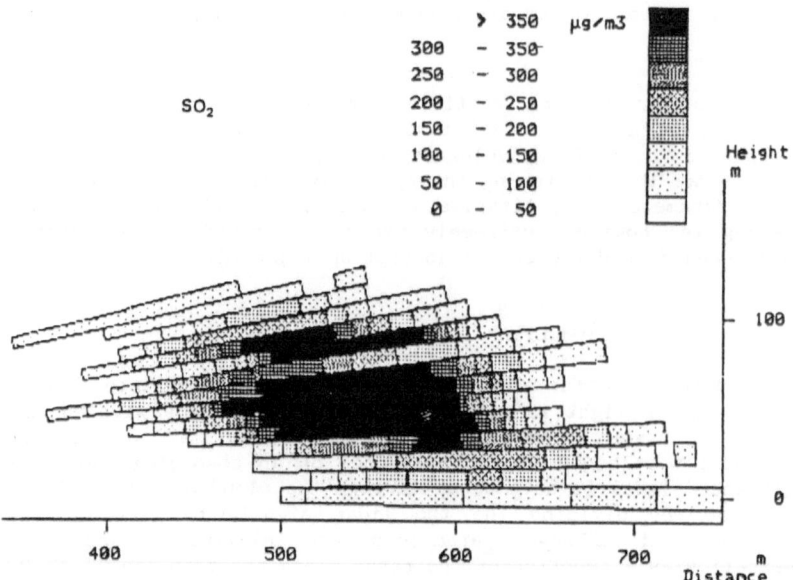

Fig. 6. Mapping of a cross section of an SO2 plume from a paper mill obtained by a 20-minute dial measurement. (From Ref. 23.)

The dial technique is operational for important pollutants such as SO2, NO2 and O3, for which molecules the appropriate wavelengths can readily be generated (see, e.g., Refs. (23-27). An example of data from a vertical scan through a spreading SO2 plume is given in Fig. 6. This type of representation can be automatically generated using the system described in Ref. 23. If the integrated concentration over the plume cross section is multiplied by the wind velocity component vertical to the measurement plane the total pollutant flux from the source is obtained. Pollutants such as NO and Hg absorb at short UV wavelengths that more recently have become accessible through nonlinear frequency conversion techniques in new materials, such as β-barium-borate. Data from a vertical lidar scan for NO, investigated at 226 nm are shown in Fig. 7[28].

Mercury is a troublesome pollutant generated from coal combustion, chlorine-alkali and refuse incineration plants. Since it is present in the atmosphere primarily in atomic form, it can be detected in very low concentrations (ppt) because its differential absorption is much higher than that of molecules[29]. Mercury data from a vertical scan over a chlorine-alkali plant are shown in Fig. 8[29]. Mercury is also an interesting geophysical tracer gas related to the occurrence of ore deposits, geothermal reservoirs and seismic activity[30]. An illustration of geophysical lidar applications, centered mainly around mercury, is shown in Fig. 9.

Great interest is presently being focused on the ozone molecule. A steadily increasing concentration of tropospheric ozone is thought to be

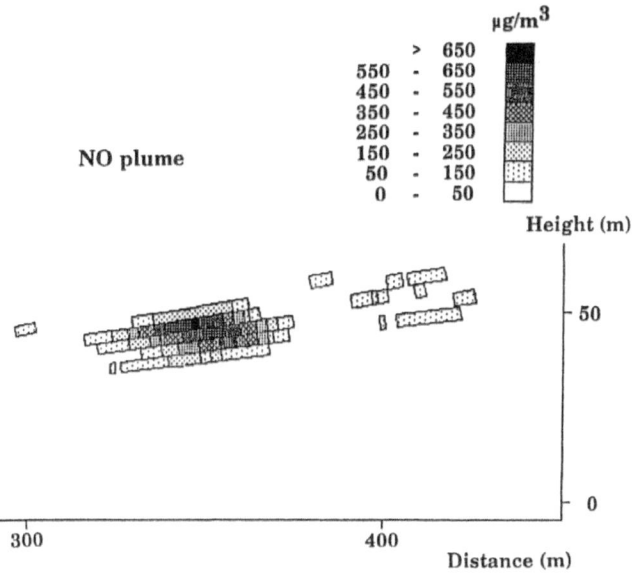

Fig. 7. NO plume mapping. (From Ref. 28.)

related to the increasing damage to forests observed throughout Europe. Ozone chemistry, dynamics and measurement techniques form an important part of the inter-European EUROTRAC research program. Within this project, a vertically sounding ozone lidar system is being developed in Lund[31]. The system incorporates a 100 Hz repetition rate 500 mJ/pulse KrF excimer system (λ=248 nm) which has an unstable resonator for high-quality beam generation. A number of suitably located frequencies within the ozone Hartley-Huggins absorption band can be generated by stimulated Raman shifting in high-pressure H_2 cells[32,33]. Our system uses a horizontal 30 cm diameter telescope placed inside the laboratory, and the transmitted and detected light beams are folded vertically by a large 45° mirror placed outside the building. A high-throughput spectrometer and multiple PMT tubes are used for simultaneous detection of two or more wavelengths. Ground-based dial systems for studies of the stratospheric ozone layer have been constructed by several groups in the search for anomalies and long-term trends[32-34]. Both NASA and ESA are planning space lidar systems providing global wind, temperature and pollution monitoring.

GAS CORRELATION LIDAR

Lidar systems are very powerful and provide unique three-dimensional information on atmospheric conditions. However, the systems also tend to be complex and expensive, which is a limiting factor in the widespread

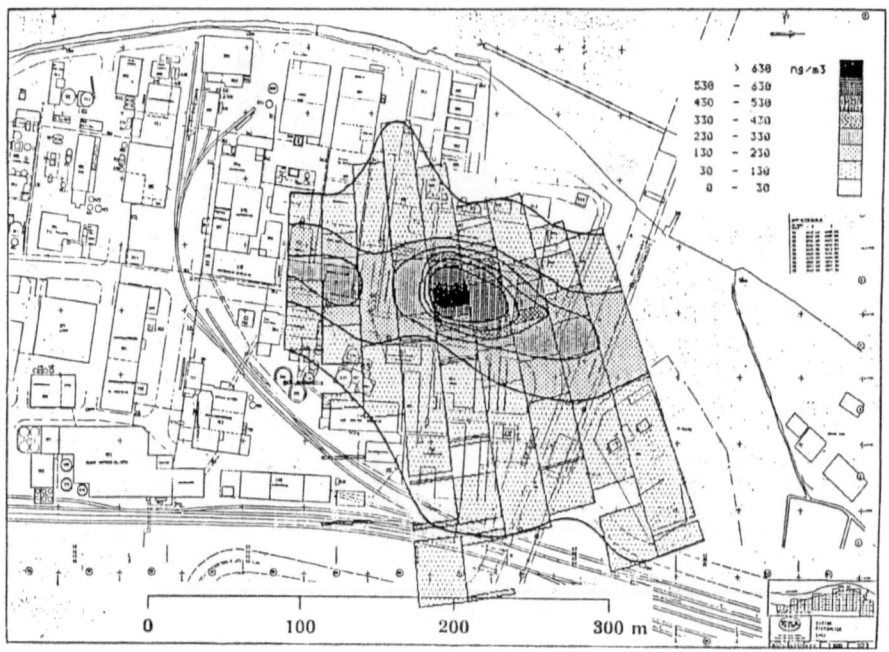

Fig. 8. Horizontal scan of Hg distribution over a chlorine-
alkali plant. (From Ref. 29.)

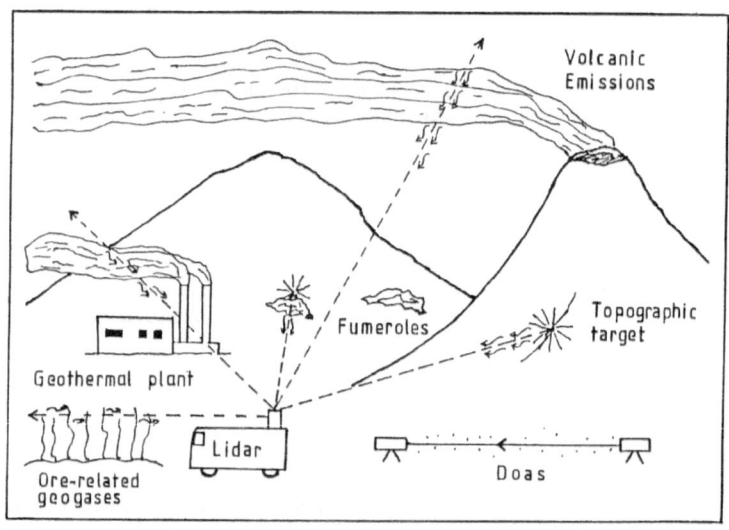

Fig. 9. Illustration of atmospheric remote sensing related to
geophysical research.

use of such systems for environmental management. Thus, there is a great need to develop simplified equipment. One such approach is gas correlation lidar[35]. Here a rather crude laser system with a comparatively broad linewidth is utilized. Since the laser wavelength is not sharp it covers both on- and off-resonance wavelengths at the same time. However, the information can be separated on the detection side by splitting the received radiation into two parts. One part is detected directly while the other part is first passed through a cell filled with an optically thick sample of the gas to be studied. In this way all the on-resonance radiation is filtered away leaving only the off-resonance radiation to be detected. In the direct channel the sum of the on- and off-resonance radiation is detected. Unknown factors are eliminated by dividing the signals. The simultaneous detection of the two signals also eliminates influences due to atmospheric turbulence and fluctuations due to changing reflectivity in airborne measurements using topographic targets. We are also investigating other aspects of gas correlation techniques, including multi-channel gas flow imaging[36,37], using ambient environmental optical radiation.

ENVIRONMENTAL FLUORESCENCE MONITORING

Fluorescence spectroscopy has long been used for analytical and diagnostic purposes[38-41]. Using UV laser sources, the techniques of laser-induced fluorescence (LIF) have become particularly powerful. The LIF process in large molecules, such as biological ones, is schematically illustrated in Fig. 10[42]. The ground as well as the excited electronic levels are broadened by vibrational motion and interactions with surrounding molecules. Thus, absorption occurs in a broad band allowing a fixed-frequency UV laser, such as a nitrogen laser (λ=337 nm), an excimer laser (XeCl, λ=308 nm; XeF, λ=351 nm) or a frequency-tripled Nd:YAG laser (λ=355 nm) to be used for the excitation. A radiationless relaxation to the bottom of the excited band then occurs on a pico-second time scale. The molecules remain here for a typical lifetime of few nanoseconds. Fluorescence light is released in a red-shifted broad band, which is frequently rather structureless. Internal conversion and transfer of energy to surrounding molecules are strongly competing radiationless processes.

Laser-induced fluorescence has an interesting potential for remote sensing of environmental parameters. For quite some time, hydrospheric pollution monitoring has been performed with airborne laser-based fluorosensors. Different kinds of oil can be identified by their fluorescence properties. Other pollutants and algal bloom patches can also be studied[43-45]. Using the blue-green transmission window of water, bathymetric measurements of sea depths can also be performed[46,47]. Some laboratory and field work performed by our group is described in Refs. 48-55. The field of laser-based hydrospheric monitoring is covered in Refs. 8 and 10. Laser-induced fluorescence has also been used by American[56-58], Italian[59] and Swedish groups for studies of land vegetation. Some work performed by our group is reported in Refs. 48 and 60.

We will now give a brief presentation of our work within the field of environmental monitoring using laser fluorescence spectroscopy. We will give examples from laboratory measurements and field work. Finally, some examples of LIF studies using similar techniques outside the environmental field will be mentioned.

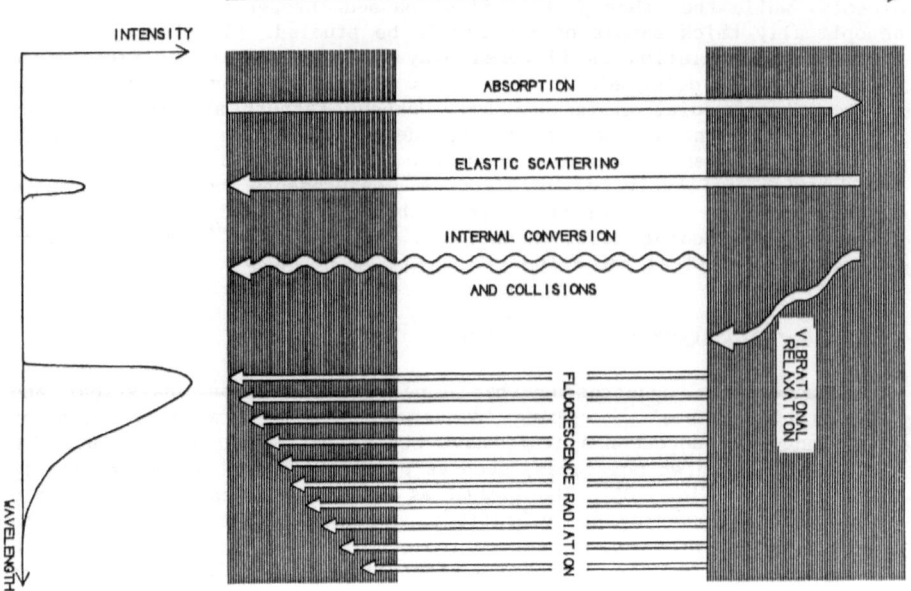

Fig. 10. Schematic diagram of the LIF process in large molecules.
(From Ref. 42.)

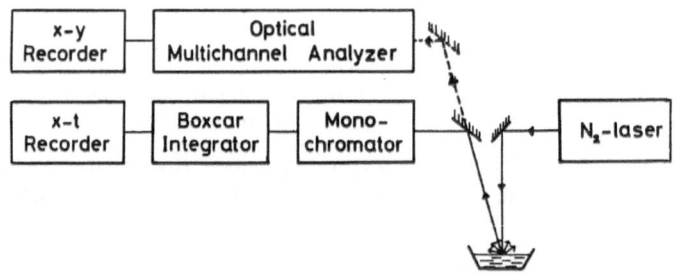

Fig. 11. Laboratory set-up for LIF studies. (From Ref. 48.)

The starting point for assessing the potential of LIF for environmental studies is laboratory studies of constituents of environmental interest, e.g. mineral oils, phytoplankton and land vegetation. A suitable experimental set-up for such studies is shown in Fig. 11[48]. An arbitrary sample is excited with a pulsed UV laser and the fluorescence is dispersed in a spectrometer and detected. A scanning monochromator can be used in conjunction with a photomultiplier tube and a boxcar integrator. Better, an optical multichannel analyzer system with a gated and intensified linear diode array is employed to capture the full LIF spectrum for each laser shot.

Examples of spectra for different substances in the aquatic environment are shown in Fig. 12[61], including data for a crude oil, polluted river water, seawater with a tracer and a green algae. Spectra of different crude oils are found in Fig. 13[48]. As a rule, light petroleum fractions exhibit blue-shifted, intense fluorescence while heavier fractions also have longer wavelength components and fluoresce more weakly. At short UV wavelengths the penetration of the exciting light into the oil is limited to micrometers. For longer wavelengths the penetration depth is larger. Thus, in order to assess the thickness of an oil film the choice of excitation wavelength is important[62]. The fluorescence characteristics of different oil products play an important role in airborne measurements and the assessment of marine oil spills in the decision regarding the correct oil-fighting counter-measures to be implemented.

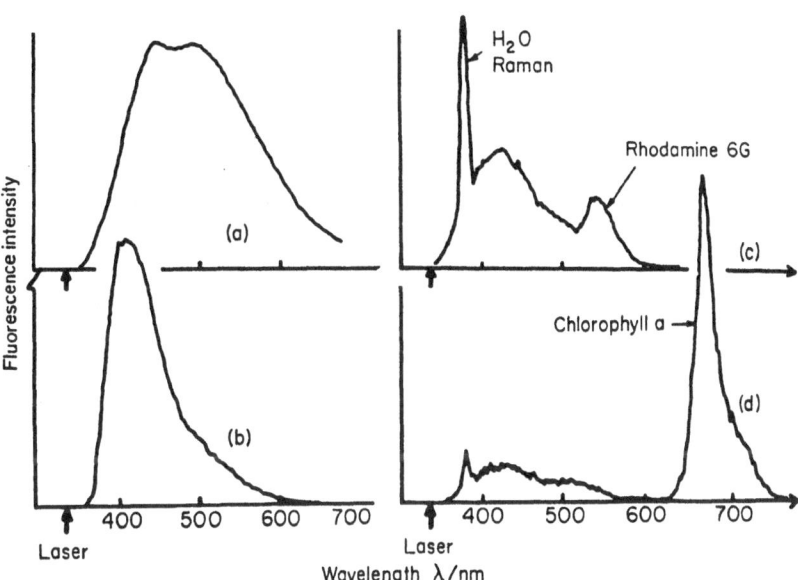

Fig. 12. Fluorescence spectra of various samples for nitrogen laser excitation (λ=337 nm). a) Crude oil, Abu Dhabi. b) River water down-stream from a sodium sulfite pulp mill. The fluorescence is about 30 times stronger than that from clean water. c) Seawater, Kattegat, containing 3 µg/l Rhodamine-6G dye, added as a tracer for hydrological studies. d) The green algae *Clorella ovalis* Butcher in seawater. (From Ref. 61.)

427

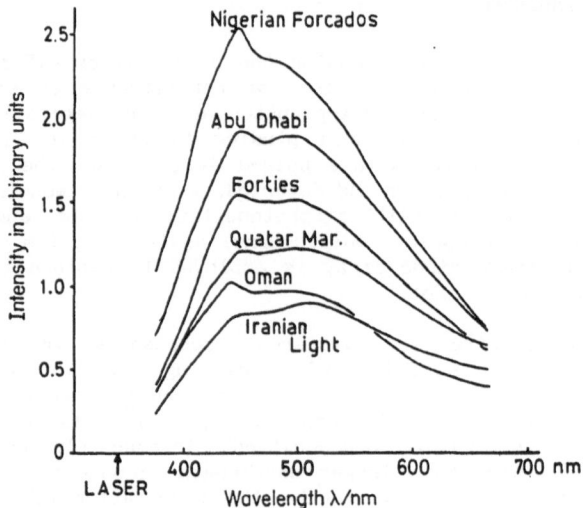

Fig. 13. LIF spectra for different crude oils. The excitation
wavelength is 337 nm. The data are spectrally corrected.
(From Ref. 48.)

Algal fluorescence monitoring can be important for measuring the
total marine productivity, which originates in the conversion of solar
energy, CO_2 and nutrients into organic matter by microscopic
phytoplankton. Recently, huge algal blooms, for instance of
Chrysochromulina polylepis, leading to devastating consequences for most
other marine life forms have occurred due to eutrophication of coastal
waters. Some classes of algae exhibit LIF spectra with certain
characteristic features[48] in addition to the dominating peak at 685 nm
due to chlorophyll *a*.

Land vegetation can be efficiently characterized by multispectral
reflectance measurements using space-borne sensors in satellites such as
LANDSAT, SPOT and ERS-1[63]. An *active* remote sensing technique such as LIF
might, in certain circumstances, complement *passive* reflectance monitoring. We have studied the LIF properties of many species under 337 nm excitation[48]. In many cases, the chlorophyll peak is absent since the
short-wavelength light is absorbed in the leaf "wax" layer, which dominates the spectrum. For longer excitation wavelengths, penetration to the
chlorophyll occurs, as shown in Fig. 14[60] for spruce needles, excited at
405 nm, in an attempt to assess forest damage due to ozone exposure using
LIF techniques.

FLUOROSENSOR CONSIDERATIONS

Once characteristic spectral features for materials of environmental
interest have been established, the successful practical application of
the LIF techniques, e.g. in airborne remote sensing, will depend on the
signal-to-noise ratio obtainable in recording faint fluorescence

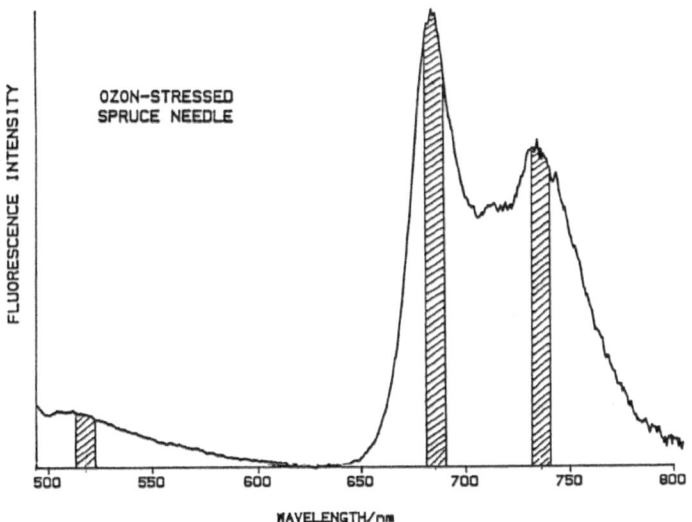

Fig. 14. LIF spectrum of spruce needles from a tree exposed to ozone. The excitation wavelength was 405 nm, and the curve is spectrally corrected. (From Ref. 60.)

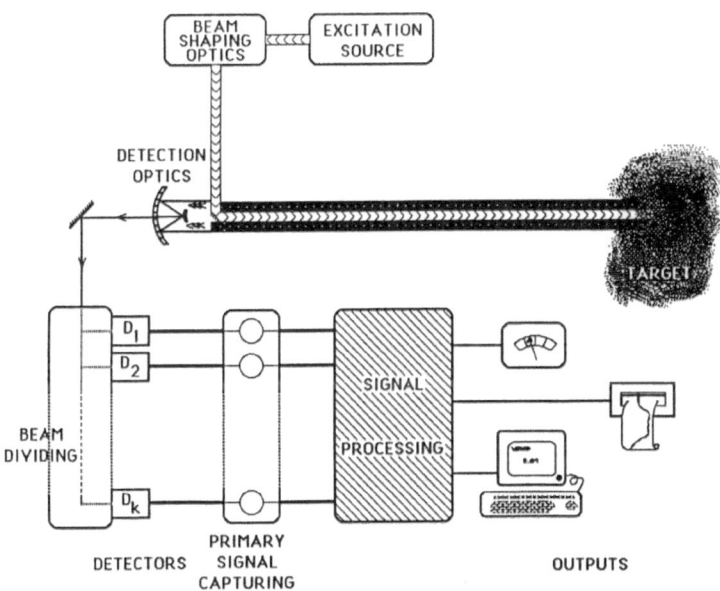

Fig. 15. General set-up for remote sample characterization based on LIF. (From Ref. 54.)

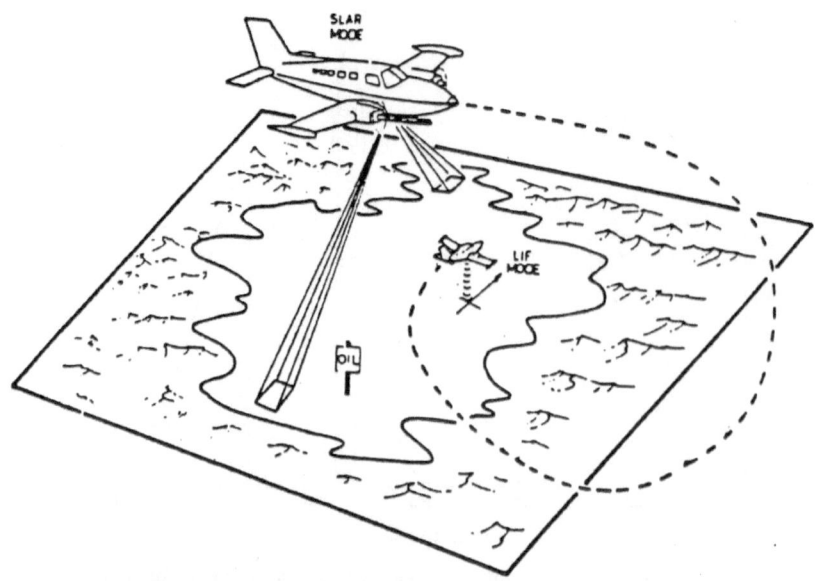

Fig. 16. Measurement scenario for an airborne marine surveillance
system. (From Ref. 54.)

phenomena in the presence of the environmental optical background
radiation. The basic elements of a fluorosensor are shown in Fig. 15[54]. A
critical discussion of the different subsystems can be found in Ref. 54,
considering transmitter, optics, electronics etc. In night-time operation
the system is basically limited only by the signal photon statistics,
while day-time operation is largely a question of background suppression,
using short gating times and narrow field-of-view optics. In certain
cases a low-cost pulsed UV lamp could be considered as a replacement for
the laser[53]. Lamps have a much longer pulse duration (>10 μs) than
lasers, but higher pulse powers. In tests we have found that the lamp is
superior during the night-time while the laser is indispensable during
the daytime.

The substance of interest must exhibit a sufficiently high contrast
in relation to the surrounding, normal material. Studies of contrast must
be performed during field tests. The LIF intensity from a thin floating
oil layer of strong intrinsic intensity will not necessarily strongly
dominate over the intrinsically weak sea water fluorescence integrated
over a layer thickness of the order of 1 m. We have found that the ratio
signal I(565 nm)/I(465 nm) for an oil discharge exhibits a contrast to
the surrounding coastal water of less than 2[50]. The fluorescence
integration of the natural water is difficult to simulate in the
laboratory. These phenomena are obviously very important when the
detection of sunken oil is considered. Experiments and conclusions on
this topic are presented in Ref. 52.

An example of a scenario for airborne oil-slick characterization is
given in Fig. 16[54]. Here, a simple laser fluorosensor to complement a
side-looking airborne radar (SLAR) system is considered. The
microwave-based system has a wide action range and all-weather capability
in oil slick detection based on a reduced sea-clutter back-scattering
signal from areas devoid of short-wavelength capillary water waves. Once
a slick is detected, the fluorosensor can characterize the the oil type
using a laser of low pulse energy at short range. Construction
considerations for such a system are presented in Ref. 55.

Fluorescence techniques can complement passive multispectral reflectance remote sensing in environmental monitoring. Several functioning demonstration systems have been constructed. For a more widespread use of LIF techniques in these contexts the development of efficient, lightweight and cheap pulsed UV lasers is highly desirable. Similarly, cheaper image-intensified array detectors would favorably influence the development in this field.

Fluorescence monitoring techniques can obviously also be used for industrial purposes. Since oils possess very strong fluorescence properties the presence or absence of thin surface layers on, e.g., sheet metal can be determined. Thus, surface cleanliness, corrosion protective measures etc., can be assessed with an industrial fluorosensor. Laboratory and field work on industrial laser-induced fluorescence is reported in Refs. 62, 64 and 65.

Finally, laser-induced fluorescence is finding many interesting applications in medicine, and many of the techniques used for environmental and industrial monitoring can be adopted for cancer diagnosis, using tumor-seeking agents, and for localizing atherosclerotic plaques in human vessels. These aspects have been pursued actively within our laboratory during the last few years and reviews of this work can be found in Refs. 66-68.

ACKNOWLEDGEMENTS

The author gratefully acknowledges fruitful cooperation with a large number of present and previous coworkers and graduate students in the field of environmental remote sensing. This work was supported by the Swedish Board for Space Activities, the National Swedish Environment Protection Board, the Swedish Natural Science Research Council, the Swedish Space Corporation and the Knut and Alice Wallenberg Foundation.

REFERENCES

1. W. Bach, J. Pankrath and W. Kellogg, eds., "Man's Impact on Climate," Elsevier, Amsterdam (1979).
2. R. Revelle, Carbon dioxide and world climate, Sci. Amer. 247/5:35 (1982).
3. J.H. Seinfeld, "Atmospheric Chemistry and Physics of Air Pollution", Wiley, New York (1986).
4. R.P. Wayne, "Chemistry of Atmospheres". Clarendon Press, Oxford (1985).
5. B.A. Trush, "The chemistry of the stratosphere," Rep. Prog. Phys. 51:1341 (1988).
6. R.S. Stolarski, "The Antarctic ozone hole," Sci. Am. 258/1:30 (1988).
7. D.A. Killinger and A. Mooradian, eds., "Optical and Laser Remote Sensing," Springer Series in Optical Sciences, vol. 39, Springer-Verlag, Heidelberg (1983).
8. R.M. Measures, "Laser Remote Sensing: Fundamentals and Applications", Wiley, New York (1984).
9. E.D. Hinkley, ed., "Laser Monitoring of the Atmosphere," Topics in Applied Physics, vol. 14, Springer-Verlag, Heidelberg (1976).
10. S. Svanberg, Lasers as probes for air and sea, Contemp. Phys. 21:541 (1980).
11. S. Svanberg, Fundamentals of atmospheric spectroscopy, in "Sur-

veillance of Environmental Pollution and Resources by Electro-magnetic Waves," T. Lund, ed., D. Reidel, Dordrecht (1978).

12. S. Svanberg, Laser technology in atmospheric pollution monitoring, in "Applied Physics - Laser and Plasma Technology," B.C. Tan, ed., World Science, Singapore (1985), p. 528.

13. W.B. Grant, Laser remote sensing techniques, in: "Laser Spectroscopy and its Applications", L.J. Radziemski, R.W. Solarz, and J.A. Paisner, eds., Marcel Dekker, New York (1987), p. 565.

14. T. Kobayashi, Techniques for laser remote sensing of the environment, Rem. Sens. Rev. 3:1 (1987).

15. R.M. Measures, "Laser Remote Chemical Analysis", Wiley-Interscience, New York (1988).

16. U. Platt, D. Perner, and H.W. Pätz, Simultaneous measurement of atmospheric CH_2O, O_3, and NO_2 by differential optical absorption, J. Geophys. Res. 84: 6329 (1979).

17. U. Platt and D. Perner, Measurements of atmospheric trace gases by long path differential UV/visible absorption spectroscopy, in Ref. (7).

18. H. Edner, A. Sunesson, S. Svanberg, L. Unéus and S. Wallin, Differential optical absorption system used for atmospheric mercury monitoring, Appl. Opt. 25: 403 (1986).

19. M.L. Chanin, Rayleigh and resonance sounding of the stratosphere and mesosphere, in Ref. (7).

20. C. Granier and G. Megie, Daytime lidar measurements of the mesospheric sodium layer, Planet. Space Sci. 30: 169 (1982).

21. K.H. Fricke and U. v. Zahn, Mesopause temperatures derived from probing the hyperfine structure of the D_2 resonance line of sodium by lidar, J. Atm. Terr. Phys. 47: 499 (1985).

22. M.P. Mc Cormick, T.J. Swisser, W.H. Fuller, W.H. Hunt, and M.T. Osborn, Airborne and groundbased lidar measurements of the El Chichon stratospheric aerosol from $90°$ N to $56°$ S, Geofisica Internacional 23-2: 187 (1984).

23. H. Edner, K. Fredriksson, A. Sunesson, S. Svanberg, L. Unéus, and W. Wendt, Mobile remote sensing system for atmospheric monitoring, Appl. Opt. 26: 4330 (1987).

24. K. Fredriksson, B. Galle, K. Nyström, and S. Svanberg, Mobile lidar system for environmental probing, Appl. Opt. 20: 4181 (1981).

25. K. Fredriksson and S. Svanberg, Pollution monitoring using Nd:YAG based lidar systems, in Ref. (7).

26. K. Fredriksson and H.M. Hertz, Evaluation of the DIAL technique for studies on NO_2 using a mobile lidar system, Appl. Opt. 23: 1403 (1984).

27. A.L. Egebäck, K. Fredriksson, and H.M. Hertz, DIAL techniques for the control of sulfur dioxide emission, Appl. Opt. 23: 722 (1984).

28. H. Edner, A. Sunesson, and S. Svanberg, NO plume mapping using laser radar techniques, Opt. Letters 12: 704 (1988).

29. H. Edner, G.W. Faris, A. Sunesson, and S. Svanberg, Atmospheric atomic mercury monitoring using differential absorption lidar techniques, Appl. Opt. 28: 921 (1989).

30. H. Edner, G.W. Faris, A. Sunesson, S. Svanberg, J.Ö. Bjarnason, H. Kristmansdòttir and K.H. Sigurdsson, Lidar search for atomic mercury in Icelandic geothermal fields, Submitted to J. Geophys. Res.

31. H. Edner, P. Ragnarsson, S. Svanberg and E. Wallinder, to appear.

32. O. Uchino, M. Togunaga, M. Maeda, and Y. Miyazoe, Differential absorption lidar measurement of tropospheric ozone with excimer-Raman hybrid laser, Opt. Lett. 8: 347 (1983).

33. J. Werner, K.W. Rothe, and H. Walther, Monitoring of the stratospheric ozone layer by laser radar, Appl. Phys. B32: 113 (1983).

34. G. Megie, G. Ancellet, and J. Pelon, Lidar measurements of ozone vertical profiles, Appl. Opt. 24: 3454 (1985).

35. H. Edner, S. Svanberg, L. Unéus, and W. Wendt, Gas correlation lidar, Opt. Lett. 9:493 (1984).

36. P.S. Andersson, S. Montán and S. Svanberg, Multi-spectral system for medical fluorescence imaging, IEEE J. Quant. Electr. **QE-23**:1798 (1987).

37. P. Ragnarsson, Spectroscopic imaging of effluent gases, Diploma paper, Lund Reports on Atomic Physics LRAP-83 (1988).

38. D.H. Hercules (ed.), "Fluorescence and Phosphorescence Analysis", Interscience, New York (1966).

39. E.L. Wehry (ed.), "Modern Fluorescence Spectroscopy", Vols 1 and 2, Plenum, New York (1976).

40. S. Udenfriend, "Fluorescence Assay in Biology and Medicine", Vol. I (1962), and Vol. II (1969), Academic Press, New York.

41. J.R. Lakowicz, "Principles of Fluorescence Spectroscopy", Plenum , New York (1983).

42. P.S. Andersson, E. Kjellén, S. Montán, K. Svanberg and S. Svanberg, Autofluorescence of various rodent tissues and human skin tumour samples, Lasers Med. Sci. **2**:41 (1986).

43. F.E. Hoge, R.N. Swift and J.K. Yungel, Active-passive airborne ocean color measurement. 2: Applications, Appl. Opt. **25**:48 (1986).

44. R.A. O'Neill, L. Buja-Bijunas and D.M. Rayner, Field performance of a laser fluorosensor for the detection of oil spills, Appl. Opt. **19**:863 (1980).

45. G.A. Capelle, L.A. Franks, D.A. Jessup, Aerial testing of a KrF laser-based fluorosensor, Appl. Opt. **22**: 3382 (1983).

46. H.H. Kim, Airborne bathymetric charting using pulsed blue-green lasers, Appl. Opt. **16**:46 (1977).

47. F.E. Hoge, R.N. Swift and E.B. Frederick, Water depth measurement using an airborne pulsed neon laser system, Appl. Opt. **19**:871 (1980).

48. L. Celander, K. Fredriksson, B. Galle and S. Svanberg, Investigation of laser-induced fluorescence with applications to remote sensing of environmental parameters, Göteborg Institute of Physics Reports GIPR-149, CTH, Göteborg (1978).

49. K. Fredriksson, B. Galle, K. Nyström, S. Svanberg and B. Öström, Underwater laser-radar experiments for bathymetry and fish-school detection, Göteborg Institute of Physics Reports GIPR-162, CTH, Göteborg (1978).

50. K. Fredriksson, B. Galle, K. Nyström, S. Svanberg and B. Öström, Marine laser probing - results form a field test, Medd. fr. Havsfiskelaboratoriet **245**, Swedish Fishery Board, Lysekil (1979).

51. B. Galle, T. Olsson and S. Svanberg, The fluorescence properties of jelly-fish, Göteborg Institute of Physics Reports GIPR-181, CTH, Göteborg (1979) (in Swedish).

52. P.S. Andersson, S. Montán and S. Svanberg, Oil-slick characterization using an airborne fluorosensor - construction considerations, Lund Reports on Atomic Physics LRAP-45, LTH, Lund (1985).

53. P.S. Andersson, S. Montán and S. Svanberg, Flashlamps for remote fluorescence characterization of oil slicks, Lund Reports on Atomic Physics LRAP-57, LTH, Lund (1986).

54. P.S. Andersson, S. Montán and S. Svanberg, Remote sample characterization based on fluorescence monitoring, Appl. Phys. **B44**:19 (1987).

55. P.S. Andersson, S. Montán and S. Svanberg, Fluorosensor for remote characterization of marine oil-slicks, Intern. Coll. on Remote Sensing of Pollution of the Sea, Oldenburg, March 31-April 3 (1987).

56. F.E. Hoge, Ocean and terrestrial lidar measurements, in "Laser Remote Chemical Analysis", R.M. Measures, ed., Wiley-Interscience, New York (1988).

57. F.E. Hoge, R.N. Swift, and J.K. Yungel, Feasiblility of airborne de-

tection of laser-induced fluorescence of green terrestrial plants, Appl. Opt. **22**:2991 (1983).

58. E.W. Chappelle, F.M. Wood, W.W. Newcomb, and J.E. McMurtrey, III, Laser-induced fluorescence of green plants. 3. LIF spectral studies of five major plant types, Appl. Opt. **24**:74 (1985).

59. F. Castagnoli et al., A fluorescence lidar for land and sea remote sensing, SPIE **663**:212 (1986).

60. S. Andersson-Engels, K. Callander and B. Galle, Investigation of the possibilities to use laser-induced fluorescence to map conifer forest damage caused by ozone, IVL Report **L88/146**, IVL, Göteborg (1988).

61. S. Svanberg, Environmental diagnostics, in "Trends in Physics", M.M. Woolfson, ed., Adam Hilger, Bristol (1979).

62. P. Herder, T. Olsson, E. Sjöblom and S. Svanberg, Monitoring of surface layers using laser-induced fluorescence, Lund Reports On Atomic Physics LRAP-9, LTH, Lund (1981).

63. H.S. Chen, "Space Remote Sensing Systems," Academic, Orlando (1985).

64. S. Montán and S. Svanberg, A system for industrial surface monitoring utilizing laser-induced fluorescence, Appl. Phys. **B38**:241 (1985).

65. S. Montán and S. Svanberg, Industrial applications of laser-induced fluorescence, L.I.A. ICALEO **47**:153 (1985).

66. S. Svanberg, Medical applications of laser-induced fluorescence, Phys. Scripta **T19**:469 (1987).

67. S. Svanberg, Medical applications of laser spectroscopy, Phys. Scripta **T26**:90 (1989).

68. S. Andersson-Engels, J. Ankerst, A. Brun, Å. Elner, A. Gustafson, J. Johansson, S.-E. Karlsson, D. Killander, E. Kjellén, E. Lindstedt, S. Montán, L.G. Salford, B. Simonsson, U. Stenram, L.-G. Strömblad, K. Svanberg and S. Svanberg, Tissue diagnostics using laser-induced fluorescence, Ber. Bunsenges. Phys. Chem. **93**:335 (1989).

REMOTE DETECTION OF ATMOSPHERIC POLLUTANTS USING DIFFERENTIAL ABSORPTION LIDAR TECHNIQUES

J.P. Wolf, H.J. Kölsch, P. Rairoux*, and L. Wöste

Institut für Experimentalphysik, Freie Universität Berlin
Arnimallee 14, D-1000 Berlin 33

*Present address: CAL, Ecole Polytechnique Fédérale de Lausanne

1015 Lausanne (Switzerland)

1. Introduction

Air pollution is an extremely dynamic phenomenon, and this makes its understanding and, therefore, its control, more elusive. This dynamic behavior appears not only in physical terms by the diffusion and transport of emitted pollutants, but also chemically, through the many reactions occuring in the atmosphere. It is therefore of outstanding importance to be able to correlate emission and immission, and thus characterize the impact of different kinds of sources of pollution (industries, vehicles, domestic heaters) on the environment. The only way to control phenomena like acid rains or hole formation in the ozone layer, is to perform a permanent and large-scale monitoring of the air pollution. Presently existing devices, however, although they may be very sensitive like Laser Induced Fluorescence (LIF) or Differential Optical Absorption Spectroscopy (DOAS), can only provide spot measurements at ground level. Three-dimensional informations, reflecting the dynamic character of pollution, are, until now, sorely lacking.

These considerations have induced, about fifteen years ago, the elaboration of a new technique for approaching air pollution phenomena: the LIDAR (Light Detection And Ranging) technique. A LIDAR apparatus allows indeed for selective measurements of pollutants concentration over several kilometers, range resolved like a RADAR (Radiowaves Detection And Ranging), and this in an interactive manner (i.e. without the usual need for samples). It is then possible to obtain "geographic maps" of concentrations, at large scale (up to 15

Applied Laser Spectroscopy, Edited by W. Demtröder and
M. Inguscio, Plenum Press, New York, 1990

km), which reflects the propagation and spread of pollution. The LIDAR approach is then to favorize fast 3D concentration measurements than a very high sensitivity or an extensively large palette of detectable pollutants. In this respect, it appears as a very powerful and complementary technique for analyzing air pollution in the future.

2. The LIDAR technique

When a laser is sent into the atmosphere, its beam is widely scattered in every direction by particles present in the air. This scattering is essentially caused by Rayleigh scattering on Nitrogen and Oxygen molecules, and Mie scattering on aerosols (dusts, water droplets, ..). At low atitudes, Mie scattering is predominant because of the higher cross-section (about 10^{-8} cm^2/sr for Mie scattering against 10^{-26} cm^2/sr for the Rayleigh one) and the high aerosols concentration. In a LIDAR arrangement, the backscattered light is collected by a telescope, usually placed coaxially with the laser emitter (see fig. 1a). The signal is then focussed on a photodetector through a spectral filter, adapted to the laser wavelength. As a pulsed laser is used, the intensity of the backscattered light can be recorded as a function of time, and then provide spatial resolution. The received signal reflects the aerosol concentration versus range, similar to an optical RADAR. More precisely, the received number of photons from a distance R at a wavelength λ is given by (assuming that each photon is scattered only once) [1, 2, 3]:

$$M(R,\lambda)= M_0(\lambda) \frac{A_0}{R^2} \beta^{\pi}(R,\lambda) \ \Delta R \ \zeta(R,\lambda) \ \exp\{-2\int_0^R \alpha(\lambda,R) \ dR\} \tag{1}$$

where: $M_0(\lambda)$ is the number of photons emitted by the laser,
A_0 the area of the telescope,
ΔR the spatial resolution of the system, limited essentially by the laser pulselength τ, $\Delta R = c\tau/2$,
$\beta^{\pi}(R, \lambda)$ the volume backscattering coefficient,
$\zeta(R, \lambda)$ the detection efficiency, and
$\alpha(R, \lambda)$ the total atmospheric extinction coefficient.

Let us now consider each of these different parameters, and their respective effect on the received signal:

(1). The A_0/R^2 factor represents the solid angle formed by the detector area, and produces a decrease in intensity with the range square.

(2). The signal dependance on the aerosol concentration N_M, which allows their detection, is implicitly included in the volume backscattering coefficient β^π:

$$\beta_\pi = \beta_\pi^R + \beta_\pi^M = \frac{\pi^2 (n_R^2 - 1)^2}{N_R \lambda^4} + \int_0^\infty \frac{d\sigma_M}{d\Omega}(\theta = \pi) \ N_M(a) \ da$$

where N_R and N_M are respectively the number densities of the air molecules and aerosols, n_R the refraction index of clean air ($N_2 + O_2$ molecules), and $d\sigma/d\Omega$ the angular differential cross-section of Mie scattering.

However, the aerosols are spread over a large size distribution (10 nm < a < 100 μm), and , as exhibited in the latter equation, β^π only reflects the combined effect of the different sizes. A quantitative measurement of the aerosols concentration becomes therefore rather complicated, unless one assumes a standard size distribution [1, 2, 3, 4]. Several numerical programs [5] have been already elaborated in order to evaluate β^π , α, and then $M(R,\lambda)$, for standard aerosols size distributions.

(3). The detection efficiency $\xi(R, \lambda)$ takes into account every geometrical and optical factors of the receiver arrangement. One usually separates $\xi(R, \lambda)$ into two different parameters, one dedicated to the spectral characteristics $\xi(\lambda)$ of the detection channels (filters, monochromator, ..), and the other, $\xi(R)$, to geometrical properties such as the overlap between the field illuminated by the laser and the telescope field of view. As described further, this latter parameter is used very often in LIDAR arrangements for compressing the large dynamic range of the signal.

(4). The exponential factor $\exp\{-2 \int \alpha(R, \lambda) \ dR\}$ represents the atmospheric attenuation of the laser beam, via Beer-Lambert law. The attenuation is caused by two different processes: the Rayleigh-Mie scattering α_{RM} (the light which is scattered in another direction than forward or backward is lost), and the specific absorption coefficient α_A of the different molecules present in the atmosphere.

437

This latter term is central in the LIDAR equation, because it allows a specific detection of a particular pollutant, using the Differential Absorption Lidar (DIAL) technique (see section 3).

The first measurements of the atmospheric backscattering by LIDAR have been performed already in the early sixties. Fiocco and Smullin [6] recorded indeed the first LIDAR echoes in 1963 up to altitudes of 140 km, using a ruby laser. More recently, several studies are undertaken in order to monitor the motions of tropospheric and stratospheric aerosols, especially arised from volcanic eruptions [7-12]. The aerosol and cloud layers play, indeed, an essential role in meteorology and in the energetic balance of the earth. Other measurements have also been performed on the structure of the clouds itself, using depolarisation effects which occur because of non-spherical droplets or multiple scattering [13-18].

3. Lidar and Differential Absorption

Besides the increasing interest about LIDAR in meteorology, a major application is constituted by the detection of traces in the atmosphere, and in particular, air pollution. Several techniques, combined to a LIDAR arrangement, have been used for this purpose:

(1). The Laser Induced Fluorescence [1, 2] Lidar has been successfully demonstrated for probing the stratosphere. A remarkable study on the remote detection of the Na- and the K- layers has been performed in particular by Mégie et al [19-21], until altitudes as high as 110 km. The utilization of this technique at tropospheric altitudes, and therefore for air pollution purposes, is however strongly limited by collision induced quenching. The fluorescence cross-section (typ. 10^{-20} cm^2/sr) is, indeed, reduced by a typical factor 10^{-5} at ground level pressure [2].

(2). In a Raman Lidar configuration, the laser is used for inducing virtual transitions in every molecules present in the air, which allows the detection of every species at the same time [2, 3, 22]. The Raman cross-section being, however, extremely small (~ 10^{-29} cm^2/sr), the detection limit is of the order of one ppm (part per million), and range-resolved measurements of pollutants are very difficult to perform. However, some remote Raman measurements have been

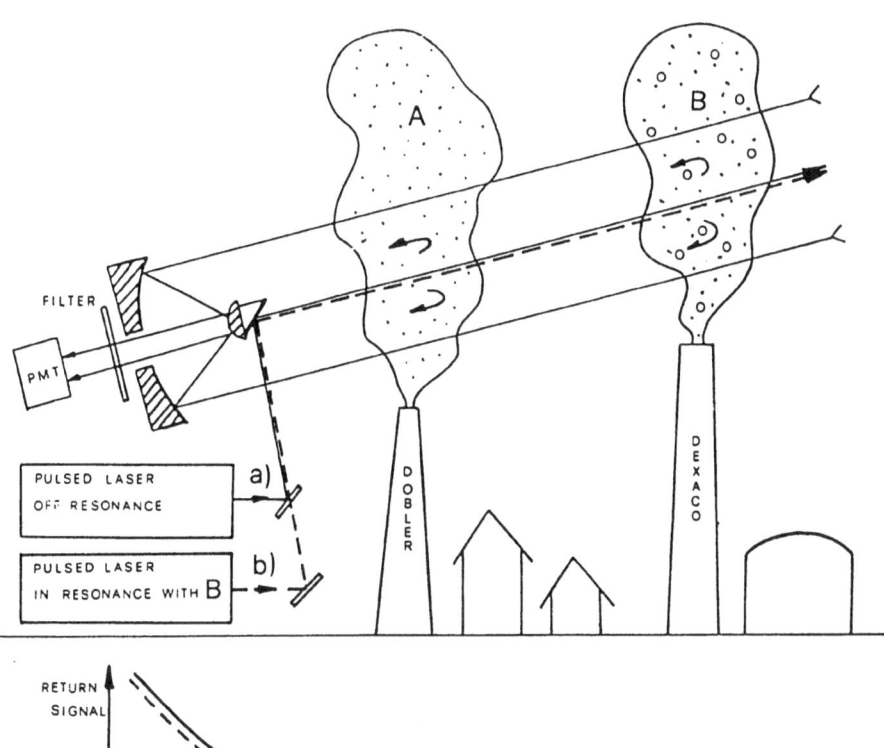

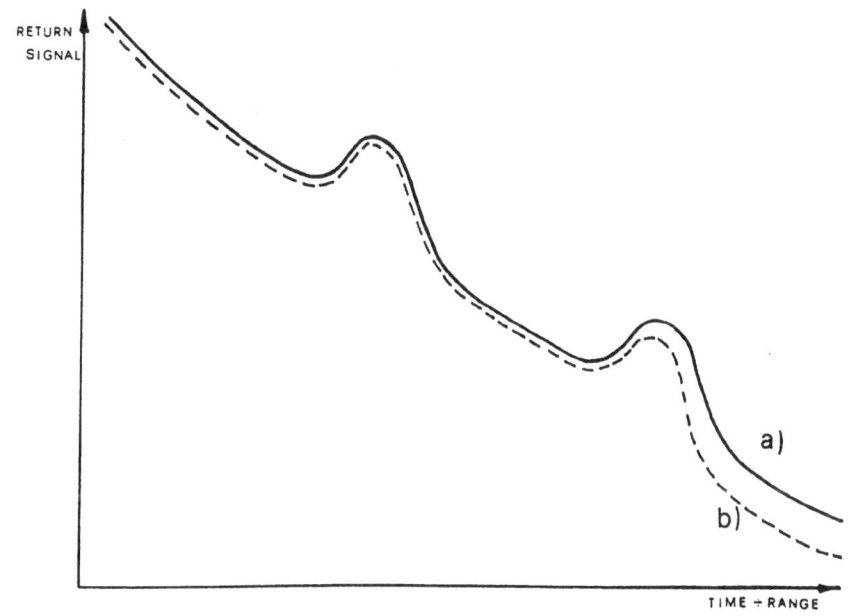

1. The DIAL principle

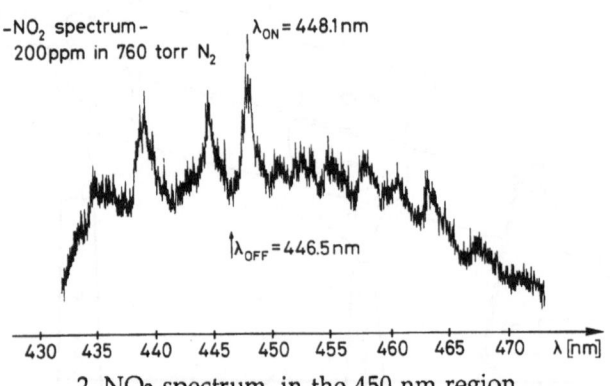

2. NO₂ spectrum, in the 450 nm region

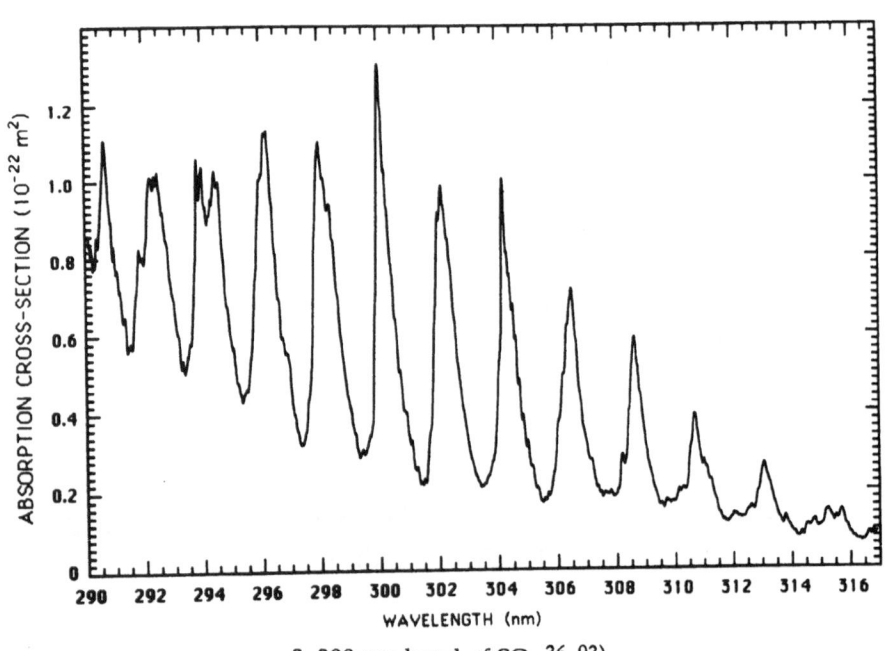

3. 300 nm band of SO₂ [36, 93]

reported recently on water vapor [23]. Pure rotational Raman effect can reach a cross-section about 10^2 times higher [24], and has allowed temperature profile measurements [25].

(3). The Differential Absorption Lidar (DIAL) is the most commonly used technique for analyzing remote air pollution. It has been demonstrated as the most sensitive at tropospheric altitudes [2, 26], and allows then a monitoring of the concentration of pollutants at both emission and immission. For these reasons, the present article will be dedicated particularly to this technique.

Absorption spectroscopy is a very powerful method for obtaining the selectivity on the measured species in Lidar experiments. Figure 2 and 3 represent the absorption spectra of two pollutants, which are most implicated in atmospheric acidification processes: NO_2 and SO_2. Both exhibit structured patterns in the visible and near UV respectively, which allows one to select a couple of wavelengths close from each other, with a large absorption coefficient difference (called λ_{on} and λ_{off}, for on-resonance and off-resonance wavelength respectively).

Let us now assume that such wavelength couple (λ_{on}, λ_{off}), selected for a substance "B", is sent simultaneously into the atmosphere (figure 1.a and 1.b). As λ_{on} and λ_{off} have been chosen close enough for exhibiting the same scattering properties, the first chimney plume (which does not contain the detected species "B") will cause an increase of the backscattered signal, because the concentration of aerosols is larger, but the same increase for both pulses. Conversely, the second chimney plume, which contains a certain quantity of the pollutant, will absorb the backscattered signal at the λ_{on}-wavelength much stronger than at the λ_{off}-one. From this difference, and using Beer-Lambert's law, one can deduce the specific concentration of the pollutant under investigation versus range: The received signals, indeed, both follow the Lidar equation (1), and by dividing each other, one obtains:

$$\frac{M(R, \lambda_{OFF})}{M(R, \lambda_{ON})} = \chi \cdot \frac{\beta(R, \lambda_{OFF})}{\beta(R, \lambda_{ON})} \frac{\xi_{OFF}(R)}{\xi_{ON}(R)} \exp\left(2 \int_{0}^{R} (\alpha(R, \lambda_{ON}) - \alpha(R, \lambda_{OFF})) \, dR\right)$$

(2)

where χ is a factor independant from the range R.

As λ_{on} is close to λ_{off}, the Rayleigh-Mie scattering can be considered as the same for both wavelengths ($\beta^{\pi}(R, \lambda_{on}) = \beta^{\pi}(R, \lambda_{off})$ and $\alpha_{RM}(R, \lambda_{on}) = \alpha_{RM}(R,$

λ_{off})), and for a perfect geometrical alignement of the two laser beams ($\xi_{ON}(R) = \xi_{OFF}(R)$), the equation above becomes:

$$\frac{M(R,\lambda_{OFF})}{M(R,\lambda_{ON})} = \chi \ \exp\{2 \int_o^R (\alpha_A(R,\lambda_{ON}) - \alpha_A(R,\lambda_{OFF})) \ dR\}$$

or, under a differential form:

$$\frac{d}{dR} \ \ln \ (\frac{M(R,\lambda_{OFF})}{M(R,\lambda_{ON})}) = 2 \ (\alpha_A(R,\lambda_{ON}) - \alpha_A(R,\lambda_{OFF}))$$

Let us now assume that, between λ_{on} and λ_{off}, only the measured species exhibits a large extinction coefficient difference (which is again assured by a proper set of wavelengths), and write $\alpha_A(R, \lambda) = N(R) \ \sigma(\lambda)$, where N is the pollutant number density, and σ the absorption cross-section at the wavelength λ. The specific concentration versus range is then expressed by:

$$N(R) = \frac{1}{2 (\sigma(\lambda_{ON}) - \sigma(\lambda_{OFF}))} \ \frac{d}{dR} \ \ln \ (\frac{M(R,\lambda_{OFF})}{M(R,\lambda_{ON})}) \qquad (3)$$

or, for a finite range resolution ΔR:

$$N(R) = \frac{1}{2 (\sigma(\lambda_{ON}) - \sigma(\lambda_{OFF}))} \ \frac{1}{\Delta R} \ \ln \ (\frac{M(R,\lambda_{OFF}) \ M(R+\Delta R,\lambda_{ON})}{M(R,\lambda_{ON}) \ M(R+\Delta R,\lambda_{OFF})})$$

The concentration of the species is then measured range resolved (see the latter mathematical treatement, summarized in the figure 4), and, by scanning the field of investigation in azimuth or elevation, one obtains 2D or 3D mappings, like a molecule-specific Radar.

Several considerations have, however, to be taken into account for performing accurate DIAL measurements with a satisfying signal to noise ratio (S/N):

(1). The atmospheric motions, inducing fluctuations in the aerosols concentration as well as in the concentration of the monitored pollutant,

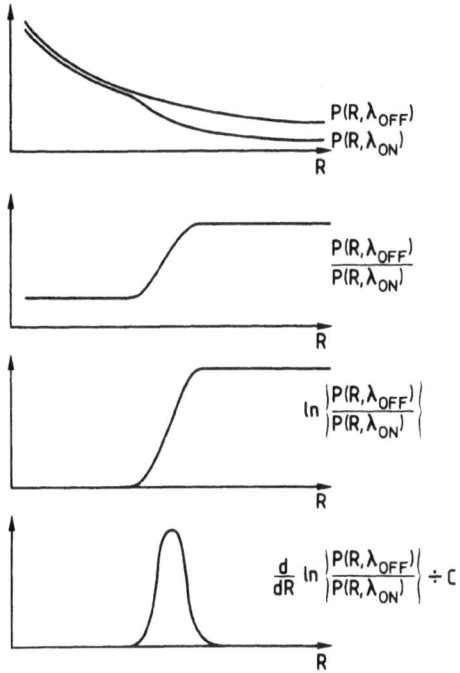

4. Concentration calculation from λ_{on} and λ_{off} returns

Legend

1   power generator
1a   cooling device

emitter:
2   excimer pump-laser
2a   gas stock
3ab   dye lasers
4ab   SHG: BBO crystal + quartz compensator
5   beam assembling: lambda/2, glan prism + chopper
6   beam expander
7   fundamental blocking filter (option)
8   scanning mirror

receptor:
8   reception mirror
9   telescope
10   video camera + monitor
11   receiver box
11a   iris diaphragm
11b   lens
11c   dichroic beamsplitter
11d   vis narrow band filter
11e   uv narrow band filter
11fg   photomultipliers
12   transient digitizer
13   computer
14   graphic display
15   data storage
16   plotter

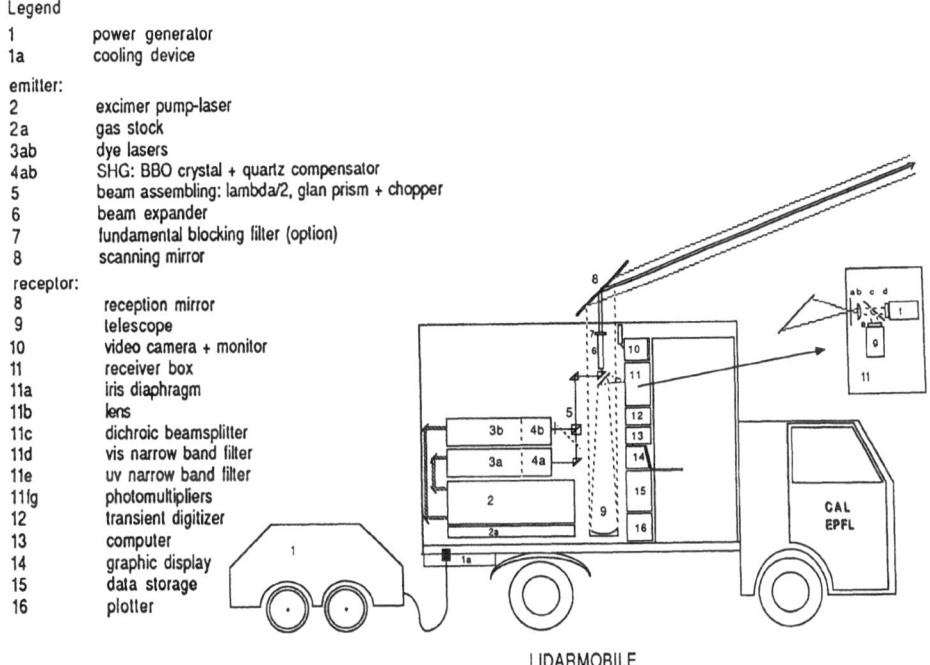

LIDARMOBILE

5. Mobile UV-DIAL system for measuring NO, NO$_2$, SO$_2$, and O$_3$

443

constitute certainly the most important source of noise in remote sensing. Most of the DIAL arrangements utilize, indeed, an alternate sending of λ_{on} and λ_{off} wavelengths, and atmospheric motions between the consecutive pulses induce an error in the concentration measurement. This error is usually reduced by averaging a large number N of laser pulse pairs. Killinger et al [27, 28] demonstrated that the atmospheric fluctuations cannot be assumed as a markovian process, and that the related standard deviation decreases much slower than $N^{-1/2}$. These experiments showed that the atmosphere is "frozen" during a timescale of about 5 mS [27, 28, 29, 30], and that the right way of reducing the fluctuations is to shorten the time, which separates λ_{on} and λ_{off} pulses, rather than to average a large number of laser shots. A solution adopted since, is to use two laser systems, one tuned on λ_{on} and the other on λ_{off} , fired at a time interval shorter than 5 mS. This solution, however, has two major disadvantages: using two different laser beams can introduce systematic errors, because of different geometries (see below), and the cost, size, and maintenance of such a laser system are limitative.

The advent of new, high repetition rate (up to 1 kHz) lasers, like Excimers, opens new perspectives for solving definitely this problem [31, 32, 33]. The time interval between λ_{on} and λ_{off} can then remain in the order of 5 mS, using only one laser system. The repetition rate has also been found of outstanding importance for averaging long term fluctuations like winds, and performing therefore realistic 2D or 3D mappings of concentration. Such measurements have been recently investigated in particularly turbulent environments [32, 33] (plume spreads, see section 7.1) .

(2). As it can be noticed from equation (2), if the geometrical form factor $\xi(R)$ is different for λ_{on} and λ_{off}, a systematic error is introduced. This error, resulting from a different investigated volume for the two wavelengths, can occur either by misalignement of both laser beams, or by a different emission mode profile. A single laser system, however, reduces in a large extent these geometrical sources of error.

(3). Several DIAL systems are using visible and near-UV radiation, where solar background level cannot be neglected. A proper spectral filtering is thus necessary to allow daytime operation. Daytime operation capability is all the more important that several chemical reactions occuring in the atmosphere are highly dependant on the solar flux. Under certain conditions, the solar radiation can even become the major source of noise.

(4). The above mathematical treatment assumes only single scattering. Multiple scattering [2, 34, 35] can occur as well, and in particular at large aerosols concentration (plumes, cloud layers). The related error can be limited in most cases by reducing the investigating field of view, or by using polarization properties [36].

(5). An accurate measurement of the pollutant absorption cross-section is of outstanding importance, because $\Delta\sigma$ appears directly in the determination of the species concentration (see equation 3). One must then care especially about the effects of saturation (Beer's law is only valid in a linear regime), temperature, pressure broadening, partial pressure broadening, and laser linewidth [37, 38, 39, 40, 41], for measuring the $\Delta\sigma$ coefficient in the laboratory. Furthermore, as $\Delta\sigma$ depends on temperature, pressure, and partial pressure (generally not below 10 ppm), corrections have to be introduced for each DIAL measurements in the atmosphere. These corrections are especially necessary for species exhibiting very narrowband transitions (such as NO), and for high concentration or high temperature environments (e.g. plumes).

(6). A major advantage of optical methods, such as DIAL, is constituted by its selectivity on the type of the detected species. However, if the interferences are rather limited in the visible and the near-UV, they can lead to a serious handicap in the middle infrared (such as for the CO_2 laser emission spectrum), where vibrational transitions are involved instead of electronic ones.

Taking into account these different sources of noise and error, typical detection limits of a differential absorption and scattering process ranges, in the UV-Vis, from about 0.1 ppm.m (part per million times meter) to 10 ppm.m. The sensitivity is expressed in ppm.m, because of the Beer-Lambert law: for a given S/N ratio, the larger the range of integration, the lower the detected concentration. For example, if the detection limit is 1 ppm.m, the lowest detectable level will be 1 ppm for a range resolution ΔR of 1 meter (emission conditions), but 5 ppb for a range resolution of 200 meters (immision conditions). The maximal range of measurement is about 3 km. The sensitivity and the absorption path can be increased in a large extent by using a topographic target (about 500 ppt for a 20 km-pathlength, see section 7.3). Nevertheless, the spatial resolution is then lost in one direction, and only average 2D mappings are performable.

4. Palette of the detectable pollutants

The DIAL technique has been tested for the first time in 1966 by Schotland [42] for evaluating remote the vertical profile of water vapor contained in the atmosphere. Several groups have used this technique since, over a large palette of molecules, as summarized in table 1. Rothe et al performed in 1974 the first real "mappings" of the NO_2 concentration over a chemical plant [43]. The nitrogen dioxide has been one of the most studied pollutant by the DIAL technique [31,33, 43-52], in reason of its absorption band at 450 nm (B^1B_2 state), a spectral region which is easily reachable with commercial laser systems. For the same reason, two other pollutants, exhibiting absorption bands near 300 nm, have been often studied as well: SO_2 [30, 31-33, 46, 47, 50, 51, 54-57] and O_3 [36, 51, 53, 57-60]. Actually, as mentioned in table 1, NO_2, SO_2, and O_3 can be detected using the same laser system (dye laser, including Second Harmonic Generation). This leaded to the apparition of very attractive mobile DIAL systems, able to measure remote the concentration of these three pollutants, which are particularly involved in the acidification processes of the atmosphere [61-63] NO has remained, for a long time, much more difficult to detect [64, 65], because of its absorption bands at 227 nm and around 5 μm. Very recently, the advent of a new frequency doubling crystal, the β-BaB_2O_4 (BBO), has allowed, however, a sensitive and range-resolved detection of NO, as well as a simultaneous detection of both NO and NO_2 [66]. Furthermore, this new crystal also provides powerful radiation at 300 nm, and allows thus a DIAL measurement of NO, NO_2, SO_2, and O_3 with a simple laser system [32] (see section 6).

Besides the above mentioned pollutants, which constitute certainly the most attractive set of molecules for the environmental protection agencies, several other species have been successfully detected using the DIAL technique. Remarkable studies have been performed, in particular, by Edner et al on the atomic mercury (at levels as low as 1 ng/m^3) [67, 68], and by Weitkamp et al on HCl [2, 69].

Hydrocarbons and Freons constitute also two groups of pollutants, which are very relevant in respect to air quality. Their electronic excited states are, however, located to high in energy for being reachable with commercially available laser systems. Furthermore, below ~ 220 nm, the atmosphere becomes almost opaque, because of molecular absorption (such as O_2) and UV-Rayleigh scattering. The detection of these species is then performed in the infrared, via vibrational excitation, especially using the CO_2-laser lines.

If the DIAL detection in the near infrared seems to be very promising [36, 70, 71], the detection in the mid-IR suffers of three major disadvantages:

(1). The backscattering cross-sections are about 1000 times smaller at 10 μm than in the UV [72], and range-resolved measurements are then rather difficult to perform. Best results are obtained, using topographic targets, and in particular for airborne applications [76, 77].

(2). Because of thermal noise, the IR detectors are much less sensitive than the usual UV-Vis photomultipliers. This detectivity problem can be partially overcome using heterodyning techniques [1, 2, 78], but remains a severe disadvantage compared to the UV-spectral region.

(3). As noticeable from table 1, several species absorb in the 10 μm - band, which represents a significant risk of interference among the different pollutants. This is especially true, as hydrocarbons are often emitted simultaneously.

Besides the above mentioned tropospheric measurements, stratospheric DIAL investigations are increasing very rapidly. An outstanding interest has been carried on the ozone layer since 1978 [7, 21, 79, 80], and has become even stronger recently, because of the discovery of the deep ozone depletion in Antarctic.

5. Mobile DIAL systems

Mobility and compactness are two major aspects for air pollution analyzers. Different groups have realized mobile UV-DIAL systems, and demonstrated the high performance level of their system in field measurements. Particularly advanced studies have been performed in Sweden by the Lund Institute of Technology, the National Swedish Environmental Protection Board, and the Chalmers University of Technology, in England by the National Physical Laboratory and the Central Electricity Research Laboratories, in Germany by the GKSS-Forschungszentrum, in France at the Service d'Aéronomie du CNRS, in the United-States at the Stanford Research Institute and at the NASA Langley Research Center, and our group in Switzerland sponsored by the Swiss National Foundation for Scientific Research. Since 1989, our group is in Berlin.

Recently, several books have been dedicated to laser remote sensing. In particular, Hinkley (1976) [1] and Measures (1984 and 1988) [2, 3] report a very detailed analysis of the subject, and can be considered as reference handbooks. For this latter reason, we prefer, presently, to concentrate on some new aspects and perspectives we found, which were allowed by stepforwards in laser technology.

Table 1. Main detectable pollutants, using the DIAL technique and commercially available lasers (from Fredriksson (1988) [36]).

SPECIES		λ	ABSORPTION CROSS-SECTION 10^{-18} CM^2	LASER
Nitric Oxide	NO	226.8 nm	4.6	DYE (SHG)
		5.215 μm	0.67	CO
		5.263 μm	0.6	CO$_2$ (SHG)
Nitrogen Dioxide	NO$_2$	448.1 nm	0.69	DYE
		6.229 μm	2.68	CO
Sulfur Dioxide	SO$_2$	300.05 nm	1.3	DYE (SHG)
		3.984 μm	0.42	DF
		9.024 μm	0.25	CO$_2$
Ozone	O$_3$	298.4 nm	1.5	DYE (SHG)
		9.505 μm	0.45	CO$_2$
Carbon monoxide	CO	4.776 μm	0.8	CO$_2$ (SHG)
Mercury	Hg	253.7 nm	$5.6 \cdot 10^4$	DYE (SHG)
Chlorine	Cl$_2$	330 nm	0.26	DYE (SHG)
Hydrogen Chloride	HCl	3.636 μm	0.2	DF
Ammonia	NH$_3$	10.333 μm	1.0	CO$_2$
Freon 11	CCl$_3$F	9.261 μm	1.09	CO$_2$
Freon 12	CCl$_2$F$_2$	10.719 μm	1.33	CO$_2$
Fluorocarbon 113	C$_2$Cl$_3$F$_3$	9.604 μm	0.77	CO$_2$
Perchlorethylene	C$_2$Cl$_4$	10.834 μm	1.14	CO$_2$
Hydrazine	N$_2$H$_4$	10.612 μm	0.18	CO$_2$
Sulfur Hexafluoride	SF$_6$	10.551 μm	30.3	CO$_2$
Methane	CH$_4$	3.715 μm	0.002	DF
		3.391 μm	0.6	HeNe
Propane	C$_3$H$_8$	3.391 μm	0.8	HeNe
Benzene	C$_6$H$_6$	250 nm	1.3	DYE (SHG)
		9.621 μm	0.07	CO$_2$
Ethylene	C$_2$H$_4$	10.553 μm	1.19	CO$_2$
Trichlorethylene	C$_2$HCl$_3$	10.591 μm	0.49	CO$_2$
Monochlorethane	C$_2$H$_5$Cl	10.275 μm	0.12	CO$_2$
1,2-Dichlorethane	C$_2$H$_4$Cl$_2$	10.591 μm	0.02	CO$_2$
Vinyl Chloride	C$_2$H$_3$Cl	10.612 μm	0.33	CO$_2$
Propylene	C$_3$H$_6$	6.069 μm	0.09	CO
1,3-Butadiene	C$_4$H$_6$	6.215 μm	0.27	CO
1-Butene	C$_4$H$_8$	10.787 μm	0.13	CO$_2$
Chloroprene	C$_4$H$_5$Cl	10.261 μm	0.34	CO$_2$
MMH	CH$_3$N$_2$H$_3$	10.182 μm	0.06	CO$_2$
UDMH	(CH$_3$)$_2$N$_2$H$_2$	10.696 μm	0.08	CO$_2$
Ethyl-Mercaptan	C$_2$H$_5$SH	10.208 μm	0.02	CO$_2$

Table 2. Characteristics of Excimer pumped dye-lasers; fundamental and frequency doubled outputs using BBO crystals

	λ_{on} / λ_{off} [nm]	Δk [1/cm.atm]	Dye used	Energy	Rep.rate
NO	226.82 / 226.83	105	Coum 2 (2v)	5 mJ	80 Hz
NO$_2$	448.1/446.5	8.0	Coum 2	30 mJ	80 Hz
SO$_2$	300.05/299.4	25.4	Rhod 6G(2v)	4 mJ	80 Hz
O$_3$	291.4/300.55	20.9	Rhod 6G(2v)	4 mJ	80 Hz

6. New possibilities allowed by Excimer lasers and BBO crystals

The UV-DIAL system we developed has for goal to approach as close as possible the concerns of atmospheric chemists and environmental protection agencies. Particular cares have then been carried on sensitivity, for allowing both emission and immission monitorings, on the choice of the detected pollutants, on the rapidity of the mapping execution, on the daytime and nighttime measurement capability, on mobility, and on compactness. The system, able to detect NO, NO_2, SO_2, and O_3 (table 2), is based on a new laser configuration.

As mentioned before, it has been found that atmospheric fluctuations constitute one of the most important source of noise in DIAL experiments. The best way to reduce these effects is to use a laser system, which is able to switch from λ_{on} to λ_{off} in a time sufficiently small for assuming the atmosphere as "frozen". The laser repetition rate appears therefore of outstanding importance for obtaining high S/N levels. Since the first DIAL measurements, performed with flashlamp-pumped-dye-lasers, the repetition rate has been increased of about one order of magnitude, using Nd:Yag-pumped-dye lasers (10 Hz instead of 1 Hz). This improvement remained, however, insufficient, because the correlation coefficient between λ_{on} and λ_{off} at 10 Hz is only around 0.2 [28].

We report, in the present article, the first mobile DIAL arrangement, based on Excimer-pumped-dye lasers (figure 5). The repetition rate of the system reaches 80 Hz, allowing a correlation of about 0.7 [28]. The repetition rate is also of prime importance for performing fast mappings of concentration, and avoiding the errors induced by slow fluctuations (winds).

The Excimer laser we use (Lambda Physik EMG 201 MSC; 0.4 J/pulse at 308 nm) pumps alternately two frequency-doubled dye-lasers (Lambda Physik FL 2002), tuned respectively on λ_{on} and λ_{off}. The dye-laser outputs are then superimposed and sent into the atmosphere through a 10 x beam expander. The possibilities of the system have been extended in a large extent by the advent of new BBO (β-BaB_2O_4) frequency doubling crystals [81]. These crystals provide highly efficient UV-generation (~ 15 %) at both 225 nm and 300 nm, and allow therefore a sensitive detection of NO, SO_2, and O_3. As NO_2 can be detected in the 450 nm-band, the dye laser system needs only two types of dyes (Coumarin 2, Rhodamin 6G) and one type of doubling crystal for the 4 investigated pollutants (table 2). The laser system becomes thus extremely simple, and in principle, only one dye-laser could be used, if a fast wavelength switch was introduced in the resonnator. Several techniques for switching from λ_{on} to λ_{off} have been reported already in the litterature [2, 36, 82].

The range-resolved NO monitoring capability is a real stepforward in the DIAL technique. It had remained difficult to perform until now, because of the low affordable laser energy at 227 nm [65] and 5 μm [64] In this respect, the BBO crystal, and its high conversion efficiency between 190 and 320 nm (at a single cut angle), must be considered as a new crucial device for detecting pollutants in the UV. Furthermore, as this UV-radiation is obtained by frequency doubling, NO and NO_2 can be monitored simultaneously by a judicious choice of the absorbed- and reference- wavelengths (see section 7.2). A direct observation of the oxidation process is then performable, which is very important for the atmospheric chemistry research.

The other major parts of the system (figure 5) can be classified in three groups: (1) The receiver optics and detectors, (2) The electronic signal capture and data processing, and (3) The scanning system and assembly in the mobile unit. Each of these sections exhibits some particular aspects, which have to be discussed in details.

(1). The receiver comprises a 40 cm diameter, 180 cm-focal length Newtonian telescope. The coating, consisting of Aluminium and MgF_2, assures a high reflectivity at both visible and UV wavelengths. The backscattered light is collected by the telescope arrangement, and focussed onto an iris for performing a geometrical compression of the signal dynamics. The LIDAR signal dynamics constitutes indeed a main problem, because the signal decreases at least with the square of the range (equation 1), or even faster in strong atmospheric attenuation conditions. This faster decrease (caused by the exponential factor of equation 1) occurs in particular at mid-UV wavelengths, and under unfavorable meteorologic situations (e.g. fog). A dynamic reduction is then essential, in particular for obtaining a satisfying precision during the digitization process. The geometrical compression of the dynamics is a widely used technique, which has been described in details by several authors [2, 83-86]. Shortly, it acts as following: an iris is placed in the focal plane of the telescope, and reduces therefore the overlap factor $\xi(R)$ (equ. 1) between the laser illuminated field and the telescope field of view. The signal, backscattered from short ranges (up to ~ 500 meters), is thus severely attenuated, reducing the dynamic range. In most cases however, the geometrical compression is insufficient, and has to be completed by a sensitivity reduction of the detector. This is usually achieved by applying a modulation voltage on the photomultiplier, during the acquisition time [87]. In the presently described system, a device has been developed to provide a modulation from 1 kV to 1.8 kV in an adjustable timescale (from 1 μS to 20 μS).

The reduction of the receiver field of view presents two other advantages: it avoids, in a certain extent, the multi-scattering effects, and it lowers the solar background light reaching the detector.

At the output of the iris, the signal is parallelized and spectrally separated by a dichroïc mirror (225 nm high reflectivity) in two different detection channels. One detection channel is dedicated to NO, and the other one to NO_2, SO_2, and O_3. As mentioned before, the rejection of the solar radiation is an essential characteristic for allowing daylight operation, and monitoring the 24-hours variations of the pollutants concentration. In the UV-region, the solar flux is rather small, because of the stratospheric ozone absorption. Therefore, standard interference filters are used (1.5 nm FWHM for SO_2 (T=10 %) and 10 nm FWHM (T=15 %) for O_3), and allow in most cases daylight operation. Conversely, the detection of NO_2 can cause some problems, the solar flux being maximum at 450 nm [88]. For solving these problems, a custom-designed birefringent etalon (Daystar Co, Pomona, Ca) is used instead of interference filters. It exhibits indeed two very narrowband (0.15 nm, T=25 %) transmission peaks (figure 6), centered on 446.5 nm and 448.1 nm, which rejects most of the solar radiation. A careful alignement has, however, to be performed, because the acceptance angle of this filter is only around 15 degrees.

After passing through the rejection filters, the signal is detected by two photomultiplier tubes (EMI 9829 QA), especially selected by the manufacturer for their linearity. The non-linearity of the photomultipliers can also become a large source of error, in particular because of the large dynamic range of the signal. Therefore, each tube must be tested carefully at different light levels before its utilization in the system.

(2) The photomultiplier output signals are preamplified in a 16 dB/150 Mhz preamplifier (Comlinear E103), and digitized in 100 MHz-8 bits transient recorders (LeCroy TR 8818, with additional memory MM 8103 A). The resulting data are treated in a LeCroy 3500 SA microcomputer. The LeCroy data handling system has been chosen for two main reasons: (1) The high precision of its ADC's (7 effective bits at 20 MHz) and (2) its very fast averaging capability (700,000 additions/second). This latter advantage is essential in the present case, as the laser repetition rate is 80 Hz. If each sweep contains 1000 points, the computer must indeed be able to perform up to 80,000 additions/second. Actually, the LeCroy system involves two separate calculation devices: a fast arithmetic unit (LeCroy PDQ), which performs only the averaging, and a microcomputer (LeCroy 3500 SA) for executing the standard operations through a high level language (Fortran). The peripherics are constituted of a floppy disks storage unit, a printer,

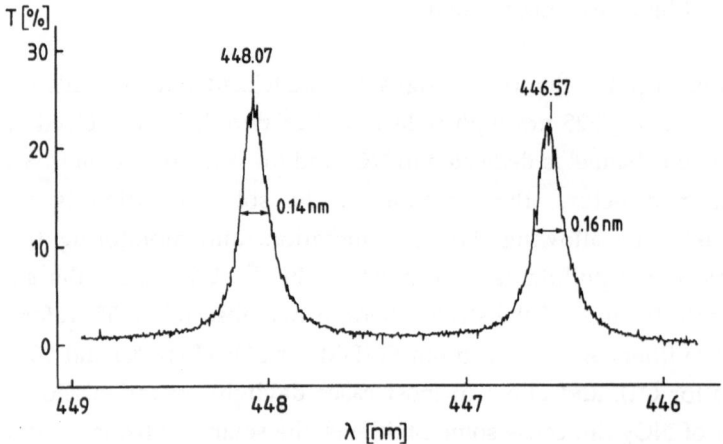

6. Transmission characteristics of a double peak birefringent filter

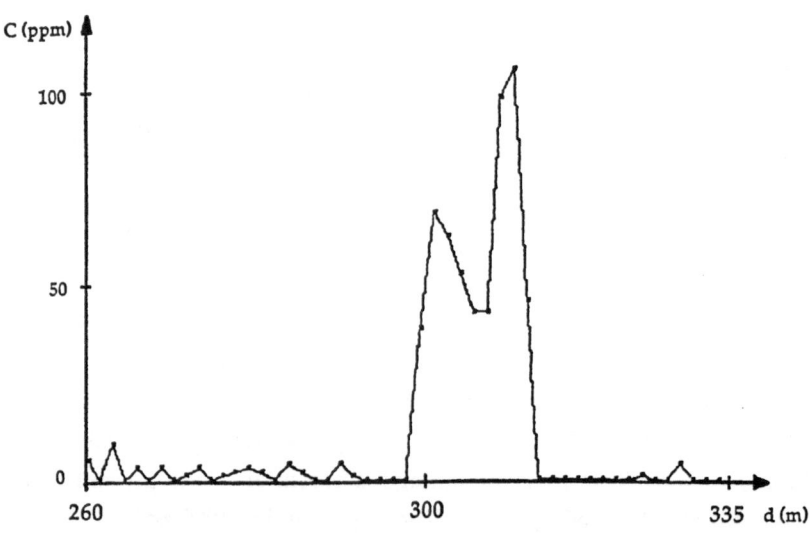

7. High resolution SO$_2$ profile in a chimney plume

and two stepping motor interfaces for driving the steering device (as described just below).

(3). The region of investigation is selected by a flat mirror (1200 mm x 500 mm x 50 mm), which directs both the laser beam and the telescope field of view. The elevation and azimuth angles are scanned by two computer controlled stepping motors (SLOSYN MO 93). A complete mapping procedure can then be performed automatically. Furthermore, a video camera monitors permanently the location of the investigated region.

The DIAL system is implanted in a van of less than 3500 kg total weight, which allows high mobility to the unit. Air mounting of the apparatus is used inside the container, for limiting misalignements during the transport. The low power consumption of the Excimer laser represents also an advantage, the total power need being as low as 380 V x 15 A, which is furnished by a transportable power unit.

7. Typical measurements performed using the DIAL technique

7.1 High resolution 2D- and 3D- mappings of concentration

Figure 7 presents a typical example of a concentration profile, obtained by the DIAL technique. The detected pollutant is SO_2. The measurement has been performed directly through the chimney plume of a power plant. This example shows the ultimate spatial resolution of the system, of about 2.5 meters, which allows the observation of the structure inside the chimney. The two concentration maxima are, indeed, due to the utilization of two separate burners in the combustion chamber. The measured SO_2 values are, furthermore, in good agreement (within ± 10%) with the standard spot measurements, performed by the company inside the chimney. A better correlation cannot be reasonably expected, because of the wide difference between both techniques. A particular attention has, however, to be carried on the optical density of the plume, while performing such kinds of analysis. The high concentration of the detected species or only of aerosols can produce, indeed, a complete extinction of the laser beam inside the plume, preventing any further concentration measurement. To avoid this problem, in case of high density plumes, the investigation must be achieved at higher altitudes, where the plume is already somewhat spread.

By scanning the measurement field in azimuth or elevation, one obtains bidimensional profiles of concentration. Figure 8 and 9 represent respectively a vertical and a horizontal mapping of the NO_2 concentration over a chemical factory. As shown in these figures, the high spatial resolution allows the monitoring of macro-scale turbulences inside the plume. The rapidity of the mapping execution has appeared necessary (in this case, some tens of seconds) for avoiding the errors, which are induced by the air motions (winds). Under sufficiently stable meteorologic situations (timescale larger than one minute), the high repetition rate of the system allows even three-dimensional analysis of the pollutant concentration, as shown in figure 10. The large quantity of informations (about 10,000 points) provides a quasi realtime 3D picture of the plume spread. A timescale of 1 - 5 minutes has been found to be maximum for obtaining realistic profiles.

7.2 Simultaneous NO and NO_2 DIAL measurements

The nitrogen oxides constitute certainly, besides ozon, the most active pollutants of the atmosphere. NO is a primary pollutant, emitted essentially by the road traffic and some other high temperature industrial processes [61]. It is then oxidized in the atmosphere during its transport in a more toxic species: NO_2. This oxidation can occur either by the molecular oxygen of the air, or, and much faster, by O_3. NO, NO_2, and O_3 are involved actually in a real photochemical cycle, which is often considered as a key process in air pollution [62, 63]. Furthermore, the nitrogen oxides react with water vapor and contribute to a large extent to the acidification of the atmosphere [63]. It is then very important to detect simultaneously NO and NO_2, during transport from emission to immission, and observe the dynamics of these oxidation processes. Such a study has recently been performed by Eishout et al [89] using classical analysers. The results demonstrate the importance of recording the oxidation rate as a function of range, ozon concentration, and solar irradiation. As range-resolved informations are needed, a LIDAR system would be the best suited tool for performing these experiments extensively, especially over urban areas.

The advent of the β-BaB_2O_4 crystal opened new perspectives in DIAL measurements of nitrogen oxides. The high energy provided at 225 nm allows not only the sensitive and range resolved detection of NO, but also the additional detection of NO_2: One can indeed correlate the B^2B_1 absorption band of NO_2 and the $A^2\Sigma^+$- state of NO, only by frequency doubling [66]. The λ_{on} of NO_2 will then

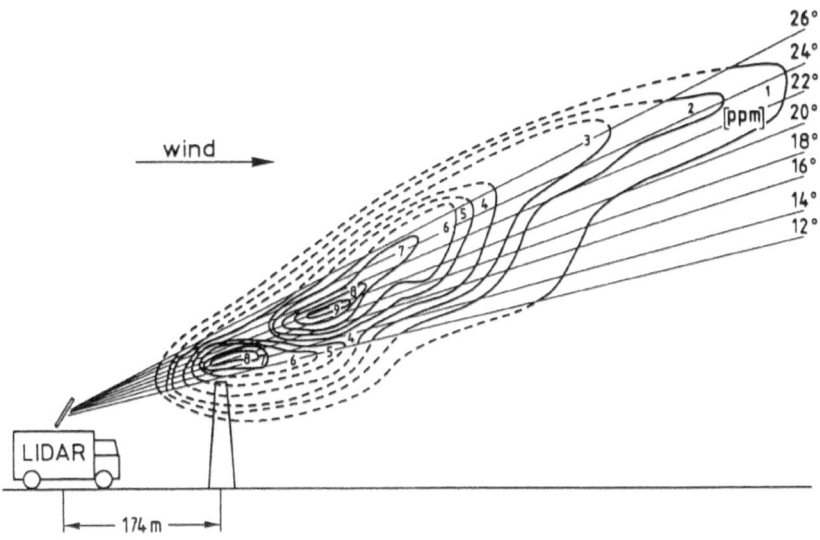

8. Vertical mapping of the NO$_2$ concentration over a chemical factory

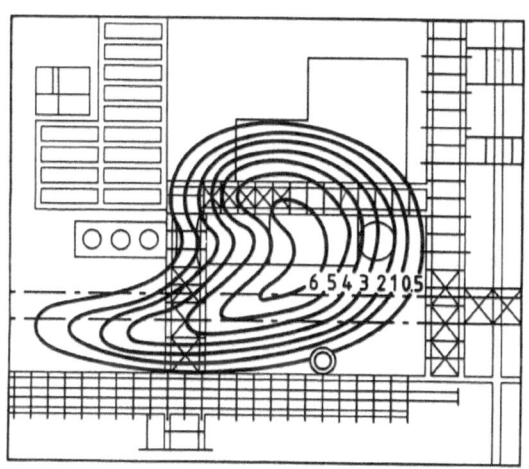

9. Horizontal mapping of the same plume, performed at 30 m. over the chimney

be twice λ_{off} of NO and the λ_{off} of NO_2 will be chosen as twice the λ_{on} of NO, i.e.

$$\lambda_{on}(NO_2) = 448.1 \text{ nm} = 2 \times 224.05 \text{ nm} = 2 \times \lambda_{off} \text{ (NO)}$$
$$\lambda_{off}(NO_2) = 453.6 \text{ nm} = 2 \times 226.8 \text{ nm} = 2 \times \lambda_{on} \text{ (NO)}$$

By frequency doubling 448.1 nm and 453.6 nm and keeping the fundamental frequencies collinear with the second harmonics, one obtains directly the four wavelengths required for detecting NO and NO_2 simultaneously.

This set of wavelengths has been selected for maximizing the differential absorption coefficient of both molecules (ΔK (NO_2) = 8.05 cm^{-1} atm^{-1}, ΔK (NO) = 105 cm^{-1} atm^{-1}, at our laser bandwidth, i.e. 0.4 cm^{-1}), and then provide highest sensitivity. It is obvious, however, that every other set of fundamental- and corresponding second harmonic- frequencies can be selected inside the two absorption bands, in order to optimize the ΔK-coefficients to the concentration ratio [NO]/[NO_2], which is present in the atmosphere. A particular attention must be paid, however, on the presence of other molecules (such as SO_2 [90)]) for avoiding interferences. Furthermore, the high thinness of the NO lines are very sensitive to broadening, and the ΔK coefficient has to be determined very carefully, under different pressure and temperature conditions.

A test experiment has been performed, in order to demonstrate the efficiency of the technique. For this purpose, a NO/NO_2 emitter was simulated by a 2 meters - long open chamber, crossed by a controlled flow of both gases. Two different calibrated gas mixings (500 ppm of the species in N_2) were used to increase the sensitivity of the flow regulation, and prevent over-emission. The LIDAR van was located at about 75 meters from the source, which represents realistic conditions for a plume spread experiment.

Figure 11 shows a typical result obtained with the system. The left side of the figure is dedicated to NO and the right one to NO_2. For each molecule, are plotted respectively: 1) The λ_{on} returns (upper), 2) The λ_{off} returns (medium), and 3) The calculated concentration (lower). The high quality of the signals, averaged over 100 shots only (i.e. 1.2 second), demonstrates the feasability of a simultaneous DIAL detection of several species. The measured NO and NO_2 concentrations are, furthermore, in good agreement with the flow rates used for both gases. The sensitivity of the measurement (about 1 ppm.m for NO_2 and 100 ppb.m for NO) is practically the same as for a DIAL detection of the individual molecules, taken separately. This means that the choice of the reference- and

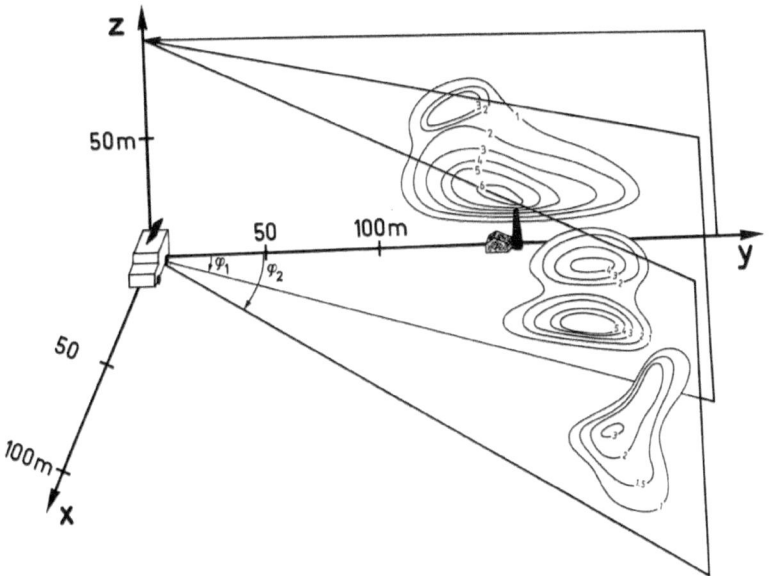

10. 3D - mapping of the NO_2 concentration

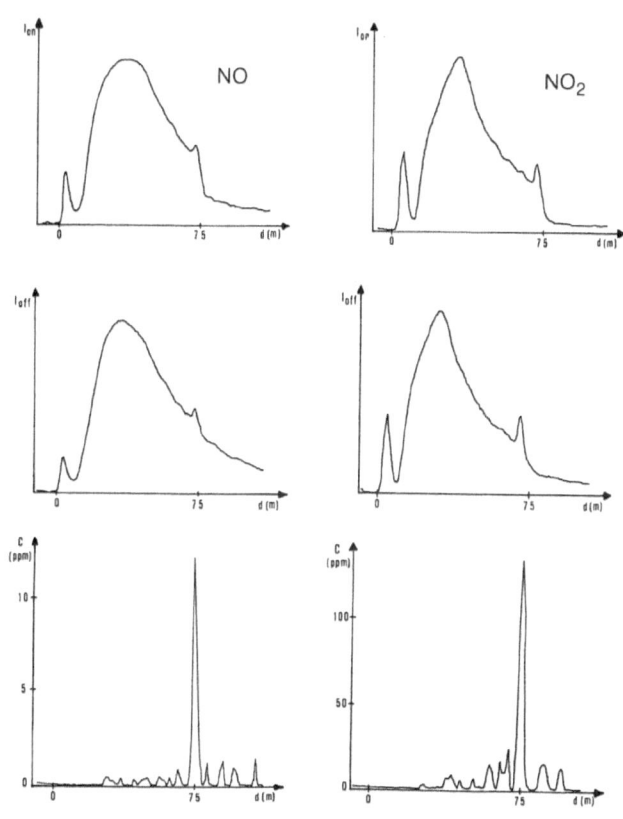

11. Simultaneous measurement of both NO and NO_2

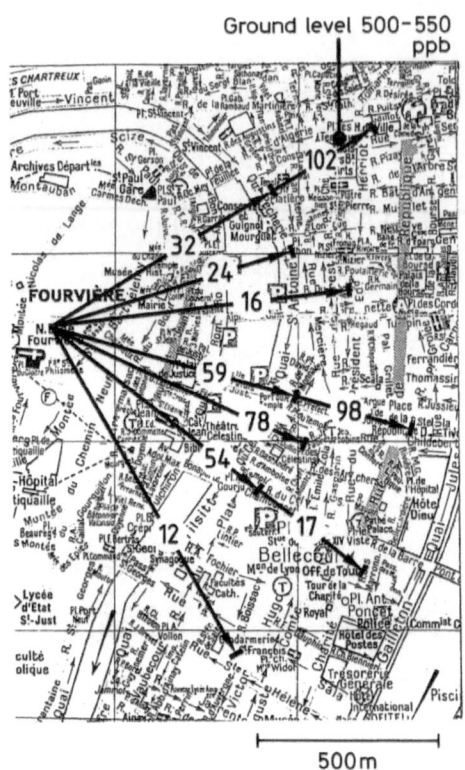

Ground level 500-550 ppb

500m

12. NO mapping over a large urban area

example, the pollution situation just after rainfall. The SO_2 concentration decreases monotonously with altitude, revealing the homogeneity of the atmosphere at this moment. Controversely, when the meteorologic situation is stable and somewhat warmer, several sublayers appear (fig. 13 b), trapping the pollution concentration at a specific altitude. The dynamics of these sublayers, due to exchanges with perpendicular colder valleys and temperature inversions, are of outstanding importance for understanding the pollution transport phenomena at tropospheric altitudes. This kind of experiments constitutes, furthermore, an interesting test for large scale models [92].
absorbed- wavelengths for NO and NO_2, which are directly connected by frequency doubling, does not induce a significant decrease in sensitivity, compared to the usually chosen set of wavelengths [32].

The maximal range of the measurement is estimated to be about 1 km, due to the 5 mJ provided by the BBO crystal. However, the effective radius of action is definitely limited by the strong Rayleigh scattering occuring in this spectral region ($\sigma_R \sim \lambda^{-4}$) and by the absorption from molecular oxygen. O_2 exhibits indeed a weak absorption band in this region [91], but sufficient to cause an attenuation of the laser beam. The performances are, furthermore, highly dependant on the visibility range, and thus very sensitive to the meteorological situation.

The high energy available at 227 nm allowed also the first NO mapping, performed over a large urban area (Lyon, France, see figure 12). In order to increase the maximal range of the measurement, both backscattering and topographic targets have been used. As exhibited in figure 12, a strong correlation is found between the high NO-concentration levels, and the density of small streets, where the traffic is slow and the pollution trapped between the buildings.

7.3 Large scale immission measurements

Mobile DIAL systems can also be used under immission conditions. The example presented here corresponds to a large scale immission campaign, which has been investigated in the Rhone Valley on SO_2. For this purpose, the unit was located at one side of the valley, and used the other side as a topographic target. The range resolution being lost in one direction, only 1D or 2D profiles can be obtained under these conditions. The goal of this particular experiment was to record the SO_2 vertical profile, as a function of time and of the meteorologic situation. Such profiles are shown in figure 13. The length of the absorption path, up to 20 km, allows a detection limit of about 500 ppt. Figure 13 a) presents, for

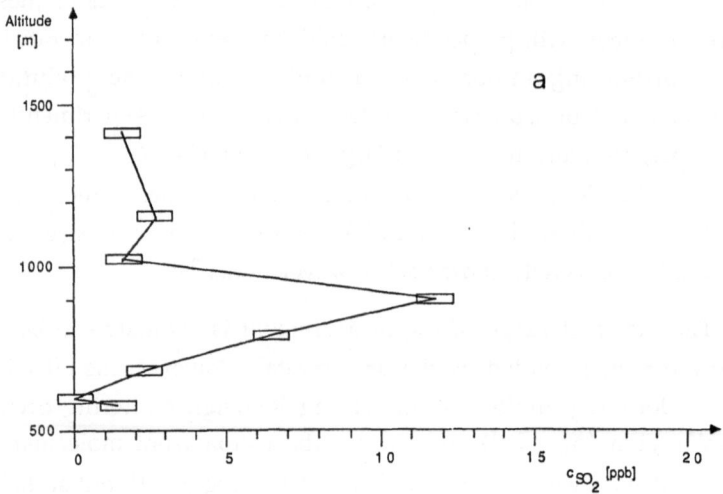

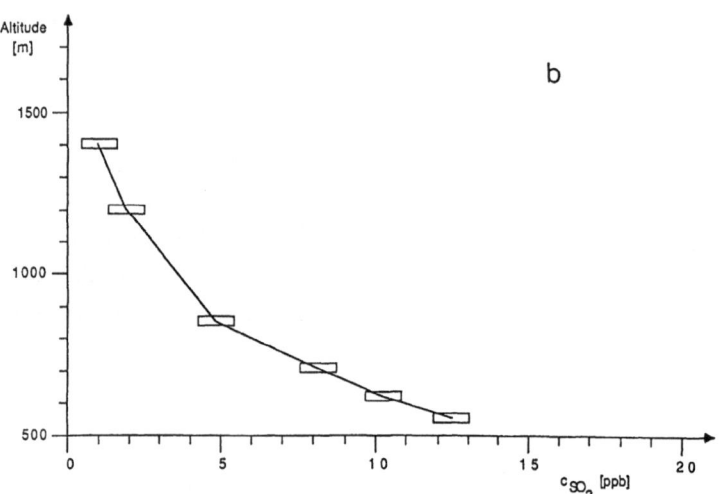

13. Vertical SO$_2$ profiles in the Rhone Valley
   a) Just after rainfall
   b) Inversion situation

8. Conclusion

Mobile DIAL systems constitute a new tool of analysis of air pollution. They allow, indeed, very selective concentration measurements, in a remote way, which prevents the well-known sampling problems. Their major advantage, however, is to give access to the pollution dynamics. The 3D cadastres, they can provide at a sequential cadence, present obviously a large number of applications. One of the most important is the urban survey, where DIAL systems could replace an impressive quantity of spot analyzers. Sequential mappings of concentration would be, in particular, very useful for detecting unknown emitters, traffic slowings, and for preventing large scal smog alarms. Some known emitters could also be controlled simultaneously, with the same arrangement. Another applications is constituted by the survey of large industrial areas, in particular for providing fast informations on the constituents and the propagation of toxic accidental exhausts.

LIDAR is also a very powerful analyzer for sounding the atmosphere, and is therefore, widely utilized in basic research. It can, in particular, provide new, and very useful informations for testing today's chemical and physical models of the atmosphere.

Acknowledgements

We wish to acknowledge the Swiss National Foundation for Scientific Research for financially supporting the present project and the Institute of Experimantal Physics of the EPFL for technical support.

References

[1]. *"Laser Monitoring of the Atmosphere"*, E.D. Hinkley, Springer Verlag, Berlin (1976)

[2]. *"Laser Remote Sensing"*, R.M. Measures, J.Wiley & Sons ed., New-York (1984)

[3]. *"Laser Remote Chemical Analysis"*, R.M. Measures, J.Wiley & Sons ed., New-York (1988)

[4]. *" Electromagnetic Scattering on Spherical Polydispersions"*, D. Deirmendjan, Am. Elsevier, New-York (1969)

[5]. " Transmission through the atmosphere", R. Beer, in *Laser Remote Chemical Analysis*", R.M. Measures, J.Wiley & Sons ed., New-York (1988)

[6]. G. Fiocco, L.D. Smullin (1963), "Detection of Scattering Layers in the Upper Atmosphere (60-140 km) by Optical Radar", *Nature* **199**:1275

[7]. J.Werner, K.W. Rothe, H. Walther (1983), Monitoring of the Stratospheric Ozone Layer by Laser Radar,*App. Phys.* **B32**:113

[8]. M.P. Mc Cormick (1982), "Lidar Measurements of Mount St. Helens Effluents",*Opt. Eng.* **21**:340

[9]. M.P. Mc Cormick, T.J. Swissler, W.P. Chu, W.H. Füller (1978), "Post-volcanic Stratospheric Aerosol Decay as Measured by Lidar",*J.Atmosph.Sciences* **35**:1296

[10]. J. Lefrère, J. Pelon, C. Cahen, A. Hauchecorne, P. Flamant (1981), "Lidar Survey of the Post-Mt. St. Helens Statospheric Aerosol at Haute Provence Observatory", *App. Optics* **20** A70:1117

[11]. M.P. McCormick, "Lidar Measurements of the El-Chichon Stratospheric Aerosol Climatology", in *Proc. of the 12th ILRC,* Aix en Provence (1984):203

[12]. H. Jäger, R. Reiter, W. Carnuth, Sun Jian, "Stratospheric Aerosols Layers during 1982 and 1983 as Observed by Lidar at Garmisch-Partenkirchen",, in *Proc. of the 12th ILRC*, Aix en Provence (1984):207

[13]. R.J. Allen, C.M.R. Platt (1977), "Lidar for Multiple Backscattering and Depolarization Observations",*App. Optics* **16**:3193

[14]. S.R. Pal, A.I. Carswell (1973), "Polarization Properties of Lidar Backscattering from Clouds",*App. Optics* **12**:1530

[15]. S.R. Pal, A.I. Carswell (1976), "Multiple Scattering in Atmospheric Clouds: Lidar Observations",*App. Optics* **15**:1990

[16]. S.R. Pal, A.I. Carswell (1978), "Polarization Properties of Lidar Scattering from Clouds at 347 nm and 694 nm", *App. Optics* **17**:2321

[17]. W.R. Mc Neil, A.I. Carswell (1975), "Lidar Polarization Studies of the Troposphere",*App. Optics* **14**:2158

[18]. K. Sassen, M.K. Griffin, and G.C. Dodd, "Subvisual Cirrus Cloud Properties Derived From Polarization Lidar, Surface Radiation Flux, and Solar Corona Measurements", in *Proc. 14th ILRC,* San Candido (1988):20

[19]. G. Mégie, F. Bos, J.E. Blamont, M.L. Chanin (1978), "Simultaneous Nighttime Lidar Measurements of Atmospheric Sodium and Potassium", *Planet Space Sci.* **26**:27

[20]. C. Granier, G.Mégie (1982), "Daytime Lidar Measurements of Mesospheric Sodium Layer",*Planet Space Sci.* **30** :169

[21]. G. Mégie, "Laser Measurements of Atmospheric Trace Constituents", in

"*Laser Remote Chemical Analysis*", Measures R.M., J.Wiley & Sons ed.,
New-York (1988)

[22]. H. Inaba, in "*Laser Monitoring of the Atmosphere*", E.D. Hinkley, Springer
Verlag, Berlin (1976)

[23]. W. Lahmann, M. Riebesell, E. Voss, C.Weitkamp, and W. Michaelis,
"Raman Lidar for Vertical Water Vapor Profiling", in *Proc. 14th ILRC*, San
Candido (1988):477

[24]. W.R. Fenner, H.A. Hyatt, J.M. Kellan, and S.P.S. Porta (1973), "Raman Cross-
sections of Some Simple Gases",*J.Opt.Soc.Am.* **63**:73

[25]. Y.F. Arshinov, S.M. Bobrovnikov, V.E. Zuev, V.M. Mitev (1983), *App. Opt.*
22:2584

[26]. R.M. Measures, G. Pilon (1972), "A Study of Tunable Laser Techniques for
Remote Mapping of Specific Gaseous Constituents of the Atmosphere",
Opto-Elec. **4**:141

[27]. N.Menyuk, D.K. Killinger, C.R. Menyuk (1982), "Limitations of Signal
Averaging due to Temporal Correlation in Laser Remote Sensing
Measurements", *App.Opt.* **21**:3377

[28]. N. Menyuk, D.K. Killinger (1981), "Temporal Correlation Measurements of
Pulsed Dual CO_2 Lidar Returns", *Opt.Lett.* **6**:301

[29]. N.Menyuk, D.K. Killinger, C.R. Menyuk (1985), "Error Reduction in Laser
Remote Sensing: Combined Effects of Cross-Correlation and Signal
Averaging", *App.Opt.* **24**:118

[30]. J.G. Hawley, L.D. Fletcher, and G.F. Wallace, "Ground-Based Ultraviolet
Differential Absorption Lidar (DIAL) System and Measurements", in
"*Optical and Laser Remote Sensing*", D.K. Killinger and A Mooradian ed.,
Springer Series in Optical Sciences, Springer Verlag, Berlin 1983

[31]. J.P. Wolf, L. Wöste (1987), "Détection Sélective et à Distance de la Pollution
Atmosphérique par LIDAR", *Helv. Phys. Acta* **60**:161

[32]. H.J. Kölsch, P. Rairoux, J.P. Wolf, L. Wöste, "New Perspectives in
Remote Sensing Using Excimer-pumped Dye Lasers and BBO Crystals" *Proc.
14th ILRC*, San-Candido (1988):484

[33]. "*Applications de la Spectroscopie Laser à la Pollution Atmosphérique*", J.P.
Wolf,Thèse de Doctorat 683, EPFL, Lausanne, Switzerland (1987)

[34]. K.N. Liou, M.R. Schotland (1971), "Multiple Backscattering and
Depolarization from Water Clouds for a Pulsed Lidar System", *J.Atmos.Sci*

[35]. R.T.H. Collis, and P.B. Russel, "Lidar Measurement of Particles and Gases by
Elastic Backscattering and Differential Absorption", in "*Laser Monitoring of
the Atmosphere*", E.D. Hinkley, Springer Verlag, Berlin (1976)

[36]. K. Fredriksson, "Differential Absorption Lidar for Pollution Mapping", in *"Laser Remote Chemical Analysis"*, R.M. Measures, J.Wiley & Sons ed., New-York (1988)

[37]. *"Laser Spectroscopy"*, W. Demtröder, Springer Series in Chemical Physics 5, Springer-Verlag, Berlin 1982

[38]. C. Cahen, G. Mégie (1981), "A Spectral Limitation of the Range Resolved Differential Absorption Lidar Technique", *J.Quant.Spectros.Radiat.Transfer* **25**:151

[39]. D.J. Brassington, R.C. Felton, B.W. Jolliffe, B.R. Marx, J.T.M. Moncrieff, W.R.C. Rowley, P.T.Woods (1984), "Errors in Spectroscopic Measurements of SO_2 due to Nonexponential Absorption of Laser Radiation, with Application to the Remote Monitoring of Atmospheric Pollutants", *App.Optics* **23**(3):469

[40]. P.T. Woods, B.W. Jolliffe (1980), "High Resolution Spectroscopy of SO_2 using Frequency-Doubled Pulsed Dye-Laser, with Application to the Remote Sensing of Atmospheric Pollutants", *Optics Comm.* **33**(3):281

[41]. T. Tajime, T. Saheki, K. Ito (1978), "Absorption Characteristics of the γ-0 Band of NO", *App.Optics* **17**(8):1290

[42]. R.M. Schotland, "Some Observations of the Vertical Profile of Water Vapor by a Laser Optical Radar", *Proc. 4th Symposium on Remote Sensing of the Environment*, Univ. Michigan, Ann Arbor, (1966):273

[43]. K.W. Rothe, U. Brinkman, H. Walther (1974), "Remote Sensing of NO2 Emission from a Chemical Factory by the Differential Absorption Technique", *App. Phys.* **4**:181

[44]. K.W. Rothe, U. Brinkman, H. Walther (1974), "Applications of Tunable Dye Lasers to Air Pollution Detection: Measurements of Atmospheric NO_2 Concentration by Differential Absorption", *App. Phys.* **3**:115

[45]. W.B. Grant, R.D. Hake, E.M. Liston, R.C. Robbins, E.K. Proctor (1974), "Calibrated Remote Measurements of NO_2 Using Differential Absorption Backscatter Technique", *App.Phys.Lett.* **24**:550

[46]. K. Fredriksson, B. Galle, K. Nystrom, S. Svanberg (1979), "Lidar System Applied in Atmospheric Pollution Monitoring", *App. Optics.* **18**:2998

[47]. K. Fredriksson, B. Galle, K. Nystrom, S. Svanberg (1981), "Mobile Lidar System for Environmental Probing", *App.Optics.* **20**:4181

[48]. K. Fredriksson, H.M. Hertz (1984), "Evaluation of the DIAL Technique for Studies on NO_2 Using a Mobile Lidar System", *App.Optics* **23**:1403

[49]. B. Galle, A. Sunesson, W. Wendt (1988), " NO2-Mapping Using Laser-Radar Techniques", *Atmos. Environm.* **22**:569

[50]. W. Staehr, W. Lahman, C. Weitkamp, W. Michaelis, "Differential Absorption Lidar System for NO_2 and SO_2 Monitoring", *Proc. 12th ILRC Conference*, Aix-en- Provence (1984)

[51]. B.W. Jolliffe, R.C. Felton, N.R. Swann, P.T. Woods, "Field Measurement Studies Using a Differential Absorption Lidar System", *Proc. 12th ILRC Conference*, Aix-en- Provence (1984)

[52]. T. Tsuji, H. Kimura, Y. Higuchi, K. Goto (1976), "NO_2 Concentration Measurement in the Atmosphere Using Differential Absorption Dye Laser Radar Technique", *Jap. J. App. Phys.* **15**:1743

[53]. W.B. Grant, R.D. Hake (1975), "Calibrated Remote Measurements of SO_2 and O_3 using Atmospheric Backscattering", *J.App.Phys.* **46**:3019

[54]. S. Adrian, D.J. Brassington, S. Sutton, R.H. Varey (1979), "The Measurement of SO_2 in Power Station Plumes with Differential Lidar", *Opt.Quant.Elec.* **11**:253

[55]. A. Marzorati, W. Corio and E. Zanzottera (1984), Remote Sensing of SO_2 During Fields Tests at Fos-Berre in June 1983. Abstr. 12th International Laser Radar Conference, Aix en Provence, Service d'Aeronomie du CNRS: 259

[56]. W. Michaelis and c. Weitkamp (1984), Sensitive Remote and in situ Detection of Air Pollutants by Laser Light Absorption Measurements, *Fresenius Z. Anal. Chem.* **317**:286

[57]. G. Ancellet, R. Capitini, D. Renaut, G. Mégie and J. Pelon (1984), DIAL Lidar Measurements of Atmospheric Pollutants (SO_2, O_3)during the Fos-Berre 83 Experiment. Abstr. 12th International Laser Radar Conference, Aix en Provence, Service d'Aeronomie du CNRS: 269

[58]. J. Pelon and G. Mégie (1982), Ozone Monitoring in the Troposphere and Lower Stratosphere: Evaluation and Operation of a Ground-Based Lidar Station, *J. Geophys. Res.* **87**:4947

[59]. E.V. Browell, A.F. Carter, S.T. Shipley, R.J. Allen, C.F. Butler, M.N. Mayo, J.H. Siviter and W.M. Hall (1983), NASA Multipurpose Airborne DIAL System and Measurements of Ozone and Aerosol Profiles, *Appl. Optics* **22**:522

[60]. G. Mégie, g. Ancellet, J. Pelon (1985), "Lidar Measurements of Ozone Vertical Profiles", *Appl. Optics* **21**:3454

[61]. C.E. Billings, in *"Industrial Pollution"*, Irving Sax ed., van Nostrand, New-York (1974)

[62]. *"Air Pollution by Photochemical Oxidants"*, R. Guderian, Ecological Studies **52**, Springer-Verlag, Berlin (1985)

[63]. *"Air Pollution"*, J.H. Seinfeld, McGraw-Hill ed., (1975)

[64]. N. Menyuk, D.K. Killinger, W.E. DeFeo (1980), "Remote Sensing of NO Using a Differential Absorption Lidar", *App.Optics*. 19:3282

[65]. M. Aldén, H. Edner, S. Svanberg (1982), "Laser Monitoring of Atmospheric NO Using Ultraviolet Differential-Absorption Techniques", *Opt.Lett.* 7 No 11: 543

[66]. H.J. Kölsch, P. Rairoux, J.P. Wolf, L. Wöste, "Simultaneous NO and NO_2 DIAL Measurement Using BBO Crystals", submitted to Applied Optics

[67]. H. Edner, G.W. Faris, A. Sunesson, S. Svanberg, "Progress in DIAL Measurements at Short UV Wavelengths", in *Proc. 14th ILRC*, San-Candido (1988):480

[68]. M. Aldén, H. Edner and S. Svanberg (1982), Remote Measurements of Atmospheric Mercury Using Differential Absorption Lidar, *Opt. Lett.* 7:221

[69]. C. Weitkamp (1981), The Distribution of Hydrogen Chloride in the Plume of Incineration Ships: Development of New Measurement Systems, *Wastes in the Ocean*, **Vol. 3**, Wiley, New York

[70]. M.J.T. Milton, R.H. Bradsell, B.W. Jolliffe, N.R.W. Swann, P.T. Woods, "The Design and Development of a Near-Infrared DIAL System for the Detection of Hydrocarbons", in *Proc 14th ILRC*, San-Candido (1988):370

[71]. E.R. Murray, J.E. van der Laan and J.G. Hawley (1976), Remote Measurement of HCI, CH_4 and NO_2 Using Single-Ended Chemical-Laser Lidar System, *Appl. Optics* **15**:3140

[72]. G. Mégie and R.T. Menzies (1980), Complementary of UV and IR Differential Absorption Lidar for Global Measurements of Atmospheric Species, *Appl. Optics* **19**:1173

[73]. R.T. Menzies and M.S. Shumate (1976) Remote Measurements of Ambient Air Pollutants with a Bistatic Laser System, *Appl. Optics* **15**:2080

[74]. D. Killinger, N. Menyuk (1981), "Remote Probing of the Atmosphere Using CO_2 DIAL System", *IEEE J.Quant.Elec* **17**(9):1918

[75]. N. Menyuk, D.K. Killinger, W.E. de Feo (1982), "Laser Remote Sensing of Hydrazine, MMH, and UDMH Using a Differential-Absorption CO_2 Lidar", *App.Optics* **21**:2275

[76]. M.S. Shumate, R.T. Menzies, W.B. Grant and D.S. Mc Dougal (1981), Laser Absorption Spectrometer: Remote Measurement of Tropospheric Ozone, *Appl. Optics* **20**:545

[77]. R.T. Menzies and S. Shumate (1978), Tropospheric Ozone Distributions Measured with an Airborne Laser Absorption Spectrometer, *J. Geophys. Res.* **83**:4039

[78]. D.K. Killinger, N. Menyuk and W.E. De Feo (1983), Experimental

Comparison of Heterodyne and Direct Detection for Pulsed Differential Absorption CO_2 Lidar, *Appl. Optics* **22**:682

[79]. O. Uchino, M. Maeda, J. Khono, T. Shibata, C. Nagasawa and M. Hirono (1978), Observation of Stratospheric Ozone Layer by a XeCl Laser Radar, *Appl. Phys. Lett.* **33**:807

[80]. O. Uchino, M. Maeda, H. Yamamura and M. Hirono (1983a), Observation of Stratospheric Vertical Ozone Distribution by a XeCl Lidar, *J. Geophys. Res.* **88**:5273

[81]. K. Miyakzaki, H. Sakai, T. Sato (1986), "Efficient Deep-UV Generation by Frequency Doubling in β-BaB_2O_4", *Opt.Lett.* **11** No 12:797

[82]. Nguyen Dai Hung, P. Brechignac (1988), "Tunable Alternative Double-Wavelength Single Grating Dye-Laser for DIAL Systems", *App.Optics* **27**:1906

[83]. J. Harms, W. Lahmann and C. Weirkamp (1978), "Geometrical Compression of Lidar Return Signals", *Appl. Optics* **17**:1131

[84]. J. Harms (1979), "Lidar return Signals for Coaxial and Noncoaxial Systems with Central Obstruction", *Appl. Optics* **18**:1559

[85]. T. Halldorsson and J. Langerholc (1978), "Geometrical Form Factors for the Lidar Function", *Appl. Optics* **17**:240

[86]. K. Sassen and G.C. Dodd (1982), Lidar Crossover Function and Misalignment Effects, *Appl. Optics* **21**:3162

[87]. R.J. Allen and W.E. Evans (1972), Laser Radar for Mapping Aerosol Structure, *Rev. Sci. Instrum.* **43**:1422

[88]. S.L. Valley, Ed. (1965), *Handbook of Geophysics and Space Environments*, Mc Graw-Hill, New York

[89]. A.J. Eishout, S. Beilke (1984), "Die Oxidation von NO zu NO_2 in Abgasfahnen von Kraftwerken", *VGB Kraftwerkstechnik* **64**:648

[90]. J. Heicklen, N. Kelly, K. Partymiller (1980), "The Photophysics and Photochemistry of SO_2", Rev.Chem.Interm. **3**:315

[91]. V. Hasson R.W. Nicholls (1971), "Absolute Spectral Absorption Measurements on Molecular Oxygen from 2640-1920 A", *J.Phys. B* **4**: 1789

[92]. *"Interactions between Energy Transformations and Atmospheric Phenomena"*, M. Beniston and R. Pielke, Reidel Publishing, Boston (1987)

[93]. D.J. Brassington (1981), Sulfur Dioxide Absorption Cross Section Measurements from 290 nm to 317 nm, *Appl. Optics* **20**:3774

LASER COOLING OF ATOMIC BEAMS AND ITS APPLICATION
TO FREQUENCY STANDARDS

N. Beverini, F. Strumia

Dipartimento di Fisica dell'Università di Pisa

THE LASER COOLING TECHNIQUE

When an atom absorbs or emits a photon, the momentum conservation law implies that its momentum must change by a quantity equal to the photon momentum $h\nu/c$. Let us consider an atom irradiated by a running wave, resonant with an atomic transition. Each absorption-spontaneous emission cycle will lead to a variation of the atom velocity of a quantity $h\nu/mc$. At contrary, an absorption followed by a stimulated emission process, with two equal photons, does not change the atomic momentum. Thus, taking into account also the saturation effects, an atom experiences an effective force given by the ratio of the photon momentum with the mean time interval between two spontaneous emission, that is

$$F = \frac{h\gamma}{\lambda} \frac{S}{1 + S + (\Delta\omega - kv)^2/\gamma^2} \tag{1}$$

where γ is the HWHM homogeneous linewidth, $\lambda = c/\nu$ is the resonance wavelength, and $S = I/I_S$ the saturation parameter of the radiation. Evidence of this force has been first demonstrated by Frische in 1933, who observed the deflection of an atomic beam of Na by the radiation of a resonance lamp[1]. The observed effect was very small. On the contrary, the interaction of resonant laser radiation with an atomic thermal beam can be efficiently used to reduce the atomic velocity [2]. Thus, atoms with an initial velocity v_0 are stopped, in a laser field of constant intensity I, after a distance

$$L(S) = \frac{v_0^2 M\lambda\tau}{h}\left(1 + \frac{1}{S}\right), \tag{2}$$

where τ is the upper level lifetime and M the atomic mass.

However, only a small fraction of atoms over the thermal velocity distribution profile can absorb the laser radiation when a monochromatic single-frequency laser beam is used, namely those atoms whose laser frequency offset is compensated, within the natural linewidth, by the Doppler shift. As soon as the deceleration process starts, the velocity is reduced, the atoms are Doppler shifted out of resonance, and the absorption becomes negligible.

Three different methods has been suggested to overcome this difficulty. The first one[3] implies the use of very high laser intensity in order to have saturation parameter $S \gg (\Delta\omega - kv)/\gamma$. It has been demonstrated for Na, whose resonant transition falls at a wavelength of the maximum efficiency of the dye lasers, but it is not effective to produce very low velocity and it is not applicable to other atomic species for which so high S values cannot be achieved.

In the other two methods the atoms are kept in resonance with the radiation during the cooling process, or by chirping quickly the laser frequency[4,5], or by tuning the atomic absorption frequency by the Zeeman effect, produced by an axial magnetic field B(z) of appropriate

Applied Laser Spectroscopy, Edited by W. Demtröder and
M. Inguscio, Plenum Press, New York, 1990

intensity along the beam[6]. Both the methods has been first demonstrated on Na atom and then applied also on other atomic species[7,8,9,10]. Frequency chirping requires a laser tuning of the order of some GHz in a fraction of ms; such a performance can be quite easily achieved, if the cooling radiation is produced by a diode laser, like in the case of Cs ($\lambda=851$ nm) [7], otherwise sofisticate electronical equipment are required. When the atomic absorption wavelength lies in the blue or in the UV region, the dye laser operation is more critical and less efficient and the magnetic tuning is the most convenient alternative. An additional advantage of the magnetic cooling method is the production of a continuous beam of decelerated atoms, while only pulsed beams can be obtain by laser chirping. The parameter relevant for laser cooling are given in Table I for different atomic species.

A scheme of the experimental apparatus is shown in fig.1, which refers to an experiment[8] for laser cooling of Ca atoms. The resonance line is the strong singlet $^1S_0$–$^1P_1$ transition (fig.2) at 422 nm ($I_S = 60$ mW/cm^2). The ground state has J=0, and the more abundant natural isotope, ^{40}Ca, does not have nuclear magnetic moment. Thus the ground state is not degenerate, and we can have a true two level system by using circularly polarized light. In Ca the presence of a metastable $^1D_2$ level at a energy lower than the $^1P_1$ level may decrease the cooling efficiency. An atom, excited in the $^1P_1$ level, that decay on this long-life metastable level, is lost for the cooling process. Recently the decay branching ratio between the $^1P_1$-$^1S_0$ and $^1P_1$-$^1D_2$ transition has been measured[11] to be of the order of 10^5, high enough to not affect significantly the cooling efficiency.

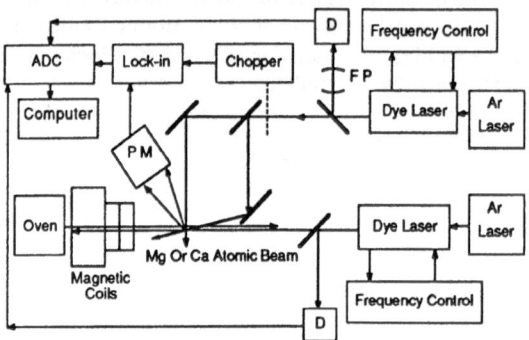

Fig.1   Scheme of the Ca levels involved in the laser cooling.

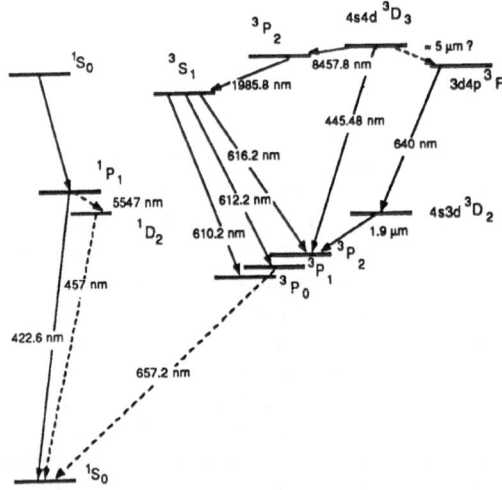

Fig. 2   Typical experimental apparatus for laser cooling

In the Ca experiment the magnetic field intensity B(z) decreases from its maximum to zero in about 25 cm, i e. the maximum cooling length. The cooling radiation at 422nm is provided by a single frequency, actively stabilized CW dye laser. The modified velocity distribution is analyzed by a second laser beam nearly collinear to the first one, and whose frequency is scanned around the resonance line. A near collinear geometry, where the analysis beam diameter is smaller than the cooling one, and, thus, of the atomic beam diameter, was preferred in order to obtain a sufficiently large analysis signal and a good resolution in the velocity profile. In fact, the analysis laser beam crosses the center of the cooling beam only in front of the photomultiplier detecting the fluorescence signal.

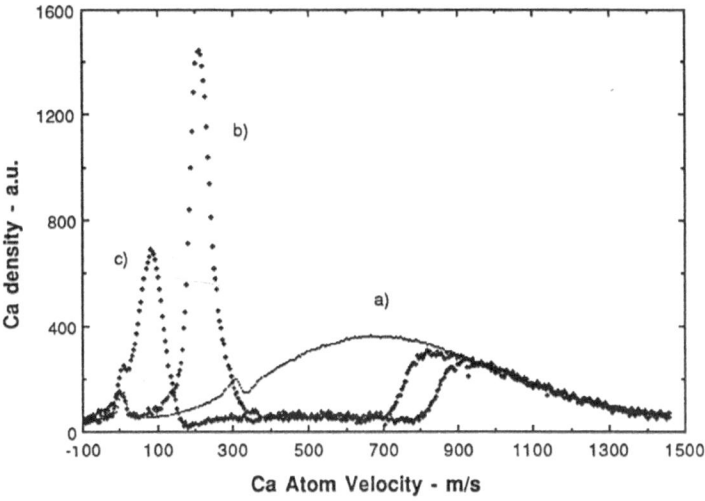

Fig.3 Evidence of laser cooling in presence of Zeeman tuning. Curve a): velocity distribution with magnetic field off and low intensity of the cooling beam; curve b): deceleration observed with an initial magnetic field B(0)= 0.07 T; curve c): same conditions but with the laser frequency shifted in correspondence of lower velocities

Fig.3 demonstrates an effective laser deceleration in presence of Zeeman tuning. Curve a) was obtained in absence of magnetic field and with the intensity of the cooling laser beam strongly reduced : the hole in correspondence of v=315 m/s shows the position of the laser frequency in the velocity distribution. Curve b) shows the modified velocity distribution obtained with an appropriate laser intensity and an initial magnetic field B(0)= 0.07 T, corresponding to a tuning of about 415 m/s. Curve c) was obtained in the same conditions, but with the laser frequency shifted in correspondence of lower velocities. Very efficient production of atomic beam at about 200 m/s, with a spread in velocity equivalent to a temperature in the center of gravity system of the order of 1 K, is demonstrated. The diffusion effect, which will be discussed in the following paragraph, reduces the efficiency at lower final velocity; however fig.4 shows that also at very low velocity, of the order of 10 m/s, the density in the cooled beam is greater than the density at the maximum of the unperturbed maxwellian distribution, while a non-negligible number of atom with reversed velocity can be observed.

The same cooling technique can be applied to cool a thermal beam of Mg, which has a level scheme similar to Ca (Fig.5). In this case, however, the resonance optical transition falls in the UV at 285 nm. This wavelength can be produced by doubling the 570 nm radiation of a Rhodamine 6G dye laser.The power is limited at a level of a few mW. Evidence of the cooling process has been observed[8], but an efficient production of cooled Mg beam requires improvement in the laser duplication efficiency.

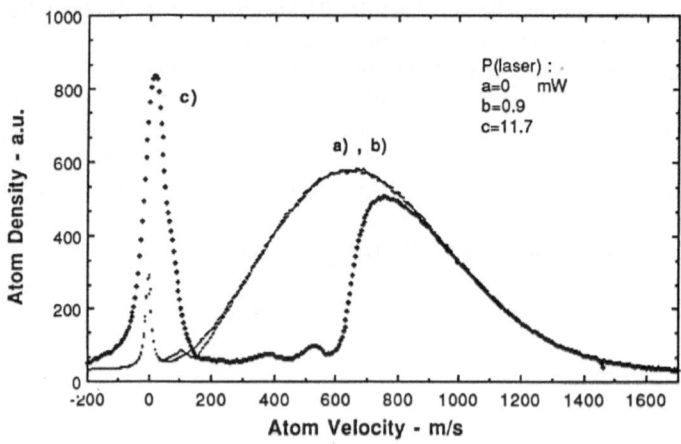

Fig.4   Laser cooling of Ca at very low velocity (≈10 m/s)

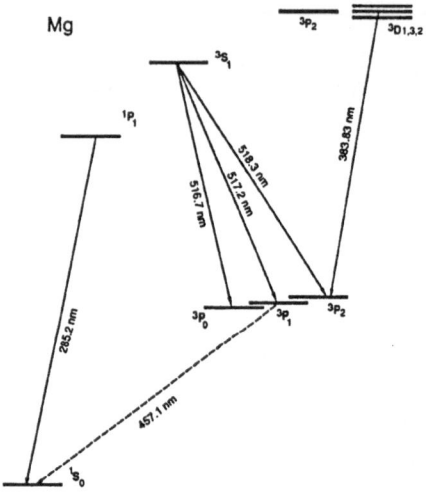

Fig.5   Scheme of the Mg levels involved in the laser cooling.

DIFFUSION PROCESSES AND BEAM COLLIMATION

The efficiency of laser cooling is limited by the diffusion process, particularly relevant for very low velocity atoms. First of all the collimation factor of the beam is worsened, if the longitudinal atomic velocity is reduced without any change in the transverse velocity. Besides, the photon absorption and emission are stochastic processes and the effective force given in (1) is only an average value. The spontaneous emission direction is a *random walk* process, and an atom, that experiences N absorption-spontaneous emission cycles, has mean square velocity $<v^2> = N (\delta v)^2$, where $\delta v = h/\lambda m$.
It is however possible to obtain laser collimation by the radiation pressure from a stationary red-shifted resonant wave, orthogonal to the atomic beam axis. Using Eq.1, it is easy to

demonstrate[12], that an atom in the beam experiences a viscosity-like force

$$F = -\beta M v_t \ , \tag{9}$$

where v_t is the transverse velocity, and

$$\beta = \frac{8\pi h}{M\lambda^2} \left|\frac{\Delta\omega}{\gamma}\right| S \left(1+\Delta\omega^2/\gamma^2\right)^{-1} \left(1+\Delta\omega^2/\gamma^2+2S\right)^{-1} \ . \tag{10}$$

The efficiency of this collimation effect has been demonstrated on a sodium beam by Balykin et al.[13] Evidence of this collimation effect has been observed also in Ca (fig.6). The efficiency in this case is limited, because the same laser beam, and thus the same laser frequency detuning, was used both for cooling and for collimation. From (10) it can be easily deduced that the viscous force is, for small saturation, proportional to $(\gamma/\Delta\omega)^3$, so that, with a proper laser detuning for cooling, the collimation is not very efficient.

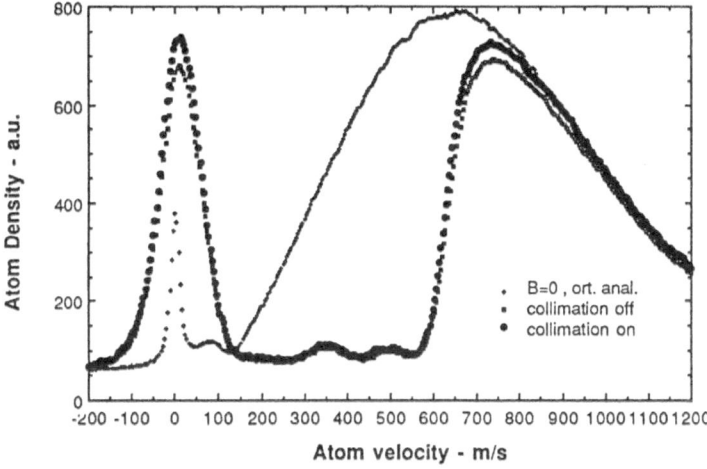

Fig.6   Effect of transverse collimation on a cooled beam of Ca.

ATOMIC FREQUENCY STANDARDS

The Atomic Frequency Standards (AFS) consist of a high spectral purity electromagnetic oscillator phase locked to a suitable atomic or molecular transition to improve either the frequency stability and frequency definition. In a primary AFS the operating frequency can be defined "a priori", and it is expected that any AFS based on that transition will reproduces the requested frequency, provided the environmental conditions are realized according to the definition.

AFS is an important tool for the scientific development (applications in radio astronomy (VLBI), general relativity, atomic and molecular spectroscopy, measurement of the continental drift etc.) and for the technological progress (telecommunications). In particular the development of more and more precise time/frequency standards has been stimulated and supported by the need to improve the timekeeping and the navigation systems. As a consequence several countries (USA, Canada, FRG, France, Italy, UK, CH, URSS, Japan, and China) support metrological national laboratories dedicated to the maintaining and development of time and frequency standards .

The measurement and comparison of frequencies is, at least in principle, a measurement of unlimited precision , the error being ± 1 count independently of the number of the counts. The ultimate precision is therefore limited only by the stability of the master oscillator.

The development of very fast diodes (MIM diodes) has permitted in the recent years the frequency synthesis up to the visible, and the direct frequency measurement of visible and near infrared actively stabilized lasers by heterodyne comparison with the microwave standards. Contrary to the wavelength measurements, the frequency measurements are not limited by diffraction and the BIPM decided at the 17th CGPM meeting on October 20th, 1983 to adopt a new definition of the meter which considers the velocity of light as a fixed number : " The meter is the length of the path travelled by light in vacuum during a time interval of 1/299792458 of a second". With this decision the two most important units , time and length , are referred to the ground state hyperfine transition $F=4$, $m_F = 0 — F=3$, $m_F = 0$ of ^{133}Cs , whose central frequency is defined as ν (Cs) = 9 192 631 770.000 Hz in absence of perturbations (13th General Conf. on Weights and Measures - 1967). Unfortunately, the precision and resolution of the frequency measurements in the microwave region is not yet fully extended to the visible region because of the phase noise introduced in the high order multiplication process and of the inadequate reproducibility of the reference oscillator in the visible and near infrared.

Today devices permit a frequency multiplication and phase locking of oscillators with the stability and resolution of the Cs primary standard only up to about 10^{13} Hz. This confines the development of new kinds of Time Standards of improved performances to the microwave and FIR (Far-infrared) spectral region. A primary frequency standard in the visible region, with a precision equal or better than that of Cs, would be also important as a practical length standard. The frequency synthesis and comparison between microwave and optical waves will then be restricted to the metrological national laboratories.

It is worth noting that the Cs AFS has been extensively studied during the last 30 years . It has reached a mature state , no significant improvement has been obtained in the past ten years and none is reasonably to be expected for the future , at least for the classical design of Stern-Gerlach states selection and Ramsey interrogation scheme on a thermal beam. For this reason new kinds of AFS are under investigation[14].

The quality of a frequency standard is defined by its Precision, Reproducibility, and Stability. The Stability, which expresses the mean square deviation of the standard frequency over an observation time τ , is a function of τ^n , n depending on the prevalent kind of noise (for the shot or white noise n=-1/2, for flicker or 1/f noise n=0). The stability, when measured following a particular procedure, is defined as the Allan Variance $\sigma(\tau)$. In the case of shot noise it is given by[15]

$$\sigma(\nu) = \frac{K}{Q \ S/N \ \sqrt{\tau}} \qquad (11)$$

where S/N is the signal to noise ratio of the detection, Q is the $Q=\nu/\Delta\nu$ of the reference transition, and K is a constant of the order of unity and depends on the actual physical structure of the standard. The Accuracy is defined as the capability to agree with the master laboratory standard, and the Reproducibility reflects the degree to which a set of standards will produce the same output frequency and the degree to which the ideal frequency of the reference atomic line is reproduced. The "primary" AFS are so called because the atomic reference sample is strongly protected against perturbations, in particular against collisions with other atoms and/or walls . The last requirement is fully satisfied by using atomic beams and the Ramsey "two field" interrogation method as is the case of the Cs AFS. On the contrary other frequently used AFS , like the Hydrogen maser, are not primary standards even if the latter has a better short term stability with respect to Cs. Secondary AFS must be calibrated more or less frequently against the Cs primary AFS.

In conclusion an AFS referred to a transition observed in an atomic beam is of particular interest for precision and reproducibility because the reference system is immune from environment perturbations to the maximum degree.

In this case there are however two crucial limitations:

i-the sample size and thus in the signal to noise ratio (S/N) of the detection

ii-the $Q=v/\Delta v$ of the reference transition since the interrogation time depends on the beam velocity and the interrogation length : $\Delta t = L / v$ and

$$Q = v/\Delta v = v \, L/v \qquad (12)$$

In the last years two different approaches towards AFS of improved performances have been proposed and demonstrated as a prototype.

The first consist in increasing the interrogation time by using as a reference atomic transitions observed in ions stored in Penning or RF traps. In this case Δt has been increased by about four order of magnitude by using hyperfine structure transition in the microwave region. Unfortunately the sample size must be small and the S/N is poor. In addition a long interrogation time make the frequency corrections of the slave oscillator very slow. It can be foreseen that the AFS based on stored ions will be, when fully developed, better than the Cs AFS as for precision, reproducibility but worse as for short term stability, thus probably requiring a combined system with a non primary AFS, like the hydrogen maser.

The second approach consists in increasing the frequency of the reference transition[16]. As stated above the spectral purity of the RF oscillators is sufficient for a direct frequency multiplication up to the THz region about 100 times higher than that of the Cs. Multi steps frequency multiplication chains can be used for the frequency synthesis up to the near infrared and visible region. In this case however the RF oscillator cannot be properly phase-locked to the reference transition.

The atoms of Magnesium and Calcium have transitions in the far-infrared and in the visible spectral regions, that can be a useful reference for AFS with superior performances[17]. In fact the metastable triplet states $^3P_0, ^3P_1,$ and $^3P_2$ and the associated magnetic dipole transitions at 0.6 and 1.2 THz for Mg and 1.5 and 3.2 THz for Ca respectively have a set of properties that are of particular interest. A scheme of the lowest energy levels is given in fig.5 and 2 for Mg and Ca respectively. The radiative decay of the ^3P levels towards the ground $^1S_0$ state is strongly forbidden with the exception of the $^3P_1$ level which has a lifetime of about 10^{-3} s as a consequence of a small mixing with the singlet $^1P_1$ state of the same configuration. It is then possible to obtain a spontaneous time of flight states selection and the SMM transition $^3P_0 - ^3P_1$ can be monitored by the fluorescence in the visible from the $^3P_1 - ^1S_0$ transition. As a consequence, the detection of the submillimeter transition with a good S/N ratio is possible, and with a Q about two order of magnitude larger than that of the Cs AFS.

In addition the reference line of an AFS must be of the kind $m_J=0 — m_J=0$, with only a residual quadratic Zeeman effect (higher order terms are negligible)

$$v = v_0 + a \, B^2 \qquad (13)$$

where $a = 427$ Hz/Gauss^2 in the case of Cs. For the submillimeter transitions of Mg and Ca, the dependence on the residual magnetic field is fortunately about four orders of magnitude smaller,being inversely proportional to the square of the energy splitting. Thus this transitions not only will have a larger Q than the Cs, but also a much smaller dependence on the residual magnetic field. This would promise an AFS of better stability, precision , and reproducibility, because the B fluctuations are one of the most important source of limitation for the Cs AFS.

Following the above criteria, at IEN (Istituto Elettrotecnico Nazionale "G.Ferraris", Torino), and in collaboration with the Pisa University, a prototype of AFS based on the Mg line at 601 277 157 860 Hz was realized in the past few years [18], by using a thermal effusive atomic beam and the optical Ramsey interrogation scheme. A linewidth of 1.2 kHz ($Q \approx 5 \times 10^8$) and a S/N = 250 was demonstrated with a distance of 30 cm between the two oscillating fields. The stability was measured by comparison with a commercial high performance Cs AFS and was estimated to be $\sigma (\tau) = 8 \times 10^{-12} \times \tau^{-1/2}$, or better. This short term stability is better than that of the present commercial Cs AFS and inferior only to that of the large laboratory Cs AFS of the National Boureau of Standards - USA and National Research Council - Canada . The Q is already larger than that of all the present Cs AFS, and can be easily further increased (L=30 cm for Mg, while L=375 and 213 cm for the NBS IV and NRC V respectively).

The red and blue intercombination lines of Mg and Ca can also be used as a high quality reference for the stabilization of visible laser source to realize a visible length/time standard. Preliminary excellent results have been recently obtained[19] at PTB(Physikalisch Technisch Bundesanstalt, Braunschweig), demonstrating the future relevance of this AFS. Also in this experiment a thermal effusive atomic beam of Ca was used.

IMPORTANCE OF THE LASER COOLING FOR THE AFS

A new and important technical approach for substantially improving the AFS will be the use of laser decelerated and collimated atomic beam. This laser cooling or control of the atom velocity may have a tremendous impact in the development of AFS with superior performances. In fact the benefits of laser cooling and collimation are crucial in many key points:

1)- By reducing v, the transition Q is increased, see eq.(12)
2)- By reducing v, a better detection of the transition is possible, thus increasing the S/N ratio. This is of particular relevance for Mg and Ca submillimeter AFS
3)- By reducing v, the RF power for interrogating the atomic transition is reduced as v^{-2}. This will strongly attenuate the technical problems for extending the synthesis of the interrogating radiation toward higher frequencies.
4)- By reducing v, the relevance of most of the small perturbations that are responsible for the maximum attainable precision and reproducibility in the passive primary AFS will be reduced. The v dependence and numerical values for the best available Cs beam AFS are summarized in Table II.
5)- By reducing the transverse velocity by laser collimation, the first order Doppler effect line broadening can be reduced to be smaller or equal to the homogeneous line width even for the very high Q expected for the reference transitions in laser cooled atomic beams.

PRODUCTION OF COOLED METASTABLE BEAMS

As stated above, the use of cooled beams can improve dramatically the performance of an atomic frequency standard. If we want to apply this technique to atomic clock, based on the transitions between the metastable levels of Mg or Ca, the problem of the production of a low velocity metastable beam must be solved. Two different technique can be used.

A thermal atomic beam can be slowed down on the ground level by using the resonance $^1P_1$ — $^1S_0$ transition, as demonstrated before, and then the atom can be excited from ground state to the metastable triplet by electronic collisions. The perturbation to the atomic motion due to the inelastic scattering between the atoms and the electrons can be controlled, if a collimated electronic beam is used. The low velocity of the atoms allows a long interaction time and thus a satisfactory efficiency.

A second way to produce low velocity metastable atomic beam is to excite in advance the thermal beam by electron collisions (more than 30% efficiency in metastable production has been obtained by a careful experimental designs [20]) and then to cool the atoms directly on the metastable states. The cooling transition must be choose in order to behave, at the best approximation, a two-level system, to avoid losses of atoms in the cooling cycle. This is fortunately possible by using the transition connecting the metastable $^3P_2$ level to a $^3D_3$ level. If a circularly polarized light beam is used, the $m_J =+2$ — $m_J =+3$ Zeeman component is an effective two-level system, provided that no other decay channel are available for the upper $^3D_3$ level. In Mg the transition lies at 383.83 nm (see fig.5), a wavelength that is now directly available from the new Exalite dye laser, or by frequency doubling high efficient Ti-sapphire laser. In Ca the lower ^3D multiplet (4s,3d) cannot be used, because the correspondent $^3P_2$ — $^3D_3$ transition lies in the infrared at 1.97 μm; but the following $^3P_2$ — $^3D_3$ (4s,4d) transition can be used. The corresponding wavelength of 445.48 nm is directly available from the high efficiency Stilbene 3 Dye laser. In this case, other decay channels are allowed, see fig.2, through the $^3P_2$ (4s,5p) level or, possibly, the ^3F (3d,4p) levels.

However, the corresponding decay rates can be estimated to be sufficiently small for allowing an efficient cooling process.

The interesting parameters for cooling in the metastable states are also reported in Table I.

Table I

Numerical data on laser cooling (N is the number of photons needed for cooling starting from the velocity v_0, and I_s is the saturation intensity of the resonance transition)

	^{23}Na	^{133}Cs	^{24}Mg $^1S_0$	^{24}Mg $^3P_2$	^{40}Ca $^1S_0$	^{40}Ca $^3P_2$
λ (nm)	589.0	852.1	285.1	383.8	422.6	445.5
τ (ns)	16	32	2.02	5.95	4.57	11.6
γ HWHM (MHz)	5	2.5	39	13.4	17.5	6.84
Δv (cm /s)	2.29	0.35	5.84	4.29	2.36	2.23
v_0 (m /s)	800	300	1000	1000	800	800
N ($\times 10^3$)	35	85.7	17.1	23.3	33.9	35.9
L_m (cm)	45	82.3	3.46	6.13	12.4	33.3
v_t (m /s)	4.3	1.02	7.6	6.55	4.34	4.2
vt_m (m /s)	0.41	0.12	1.13	0.66	0.58	0.37
I_s (mw /cm^2)	6.4	1.05	444	61.7	60	20.2

Table II

Accuracy and precision limitations of the NBS-6 and NRC-CsVI Cs atomic beam frequency standards (the data are in units of 10^{-14}). The dependence of the effects on the atoms velocity is also shown.

Effect	Velocity dependence	Value NBS	Uncertainty NBS	Uncertainty NRC
1- B-field		5335	3	5
2- B-field inhomogeneity			0.2	2
3- Majorana effect	v^2		0.3	
4- tail pulling	v^4		2	
5- blackbody shift		1.7	0.0	
6- cavity pulling	v^2		0.1	
7- second order Doppler	v^2	26	1.0	2
8- cavity phase shift	v	36	8.0	3
9- uncertainty in phase due to n(v)	v		1.0	
10- RF spectrum	v^2		1.0	2
11- amplifier offset	v		1.0	
12- second harmonic distortion	v		2.0	

References

1. O.R. Frisch :Z. Phys. 86, 42 (1933)
2. T.W. Hänsch, A.W. Schawlow : Opt. Commun. 13, 68 (1975)
3. V.I. Balykin, V.S. Letokhov, A.I. Sidorov : Opt. Commun. 49, 248 (1984) and Zh. Eksp.Teor. Fiz. 86, 2016 (1984)
4. V.S. Letokhov, V.G. Minogin, B.D. Pavlik : Opt. Commun. 19, 72 (1976)
5. W. Ertmer, R. Blatt, J. Hall, M. Zhu : Phys.Rev.Lett. 54,996 (1985)
6. W.D. Phillips, J.V. Prodan, H.J. Metcalf: J.Opt.Soc.Am., B2, 1751-1767 (1985)
7. R.N.Watt, C.E. Wieman : Opt.Lett. 11,291 (1986)
8. N. Beverini,S. De Pascalis, E. Maccioni, D. Pereira, F. Strumia, G. Vissani, Y.Z. Wang, C. Novero : Opt. Lett., 14, 350 (1989)
9. N. Beverini, E. Maccioni, D. Pereira, F. Strumia, G. Vissani : Ital. Phys. Soc. Conf. Proc., 21, 205 (1989)
    N. Beverini, E. Maccioni, D. Pereira, F. Strumia, G. Vissani : " Frequency Standards. and Metrology", A. De Marchi ed. , Springer Verlag, pag. 282 (1989)
    N. Beverini, E. Maccioni, D. Pereira, F. Strumia, G. Vissani : Proc. IV Conf. Laser Science, 1988, AIP Conf. Proc. 1989, in press
10. F. Shimizu, K. Shimizu, H. Takuma : Jap.J.Appl.Phys. 26, L1847-49 (1987)
11. N. Beverini, F. Giammanco, E. Maccioni, D. Pereira, F. Strumia, G. Vissani : J.Opt. Soc.Am. B (1989)
12. V.G. Minogin, V.S. Letokhov: "Laser light pressure on atoms", Gordon and Breach Sc. Publ., New York 1987
13. V.I. Balykin, V.S. Letokhov, V.G. Minogin, Yu.V. Rozhdestveensky, A.I. Sidorov: J. Opt. Soc. Am. B 2, 1776-1783 (1985)
14. Proc. Fouth Symp. on Frequency Standards and Metrology, A. DeMarchi ed., Springer-Verlag, 1989
15. J. Vanier, and C. Audoin, " The Quantum Physics of Atomic Frequency Standards", Adam Hilger,1989
    " Metrology and Foundamental Constants", A. Ferro Milone, P. Giacomo, S. Leschiutta eds., North Holland, 1980
16. F. Strumia: "Application of laser cooling to the atomic frequency standards", in "Laser Science and Technology", Ed. da A.N. Chester e S. Martellucci, Plenum Press, N.Y. 1988, pagg. 367-401.
17. F.Strumia : Metrologia,8, 85-90,(1972)
18. E.Bava, A.Godone, C.Novero : Appl.Phys., B48, 495, (1989)
    E.Bava, A.Godone, C.Novero : Metrologia , 24, 133, (1987)
19. A. Morinaga, F. Riehle, J. Ishikawa, J. Helmcke: Appl.Phys., B48, 165, (1989)
20. G.Giusfredi, P.Minguzzi, F.Strumia, M.Tonelli : Z. Physik, A274, 279-287, (1975)

POSSIBLE PRECISION FAR-INFRARED SPECTROSCOPY ON TRAPPED IONS

Günter Werth

Institut für Physik
Universität Mainz
D-6500 Mainz, Fed.Rep.Germany

One of the necessary conditions for very high resolution spectroscopy is the existence of long-living energy levels between
which transitions can be induced by tunable electromagnetic
radiation. Moreover the observation time of the particles under investigation should be sufficiently long, and perturbing
effects like Doppler broadening and collisions should be very
small. All this leads to the technique of trapped ion spectroscopy which offers all the required conditions. It uses
the confinement of charged particles to a very small volume
in space under ultrahigh vacuum conditions either by application of a r.f. voltage and a small additional d.c. bias to an
arrangement of three electrodes as shown in Fig. 1 (r.f. trap
or Paul trap), or by application of a d.c. voltage to the
same electrode configuration and a superimposed magnetic
field directed along the axis of symmetry (Penning trap). Details of the operation of such traps can be found in the
literature[1,2].

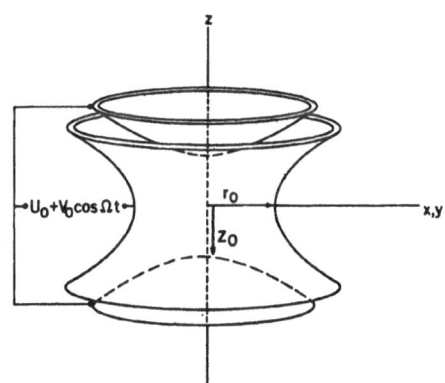

Fig. 1. Radio frequency ion trap (Paul trap)

Applied Laser Spectroscopy, Edited by W. Demtröder and
M. Inguscio, Plenum Press, New York, 1990

Many singly charged ions have long-living metastable states, which - apart from the stable ground state - may be used for high resolution spectroscopy. Among them the ions of the earth alkalines are particularly well suited as it is evident from the general level diagram of such ions shown in Fig. 2.

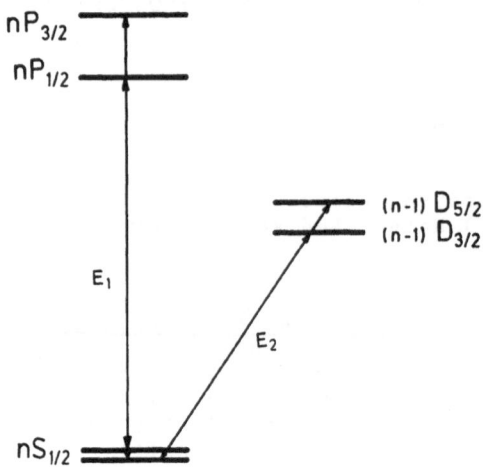

Fig. 2. General partial level diagram of alkali-like ions. Examples of such ions are Ca^+, Sr^+, Ba^+, Yb^+, Hg^+.

It is similar to that of alkali atoms, having an $S_{1/2}$ ground state and electric dipole resonance transitions to $P_{1/2}$ and $P_{3/2}$ states. In contrast to the neutral alkali atoms they have additionally low-lying $D_{3/2}$ and $D_{5/2}$ states which decay by quadrupole radiation into the ground state and hence are metastable. Examples of such ions are Ca^+, Sr^+, Ba^+, Yb^+ and Hg^+. Including ground state hyperfine splittings of odd isotopes of these ions, the level scheme offers different possibilities of high resolution spectroscopy which have in part been exploited in recent experiments:

(a) Ground state hyperfine transitions in the microwave spectral region

Since the lifetime of ground state hyperfine levels approaches infinity, the achievable resolution is determined by the available frequency reference and by the phase coherence time of the trapped particles. The latter may exceed several minutes under UHV conditions. Thus linewidths in the mHz-range have been obtained in different experiments[3,4].

(b) Quadrupole transitions in the optical spectral region

A direct excitation of one of the metastable levels from the ground state has been demonstrated in an experiment on a single confined Hg^+ ion, performed at the NBS Boulder. It was detected by the "quantum jump" method where a S-D transition is indicated by the absence of otherwise very strong fluores-

cence from a simultaneously excited S-P resonance transition[5].
The linewidth is entirely determined by the frequency stability
of the tunable laser. Technical improvements of the laser
have led meanwhile to a linewidth of about 50 Hz, which is
not far from the natural linewidth of 16 Hz, given by the
9 ms lifetime of the excited D-state[6].

(c) A third and so far unexploited possibility of high resolution spectroscopy offers the finestructure separation
between the metatstable D-states. The natural line-Q of those
transitions would in general exceed 10^{13}. Table 1 summarizes
the frequency differences for the above mentioned sample of
ions as well as $S_{1/2}$-$P_{1/2}$ resonance wavelengths to populate
the $D_{3/2}$ states via spontaneous decay of the excited $P_{1/2}$
state. Additionally, P-D wavelengths are mentioned which may
serve for resonance detection and ion cooling as explained
further below.

Table 1.   Fine structure splittings ΔE_{Fs} between metastable
           D-states of different ions. τ is the shorter of the
           $D_{3/2}$ or $D_{5/2}$ lifetimes which determines the natural
           linewidth. Q_{max} is the maximum obtainable line-Q
           calculated from ΔE_{Fs} and τ. λ_1 is the $S_{1/2}$-$P_{1/2}$
           resonance wavelength, used to excite the ion, λ_2
           and λ_2' correspond to the $P_{1/2}$-$D_{3/2}$ and $P_{3/2}$-$D_{5/2}$
           transitions, respectively, (see Fig. 2)

Ion	ΔE_{Fs} [THz]	τ [s]	Q_{max}	λ_1 [nm]	λ_2 [nm]	λ_2' [nm]
Ca^+	1.89	1	10^{13}	393	866	854
Sr^+	8.40	0.39	$2 \cdot 10^{13}$	422	1091	1033
Ba^+	24.0	17	$2 \cdot 10^{15}$	493	650	614
Yb^+	41.1	0.052	10^{13}	369	2438	1650
Hg^+	446.0	0.009	$2 \cdot 10^{13}$	193	-	-

The choice of an ion for visible-infrared double resonance depends on the availability of narrow band tunable
lasers for ion excitation and cooling and on the other hand
of tunable low noise infrared sources. Both considerations
lead to Ba^+ and Ca^+ as good candidates. While Ba^+ is presently
investigated by a group at NRC Ottawa, Canada, using a stable
ammonia laser as FIR radiation source[7], we propose to perform
a double resonance experiment on Ca^+. As evident from the
level diagram (Fig. 3), Ca^+ offers the particular feature that
all the wavelengths connecting S,P and D states can be produced
by tunable diode lasers if we include frequency doubling of a
786 nm diode laser to obtain the S-P resonance wavelength at
393.4 nm. This seems possible by using new powerful commercially available diode lasers. It should be noted that trapped

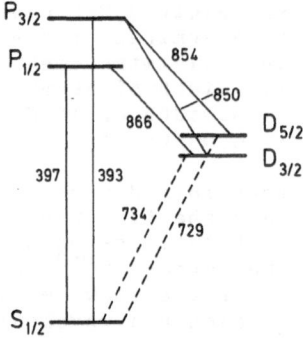

Fig. 3. Level diagram of Ca$^+$. All wavelengths are in nm

Ca$^+$ ions also might provide the possibility to stabilize diode laser on a narrow quadrupole transition, similar to the above mentioned experiment on Hg$^+$ [5].

The FIR radiation may be produced in two different ways[8]: A near coincidence of the Ca$^+$ D-state fine structure splitting is given by the 1.89828 THz line of a $^{13}CH_3^{16}OH$ laser, pumped by the 9P12 line of a CO_2 laser. A more careful investigation of the Ca$^+$ FS separation will show how large the actual displacement is. Since the alcohol laser can be Stark-modulated, it may be possible to stabilize it to the fine structure line. A tunable FIR source is provided by the mixing of different CO_2 laser lines in a MIM-diode with additional microwave radiation from a synthesizer. Although the available output power of such devices is only in the µW range, it would be sufficient to drive a narrow fine structure transition.

The detection of fine structure transitions can be performed in different ways as indicated in Fig. 4: Most simply from the experimental point of view would be a detection of the quadrupol radiation originating from the $D_{5/2}$ state which can be populated only by a transition from the $D_{3/2}$ state. The transition wavelength is different from the excitation wavelength which reduces background problems. On the other hand the number of photons is rather small considering the long D-state lifetime of about 1 second as predicted by calculations[9]. Earlier measurements indicate, however, that such transitions may be observed with reasonable signal-to-noise ratio as shown in Fig. 5, where we observed the $D_{5/2}-S_{1/2}$ quadrupole transition in Sr$^+$, which occurred when the $D_{5/2}$ state was populated by collisions from the $D_{3/2}$ state (fine-structure mixing)[10]. In the case of collisions we have the competing process of direct de-excitation of the $D_{3/2}$ level into the ground state which reduces the observed signal. Such competing processes would be absent in the case of FIR reso-

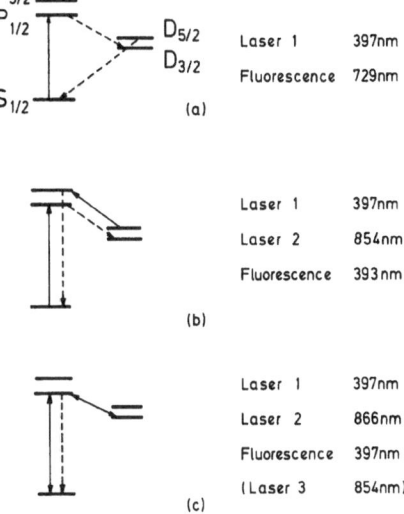

Laser 1	397nm	
Fluorescence	729nm	
(a)		

Laser 1	397nm	
Laser 2	854nm	
Fluorescence	393nm	
(b)		

Laser 1	397nm	
Laser 2	866nm	
Fluorescence	397nm	
(Laser 3	854nm)	
(c)		

Fig. 4   Possible detection schemes for FIR fine structure
transitions between metastable D-states.
(a) Observation of weak $D_{5/2}$-$S_{1/2}$ fluorescence after
$S_{1/2}$-$P_{1/2}$ laser excitation to populate the $D_{3/2}$ level
and FIR $D_{3/2}$-$D_{5/2}$ transition.
(b) Laser depopulation of $D_{5/2}$ state and observation
of $P_{3/2}$-$S_{1/2}$ fluorescence.
(c) Population of the $D_{3/2}$ level by $S_{1/2}$-$P_{1/2}$ and
$P_{1/2}$ -$D_{3/2}$ laser excitation. Observation of $P_{1/2}$-
$S_{1/2}$ fluorescence. In case of a single ion this
fluorescence will vanish completely, when a $D_{3/2}$-$D_{5/2}$
FIR transition takes place. The lasers may be used
to cool the ions. A third laser, tuned to the $D_{5/2}$-
$P_{3/2}$ transition, would drive the ion back to the
pumping cycle after a FIR transition and thus increase the signal-to-noise ratio.

nance transitions. The detection mechanism, however, requires
a large (10^5) number of trapped ions because of solid angle
losses which makes it very difficult to cool the ions as
required for very high precision (see below).

Instead of observing the quadrupole photons one might
detect the fine structure transitions by a second laser,
tuned to the $D_{5/2}$-$P_{3/2}$ transition, which results in fluorescence photons at 396.8 nm, again different from the excitation
wavelength of 393 nm (Fig. 4b). The number of photons per
unit time would be substantially larger than in the first
case, since the D-state lifetime is no longer a limiting
factor, and consequently the number of ions could be reduced.

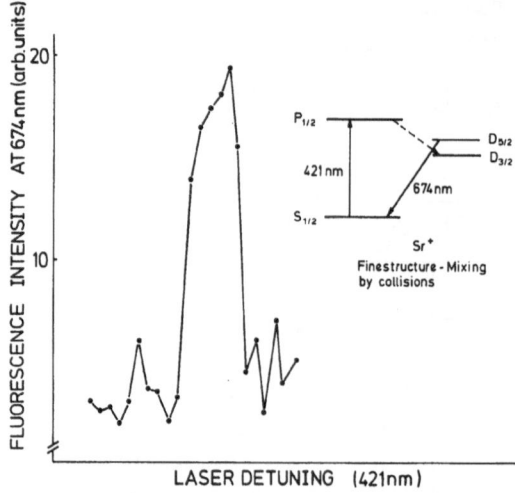

Fig. 5 Observation of $D_{3/2}-D_{5/2}$ transitions in Sr$^+$ induced
        by collisions of trapped ions with buffer gas atoms.
        The detection scheme is like in Fig.4(a).(from Ref.10).

A third and most effective detection mechanism would be
the "quantum jump" scheme as demonstrated by several
groups[11,12,13]: Two narrow beam lasers, tuned to the low energy
side of the $S_{1/2}-P_{1/2}$ and $P_{1/2}-D_{3/2}$ transitions, respectively,
cool the ions by photon recoil and give rise to a strong
(about 10^8 photons per second per ion) fluorescence at 393 nm.
This fluorescence decreases or in case of a single trapped
ion even goes to zero when the ion is driven to the $D_{5/2}$
state by FIR radiation. Thus every single transition may be
detected with an efficiency approaching unity. At the same time
the ions' temperature may be reduced to below 1 K which substantially increases the spectral resolution of the fine structure
transition.

Two main reasons limit the experimentally achievable resolutions: Uncooled ion clouds in r.f. traps have a typical
average kinetic energy of the order of 1 eV. This gives
rise to a second order Doppler shift and due to the near
Maxwellian distribution of energies to an associated broadening. The order of magnitude is given by E_{kin}/mc^2, which is
for Ca$^+$ $2 \cdot 10^{-11}$, well above the natural linewidth of 10^{-13}.
Reduction of the ion temperature to 1 K would reduce the
shift to less than 10^{-13}.

A residual magnetic field in the trap gives a Zeeman
shift of the fine structure transition, which is of the order
of 10^{-9} for residual fields in the mG range, since all Zeeman
sublevels depend linearly on the magnetic field for even isotopes. The odd isotope ^{41}Ca$^+$, however, has a nuclear spin of
5/2, which couples to the D state J-values of 3/2 and 5/2 to an
integer total angular momentum. It has magnetic sublevels $m_F=0$
which are independent of the magnetic field to first order.

The residual second order shift is about $10^{-3}B^2(Hz/G^2)$, which for moderate shielding gives very small shifts. In addition it can be measured by magnetic field determination from $\Delta m_F=1$ transitions. The use of odd isotopes, however, requires measures to avoid ground state hyperfine pumping. This could be done either by modulating the 393 nm laser at the hyperfine frequency or by driving the ground state hyperfine transition by an additional microwave source.

In summary trapped Ca^+ ions offer a good possibility of precision fine structure spectroscopy in the FIR spectral range. The required radiative sources are either commercially available or can be constructed by proven techniques. Attempts have been started to set up an experiment at the recently opened European Laser Laboratory, L.E.N.S., at Florence, Italy, in cooperation with M. Inguscio (Napoli/ Firenze) and his coworkers.

Discussions with M. Inguscio, who provided the information about the FIR sources, and K. Siemsen are gratefully acknowledged. Basic experimental work has been started at Mainz by B. Abel and F. Marin.

References

1. D.J. Wineland, W.M. Itano, and R.S.van Dyck, High resolution spectroscopy of stored ions, in: "Advances in Atomic and Molecular Physics", B. Bederson, ed., Academic Press, New York (1983).
2. L.S. Brown and G. Gabrielse, Rev.Mod.Phys. 58:233 (1986).
3. W.M. Itano and D.J. Wineland, Precision Measurement of the ground-state hyperfine constant of $^{25}Mg^+$, Phys.Rev. A24:1364 (1981).
4. R. Blatt, H. Schnatz, and G. Werth, Precise Determination of the $^{171}Yb^+$ Ground State Hyperfine Separation, Z.Phys. A312:143 (1983).
5. J.C. Bergquist, W.M. Itano, and D.J. Wineland, Recoillers Optical Absorption and Doppler Sidebands of a Single Trapped Ion, Phys.Rev. A36:428 (1987).
6. L. Holberg, private communication.
7. K. Siemsen, private communication. See also contribution in this volume.
8. M. Inguscio, private communication.
9. B. Warner, Atomic Oscillator Strengths-III, Mon.Nat.R. astr.Soc. 139:115 (1968).
10. Ch. Gerz, Th. Hilberath, and G. Werth, Lifetime of the $4D_{3/2}$ and $4D_{5/2}$ Metastable States in Sr II, Z.Phys. D5:97 (1987).
11. W. Nagourney, J. Sandberg, and H. Dehmelt, Shelved optical amplifier:Observation of quantum jumps, Phys.Rev.Lett. 56:2797 (1986).
12. Th. Sauter, R. Blatt, W. Neuhauser, and P. Toschek, Observation of Quantum Jumps, Phys.Rev.Lett. 57:1696 (1986).
13. J.C. Berquist, R. Hulet, W.M. Itano, and D.J. Wineland, Observation of Quantum Jumps in a Single Atom, Phys.Rev. Lett. 7:1699 (1986).

STABLE AMMONIA LASER FOR COOLED SINGLE ION FREQUENCY STANDARD

K.J. Siemsen [1], A.A. Madej [1], and G. Magerl [2]

1 Time and Length Standards, National Research Council
of Canada, Ottawa, CANADA K1A OR6

2 Technische Universität Wien, A1040 Wien, AUSTRIA

It has been demonstrated recently that an optically pumped NH_3 laser operating in the mid-infrared can provide a sub-kilohertz linewidth source when run under proper pumping conditions [1]. Although such a laser possesses an excellent spectral purity, long term stabilization must be referenced to an absolute frequency, typically provided by an atomic or molecular absorber. In the present project, it is proposed to use the NH_3 laser for the required frequency source to stabilize to the 12.5 μm resonance of a trapped, laser cooled single ion of Barium. The trapped single ion system provides the beneficial features of an absorber system in which motional Doppler shifts and broadenings can be eliminated to very good accuracy via confinement and laser cooling [2], together with minimal perturbation by external fields. The ion's 5d $^2D_{5/2}$ - 5d $^2D_{3/2}$ magnetic dipole transition at 800.974 cm^{-1} has an ultimate natural linewidth below 0.1 Hz and is situated near the NH_3 laser transitions at 12.5 μm, thus it has the interesting possibility for stabilization to an extremely high level. Such a weakly allowed transition can be detected for single ion systems by observing interruption in fluorescence from the ion cycling on a strongly allowed transition. The stability of the source may then be evaluated by comparing the ammonia laser frequency with the NRC primary cesium beam frequency standard via a frequency chain [3].

In order to obtain a NH_3 laser source which can provide radiation to be stabilized on the $^{138}Ba^+$ transition, a study of available laser transitions

Applied Laser Spectroscopy, Edited by W. Demtröder and
M. Inguscio, Plenum Press, New York, 1990

was performed. The sP(8,6) laser transition in $^{15}NH_3$ was found in close proximity to the Ba^+ magnetic dipole transition. The NH_3 laser is pumped by a ^{16}O ^{12}C ^{18}O laser operating on the $R(40)_{II}$ transition delivering 2 W of pump power. The radiation from the pump laser is downshifted by 72 MHz by an acousto-optic modulator into resonance with the sR(6,6) absorption in $^{15}NH_3$ (Fig.1). In order to prevent frequency drifts of the pump radiation, a small fraction of the CO_2 laser beam is used to give a saturation absorption dip in a cell filled with $^{15}NH_3$ at 1 mTorr pressure. The half width at half maximum of this dip is 150 kHz and the shifted pump laser frequency is locked via servo electronics to this dip. The ammonia laser cavity consists of a 30 cm long, dry ice cooled quartz envelope filled with 50 mTorr of $^{15}NH_3$, and provides 30 mW of power on the sP(8,6) transition. We have recently measured the laser frequency by heterodyning the laser output with two Lamb-dip stabilized CO_2 lasers in a tungsten antimony point-contact diode. The sP(8,6) center frequency was thus determined to be f= 24 029 248.2 MHz with a tunability of $\pm$ 2 MHz.

With the Ba^+ reference transition at 24 012 660. $\pm$ 180 MHz [4], a frequency shift of 16 GHz is required to allow interaction with the NH_3 laser. For this purpose, it is planned to use an electro-optic modulator of the type described by Magerl and coworkers [5]. By employing CdTe as the modulator medium, a sideband efficiency of approximately 130 μW per watt laser power and per watt microwave power is expected. For a 1 mm beamsize, the sideband power required to excite the Ba^+ reference transition on a time scale of a few seconds was calculated to be on the order of a few nanowatts.

The overall arrangement for the experiment is shown in Fig.2. Detection of excitation of the magnetic dipole transition is obtained indirectly by observing the fluorescence from the ion excited on the strongly allowed 6p $^2P_{1/2}$ - 6s $^2S_{1/2}$ (λ=494 nm) and 6p $^2P_{1/2}$ - 5d $^2D_{3/2}$ (λ=650 nm) transitions (Fig.3). Laser excitation on these two transitions is detuned slightly to the red in order to provide laser cooling of the ion. Detection of whether the ion has been excited into the 5d $^2D_{5/2}$ metastable level is provided by observing whether there is an interruption of the 494 nm fluorescence. In order to prevent power broadening and shifts of the reference transition by the visible laser light, alternate illumination by the visible (cooling and detection) and infrared (excitation of reference transition) is provided.

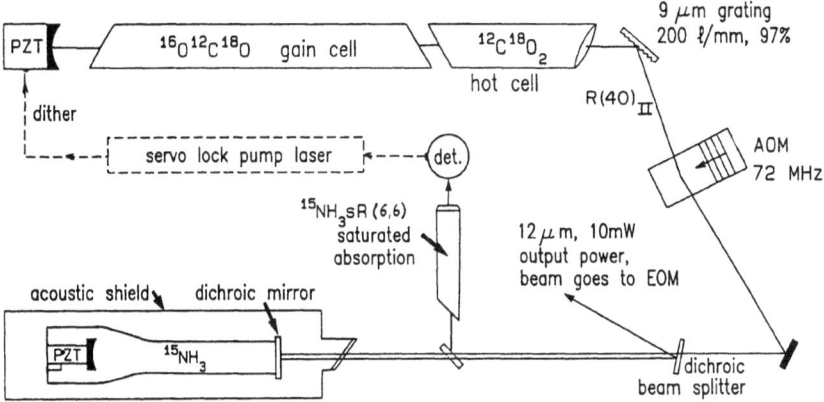

Fig.1. $^{15}NH_3$ laser operating on the sP(8,6) transition.

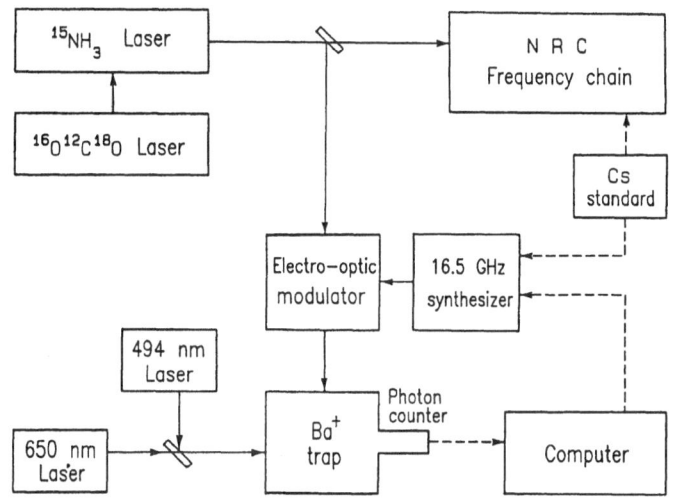

Fig.2. Overview of proposed measuring scheme.

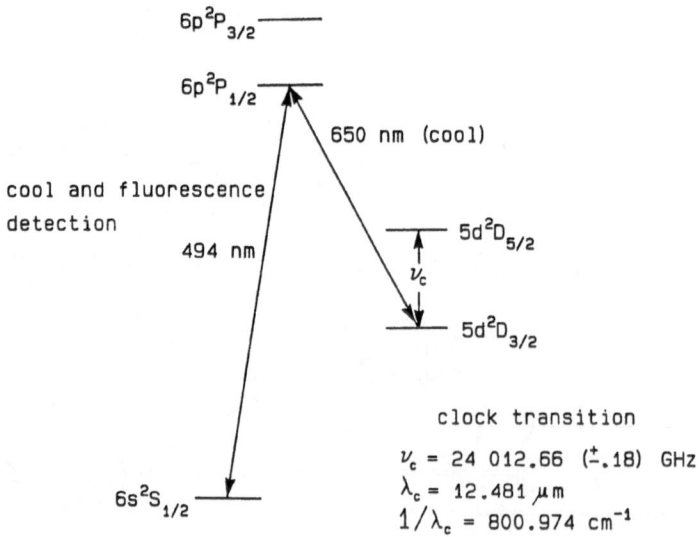

Fig.3. Simplified energy level diagram of $^{138}Ba^+$.

References

[1] K.J. Siemsen, E. Williams, and J. Reid, Opt. Lett. 12, 879 (1987).

[2] J.C. Bergquist, Wayne M. Itano, and D.J. Wineland, Phys. Rev. A 36, 428 (1987).

[3] B.G. Whitford, Appl. Phys. B 35, 119 (1984).

[4] C.E. Moore, (in "Atomic Energy Levels", vol.III, N.B.S. Washington, 1958).

[5] G. Magerl, W. Schupita, and E. Bonek, IEEE QE-18, 1214 (1982).

FREQUENCY STABILIZATION OF INFRARED LASERS:

THE OPTOGALVANIC EFFECT IN CO_2

Lyndon Zink

European Laboratory for
Nonlinear Spectroscopy (LENS)
Largo Enrico Fermi, 2
Florence, Italy

INTRODUCTION

CO_2 and CO lasers provide accurate frequency standards in
the infrared. CO_2 laser frequencies have been measured to an
accuracy of 3 kHz[1] and are frequently used in heterodyne
experiments to determine molecular transition frequencies in the
infrared[2] and far-infrared[3]. However, to fully realize this
accuracy the laser must be frequency stabilized.

Several stabilization schemes for CO_2 and CO lasers have
been employed and are listed in Table 1. The simplest scheme and
one which is applicable to all the laser lines is to stabilize on
the peak of the Doppler broadened gain curve[4]. But, this only
provides a stability of 1 MHz at best. The optogalvanic (OG)
effect[5] and opto-hertzian effect[6] (rf equivalent of the OG
effect) have also been used for all the lasing lines, but again
the signal is stabilized on the peak of a Doppler broadened
curve. To achieve higher frequency stability requires a sub-
Doppler signal.

The most commonly used sub-Doppler signal for CO_2
stabilization is the saturation dip in the 4.3 micron CO_2
fluorescence[7,8]. The energy levels involved in this scheme are
illustrated in Fig. 1. Similarly the saturation dip in the
absorbed CO_2 laser power can also be monitored, but with reduced
signal to noise. Both of these techniques are limited to the
regular bands or hot bands because they require a population of
the lower laser level (which absorbs the laser radiation) at room
temperature or slightly higher. To observe the fluorescence
signal from the sequence bands (see Fig. 1) requires a much
higher temperature of the absorbing gas, which then increases the
background radiation at 4.3 microns, overwhelming the laser
induced signal. Saturated absorption in other gases, for example
OsO_4[9,10] and SF_6[11] can provide stronger absorption and narrower
lines than CO_2 itself. These promise the highest stability[12] for

Applied Laser Spectroscopy, Edited by W. Demtröder and
M. Inguscio, Plenum Press, New York, 1990

Table 1. Stabilization techniques for
CO_2 and CO lasers.

Stabilization Signal	Laser	Frequency Stability	Reference
Peak Power	CO_2	1 MHz	4
	CO	5 MHz	
Optogalvanic	CO_2	1 MHz	5
Optohertzian	CO	5 MHz	6
Sub-Doppler			
Fluorescence-CO_2	CO_2	10 kHz	7,8
Absorption-OsO_4	CO_2	100 Hz	9,10
Absorption-SF_6	CO_2	100 Hz	11
Optogalvanic	CO_2	30 kHz	14,15
	CO	100 kHz	13

CO_2 lasers but rely on an accidental coincidence between the CO_2 laser frequency and the molecular frequency and therefore are not applicable to all the laser lines.

The stabilization of a CO laser to absorption in room temperature CO is not possible because, like the CO_2 sequence bands, the lower laser levels are not sufficiently populated. Therefore to see saturated absorption in these lines the laser levels are populated in a discharge and the signal detected via the OG effect[13]. This detection technique provides superior signal to noise than detecting the absorption directly. Similarly this method has been applied to the regular[14] and sequence[15] bands of a CO_2 laser. Lamb-dip optogalvanic stabilization in CO_2 and measurement of the stability achieved is discussed below. The complex processes involved in the OG effect in CO_2 is discussed in references 5 and 16.

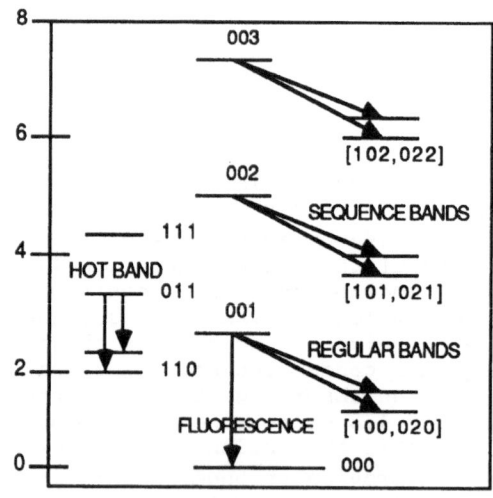

Figure 1. Energy level diagram of CO_2.

EXPERIMENTAL

Figure 2 is a diagram of the experiment. In brief, the output radiation from CO_2 laser 1 is focused into the external OG cell with a telescope and reflected back through the cell for observation of the saturated signal. The return beam is combined with radiation from a reference laser and detected on an infrared detector. This detector generates the difference frequency (beat note) between the two lasers. Laser 1 is stabilized on the Lamb dip observed in the OG cell, the reference laser is stabilized on the saturated fluorescence signal and the beat note measures the frequency and stability of the OG stabilized laser. The previous report of this technique[14,15] utilized an internal OG cell, yet to our knowledge this is the first demonstration with an OG cell outside the laser cavity.

The two CO_2 lasers are identical in design. The 1.62 m long resonator cavity consists of a 150 lines/mm grating and a totally reflecting end mirror with 10 m radius of curvature. The end mirror is mounted on a PZT for cavity length adjustment and modulation. Radiation is coupled out via the zeroth order of the grating, which was specially constructed by Hyperfine, Inc. to give 5 % output coupling. The laser excitation is achieved by a dc discharge in a flowing gas mixture of 15 % CO_2, 15 % N_2, 70 % He confined in a 13 mm bore, water cooled Pyrex tube. The resonator elements and the laser tube are supported by three INVAR rods, irises at each end of the tube provide mode selection. At typical operating conditions of 2500 Pa and 25 mAmps these lasers yield an average of three Watts in TEM_{00} from 9R(60) through to the 11 micron hot band transitions. They also

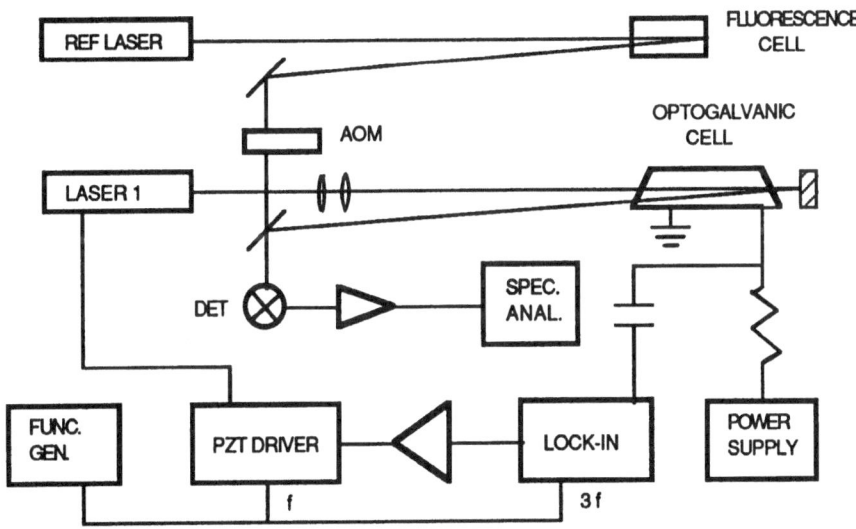

Figure 2. Experimental set-up for the stabilization of a CO2 laser to an external OG cell.

lase on many of the sequence lines without an internal hot cell. These lasers were designed by Dr. K. M. Evenson at NIST, Boulder, Colorado and constructed by Karl Gebert and Cole Briggs.

To observe the saturated signal in the OG cell it was necessary to increase the power density with the telescope. Estimated power density in the cell after the telescope is 40 W/cm^2. The OG cell was placed approximately 4 meters from the laser output window to reduce the necessary angular displacement of the return beam and optimize the saturated signal.

Two OG cells were used in the experiment. The first is a water cooled Pyrex tube of internal diameter 20 mm sealed at both ends with ZnSe Brewster windows. The total discharge length is 150 mm and the gas mixture was slowly flowed through the cell. The second cell is identical in design except the return mirror replaced one of the Brewster windows and the gas was not flowed. A current stabilized high voltage power supply provides the energy for the dc discharge. The OG effect manifests itself as a change in the voltage across the discharge and this signal is directly coupled via a capacitor and diode protection circuit into a lock-in amplifier.

For stabilization the laser frequency is modulated by applying a small ac voltage onto the end mirror PZT. Three times this modulation frequency serves as the reference for the lock-in amplifier. 3f detection reduces any offset problems caused by a sloping baseline in the OG signal[17,18]; optimum modulation frequency (1f) was 275 Hz. The signal from the lock-in is then averaged in an integrator which provides a correction voltage (after amplification) onto the PZT.

The laser frequency and stability are measured by heterodyning the OG stabilized laser and the reference laser. In the regular CO_2 band measurements a HgCdTe detector was used to generate the beat note. The two lasers operated on the same laser transition and the reference laser passed through a 90 MHz acousto-optic modulator which shifts the frequency. The beat note frequency (f) is given by $f_{bn}=(f_{ref}-f_{og}) \pm f_{aom}$. In the sequence band measurement a tungsten-nickel MIM diode was used to generate the beat note. In this case microwave radiation was also coupled onto the diode and the beat note is given by $f_{bn}=(f_{ref}-f_{og}) \pm nf_{mw}$, where n is a whole number.

RESULTS

The frequency stability achieved with the 3f OG Lamb-dip technique is listed in Table 2. Note that the technique is also applicable to sequence band lines. Each of these measurements is the fluctuation of the frequency about the line center over a period of five minutes, these fluctuations did not increase after this time. The linewidth of the OG Lamb-dip was about 3.1 MHz FWHM, with linear dependence on the gas pressure and independent of the discharge current.

Signal-to-noise (S/N), and hence stability, depended critically on the discharge pressure, gas mixture, and current.

Table 2. Frequency stability achieved with
the 3f optogalvanic technique.

CO_2 laser line	Stability (kHz)
I P(30)	70
I R(26)	140
II P(10)	105
II R(20)	200
Seq I P(33)	600*

* Stabilized on 1f signal.

The best S/N ratio of 40 was achieved with a 1:1 mixture of
CO_2/N_2 at a pressure of 60 Pa (500 mTorr) and 20-23 mAmps
current. Optimum temperature for the cooling water was 10-15° C,
but this was less critical. The flowing gas cell and the sealed
off cell gave similar results, however the signal from the sealed
off system decayed after several hours because of CO_2
dissociation.

On the regular band transitions we did not observe any
difference in the center frequency of the OG stabilized laser and
the reference laser, outside of the measured fluctuations.
Therefore, the plasma does not induce any shift in the frequency
of the molecular transition at this level of accuracy.

Since this technique gives improved stability for the
sequence band lines over the previous method of stabilizing to
the top of the gain curve, we have remeasured the frequency of
the I P(33) sequence band line. As noted in Table 2, the S/N for
this transition was too weak to allow a 3f lock (in which the
signal decreases by a factor of 2 compared to 1f), so a 1f lock
was used for the measurement. The RF beat note between seq. I
P(33), reg. I P(38), and microwave radiation was generated in a
MIM diode. The beat note was averaged for 90 seconds on a
spectrum analyzer and the center frequency measured with an
accurate frequency marker. Ten measurements were performed and
the new measured frequency is 27 875 546.157 (165) MHz. The
number in parenthesis is one sigma of the ten measurements. This
value agrees with the previous value of Siemsen and Whitford[4] to
within the uncertainty of the measurements, but is three times
more accurate.

DISCUSSION

Since our reference CO_2 laser is stabilized via the
saturated absorption technique, it provides an easy comparison
between the OG Lamb-dip technique and the fluorescence technique.
The linewidth of the fluorescence signal is 1.5 MHz, with a S/N
of about 300, compared to 3.1 MHz and 40 for the OG signal. This
results in a much tighter lock for the fluorescence technique.
In fact, when locked with the fluorescence technique, laser 1 has
a five minute fluctuation of 20 kHz; a factor of five to ten
times better than the OG technique.

Also the fluorescence technique does not require a dc discharge and is therefore easier to realize experimentally. In reference 14 it was stated that the OG technique is less expensive than the fluorescence technique, but we feel the costs are about equal if you purchase the current stabilized power supply for the OG discharge.

However, the fluorescence technique is not usable, at present, for the sequence bands. Although other molecules, for example SF_6 or OsO_4, may be used to stabilize sequence band lines to higher accuracy, they do not stabilize to the center of the CO_2 molecular line. Hence, the OG Lamb-dip stabilization technique is the best method for stabilizing a CO_2 laser to the line center of the sequence band transitions.

SUMMARY

Frequency stabilization of CO_2 and CO lasers has been briefly reviewed and a new technique for stabilizing CO_2 lasers described. This technique, using the saturated optogalvanic signal, gives a stability of 100 kHz over five minutes. More importantly it is applicable to the sequence band lines. Using this technique of stabilization we have remeasured the frequency of the laser line seq. I P(33), improving the accuracy by a factor of three over the previous measurement.

REFERENCES

1. F.R. Petersen, E.C. Beatty, and C.R. Pollock, J. Mol. Spectrosc. 102, 112 (1983).
2. L.R. Zink, J.S. Wells, and A.G. Maki, J. Mol. Spectrosc. 123, 426 (1987).
3. See for example; K.M. Evenson, this publication.
4. K.J. Siemsen and B.G. Whitford, Opt. Commun. 22, 11 (1977).
5. A.L.S. Smith and S. Moffatt, Opt. Commun. 30, 213 (1979).
6. G.N. Pearson and D.R. Hall, IEEE J. Quantum Electron. 25, 245 (1989).
7. C. Freed and A. Javan, Appl. Phys. Lett. 17, 53 (1970).
8. C. Freed and R.G. O'Donnell, Metrologia 13, 151 (1977).
9. O.N. Kompanets, A.R. Kukudzhanov, V.S. Letokhov, V.G. Minogin, and E.L. Mikhailov, Sov. Phys.-JETP 42, 15 (1976).
10. A. Clairon, B. Dahmani, A. Filimon, and J. Rutman, IEEE Trans. Instrum. Meas. IM-34, 265 (1985).
11. M. Ouhayoun and C.J. Borde, Metrologia 13, 149 (1977).
12. A. Godone, M.P. Sassi, and E. Bava, Metrologia 26, 1 (1989).
13. M. Schneider, A. Hinz, A. Groh, K.M. Evenson, and W. Urban, Appl. Phys. B 44, 241 (1987).
14. J.T. Shy and T.C. Yen, Opt. Commun. 60, 306 (1986).
15. J.T. Shy and T.C. Yen, Opt. Lett. 12, 325 (1987).
16. F.O. Shimizu, K. Sasaki, and K. Ueda, Jap. J. Appl. Phys. 22, 1144 (1983).
17. C. Audoin, in Frequency Metrology, ed. A.F. Milone and P. Giacomo, 195 (1976).
18. J.L. Hall, H.G. Robinson, T. Baer, and L. Hollberg, in Advances in Laser Spectroscopy, eds. F.T. Arecchi, F. Strumia, and H. Walther (Plenum Press, New York, 1983).

INDEX

Absorption coefficient, 405
Anomalous Doppler–broadening, 249
Antibunching of Photons, 18
Atmospheric monitoring, 418, 435
Atomic discharges, 173
Atomic frequency standards, 473

Barium borate, 101f
Bloch equation, 295
Bolometer, 218, 236
Box CARS, 378
Breit–Rabi diagram, 151
Brillouin gain spectroscopy, 307
Brillouin scattering, 308
CARS, 272, 302, 313ff, 365 ff
    collinear, 385
    resonance enhanced, 389
Chaos, 24, 31
Coherence, 73
    temporal, 73
Coherent states, 22
Clusters, 319
CO laser, 127
CO_2 laser, frequency
    stabilization, 493
Collection efficiency, 2
Collinear laser spectroscopy, 341
Collinear CARS, 385
Combustion spectroscopy, 393
Concentration mapping, 457
Conversion efficiency, 66, 103, 105

Density matrix, 228
Detectable pollutants, 448
Detection sensitivity, 3
Deterministic Chaos, 32f
Difference frequency mixing, 65
Differential absorption, 419ff, 435

Dimension of fractals, 37
Diode lasers, 117
Dipole moment, 48
Discharges, 251
    holllow cathode, 195
    Doppler–tuning, 199, 205

Einstein–Podolski–Rosen
    paradoxon, 14
Electric discharges, 193, 251,
Electron storage ring, 73
Enviromental monitoring, 417, 435

Far–infrared spectrometer, 142
    spectroscopy, 141, 479
Fine structure of oxygen, 181
FIR spectrometer, 142
Fluorescence monitoring, 431
Fourwave mixing, 78
Fractal, 36
Fractal, dimension, 37
Free electron laser, 74, 85ff, 94
Frequency mixing, 63
    tripling, 64
    stabilization, 491
    standards, 469

Gas correlation LIDAR, 423,
Hanle effect, 164
Harmonic generation, 84
Holeburning, 240
Hollow cathode discharge, 195
Hydrogen spectroscopy, 167ff

Idler wave, 104
Immision measurements, 459
Infrared lasers, 191
    laser spectroscopy, 189
Intermodulated spectroscopy, 176

Intermodulated spectroscopy, 176
Intermolecular transfer, 262
Intracavity absorption, 8
Intermolecular dynamics, 70
Ion traps, 24f, 479
Isotope shift
    spectroscopy, 331, 339

Kolmogrov entropy, 37

Lamb–dip, 7, 9
Laser, infrared, 191
    VUV, 77
Laser cooled ions, 24
Laser cooling, 469
Laser desorption, 357
Laser induced fluorescence, 1,4,
        261, 426, 429
Laser ion sourcre, 355
Laser spectroscopy, 1, 117, 137,
        149 173, 206, 271, 293,
        339, 341, 393, 408, 479
Level crossing, 152, 161
LIDAR, 419, 435
    gas correlation, 423
    LIF, 429
    detectable pollutants, 448
Light–induced kinetic effects,
        241
    drifts, 245
Line shape, 253
Lyapunov exponent, 33f

Mass spectroscopy, 343, 349
Metastable beams, 476
Micro clusters, 321
Microwave modulation, 137
    spectroscopy, 257ff
MIM–diodes, 144
Multiphoton spectroscopy, 222,
        227

One–atom maser, 15
Optical Bloch equation, 229
    double resonance, 11, 150, 163
    klystron, 80, 83
    parametric oscillator, 101ff
    Stark effect, 317
    optoacoustic spectroscopy, 6
Optogalvanic effect, 491
Optothermal spectroscopy, 8, 215,
        225
Overtone laser, 127ff
Overtone spectroscopy, 224

Oxygen fine structure, 181

Parametric oscillator, 101ff
    process, 376
Partial inversion, 129
Phonon decay, 291
    lifetimes, 269
Photoacoustic resonator, 408
    spectroscopy, 6
Photo dissociation, 79
Photo ionisation, 80, 333
Photon–antibunching, 18
statistics, 19
Polarization spectroscopy, 10, 178
Pollutant mapping, 448, 457
Predissociation limited LIF, 4
Probably amplitude, 48
Pulse response, 55

Quantum beats, 20f, 150
Quantum efficiency, 2
Quantum jumps, 14, 158
Quantum optics, 31

Rabi oscillations, 229
Radiation pressure, 243
Radicals, 106ff
Radioactive isotopes, 329
Raman interactions, 48
Raman spectroscopy, 271
Relaxation measurements, 234
    processes, 294
Remote detection, 435
REMPI, 4
Resonance ionisation spectroscopy, 342, 349ff
Resonant frequency mixing, 565
Response amplitude, 55
Return time, 39, 44
Rotational CARS, 315
Rydberg atoms, 17, 23
    constant, 167

Saturation spectroscopy, 9
Semiconductor lasers, 117ff
Spectroscopy, 6
Brioullin, 307
    CARS, 274, 302, 313, 365
    FIR, 479
    LMR, 209, 274, 302, 313, 365
    microwave, 257ff
    multiphoton, 222, 227
    nonlinear, 293
    optogalvanic, 176
    optothermal, 8, 215, 225

Spectroscopy (continued)
  photoacoustic, 408
  polarization, 10, 178
  Raman, 271, 302, 365
  resonance ionisation, 341, 349
  sideband, 137, 140
  Stark, 206
  time–resolved, 293, 297
  velocity modulation, 198
Spiking, 39
Spin Echo, 53
Stability analysis, 40
Sum frequency generation, 64,
  153
Super–conducting bolometer, 219f
  cavity, 16
Supersonic expansion, 197, 217
Susceptibility, 369
Synchrotron radiation, 70–76

Thermal detectors, 218
Time delayed pulses, 111,
Time standard, 487
Trace analysis, 354f
  gas detector, 403
Transient optical grating, 302
Trapped atoms, 53
Tunable FIR spectrometer, 144
Two–photon excitation, 235
  Raman–interaction, 48
  spectra, 156
  transitions, 109, 168

Ultrashort light pulses, 49
Undulators, 80f, 154
Unstable molecules, 192
  generation in,
  discharges, 194
  supersonic expansions, 197

Vacuum ultraviolet lasers, 77
  radiation, 63
  spectroscopy, 149, 155ff
Van der Waals complexes, 197, 204
Velocity modulation spectroscopy, 198
Vibrational energy transfer, 261
Vibronic bandwidth, 284
VUV–lasers, 77
  radiation, 63
  spectroscopy, 149, 155ff

Waveguide modulator, 138
Wigner crystal, 26

Zeeman tuning, 471